HANDBOOK OF

MOLECULAR PLASMONICS

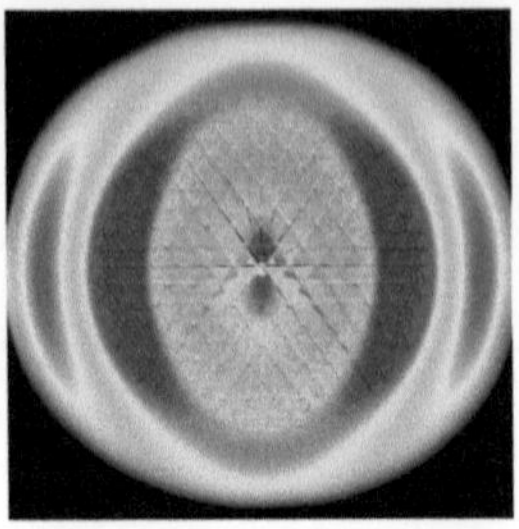

HANDBOOK OF

MOLECULAR PLASMONICS

edited by Fabio Della Sala and Stefania D'Agostino

PAN STANFORD PUBLISHING

Published by

Pan Stanford Publishing Pte. Ltd.
Penthouse Level, Suntec Tower 3
8 Temasek Boulevard
Singapore 038988

Email: editorial@panstanford.com
Web: www.panstanford.com

British Library Cataloguing-in-Publication Data
A catalogue record for this book is available from the British Library.

Handbook of Molecular Plasmonics

ISBN 978-981-4303-20-0 (Hardcover)
ISBN 978-981-4303-21-7 (eBook)

Printed in the USA

Contents

Preface

The term *molecular plasmonics* refers to a rapidly growing interdisciplinary science which aims at investigating the coupling, at the nanoscale level, between emitting molecules and metallic nanostructures. Collective oscillations of the conduction electrons, named *surface plasmons*, can strongly enhance the electromagnetic field around metallic nanoparticles and nano-patterned surfaces: the optical responses (absorption, fluorescence, lifetime, and Raman scattering) of molecules close to the metal are thus strongly modified.

Despite these effects have been known since the seventies, it is only with the recent progress in the synthesis and fabrication of nanosystems as well as innovations in the characterization procedures and spectroscopies that interest in molecular plasmonics has been boosted. Surface-enhanced Raman spectroscopy (SERS), localized surface plasmon resonance (LSPR) spectroscopy, and metal-enhanced fluorescence (MEF) find large applications in biology, to realize plasmonic biosensors or to detect molecular-binding events, as well as in medicine, for molecular-specific imaging, detection, and photothermal therapy of cancer. Surface plasmons and MEF are also widely used in organic opto-electronics, photonics, and energy-conversion applications. In addition, different theoretical approaches and modeling tools have been developed in recent years to describe both organic molecules and metal nanoparticles as well as their interactions, with increased accuracy and efficiency. Molecular plasmonics thus has great interdisciplinary appeal, attracting researchers from fields as diverse as telecommunication engineering (as emitting molecules behave like electromagnetic antennas), inorganic chemistry (to synthesize metal nanoparticles), quantum mechanics (to describe optical properties of molecules and metals), nano-photonics (to manipulate light at a length scale below the diffraction limit), and optical microscopy (to measure the near-field around metallic objects).

Handbook of Molecular Plasmonics is intended for a broad readership and contains both high-level specialized chapters and introductory chapters as well as theoretical and experimental reviews. The main idea underlying this project is to create a useful feedback between theory and experiments, giving a theoretical reference to experimentalists and, at the same time, new inputs to theoreticians for further developments.

This handbook is organized in 10 chapters that reflect the current status of this evolving scientific field, discuss the most recent developments, and identify the directions of future research.

Chapter 1 introduces the basic foundations of molecular plasmonics. It is a self-contained chapter, starting with Maxwell's equations and concluding with the derivation of the radiative and non-radiative decay rates of emitting molecules near metal surfaces and nanoparticles. After this introductory chapter, the handbook is subdivided in two parts: the first one describes the computational and theoretical methods of interest in molecular plasmonics, while the second is entirely dedicated to the most relevant applications and experimental techniques. Both parts contain precious contributions from international experts to ensure a plurality of points of view.

Part I, *Theory and Computational Methods*, opens with a chapter by M. A. Yurkin (Russia) who describes in detail the Discrete Dipole Approximation (DDA) approach, which is an efficient method to study the absorption and scattering of metal nanoparticles of arbitrary shapes. This chapter will serve as an important reference for theoreticians to model metal nanoparticles. Chapter 3 reports DDA results for nanoparticles of different sizes and shapes. This systematic analysis, inspired by recent literature, should represent an important reference for both experimentalists and theoreticians to verify and compare the absorption and scattering spectra of different nanoparticles. While these first two chapters are completely dedicated to metal nanoparticles, Chapter 4 introduces the discussion about the molecular counterpart. In this chapter E. Fabiano (Italy) sheds light on the optical and photophysical properties of organic molecules as well as theoretical methods of quantum mechanics. The interaction between metals and organic molecules is then analyzed in detail in Chapter 5 by Stefano Corni (Italy). This chapter represents the synthesis of the concepts of all previous chapters.

Part II, *Applications and Experimental Aspects*, starts with a chapter by P. K. Jain (Illinois) and C. Deeb (Illinois). They describe the coupling between two plasmonic nanostructures and concepts and applications of the optical near-field. This chapter serves as a link between theory and experiments. In Chapter 7, R. J. Phaneuf (Maryland) discusses recent results on MEF. Both localized and delocalized surface plasmon resonances are presented, including comparison between theory and experiments. Chapter 8 is dedicated to SERS, and M. Sun (China) gives special emphasis to theoretical calculations of charge-transfer effects and to experimental results on remote SERS excitation, a treatment which should be useful for both theoreticians and experimentalists. In Chapter 9, D. Zhang and A. J. Meixner (Germany) describe near-field microscopy and the techniques based on the parabolic mirror confocal microscope. Different types of gap modes are discussed and relevant experimental images and results are presented. The last chapter is devoted to the colloidal synthesis of metal nanoparticles. L. Carbone (Italy) presents a complete review on different synthetic approaches, ranging from convective to microwave heating. This chapter should prove very useful not only to chemists but also to readers from other backgrounds as a reference for all types of nanoparticles that can be realized experimentally.

It has been a great pleasure to work with all these leading scientists of this field, and we thank all of them for their support. We hope this handbook proves to be a helpful reference manual and a useful tool for both students and experts in this field, and we encourage readers to give their valuable comments/suggestions so that the book can be improved further.

Fabio Della Sala
Stefania D'Agostino
Summer 2013

Acknowledgment

We would like to thank all our collegues of the National Nanotechnology Laboratory who have collaborated with us in the field of molecular plasmonics.

F. D. S. thanks his wife Erika for her constant support and for proofreading the book.

S. D. thanks the Photonics and Nanostructures group of the Physics Department of Pavia for the precious friendship and support received during the period of the project.

Chapter 1

Foundations of Molecular Plasmonics

F. Della Sala[a,b]

[a] *National Nanotechnology Laboratory, Istituto Nanoscienze CNR, Via Arnesano 16, I-73100 Lecce, Italy*
[b] *Center for Biomolecular Nanotechnologies @UNILE, Istituto Italiano di Tecnologia, Via Barsanti, I-73010 Arnesano (LE), Italy*
fabio.dellasala@unisalento.it

In this chapter the physics of the interaction between metallic systems (bulk, surfaces and nanoparticles) and the electromagnetic radiation will be considered. The light-matter interaction will be analyzed for a plane-wave incident field as well as for a dipolar radiation, which represents the electromagnetic field emitted by molecules. We will start discussing the Maxwell's equations, the dielectric function of metals and the localized surface plasmon resonances of metallic nanospheres. Then the Green's function formalism and the electromagnetic scattering theory will be introduced. Finally the radiative and non-radiative decay rates of emitting molecules near metal surfaces and nanoparticles will be discussed.

This chapter also introduces all the definitions and the theoretical formalism which will be used in all the other chapters.

Handbook of Molecular Plasmonics
Edited by Fabio Della Sala and Stefania D'Agostino

ISBN 978-981-4303-20-0 (Hardcover), 978-981-4303-21-7 (eBook)
www.panstanford.com

1.1 Electromagnetic Waves

The optics of metallic systems described in this handbook falls within the framework of the classical electromagnetic theory based on the Maxwell's equations [1]. This description is valid in the case of nanostructures down to sizes of the order of a few nanometers. In fact there is no need to resort to a more accurate treatment based on quantum mechanics when the spacing of the electron energy levels is smaller than the thermal excitations of energy ($k_B T$).

The macroscopic Maxwell's equations in matter are (in the *Gaussian unit system* used throughout this handbook):

$$\nabla \cdot \mathbf{D} = 4\pi \rho_e \,, \tag{1.1}$$

$$\nabla \cdot \mathbf{B} = 0 \,, \tag{1.2}$$

$$\nabla \times \mathbf{E} = -\frac{1}{c}\frac{\partial \mathbf{B}}{\partial t} \,, \tag{1.3}$$

$$\nabla \times \mathbf{H} = \frac{1}{c}\frac{\partial \mathbf{D}}{\partial t} + \frac{4\pi}{c}\mathbf{J}_e \,, \tag{1.4}$$

where c is the speed of light, $\mathbf{E}$ and $\mathbf{H}$ are the *electric field* and the *magnetizing field*, respectively, and $\mathbf{D}$ and $\mathbf{B}$ are the *electric displacement field* and the *magnetic induction*, respectively. All quantities depend on time and on spatial coordinates (not explicitated for simplicity). $\mathbf{E}$, $\mathbf{H}$, $\mathbf{D}$, $\mathbf{B}$, and $\mathbf{J}_e$ are vectors in the three-dimensional space, and throughout this handbook are indicated in **boldface**.

The Maxwell's equations link the four macroscopic fields $\mathbf{D}$, $\mathbf{E}$, $\mathbf{H}$ and $\mathbf{B}$ with the *external charge density* ρ_e and the *external current density* $\mathbf{J}_e$. The term $(\partial/\partial t)\mathbf{D}$ is called *displacement current* and it is added to the external current density term in Eq. (1.4) to satisfy the charge conservation, i.e. the *continuity equation*:

$$\nabla \cdot \mathbf{J}_e + \frac{\partial \rho_e}{\partial t} = 0 \,. \tag{1.5}$$

The displacement current can be regarded as arising from the polarization changing of the medium under the effect of time-varying electric fields.

In macroscopic media the Maxwell's equations must be supplemented by the *constitutive relations* which allow to describe the average effect of a large number of atoms without requiring detailed knowledge of the response of individual atoms to electric and magnetic fields. These relations are:

$$\mathbf{D} = \mathbf{E} + 4\pi \mathbf{P}\,, \tag{1.6}$$

$$\mathbf{B} = \mathbf{H} + 4\pi \mathbf{M}\,, \tag{1.7}$$

where **P** and **M** are the *polarization field* and the *magnetization field*, respectively. In this handbook, we will limit the description to *non-magnetic* media: the magnetic response represented by **M** will be disregarded, and we will focus on the analysis of electric polarization effects. Note in fact that noble metals have a magnetic permeability very close to unity [2].

P describes the electric dipole moment per unit volume inside the material, caused by the alignment of microscopic dipoles by the electric field. It is related to the *bound charge density* via:

$$\nabla \cdot \mathbf{P} = -\rho_b\,. \tag{1.8}$$

A continuity equation also holds for the bound charge density:

$$\mathbf{J}_b = \frac{\partial \mathbf{P}}{\partial t}\,, \tag{1.9}$$

where $\mathbf{J}_b$ is the *polarization current density.*

The relation between **P** and **E** defines the dielectric properties of the material. As the optical response of metals is in general non-local in space and in time and anisotropic, we have:

$$\mathbf{P}(\mathbf{r}, t) = \int dt' \int \mathrm{d}^3\mathbf{r}'\, \mathbb{\chi}\left(\mathbf{r}; \mathbf{r} - \mathbf{r}'; t - t'\right) \cdot \mathbf{E}\left(\mathbf{r}', t'\right)\,, \tag{1.10}$$

where $\mathbb{\chi}$ is the *dielectric susceptibility tensor*. In this handbook tensors, 3×3 matrices and dyadics, will be indicated with blackboard bold face ($\mathbb{A}, \mathbb{B}, \ldots \mathbb{\alpha}, \mathbb{\beta}, \ldots$). Equation (1.10) is valid in a linear regime and for a material uniform in time, e.g. memory-less: in fact Eq. (1.10) indicates that the material response function does not depend on absolute temporal coordinates, but only on their differences. In this handbook we further assume that the medium is *isotropic* and has a *locality in space*. The non-locality in space can be quite strong for metals, but a local approximation still yields a quite accurate description (see Sec. 5.3.4 for further details). In fact this simplification is valid as long as the wavelength λ in the material is significantly longer than all other characteristic dimensions such as the mean free path of the electrons, which is usually the case up to ultraviolet frequencies.

With the above assumption Eq. (1.10) simplifies into:

$$\mathbf{P}(\mathbf{r}, t) = \int dt' \chi\left(\mathbf{r}; t - t'\right) \mathbf{E}\left(\mathbf{r}, t'\right) , \tag{1.11}$$

where χ is the *scalar dielectric susceptibility*. By taking the Fourier transform with respect to time coordinates, i.e. $f(t) = \mathrm{Re}\left[\int d\omega f(\omega) e^{-\mathrm{i}\omega t}\right]$, Eq. (1.11) becomes:

$$\tilde{\mathbf{P}}(\mathbf{r}, \omega) = \tilde{\chi}(\mathbf{r}, \omega) \tilde{\mathbf{E}}(\mathbf{r}, \omega) . \tag{1.12}$$

Thus, for linear, isotropic media $\tilde{\mathbf{P}}(\mathbf{r}, \omega)$ and $\tilde{\mathbf{E}}(\mathbf{r}, \omega)$ are linearly related. Note that $\tilde{\mathbf{P}}(\mathbf{r}, \omega)$, $\tilde{\mathbf{E}}(\mathbf{r}, \omega)$, and $\tilde{\chi}(\mathbf{r}, \omega)$ are *complex* quantities and throughout this handbook will be indicated with the symbol $\tilde{}$.

We can also rewrite the dielectric constitutive relation in Eq. (1.12) as:

$$\tilde{\mathbf{D}}(\mathbf{r}, \omega) = \tilde{\varepsilon}(\mathbf{r}, \omega) \tilde{\mathbf{E}}(\mathbf{r}, \omega) , \tag{1.13}$$

where $\tilde{\varepsilon}(\mathbf{r}, \omega)$ is the complex *dielectric relative permittivity* of the medium. Inserting Eq. (1.6) and Eq. (1.12) into Eq. (1.13) we obtain:

$$\tilde{\varepsilon}(\mathbf{r}, \omega) = 1 + 4\pi \tilde{\chi}(\mathbf{r}, \omega) , \tag{1.14}$$

which relates the dielectric susceptibility to the relative dielectric permittivity.

In metallic systems, another important constitutive linear relationship has to be taken into account: the *free current density* $\mathbf{J}_f$ which is generated by the electric field $\mathbf{E}$. The free current density is added to the external current density in Eq. (1.4). For linear, memory-less, and isotropic media we have

$$\mathbf{J}_f(\mathbf{r}, t) = \int dt' \int \mathrm{d}^3\mathbf{r}' \sigma\left(\mathbf{r}; \mathbf{r} - \mathbf{r}'; t - t'\right) \mathbf{E}\left(\mathbf{r}', t'\right) , \tag{1.15}$$

where σ is the *electrical conductivity*. Considering again the Fourier representation and neglecting the space-non-locality we obtain:

$$\tilde{\mathbf{J}}_f(\mathbf{r}, \omega) = \tilde{\sigma}(\mathbf{r}, \omega) \tilde{\mathbf{E}}(\mathbf{r}, \omega) . \tag{1.16}$$

We now can rewrite Eq. (1.4) in the Fourier space as:

$$\nabla \times \tilde{\mathbf{H}} = -\frac{\mathrm{i}\omega}{c}\tilde{\mathbf{D}} + \frac{4\pi}{c}\left(\tilde{\mathbf{J}}_f + \tilde{\mathbf{J}}_e\right) \tag{1.17}$$

$$= -\frac{\mathrm{i}\omega}{c}\left(\tilde{\varepsilon}\left(\mathbf{r},\omega\right) + \frac{4\pi\mathrm{i}}{\omega}\tilde{\sigma}\left(\mathbf{r},\omega\right)\right)\tilde{\mathbf{E}} + \frac{4\pi}{c}\tilde{\mathbf{J}}_e\,, \tag{1.18}$$

where the complex permittivity with conductivity is

$$\tilde{\varepsilon}_c\left(\mathbf{r},\omega\right) = \tilde{\varepsilon}\left(\mathbf{r},\omega\right) + \frac{4\pi\mathrm{i}}{\omega}\tilde{\sigma}\left(\mathbf{r},\omega\right) = \varepsilon_1\left(\mathbf{r},\omega\right) + \mathrm{i}\varepsilon_2\left(\mathbf{r},\omega\right)\,. \tag{1.19}$$

The imaginary part of the dielectric constant ($\varepsilon_2\left(\mathbf{r},\omega\right)$) takes into account the dissipation inside the material (see Sec. 1.1.1.1 below). When $\tilde{\varepsilon}_c\left(\mathbf{r},\omega\right)$ is real then the medium is *non-dissipative*. Often, and in the following, $\tilde{\varepsilon}_c$ is simply indicated with $\tilde{\varepsilon}$.

1.1.1 *Propagation in a Homogeneous Medium*

If we apply the curl to both the members of Eq. (1.3) we obtain in the Fourier space:

$$\nabla \times \nabla \times \tilde{\mathbf{E}} = \mathrm{i}\frac{\omega}{c}\nabla \times \tilde{\mathbf{H}}\,, \tag{1.20}$$

where we also used that in non-magnetic media $\mathbf{B} = \mathbf{H}$, see Eq. (1.7). Using Eq. (1.18) we then obtain:

$$\nabla \times \nabla \times \tilde{\mathbf{E}} = \frac{\omega^2}{c^2}\tilde{\varepsilon}\left(\mathbf{r},\omega\right)\tilde{\mathbf{E}} + \mathrm{i}\frac{4\pi\omega}{c^2}\tilde{\mathbf{J}}_e\,. \tag{1.21}$$

In the absence of external stimuli (i.e. when $\rho_e = 0$ and $\mathbf{J}_e = 0$), and restricting the analysis to homogeneous media ($\tilde{\varepsilon}\left(\mathbf{r},\omega\right) = \tilde{\varepsilon}(\omega)$) so that $\nabla \cdot (\tilde{\varepsilon}(\omega)\tilde{\mathbf{E}}) = \tilde{\varepsilon}(\omega)\nabla \cdot \tilde{\mathbf{E}} = 0$, we finally obtain:

$$\nabla^2\tilde{\mathbf{E}}\left(\mathbf{r},\omega\right) = -\frac{\omega^2}{c^2}\tilde{\varepsilon}(\omega)\tilde{\mathbf{E}}\left(\mathbf{r},\omega\right)\,, \tag{1.22}$$

where we used the vector identity

$$\nabla \times \nabla\mathbf{X} \equiv \nabla(\nabla \cdot \mathbf{X}) - \nabla^2\mathbf{X}\,. \tag{1.23}$$

Equation (1.22) is called *homogeneous Helmholtz equation*. This equation has a plane-wave solution of form:

$$\tilde{\mathbf{E}}\left(\mathbf{r},\omega\right) = \tilde{\mathbf{E}}^0\exp\left(\mathrm{i}\tilde{\mathbf{k}}\cdot\mathbf{r}\right) = \tilde{\mathbf{E}}^0\exp\left(\mathrm{i}\boldsymbol{\beta}\cdot\mathbf{r}\right)\exp\left(-\boldsymbol{\alpha}\cdot\mathbf{r}\right)\,, \tag{1.24}$$

being $\tilde{\mathbf{k}} = \boldsymbol{\beta} + \mathrm{i}\boldsymbol{\alpha}$ the *complex propagation vector* with

$$\tilde{\mathbf{k}}\cdot\tilde{\mathbf{E}}^0 = 0\,, \tag{1.25}$$

which follows from $\nabla \cdot \tilde{\mathbf{E}} = 0$. Note that both $\tilde{\mathbf{E}}^0$ and $\tilde{\mathbf{k}}$ are *complex* vectors, while $\boldsymbol{\beta}$ and $\boldsymbol{\alpha}$ are real vectors. The vector $\boldsymbol{\beta}$ describes the propagation of the wave, whereas $\boldsymbol{\alpha}$ its attenuation (or amplification).

The vector $\tilde{\mathbf{E}}^0 = \mathbf{E}_R^0 + \mathrm{i}\mathbf{E}_I^0$ is called *phasor*. The full time dependence of the field can be written as:

$$\mathbf{E}(t) = \mathrm{Re}\left[\tilde{\mathbf{E}}^0 \exp\left(-\mathrm{i}\omega t\right)\right] \tag{1.26}$$

$$= \mathbf{E}_R^0 \cos\left(\omega t\right) + \mathbf{E}_I^0 \sin\left(\omega t\right). \tag{1.27}$$

The two vectors $\mathbf{E}_R^0$ and $\mathbf{E}_I^0$ are in general different and define a plane where the field oscillates and describes an ellipse [2]. In this general case the plane-wave has an *elliptical polarization*. When the two vectors are orthogonal and with the same magnitude then we have a *circular polarization*. Finally when the two vectors are parallel or one of them is null then the plane-wave has a *linear polarization*.

Coming back to the plane-wave propagation, if we insert Eq. (1.24) into Eq. (1.22), we obtain the relation

$$\tilde{\mathbf{k}} \cdot \tilde{\mathbf{k}} = \frac{\omega^2}{c^2}\tilde{\varepsilon}(\omega), \tag{1.28}$$

which is called *dispersion relation* between wavevector $\tilde{\mathbf{k}}$ and the frequency ω in an homogeneous medium. We then introduce the *complex refractive index*, $\tilde{n} = n + \mathrm{i}\kappa$, which is defined by the equation:

$$\tilde{\varepsilon} = \varepsilon_1 + \mathrm{i}\varepsilon_2 = \tilde{n}^2 = (n + \mathrm{i}\kappa)^2. \tag{1.29}$$

The imaginary part of complex refractive index (κ) is named *extinction coefficient*, and describes the attenuation of light inside the medium. The values ε_1 and ε_2 are interrelated by the following relations:

$$\varepsilon_1 = n^2 - \kappa^2, \tag{1.30}$$

$$\varepsilon_2 = 2n\kappa, \tag{1.31}$$

$$n = +\sqrt{\frac{\sqrt{\varepsilon_1^2 + \varepsilon_2^2} + \varepsilon_1}{2}}, \tag{1.32}$$

$$\kappa = \mathrm{Sgn}(\varepsilon_2)\sqrt{\frac{\sqrt{\varepsilon_1^2 + \varepsilon_2^2} - \varepsilon_1}{2}}. \tag{1.33}$$

Note that $n > 0$ by definition (exceptions are meta-materials [3]) even if in metals it is often $\varepsilon_1 < 0$. The sign of κ is also positive;

for systems that can amplify the light (e.g. lasers or any system with gain), $\kappa < 0$ and $\varepsilon_2 < 0$. If the medium is non-dissipative then we have simply:

$$n = \sqrt{\varepsilon_1} \text{ and } \kappa = 0 . \tag{1.34}$$

If $\boldsymbol{\beta}$ and $\boldsymbol{\alpha}$ have the same direction $\widehat{\mathbf{k}}$ (i.e. *uniform* plane-wave), then from Eq. (1.28) and Eq. (1.29) we have:

$$\tilde{\mathbf{k}} = \frac{\omega}{c}(n + \mathrm{i}\kappa)\widehat{\mathbf{k}} , \tag{1.35}$$

i.e.

$$\boldsymbol{\beta} = \frac{\omega}{c} n\widehat{\mathbf{k}} \text{ and } \boldsymbol{\alpha} = \frac{\omega}{c}\kappa\widehat{\mathbf{k}} . \tag{1.36}$$

Equation (1.36) states that the real part of the refractive index (n) describes the propagation of electromagnetic waves whereas the imaginary part of the refractive index (κ) its attenuation (or amplification). In this case the condition in Eq. (1.25) becomes:

$$\mathbf{E}_R^0 \cdot \widehat{\mathbf{k}} = 0 \text{ and } \mathbf{E}_I^0 \cdot \widehat{\mathbf{k}} = 0 , \tag{1.37}$$

which means that the polarization of the electric field is in the plane perpendicular to the propagation direction, i.e. the plane-wave is *transverse electric (TE)*. From Eq. (1.3) we have:

$$\tilde{\mathbf{H}}(\mathbf{r}, \omega) = \frac{c}{\mathrm{i}\omega}\nabla \times \tilde{\mathbf{E}}(\mathbf{r}, \omega) = \frac{c}{\omega}\tilde{\mathbf{k}} \times \tilde{\mathbf{E}}^0 \exp(\mathrm{i}\tilde{\mathbf{k}} \cdot \mathbf{r}) \tag{1.38}$$

$$= (n + \mathrm{i}\kappa)\widehat{\mathbf{k}} \times \tilde{\mathbf{E}}^0 \exp(\mathrm{i}\tilde{\mathbf{k}} \cdot \mathbf{r}) , \tag{1.39}$$

and thus the plane-wave is also *transverse magnetic (TM)*. As a whole we have that a plane-wave in an homogeneous space is *transverse electro-magnetic (TEM)*.

1.1.1.1 Poynting vector

The *Poynting vector*

$$\mathbf{S}(\mathbf{r}, t) = \frac{c}{4\pi}\mathbf{E}(\mathbf{r}, t) \times \mathbf{H}(\mathbf{r}, t) \tag{1.40}$$

represents the energy flux of the electromagnetic field. In the Fourier space we can introduce the *complex Poynting vector* which is given by:

$$\tilde{\mathbf{S}}(\mathbf{r}, \omega) = \frac{c}{8\pi}\tilde{\mathbf{E}}(\mathbf{r}, \omega) \times \tilde{\mathbf{H}}^*(\mathbf{r}, \omega) , \tag{1.41}$$

where the symbol * indicated the complex conjugate and the factor 1/2 is introduced so that the real part of $\tilde{\mathbf{S}}(\mathbf{r},\omega)$ represents the transferred power per unit period ($T = 2\pi/\omega$). In fact it can be easily shown that:

$$< \mathbf{S}(\mathbf{r},t) >_T = \frac{1}{T}\int_0^T \mathrm{d}t \frac{c}{4\pi}\mathbf{E}(\mathbf{r},t)\times\mathbf{H}(\mathbf{r},t) \tag{1.42}$$

$$= \frac{1}{2}\mathrm{Re}\left[\frac{c}{4\pi}\tilde{\mathbf{E}}(\mathbf{r},\omega)\times\tilde{\mathbf{H}}^*(\mathbf{r},\omega)\right] = \mathrm{Re}\left[\tilde{\mathbf{S}}(\mathbf{r},\omega)\right]. \tag{1.43}$$

Considering a volume V with a surface A, the following *energy conservation theorem* holds [1]:

$$-\int_V \frac{1}{2}\mathrm{Re}\left[\tilde{\mathbf{E}}\cdot\tilde{\mathbf{J}}_e^*\right]\mathrm{d}^3\mathbf{r} = \oint_A \mathrm{Re}\left[\tilde{\mathbf{S}}\right]\cdot\hat{\mathbf{n}}\mathrm{d}A + \int_V \frac{1}{2}\mathrm{Re}\left[\sigma\tilde{\mathbf{E}}\cdot\tilde{\mathbf{E}}^*\right]\mathrm{d}^3\mathbf{r}. \tag{1.44}$$

The power furnished to the external current source (left-hand side) in the volume V equals the power radiated out from the surface A (first term at the right-hand side) plus the power dissipated in the volume V by the *Joule effect* (second term at the right-hand side). The Joule effect dissipation can also be rewritten as:

$$\int_V \frac{1}{2}\mathrm{Re}\left[\tilde{\mathbf{J}}_b\cdot\tilde{\mathbf{E}}^*\right]\mathrm{d}^3\mathbf{r} = \int_V \frac{1}{2}\mathrm{Re}\left[-\mathrm{i}\omega\left(\frac{\tilde{\varepsilon}-1}{4\pi}\right)\tilde{\mathbf{E}}\cdot\tilde{\mathbf{E}}^*\right]\mathrm{d}^3\mathbf{r} \tag{1.45}$$

$$= \int_V \frac{1}{2}\frac{\omega}{4\pi}\mathrm{Im}[\tilde{\varepsilon}]\left|\left|\tilde{\mathbf{E}}\right|\right|^2\mathrm{d}^3\mathbf{r}, \tag{1.46}$$

where we used Eq. (1.9) and Eq. (1.14).

The Poynting vector for an uniform plane-wave reads:

$$\tilde{\mathbf{S}}(\mathbf{r},\omega) = \frac{c}{8\pi}(n-\mathrm{i}\kappa)\tilde{\mathbf{E}}^0\times\widehat{\mathbf{k}}\times\tilde{\mathbf{E}}^{0*} \tag{1.47}$$

$$= \frac{c}{8\pi}(n-\mathrm{i}\kappa)\widehat{\mathbf{k}}\left|\left|\tilde{\mathbf{E}}^0\right|\right|^2\exp\left(-2\boldsymbol{\alpha}\cdot\mathbf{r}\right), \tag{1.48}$$

i.e. the power is attenuated (or constant if $\kappa = \boldsymbol{\alpha} = 0$) in the propagation direction. For a plane-wave in the vacuum we have simply

$$\tilde{\mathbf{S}}(\mathbf{r},\omega) = \frac{c}{8\pi}\widehat{\mathbf{k}}\left|\left|\tilde{\mathbf{E}}^0\right|\right|^2. \tag{1.49}$$

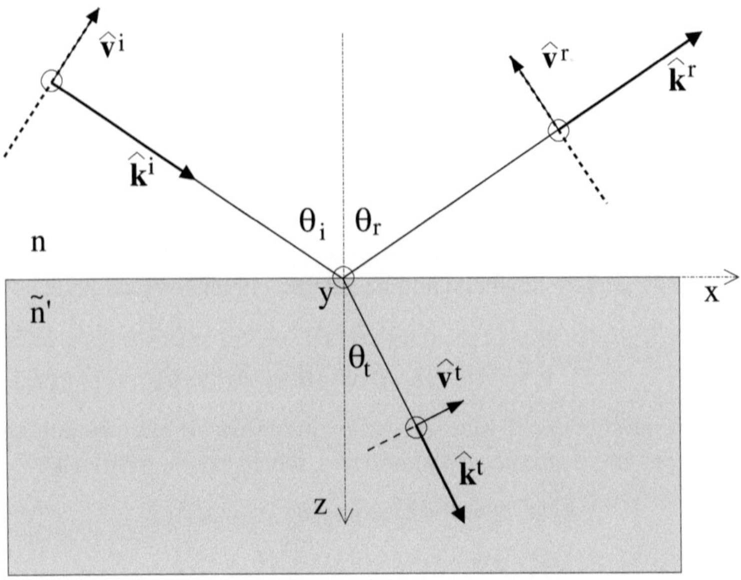

Figure 1.1 Incident (i), reflected (r) and transmitted (t) plane-waves at an interfaces between media with different refractive indexes (n above and $\tilde{n}'$ below. For each plane-wave, $\widehat{\mathbf{k}}$ indicates the propagation direction, θ is the incident angle (with respect the surface normal) and $\widehat{\mathbf{v}}$ the electric-field direction for a p-polarized incidence (for the s-polarized incidence all electric-field directions are along the y-axis, outward the figure plane).

1.1.2 *Reflection and Transmission at an Interface*

A very important case of the plane-wave propagation is the reflection at a surface. We consider a plane-wave traveling from a non-dissipative medium with a real refractive index n to second medium with a general complex refractive index $\tilde{n}'$. The incident wavevector is:

$$\mathbf{k}^i = \boldsymbol{\beta}^i = \widehat{\mathbf{k}}^i \frac{\omega}{c} n \tag{1.50}$$

with $\widehat{\mathbf{k}}^i$ being in the xz plane (the plane of incidence) and forming an angle θ_i with the surface normal $\widehat{\mathbf{z}}$, see Fig. 1.1.

Without loss of generality it is possible to consider only two types of incident plane-waves with linear polarization: i) s-*polarized* i.e. with a *real* vector $\mathbf{E}_i^0$ *perpendicular* (parallel) to the plane of incidence (to the interface plane); ii) *p-polarized* with $\mathbf{E}_i^0$ *parallel* to the plane of incidence. According to the axes in Fig. 1.1 the perpendicular case has $\mathbf{E}_i^0 = E_i^0\hat{\mathbf{y}}$, the parallel case has $\mathbf{E}_i^0 = E_i^0\hat{\mathbf{v}}_i$, with $\hat{\mathbf{v}}_i$ in the xz plane. All other types of plane-waves can be written as a linear combination of these two cases.

Thus the *incident* plane-wave can be written as:

$$\mathbf{E}_i(\mathbf{r},t) = \mathbf{E}_i^0 \mathrm{Re}\left[\exp\left(\mathrm{i}\left(\boldsymbol{\beta}^i \cdot \mathbf{r} - \omega t\right)\right)\right]. \tag{1.51}$$

Recalling the *boundary conditions* [1] for an interface (characterized by a surface normal $\hat{\mathbf{n}}$ and no surface current or charges):

$$\hat{\mathbf{n}} \times (\tilde{\mathbf{E}}_1 - \tilde{\mathbf{E}}_2) = \hat{\mathbf{n}} \cdot (\tilde{\mathbf{D}}_1 - \tilde{\mathbf{D}}_2) = 0, \tag{1.52}$$

$$\hat{\mathbf{n}} \times (\tilde{\mathbf{H}}_1 - \tilde{\mathbf{H}}_2) = \hat{\mathbf{n}} \cdot (\tilde{\mathbf{B}}_1 - \tilde{\mathbf{B}}_2) = 0, \tag{1.53}$$

it can be shown [2] that two other plane-waves are present, the *reflected* and *transmitted* plane-waves, which can be written as:

$$\mathbf{E}_r(\mathbf{r},t) = \mathrm{Re}\left[\tilde{\mathbf{E}}_r^0 \exp\left(\mathrm{i}\left(\boldsymbol{\beta}^r \cdot \mathbf{r} - \omega t\right)\right)\right], \tag{1.54}$$

$$\mathbf{E}_t(\mathbf{r},t) = \mathrm{Re}\left[\tilde{\mathbf{E}}_t^0 \exp\left(\mathrm{i}\left(\tilde{\mathbf{k}}^t \cdot \mathbf{r} - \omega t\right)\right)\right]. \tag{1.55}$$

Here the fields $\tilde{\mathbf{E}}_r^0$ and $\tilde{\mathbf{E}}_t^0$ are now complex as they can have a phase shift with respect the incident field. The wavevector $\boldsymbol{\beta}^r$ describes a reflected plane-wave propagating in the same non-dissipative medium as the incident field and thus:

$$\boldsymbol{\beta}^r = \hat{\mathbf{k}}^r \frac{\omega}{c} n. \tag{1.56}$$

The wavevector $\tilde{\mathbf{k}}^t = \boldsymbol{\beta}^t + \mathrm{i}\boldsymbol{\alpha}^t$ is in general a complex vector (it is real only if the second medium is non-dissipative too).

The boundary condition for the electric field at the plane $z = 0$ (see Eq. (1.52)) implies:

$$\hat{\mathbf{z}} \times \mathbf{E}_i^0 \exp\left(\mathrm{i}\boldsymbol{\beta}^i \cdot \mathbf{r}\right) + \hat{\mathbf{z}} \times \tilde{\mathbf{E}}_r^0 \exp\left(\mathrm{i}\boldsymbol{\beta}^r \cdot \mathbf{r}\right) = \hat{\mathbf{z}} \times \tilde{\mathbf{E}}_t^0 \exp\left(\mathrm{i}\tilde{\mathbf{k}}^t \cdot \mathbf{r}\right). \tag{1.57}$$

At the origin ($\mathbf{r} = 0$) we have simply:

$$\hat{\mathbf{z}} \times \mathbf{E}_i^0 + \hat{\mathbf{z}} \times \tilde{\mathbf{E}}_r^0 = \hat{\mathbf{z}} \times \tilde{\mathbf{E}}_t^0. \tag{1.58}$$

Thus along the y-axis we have:

$$\hat{\mathbf{z}} \times \mathbf{E}_i^0 + \hat{\mathbf{z}} \times \tilde{\mathbf{E}}_r^0 \exp\left(\mathrm{i}\beta_y^r y\right) = \left(\hat{\mathbf{z}} \times \mathbf{E}_i^0 + \hat{\mathbf{z}} \times \tilde{\mathbf{E}}_r^0\right) \exp\left(\mathrm{i}\beta_y^t y - \alpha_y^t y\right). \tag{1.59}$$

The above equation can be satisfied for every y only if:

$$\beta_y^r = 0 \ ; \ \beta_y^t = 0 \ ; \ \alpha_y^t = 0. \tag{1.60}$$

These conditions simply mean that $\tilde{\mathbf{k}}^r$ and $\tilde{\mathbf{k}}^t$ are in the incident plane, as already shown in Fig. 1.1. Along the x-axis we have:

$$\begin{aligned} &\widehat{\mathbf{z}} \times \mathbf{E}_i^0 \exp\left(\mathrm{i}\beta_x^i x\right) + \widehat{\mathbf{z}} \times \tilde{\mathbf{E}}_r^0 \exp\left(\mathrm{i}\beta_x^r x\right) \\ &= (\widehat{\mathbf{z}} \times \mathbf{E}_i^0 + \widehat{\mathbf{z}} \times \tilde{\mathbf{E}}_r^0) \exp\left(\mathrm{i}\beta_x^t x - \alpha_x^t x\right) . \end{aligned} \tag{1.61}$$

The above equation can be satisfied for every x only if:

$$\beta_x^r = \beta_x^i \ ; \ \beta_x^t = \beta_x^i \ ; \ \alpha_x^t = 0 . \tag{1.62}$$

The first two conditions in Eq. (1.62) can be rewritten as:

$$\left|\left|\boldsymbol{\beta}^i\right|\right| \sin\theta_i = \left|\left|\boldsymbol{\beta}^r\right|\right| \sin\theta_r , \tag{1.63}$$

$$\left|\left|\boldsymbol{\beta}^i\right|\right| \sin\theta_i = \left|\left|\boldsymbol{\beta}^t\right|\right| \sin\theta_t , \tag{1.64}$$

and imply that $\theta_r = \theta_i$.

If the second-medium is non-dissipative and the incident angle is below then *critical angle* θ_c, defined by:

$$\sin\theta_c = \frac{n'}{n} , \tag{1.65}$$

then we have:

$$\boldsymbol{\beta}^t = \widehat{\mathbf{k}}^t \frac{\omega}{c} n' , \tag{1.66}$$

and the well-known *Snell's law* [1] is:

$$n \sin\theta_i = n' \sin\theta_t . \tag{1.67}$$

Otherwise the transmitted plane-wave is a *non-uniform* plane-wave with:

$$\tilde{\mathbf{k}}^t = \boldsymbol{\beta}^t + \mathrm{i}\widehat{\mathbf{z}}\alpha^t , \tag{1.68}$$

i.e. with an attenuating component in the $\widehat{\mathbf{z}}$ direction. In this case we have from Eqs. (1.28, 1.30, 1.31):

$$\left|\left|\boldsymbol{\beta}^t\right|\right|^2 - (\alpha^t)^2 = \frac{\omega^2}{c^2}(n'^2 - \kappa'^2) , \tag{1.69}$$

$$2\alpha_t\beta_z^t = 2\alpha^t \left|\left|\boldsymbol{\beta}^t\right|\right| \sqrt{1 - \sin^2(\theta_t)} = \frac{\omega^2}{c^2} 2n'\kappa' . \tag{1.70}$$

If the second-medium is non-dissipative ($\kappa' = 0$) but $\theta_i > \theta_c$, Eqs. (1.64, 1.69, 1.70) simplify to:

$$(\alpha_t)^2 = (\beta_x^t)^2 - \frac{\omega^2}{c^2} n'^2 , \tag{1.71}$$

$$\beta_x^t = (\beta_x^i) \sin\theta_i \ , \ \beta_z^t = 0 , \tag{1.72}$$

which represents a transmitted non-uniform plane-wave propagating in the x direction and attenuating in the z direction. In the most general case Eqs. (1.64, 1.69, 1.70) have to be solved together to obtain $\boldsymbol{\beta}^t$ and α^t as a function of θ_i.

From the boundary conditions (Eqs. 1.52, 1.53), the following *Fresnel coefficients* can be obtained [4]:

$$\tilde{r}_\perp = \left(\frac{\tilde{E}_r^0}{E_i^0}\right)_\perp = \frac{k_z^r - \tilde{k}_z^t}{k_z^r + \tilde{k}_z^t} = \frac{\beta^i \cos\theta_i - \beta^t \cos\theta_t - i\alpha^t}{\beta^i \cos\theta_i + \beta^t \cos\theta_t + i\alpha^t}, \quad (1.73)$$

$$\tilde{t}_\perp = \left(\frac{\tilde{E}_t^0}{E_i^0}\right)_\perp = \frac{2k_z^r}{k_z^r + \tilde{k}_z^t} = \frac{2\beta^i \cos\theta_i}{\beta^i \cos\theta_i + \beta^t \cos\theta_t + i\alpha^t}, \quad (1.74)$$

$$\tilde{r}_\parallel = \left(\frac{\tilde{E}_r^0}{E_i^0}\right)_\parallel = \frac{k_z^r \tilde{\varepsilon}' - \tilde{k}_z^t \varepsilon}{k_z^r \tilde{\varepsilon}' + \tilde{k}_z^t \varepsilon}$$

$$= \frac{(\beta^i \cos\theta_i)\tilde{\varepsilon}' - (\beta^t \cos\theta_t + i\alpha^t)\varepsilon}{(\beta^i \cos\theta_i)\tilde{\varepsilon}' + (\beta^t \cos\theta_t + i\alpha^t)\varepsilon}, \quad (1.75)$$

$$\tilde{t}_\parallel = \left(\frac{\tilde{E}_t^0}{E_i^0}\right)_\parallel = \frac{2k_z^r \tilde{\varepsilon}'}{k_z^r \tilde{\varepsilon}' + \tilde{k}_z^t \varepsilon}\sqrt{\frac{\varepsilon}{\tilde{\varepsilon}'}}$$

$$= \frac{2(\beta^i \cos\theta_i)\tilde{n}' n}{(\beta^i \cos\theta_i)\tilde{\varepsilon}' + (\beta^t \cos\theta_t + i\alpha^t)\varepsilon}, \quad (1.76)$$

where $\tilde{r}$ is the *amplitude reflection coefficient* and $\tilde{t}$ is the *amplitude transmission coefficient*, for the two different polarizations.

If the second medium is non-dissipative (i.e. $\alpha^t = 0, \tilde{\varepsilon}' = (n')^2$) then all the expressions simplify in [2]:

$$\tilde{r}_\perp = \frac{n \cos\theta_i - n' \cos\theta_t}{n \cos\theta_i + n' \cos\theta_t}, \quad (1.77)$$

$$\tilde{t}_\perp = \frac{2n \cos\theta_i}{n \cos\theta_i + n' \cos\theta_t}, \quad (1.78)$$

$$\tilde{r}_\parallel = \frac{n' \cos\theta_i - n \cos\theta_t}{n' \cos\theta_i + n \cos\theta_t}, \quad (1.79)$$

$$\tilde{t}_\parallel = \frac{2n \cos\theta_i}{n' \cos\theta_i + n \cos\theta_t}. \quad (1.80)$$

If Fig. 1.2 we report the Fresnel coefficients for two different conditions: incidence from vacuum to a more dense medium ($n = 1$, $n' = 2$) and to a less dense medium ($n = 1, n' = 0.5$).

The incident angle for which $\tilde{r}_\parallel = 0$ is called *Brewster's angle* (θ_B). In the first case (panel a) the Fresnel coefficients are defined

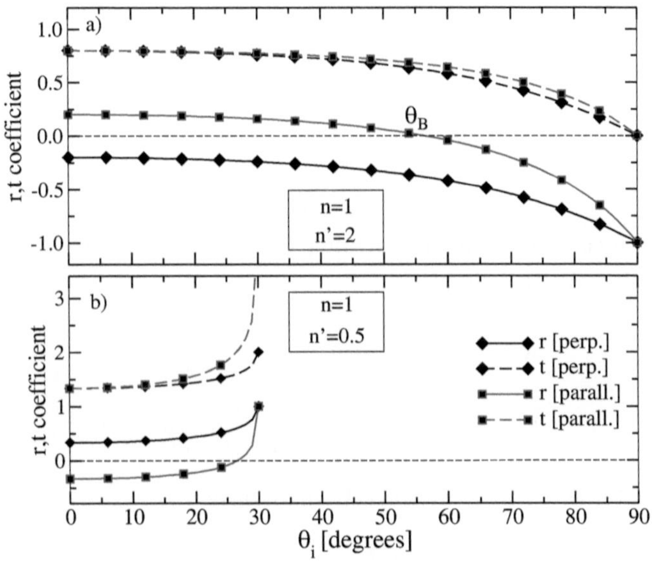

Figure 1.2 Reflection (r) and transmission (t) coefficients from propagation into a) denser medium (i.e. from $n = 1$ to $n' = 2$) or b) lighter medium (i.e. from $n = 1$ to $n' = 0.5$), for different incident angles (θ_i). The Brewster's angle (θ_B) is also indicated.

for all incident angles, because the transmitted (refracted) angle is always smaller than the incident angle. For increasing θ_i both reflection coefficients approaches -1, i.e. the surface is perfectly reflecting (mirror-like). In the second case (panel b), the critical angle is $\theta_c = 30°$, and above it the coefficients become complex-quantities (the reflection coefficients are uni-modular).This effect is called *total internal reflection* and it is characterized by an evanescent wave in the second medium, see Eqs. (1.71, 1.72).

1.1.2.1 Normal incidence

For normal incidence, we have that $\theta_i = \theta_r = \theta_t = 0$. In this case the parallel or perpendicular orientations coincide and the plane-wave transmitted in the second medium is uniform. Considering the power (i.e. from the Poynting vector) instead of the amplitude coefficients, we obtain the *reflectance* (R) and the *transmittance* (T) defined as:

$$R = \left| \frac{\tilde{E}_r^0}{E_i^0} \right|^2 = |\tilde{r}|^2 , \tag{1.81}$$

$$T = \frac{\mathrm{Re}[\tilde{n}']}{n} \left| \frac{\tilde{E}_t^0}{E_i^0} \right|^2 = \frac{n'}{n} \left|\tilde{t}\right|^2 , \tag{1.82}$$

where $n' = \mathrm{Re}[\tilde{n}']$. If we also assume that $n = 1$, i.e. the light is incoming from the vacuum, then the reflectance and the transmittance are simply:

$$R = \left| \frac{1 - \tilde{n}'}{1 + \tilde{n}'} \right|^2 = \frac{(1 - n')^2 + \kappa'^2}{(1 + n')^2 + \kappa'^2} , \tag{1.83}$$

$$T = n' \left| \frac{2}{1 + \tilde{n}'} \right|^2 = \frac{4n'}{(1 + n')^2 + \kappa'^2} . \tag{1.84}$$

Note that due to energy conservation, we have:

$$R + T = 1 . \tag{1.85}$$

1.1.3 *Current and Charges as Sources of Fields*

So far we discussed the propagation of electromagnetic waves without external sources (i.e. when $\rho_e = 0$ and $\mathbf{J}_e = 0$). To solve the Maxwell's equations in presence of sources, we can introduce the *magnetic vector potential* $\mathbf{A}$ and the *electric scalar potential* ϕ so that:

$$\mathbf{H} = \nabla \times \mathbf{A} , \tag{1.86}$$

$$\mathbf{E} = -\nabla \phi - \frac{1}{c} \frac{\partial \mathbf{A}}{\partial t} . \tag{1.87}$$

The magnetic vector potential always exists as $\mathbf{H}$ is divergence-less (see Eq. (1.2), for non magnetic media) and thus it can be written as Eq. (1.86). The electric scalar potential also always exists: in fact from Eq. (1.3) we obtain:

$$\nabla \times \left(\mathbf{E} + \frac{1}{c} \frac{\partial \mathbf{A}}{\partial t} \right) = 0 , \tag{1.88}$$

and thus $\mathbf{E} + \frac{1}{c}\frac{\partial \mathbf{A}}{\partial t}$ is curl-less and it can be written as the gradient of a scalar potential.

We now simply consider the vacuum and we use Eqs. (1.86, 1.87) in Eqs. (1.1, 1.4):

$$-\nabla^2\phi - \frac{1}{c}\frac{\partial \nabla\cdot\mathbf{A}}{\partial t} = 4\pi\rho_e\,, \tag{1.89}$$

$$\nabla\left(\nabla\cdot\mathbf{A} + \frac{1}{c}\frac{\partial\phi}{\partial t}\right) - \nabla^2\mathbf{A} + \frac{1}{c^2}\frac{\partial^2\mathbf{A}}{\partial t^2} = \frac{4\pi}{c}\mathbf{J}_e\,, \tag{1.90}$$

where we also used Eq. (1.23).

The magnetic vector potential and the electric scalar potential are not uniquely defined. A particular form of the **A** and ϕ is called *gauge*. Here we will consider the *Lorentz gauge* which is defined by

$$\nabla\cdot\mathbf{A} + \frac{1}{c}\frac{\partial\phi}{\partial t} = 0\,. \tag{1.91}$$

With this gauge we finally obtain:

$$\frac{1}{c^2}\frac{\partial^2\phi}{\partial t^2} - \nabla^2\phi = 4\pi\rho_e\,, \tag{1.92}$$

$$\frac{1}{c^2}\frac{\partial^2\mathbf{A}}{\partial t^2} - \nabla^2\mathbf{A} = \frac{4\pi}{c}\mathbf{J}_e\,. \tag{1.93}$$

The solutions of these equations can be readily obtained [1]:

$$\phi(\mathbf{r}, t) = \int \mathrm{d}^3\mathbf{r}'\frac{\rho_e(\mathbf{r}', t-t')}{||\mathbf{r}-\mathbf{r}'||}\,, \tag{1.94}$$

$$\mathbf{A}(\mathbf{r}, t) = \int \mathrm{d}^3\mathbf{r}'\frac{\mathbf{J}_e(\mathbf{r}', t-t')}{c\,||\mathbf{r}-\mathbf{r}'||}\,, \tag{1.95}$$

where $t' = \left|\left|\mathbf{r}-\mathbf{r}'\right|\right|/c$. Eqs. (1.94, 1.95) allow to directly compute the magnetic vector potential and the electric scalar potential in terms of the current and charge sources. Eqs. (1.94, 1.95) are named *retarded potentials* as the values of the electric scalar potential at time t depend on the source values at time $t - t'$.

1.1.3.1 Green's function

In the Fourier space Eqs. (1.92, 1.93) become two *inhomogeneous Helmholtz equations*:

$$\nabla^2\tilde{\phi} + \frac{\omega^2}{c^2}\tilde{\phi} = -4\pi\tilde{\rho}_e\,, \tag{1.96}$$

$$\nabla^2\tilde{\mathbf{A}} + \frac{\omega^2}{c^2}\tilde{\mathbf{A}} = -\frac{4\pi}{c}\tilde{\mathbf{J}}_e\,. \tag{1.97}$$

We now introduce the scalar *Green's function* $g(\mathbf{r})$ as a solution of

$$\nabla^2 g(\mathbf{r}) + k^2 g(\mathbf{r}) = -\delta^{(3)}(\mathbf{r}) \,. \tag{1.98}$$

Equation (1.98) indicates that $g(\mathbf{r})$ represents the potential generated by a delta source at the origin. It can be easily shown that [4]:

$$g(\mathbf{r}) = \frac{\mathrm{e}^{\mathrm{i}kr}}{4\pi r} \tag{1.99}$$

which represents a *spherical wave* (and $r = ||\mathbf{r}||$).

Considering an arbitrary source distribution

$$f(\mathbf{r}) = \int \mathrm{d}^3\mathbf{r}' f(\mathbf{r}')\delta(\mathbf{r} - \mathbf{r}') \,, \tag{1.100}$$

the general solution of the inhomogeneous Helmholtz equation

$$\nabla^2 A(\mathbf{r}) + k^2 A(\mathbf{r}) = -f(\mathbf{r}) \tag{1.101}$$

is:

$$A(\mathbf{r}) = \int \mathrm{d}^3\mathbf{r}' f(\mathbf{r}')g(\mathbf{r} - \mathbf{r}') \,, \tag{1.102}$$

due to the linear superposition principle. It follows that the solutions of Eqs. (1.96, 1.97) are:

$$\tilde{\phi}(\mathbf{r}) = 4\pi \int \mathrm{d}^3\mathbf{r}' \, \tilde{\rho}_e(\mathbf{r}')g(\mathbf{r} - \mathbf{r}') \tag{1.103}$$

$$= \int \mathrm{d}^3\mathbf{r}' \frac{\tilde{\rho}_e(\mathbf{r}') \exp\left(\mathrm{i}k\,||\mathbf{r} - \mathbf{r}'||\right)}{||\mathbf{r} - \mathbf{r}'||} \,, \tag{1.104}$$

$$\tilde{\mathbf{A}}(\mathbf{r}) = \frac{4\pi}{c} \int \mathrm{d}^3\mathbf{r}' \, \tilde{\mathbf{J}}_e(\mathbf{r}')g(\mathbf{r} - \mathbf{r}') \tag{1.105}$$

$$= \int \mathrm{d}^3\mathbf{r}' \frac{\tilde{\mathbf{J}}_e(\mathbf{r}') \exp\left(\mathrm{i}k\,||\mathbf{r} - \mathbf{r}'||\right)}{c\,||\mathbf{r} - \mathbf{r}'||} \,, \tag{1.106}$$

with $k = \omega/c$. Once the potentials $\mathbf{A}$ and ϕ are obtained from the sources, the electric and magnetic fields can be obtained from:

$$\tilde{\mathbf{H}}(\mathbf{r}) = \nabla \times \tilde{\mathbf{A}}(\mathbf{r}) \,, \tag{1.107}$$

$$\tilde{\mathbf{E}}(\mathbf{r}) = -\nabla\tilde{\phi}(\mathbf{r}) + \mathrm{i}\frac{\omega}{c}\tilde{\mathbf{A}}(\mathbf{r}) \,. \tag{1.108}$$

From the Lorentz condition in the Fourier space we have:

$$\nabla(\nabla \cdot \tilde{\mathbf{A}}) = \mathrm{i}\frac{\omega}{c}\nabla\tilde{\phi} \,, \tag{1.109}$$

and thus the electric field can be rewritten as:

$$\tilde{\mathbf{E}} = \frac{\mathrm{i}c}{\omega}\left(\nabla(\nabla \cdot \tilde{\mathbf{A}}) + \frac{\omega^2}{c^2}\tilde{\mathbf{A}}\right) \tag{1.110}$$

$$= \frac{\mathrm{i}c}{\omega}(\nabla \times \nabla \times \tilde{\mathbf{A}}), \tag{1.111}$$

where Eq. (1.111) holds only in the region without sources and we used Eq. (1.97) and Eq. (1.23).

Finally we note that without sources in an homogeneous space, Eq. (1.97) is equivalent to Eq. (1.22), and thus a plane-wave solution can be expressed using the magnetic vector as:

$$\tilde{\mathbf{A}}(\mathbf{r}) = \tilde{\mathbf{A}}^0 \exp(\mathrm{i}\tilde{\mathbf{k}} \cdot \mathbf{r}). \tag{1.112}$$

The electric field can be obtained from Eq. (1.111) and the magnetic field from Eq. (1.107), while the the Poynting vector in the vacuum is:

$$\tilde{\mathbf{S}}(\mathbf{r}) = \frac{\omega^2}{8\pi c}\widehat{\mathbf{k}}\,\left|\left|\tilde{\mathbf{A}}^0\right|\right|^2. \tag{1.113}$$

1.2 Dielectric Function of Metals

1.2.1 *The Drude Model*

The dielectric function of a metal can be obtained considering the dielectric response of a *plasma sea* of electrons with electron concentration N. The optical properties of metals are in fact determined mainly by the response of free electrons. The role of the crystal lattice can be reduced to the modification of the electron mass to give the *effective mass* m_e^* instead of the free electron mass m_e, and to the development of states of high energy for optical transitions.

In this plasma sea model the starting point thus consists in a single particle equation of motion including a damping factor to account for collisions with fixed ions described by the damping constant, $\Gamma = \tau_{coll}^{-1}$, with τ_{coll} being the *mean relaxation time*, i.e. the averaged time between successive collisions. If we assume again an harmonic time dependence of the driving field along the x-axis, ($E_x(t) = \mathrm{Re}\left[E_0 e^{-\mathrm{i}\omega t}\right]$) the equation of motion for the position $x(t)$

of an electron in the plasma sea assumes the simple form:

$$m_e^* \frac{d^2x(t)}{dt^2} + m_e^* \Gamma \frac{dx(t)}{dt} = (-e)\mathrm{Re}\left[E_0 e^{-\mathrm{i}\omega t}\right], \qquad (1.114)$$

where the second term at the left-hand side is due to the damping due to the collisions and the term at the right-end side is the external driving force.

A particular solution of this equation, describing the oscillation of the electron, is $x(t) = \mathrm{Re}\left[\tilde{x}_0 e^{-\mathrm{i}\omega t}\right]$ with [5]:

$$\tilde{x}_0 = \frac{e}{m_e^* \left(\omega^2 + \mathrm{i}\Gamma\omega\right)} E_0 \,. \qquad (1.115)$$

The amplitude $\tilde{x}_0$ is complex to account for possible phase shift between the driving field and the medium response.

The displaced electrons contribute to a macroscopic polarization P_x given by:

$$P_x(t) = (-e)Nx(t) = -\mathrm{Re}\left[\frac{Ne^2}{m_e^* \left(\omega^2 + \mathrm{i}\Gamma\omega\right)} E_0 e^{-\mathrm{i}\omega t}\right]. \qquad (1.116)$$

From Eqs. (1.116, 1.12, 1.14) we can arrive at the expression for the dielectric function for the *Drude free-electron model* [6]:

$$\tilde{\varepsilon}(\omega) = 1 - \frac{4\pi Ne^2}{m_e^* \left(\omega^2 + \mathrm{i}\Gamma\omega\right)} = 1 - \frac{\omega_p^2}{\omega^2 + \mathrm{i}\Gamma\omega}, \qquad (1.117)$$

with the common notation of the *volume plasma frequency* ω_p:

$$\omega_p^2 = \frac{4\pi Ne^2}{m_e^*}\,. \qquad (1.118)$$

In a similar way we can consider the current given by $J(t) = (-e)N\frac{\partial x(t)}{\partial t}$ so that the Drude conductibility is [5]:

$$\tilde{\sigma}(\omega) = \frac{e^2 N}{m_e^* (\Gamma - \mathrm{i}\omega)} \qquad (1.119)$$

The physical meaning of the plasma frequency can be easily understood in the following way: at $\omega = \omega_P$, we have $\tilde{\varepsilon}(\omega) = 0$ so that $\tilde{\mathbf{D}} = \tilde{\varepsilon}(\omega)\tilde{\mathbf{E}} = 0 = \tilde{\mathbf{E}} + 4\pi\tilde{\mathbf{P}}$, i.e.

$$\tilde{\mathbf{P}} = -\frac{\tilde{\mathbf{E}}}{4\pi}\,. \qquad (1.120)$$

Thus a pure depolarization field is present in the medium. If all the free electrons are slightly displaced in one direction with respect to the ions a Coulomb force is created due to macroscopic displacement of the electrons with the respect to the positive ions lattice: this force 'pulls' the electrons back. The *bulk plasmon* is thus a free (i.e. without external field) oscillation of the metal, and it is characterized by a frequency equal to ω_p and a null wavevector: all electrons moves in phase. These oscillations have a longitudinal character, thus cannot be excited in a direct way by the standard optical excitation, as a longitudinal wave is not able to couple with a transverse one. From an experimental point of view, plasmon excitations can be observed by means of the oscillations registered in EELS (electron energy loss spectroscopy) measurements [7].

Equation (1.117) can be also written separating the real and the imaginary part of the dielectric function,

$$\tilde{\varepsilon}(\omega) = 1 - \frac{\omega_p^2}{\omega^2 + \Gamma^2} + \mathrm{i}\frac{\omega_p^2 \Gamma}{\omega\left(\omega^2 + \Gamma^2\right)} \equiv \varepsilon_1(\omega) + \mathrm{i}\varepsilon_2(\omega) . \quad (1.121)$$

The functions $\varepsilon_1, \varepsilon_2$ are shown in Fig. 1.3 together with the associate complex refractive index, see Eqs. (1.32, 1.33), and the normal reflectance from Eq. (1.83). In the plot we considered the case $\Gamma \ll \omega_p$ which holds for typical noble metals.

For a wide range of frequencies $\omega \gg \Gamma$, the dielectric constant is predominantly real and can be well described as:

$$\tilde{\varepsilon}(\omega) \approx 1 - \frac{\omega_p^2}{\omega^2} , \quad (1.122)$$

which corresponds to the expression of the dielectric function for a *pure free electron gas* (i.e. without dissipation). The nearly real value of the dielectric function means negligible absorption. The bottom panel of Fig. 1.3 shows that the ideal free electron gas is a perfect mirror, i.e. reflects all the light for $\omega < \omega_p$, where the real part of the refractive index is close to zero and the extinction coefficient is positive. On the other hand the metal is transparent ($R \approx 0$) for $\omega > \omega_p$, because the electrons in the metals cannot follow the too rapid oscillating incident radiation.

Inserting the free electron gas dielectric function Eq. (1.122) in Eq. (1.28) we obtain the *bulk-plasmon dispersion relation*:

$$\omega^2 = c^2 \left\|\mathbf{k}\right\|^2 + \omega_p^2 , \quad (1.123)$$

implying propagation of transverse electromagnetic waves with frequency $\omega > \omega_p$.

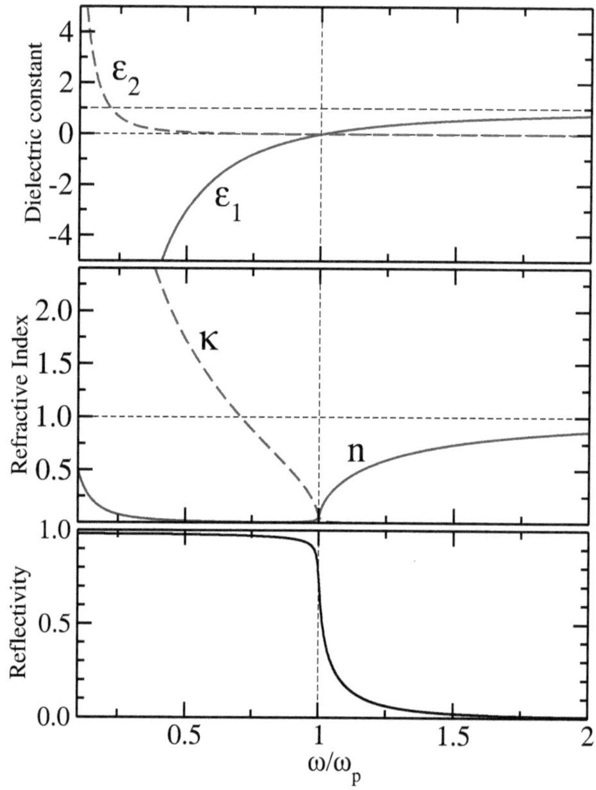

Figure 1.3 Dielectric constant, refractive index and reflectivity from the Drude model ($\hbar\omega_p = 10$ eV, $\hbar\Gamma = 0.1$ eV).

Despite of its simplicity the Drude model correctly describes the optical properties of simple metals. In Fig. 1.4 we report the experimental dielectric constant, the refractive index and the reflectance for aluminum. Figure 1.4 closely resembles Fig. 1.3 with $\omega_p \approx 15$ eV. The best agreement of the dielectric function of metal with that obtained in the framework of the free electron model can be obtained for the alkali metals (Li, Na, K, Cs, Rb), whose response seems to be weakly affected by the contribution from the core electrons. Notably, alkali metals, such as sodium, have an almost free-electron-like response and thus, in accordance

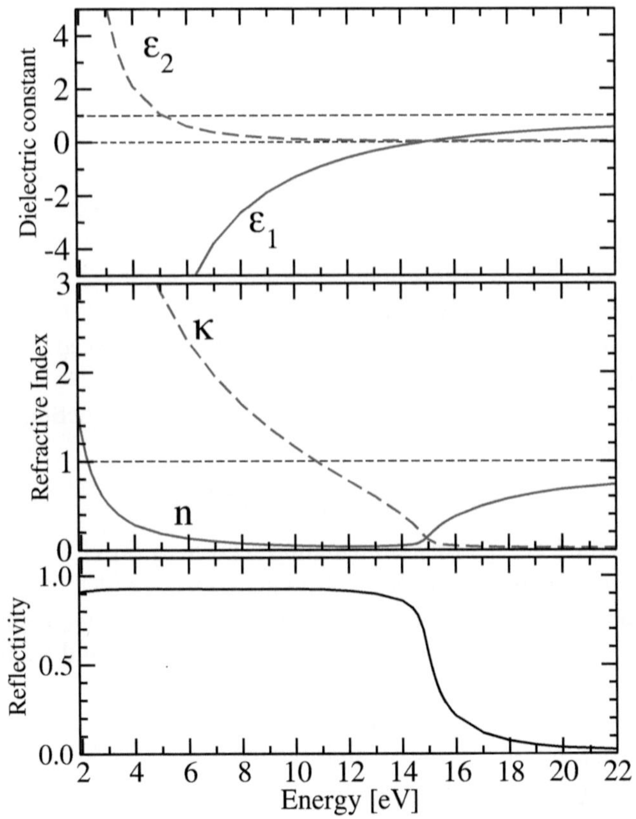

Figure 1.4 Dielectric constant, refractive index and reflectivity of aluminum (from experiments [9]).

with the free electron gas model which predicts propagation of electromagnetic waves for high frequencies $\omega > \omega_p$, exhibit an ultraviolet transparency. Moreover, since for alkali metals the plasma frequency values are quite small ($\hbar\omega = 3.8$ eV for K, 5.7 eV for Na, 7.12 eV for Li [8]), the experimental observation of their transparency is easy since it falls in the near-UV spectral range.

However, at high frequencies in real metals, inter-band transitions (not included in the Drude model) contribute to the imaginary part of $\tilde{\varepsilon}(\omega)$ resulting in absorptive losses.

1.2.2 *Noble Metals*

In *noble or coinage metals* (Cu, Ag, Au), the optical response does not reduce to the response of the free electron gas. Noble metals consist of atoms with completely filled $3d$, $4d$, and $5d$ shells and just a single electron in the $4s$, $5s$, and $6s$ bands, respectively: this last electron in not completely free to move and the dielectric response is essentially influenced by optical transitions of electrons in deeper (e.g. core) levels. These *inter-band excitations* alter the dielectric function considerably. This contribution can be described using a full quantum mechanical treatment, which is introduced in the next section.

For electrons in a real metal, the dielectric function can be written as:

$$\tilde{\varepsilon}(\omega) = \tilde{\varepsilon}^{free}(\omega) + \delta\tilde{\varepsilon}^{inter}(\omega)\,, \tag{1.124}$$

where the free electrons contribution is given by Eq. (1.117) and the inter-band contribution $\delta\tilde{\varepsilon}^{inter}$ is in general a complex function. Its imaginary part describes the direct energy dissipation (absorption) and it is present only for frequencies where inter-band transitions occur.

Figure 1.5 presents the experimentally observed dielectric function, refractive index and reflectance of bulk solid silver. Deviations from the Drude model are evident. Inter-band transitions drastically alter the dielectric function, with absorption losses in the range $5-9$ eV.

In addition for the noble metals further extension of the free electron gas model is necessary for higher frequencies $\omega > \omega_p$. Though the response is essentially determined by free s-electrons, the filled d-band close to the Fermi surface represent a highly polarizable background which can be described by a dielectric constant ε_∞ (typically $1 \leq \varepsilon_\infty \leq 10$), and we can write:

$$\tilde{\varepsilon}\,(\omega) = \varepsilon_\infty - \frac{\omega_p^2}{\omega^2 + \mathrm{i}\omega\Gamma}\,. \tag{1.125}$$

For silver, the parameters in Eq. (1.125) are $\varepsilon_\infty \approx 5$, $\hbar\omega_p \approx 9$ eV, $\hbar\Gamma \approx 0.02$ eV [10].

Finally we recall that in real metals, plasmon oscillations occur not only at the plasma frequency ω_p but in every case when the

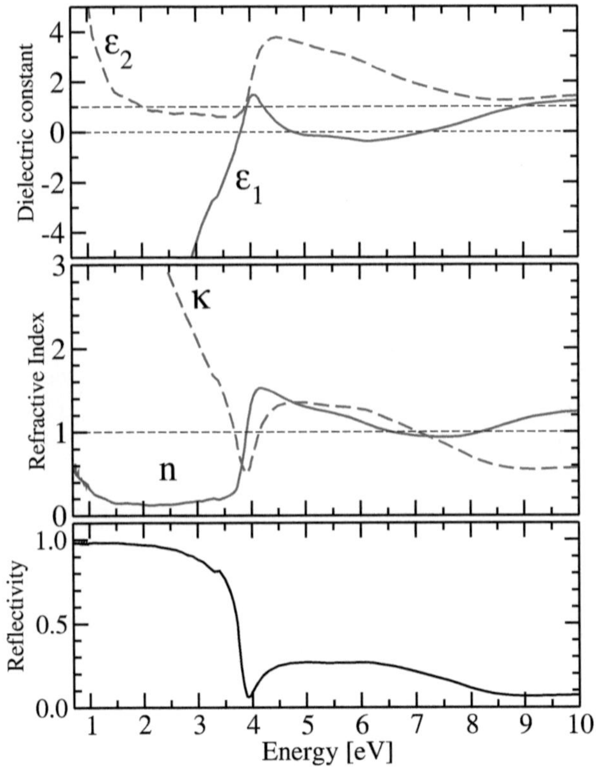

Figure 1.5 Dielectric constant, refractive index and reflectivity of silver (from experiments [9]).

dielectric function equals zero: thus two or more solutions become possible because of the complicated character of the $\tilde{\varepsilon}(\omega)$ function (see Fig. 1.5).

1.2.3 *Linear Response Theory*

The *linear response theory* of quantum mechanics allows to describe how an interacting system of electron respond to an external perturbation [5, 11]. The inverse of the *longitudinal dielectric response function* ($\tilde{\epsilon}^L$) is defined from:

$$\delta\tilde{V}^{tot}(\mathbf{r}) = \int \tilde{\epsilon}^L(\mathbf{r}, \mathbf{r}', \omega)^{-1}\delta\tilde{V}^{ext}(\mathbf{r}')\mathrm{d}^3\mathbf{r}', \qquad (1.126)$$

where the perturbing external potential:

$$\delta V^{ext}(\mathbf{r},t) = \mathrm{Re}\left[\delta \tilde{V}^{ext}(\mathbf{r})e^{-\mathrm{i}\omega t}\right], \quad (1.127)$$

induces a total potential $\delta \tilde{V}^{tot} = \delta \tilde{V}^{ind} + \delta \tilde{V}^{ext}$.

The quantity $\tilde{\epsilon}^L(\mathbf{r},\mathbf{r}',\omega)$ is called *longitudinal* because it is related to the electrostatic potential and not the electric field. An electric field can always be written as:

$$\tilde{\mathbf{E}}(\mathbf{r}) = \tilde{\mathbf{E}}^L(\mathbf{r}) + \tilde{\mathbf{E}}^T(\mathbf{r}), \quad (1.128)$$

where the longitudinal component can be written as $\tilde{\mathbf{E}}^L(\mathbf{r}) = -\nabla V(\mathbf{r})$ and satisfies $\nabla \times \tilde{\mathbf{E}}^L(\mathbf{r}) = 0$, while for the transverse component we have $\nabla \cdot \tilde{\mathbf{E}}^T(\mathbf{r}) = 0$. In addition Eq. (1.126) holds only for isotropic systems.

The function $\tilde{\epsilon}^L(\mathbf{r},\mathbf{r}',\omega)^{-1}$ can be rewritten as:

$$\tilde{\epsilon}^L(\mathbf{r},\mathbf{r}',\omega)^{-1} = \delta(\mathbf{r}-\mathbf{r}') + \int \frac{e^2}{||\mathbf{r}-\mathbf{r}''||}\tilde{\Pi}(\mathbf{r}'',\mathbf{r}',\omega)\mathrm{d}^3\mathbf{r}'', \quad (1.129)$$

where the *linear density response* $\tilde{\Pi}$ (also called *reducible polarizability propagator*) relates the induced electron density $\delta\tilde{n}^{ind}$ to the $\delta\tilde{V}^{ext}$:

$$\delta\tilde{n}^{ind}(\mathbf{r}) = \int \tilde{\Pi}(\mathbf{r},\mathbf{r}',\omega)\delta\tilde{V}^{ext}(\mathbf{r}')\mathrm{d}^3\mathbf{r}'. \quad (1.130)$$

The non-local operator $\tilde{\Pi}$ is a quite complex object, which describes the response of an interacting system of electrons to an external potential. A related definition connects the $\delta\tilde{n}^{ind}$ to the total potential $\delta\tilde{V}^{tot}$:

$$\delta\tilde{n}^{ind}(\mathbf{r}) = \int \tilde{\pi}(\mathbf{r},\mathbf{r}',\omega)\delta\tilde{V}^{tot}(\mathbf{r}')\mathrm{d}^3\mathbf{r}'. \quad (1.131)$$

where $\tilde{\pi}$ is the *irreducible polarizability propagator.* It can be easily shown that [11]:

$$\tilde{\pi}(\mathbf{r},\mathbf{r}',\omega) = \int \tilde{\Pi}(\mathbf{r},\mathbf{r}',\omega)\tilde{\epsilon}^L(\mathbf{r}',\mathbf{r},\omega)^{-1}\mathrm{d}^3\mathbf{r}', \quad (1.132)$$

and

$$\tilde{\epsilon}^L(\mathbf{r},\mathbf{r}',\omega) = \delta(\mathbf{r}-\mathbf{r}') - \int \frac{e^2}{||\mathbf{r}-\mathbf{r}''||}\tilde{\pi}(\mathbf{r}'',\mathbf{r}',\omega)\mathrm{d}^3\mathbf{r}''. \quad (1.133)$$

The exact linear density response can be related to the linear density response of a *non-interacting reference system* of electrons $\tilde{\Pi}_s$ using the *Dyson equation*:

$$\tilde{\Pi}\left(\mathbf{r},\mathbf{r}',\omega\right)=\tilde{\Pi}_s\left(\mathbf{r},\mathbf{r}',\omega\right)+\int \mathrm{d}^3\mathbf{r}''\tilde{\Pi}_s\left(\mathbf{r},\mathbf{r}'',\omega\right) \tag{1.134}$$

$$+\int \mathrm{d}^3\mathbf{r}'''\left(\frac{e^2}{||\mathbf{r}''-\mathbf{r}'''||}+f_{xc}\left(\mathbf{r}'',\mathbf{r}''',\omega\right)\right)\tilde{\Pi}\left(\mathbf{r}''',\mathbf{r}',\omega\right), \tag{1.135}$$

where f_{xc} is the *exchange-correlation kernel* which describes all the quantum effects between electrons. More details will be given in Chapter 4. Neglecting f_{xc}, the dielectric response can be directly written as:

$$\tilde{\epsilon}^L(\mathbf{r},\mathbf{r}',\omega)\approx\delta(\mathbf{r}-\mathbf{r}')-\int\frac{e^2}{||\mathbf{r}-\mathbf{r}''||}\tilde{\Pi}_s(\mathbf{r}'',\mathbf{r}',\omega)\mathrm{d}^3\mathbf{r}''. \tag{1.136}$$

Equation (1.136) is known as the *Random Phase Approximation (RPA)* and it is widely used in solid-state theoretical physics [5, 11]. From Eq. (1.133) it follows that in the RPA, $\tilde{\pi}(\mathbf{r},\mathbf{r}',\omega)\approx\tilde{\Pi}_s(\mathbf{r},\mathbf{r}',\omega)$.

For a homogeneous bulk solid one can use the Fourier transform in the space, i.e. assuming a spatial dependent of the type $e^{\mathrm{i}\mathbf{q}\cdot\mathbf{r}}$ and thus we have:

$$\tilde{\epsilon}^L(\mathbf{q},\omega)\approx 1-\frac{4\pi e^2}{||\mathbf{q}||^2}\tilde{\Pi}_s(\mathbf{q},\omega)\,. \tag{1.137}$$

In quantum mechanics, a periodic bulk solid can be described as [5, 8, 11]:

$$\hat{\mathcal{H}}\left|n\mathbf{k}\right\rangle=\mathcal{E}_{n,\mathbf{k}}\left|n\mathbf{k}\right\rangle\,, \tag{1.138}$$

where $\hat{\mathcal{H}}$ in the bulk Hamiltonian operator and $|n\mathbf{k}\rangle$ are the *Bloch* states. The RPA dielectric constant can be written as [5, 11]:

$$\tilde{\epsilon}^L(\mathbf{q},\omega)=1-\frac{4\pi e^2}{||\mathbf{q}||^2 V}\sum_{\mathbf{k},n,n'}\left|\left\langle n',\mathbf{k}+\mathbf{q}|n,\mathbf{k}\right\rangle\right|^2$$
$$\times\frac{f_{n',\mathbf{k}+\mathbf{q}}-f_{n,\mathbf{k}}}{\mathcal{E}_{n',\mathbf{k}+\mathbf{q}}-\mathcal{E}_{n,\mathbf{k}}-\hbar\omega-\mathrm{i}\eta^+}, \tag{1.139}$$

where $f_{n,\mathbf{k}}$ is the Fermi occupation number of the Bloch state (from 0 to 2, due to spins) and η^+ is an infinitesimal small positive number. Equation (1.139) describes the complex dielectric response function for an external perturbation of the type $e^{\mathrm{i}\mathbf{q}\cdot\mathbf{r}}$ as a sum over all $\mathbf{k}$ and all bands n, n'. Thus it includes both *intra-band excitations*, i.e. with $n = n'$, as well as *inter-band excitations*, i.e. with $n \neq n'$.

Here, it is interesting to consider intra-band transition and to verify that Eq. (1.139) recovers the Drude model [5]. We are interested for perturbation with small $\mathbf{q}$, i.e. for wavelength much larger than the bulk unit cell. In this long wavelength limit we have:

$$\mathcal{E}_{n,\mathbf{k}+\mathbf{q}} \approx \mathcal{E}_{n,\mathbf{k}} + \nabla_{\mathbf{k}}\mathcal{E}_{n,\mathbf{k}} \cdot \mathbf{q} + \frac{1}{2}\mathbf{q} \cdot \nabla_{\mathbf{k}}\nabla_{\mathbf{k}}\mathcal{E}_{n,\mathbf{k}} \cdot \mathbf{q} \tag{1.140}$$

$$\approx \mathcal{E}_{n,\mathbf{k}} + \nabla_{\mathbf{k}}\mathcal{E}_{n,\mathbf{k}} \cdot \mathbf{q} + \frac{1}{2}||\mathbf{q}||^2 \frac{\hbar^2}{m_n^*}, \tag{1.141}$$

where we considered an isotropic medium and define the effective mass as:

$$\frac{1}{m_n^*} = \frac{1}{\hbar^2}\frac{\mathrm{d}^2\mathcal{E}_{n,\mathbf{k}}}{\mathrm{d}k^2}. \tag{1.142}$$

For intra-band transition $n = n'$ so that the matrix element in Eq. (1.139) is simply

$$\langle n, \mathbf{k}+\mathbf{q}|n, \mathbf{k}\rangle \rightarrow 1, \tag{1.143}$$

due to the limit $\mathbf{q} \rightarrow 0$. The intra-band dielectric constant can be rewritten as:

$$\tilde{\epsilon}^{intra}(\mathbf{q} \rightarrow 0, \omega) = 1 - \frac{4\pi e^2}{||\mathbf{q}||^2 V}\sum_{\mathbf{k},n} f_{n,\mathbf{k}}\left(\frac{1}{\mathcal{E}_{n,\mathbf{k}} - \mathcal{E}_{n,\mathbf{k}-\mathbf{q}} - \hbar\omega - i\eta^+} - \frac{1}{\mathcal{E}_{n,\mathbf{k}+\mathbf{q}} - \mathcal{E}_{n,\mathbf{k}} - \hbar\omega - i\eta^+}\right). \tag{1.144}$$

For intra-band transitions with energy $\hbar\omega$ much larger than the $|\mathcal{E}_{n,\mathbf{k}+\mathbf{q}} - \mathcal{E}_{n,\mathbf{k}}|$: the real part of the dielectric constant becomes:

$$\tilde{\epsilon}_1^{intra}(\mathbf{q} \rightarrow 0, \omega) \approx 1 - \frac{4\pi e^2}{V}\frac{1}{\omega^2}\sum_{\mathbf{k},n} f_{n,\mathbf{k}}\frac{1}{m_n^*} \tag{1.145}$$

where we used Eq. (1.141).

For a simple metal with just one band we have

$$\tilde{\epsilon}_1^{intra}(\mathbf{q} \rightarrow 0, \omega) \approx 1 - \frac{4\pi e^2}{V}\frac{N}{\omega^2 m^*} \tag{1.146}$$

which coincides with the result in Eq. (1.117).

The imaginary part of the dielectric constant for intra-band transition is obtained directly from Eq. (1.144):

$$\tilde{\epsilon}_2^{intra}(\mathbf{q} \to 0, \omega) = \frac{4\pi^2 e^2}{||\mathbf{q}||^2 V} \sum_{\mathbf{k},n} f_{n,\mathbf{k}} \Big(\delta(\mathcal{E}_{n,\mathbf{k+q}} - \mathcal{E}_{n,\mathbf{k}} - \hbar\omega) - \delta(\mathcal{E}_{n,\mathbf{k-q}} - \mathcal{E}_{n,\mathbf{k}} + \hbar\omega) \Big), \tag{1.147}$$

using that $\mathrm{Im}\left[1/(X - \mathrm{i}\eta^+)\right] = \pi\delta(X)$. The term $\tilde{\epsilon}_2^{intra}$ correspond to absorption losses at low frequencies.

Inter-band transitions can be considered as well, directly from Eq. (1.139) [5].

1.3 Delocalized Surface Plasmon Resonances

In the previous sections we have considered electromagnetic waves which propagate in an homogeneous material or which are reflected at an interface. A different type of electromagnetic waves which are important in metal optics are the *delocalized surface plasmon resonances* or *surface plasmon polaritons* (SPPs), which consist in propagating, dispersive electromagnetic waves coupled to the electron plasma of a metal at a dielectric interface, but evanescently confined in the perpendicular direction [7, 12, 13].

In order to investigate the origin and the physical properties of SPPs, we consider a metal-dielectric interface described by the plane $z = 0$ (see Fig. 1.6). The local frequency-dependent dielectric function is supposed to change in a stepwise manner from the dielectric with $\tilde{\varepsilon}(z) = \varepsilon_d$ (for $z > 0$) to the metal with $\tilde{\varepsilon}(z) = \tilde{\varepsilon}_m(\omega)$ (for $z < 0$).

In absence of external current Eq. (1.21) becomes:

$$\nabla \times \nabla \times \tilde{\mathbf{E}}(x, y, z) = \frac{\omega^2}{c^2} \tilde{\varepsilon}(z) \tilde{\mathbf{E}}(x, y, z) \,. \tag{1.148}$$

We can assume propagation in the x direction and search for a solution of the type:

$$\tilde{\mathbf{E}}(x, y, z) = \tilde{\mathbf{E}}^0(z) e^{\mathrm{i}\tilde{k}_x x}, \tag{1.149}$$

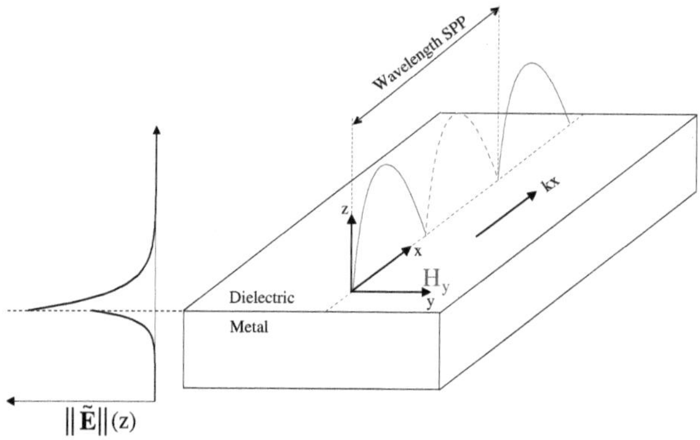

Figure 1.6 Geometry for surface plasmon polariton (SPP) propagation (along the x-axis) at a single interface between a dielectric ($z > 0$) and a metal ($z < 0$). The SPP wavelength is also shown. On the left, the norm electric field $||\tilde{\mathbf{E}}||$ for a SPP, which depends only on the z coordinate and it is confined at the interface.

(due to homogeneity in the y direction) so that:

$$\frac{\mathrm{d}^2\tilde{\mathbf{E}}^0(z)}{\mathrm{d}z^2} + \left(\frac{\omega^2}{c^2}\varepsilon_d - \tilde{k}_x^2\right)\tilde{\mathbf{E}}^0(z) = 0 \text{ for } z > 0\,, \qquad (1.150)$$

$$\frac{\mathrm{d}^2\tilde{\mathbf{E}}^0(z)}{\mathrm{d}z^2} + \left(\frac{\omega^2}{c^2}\tilde{\varepsilon}_m(\omega) - \tilde{k}_x^2\right)\tilde{\mathbf{E}}^0(z) = 0 \text{ for } z < 0\,. \qquad (1.151)$$

Eqs. (1.150, 1.151) can have solutions *confined* at the interface (see Fig. 1.6) of the type:

$$\tilde{\mathbf{E}}^0(z) = \tilde{\mathbf{E}}^d \exp(-\tilde{k}_z^d(\omega)\,z) \text{ for } z > 0\,, \qquad (1.152)$$

$$\tilde{\mathbf{E}}^0(z) = \tilde{\mathbf{E}}^m \exp(+\tilde{k}_z^m(\omega)\,z) \text{ for } z < 0\,, \qquad (1.153)$$

with

$$\tilde{k}_z^d(\omega)^2 = \tilde{k}_x^2 - \frac{\omega^2}{c^2}\varepsilon_d \text{ for } z > 0\,, \qquad (1.154)$$

$$\tilde{k}_z^m(\omega)^2 = \tilde{k}_x^2 - \frac{\omega^2}{c^2}\tilde{\varepsilon}_m(\omega) \text{ for } z < 0\,. \qquad (1.155)$$

Eqs. (1.149, 1.152, 1.153) define the form of a SPP solution, which must propagate along the x direction and must be confined along the z direction; such requirements imply the following conditions:

$$\tilde{k}_z^d(\omega) \approx \mathrm{Re}\left[\tilde{k}_z^d(\omega)\right] > 0\,, \qquad (1.156)$$

$$\tilde{k}_z^m(\omega) \approx \mathrm{Re}\left[\tilde{k}_z^m(\omega)\right] > 0\,, \qquad (1.157)$$

$$\tilde{k}_x \approx \mathrm{Re}\left[\tilde{k}_x\right]. \qquad (1.158)$$

In addition we have the condition $\nabla \cdot \tilde{\mathbf{E}}(x, y, z) = 0$ that implies:

$$i\tilde{k}_x \tilde{E}_x^d - \tilde{k}_z^d(\omega)\tilde{E}_z^d = 0 \quad \text{for} \quad z > 0\,, \tag{1.159}$$

$$i\tilde{k}_x \tilde{E}_x^m + \tilde{k}_z^m(\omega)\tilde{E}_z^m = 0 \quad \text{for} \quad z < 0\,. \tag{1.160}$$

No condition is found for the y component of the field which can be set to zero. A solution with a non zero y component of the field will correspond to a TE (s-polarized) electromagnetic wave. However it can be easily shown from boundary conditions that such solution cannot exist [7]. Thus surface plasmon polaritons localized at the interfaces exist only in the TM (p-polarized) mode.

From the boundary conditions (see Eq. (1.52)) we have that:

$$\tilde{E}_x^d = \tilde{E}_x^m\,, \tag{1.161}$$

$$\varepsilon_d \tilde{E}_z^d = \tilde{\varepsilon}_m(\omega)\tilde{E}_z^m\,. \tag{1.162}$$

Eqs. (1.150, 1.151, 1.159, 1.160, 1.161, 1.162) form a system of equations which can be solved as a function of $\tilde{E}_x^d$ obtaining:

$$\tilde{k}_x^2 = \frac{\omega^2}{c^2}\frac{\tilde{\varepsilon}_m(\omega)\varepsilon_d}{\tilde{\varepsilon}_m(\omega) + \varepsilon_d}\,, \tag{1.163}$$

$$\tilde{k}_z^d(\omega)^2 = -\frac{\omega^2}{c^2}\frac{\varepsilon_d^2}{\tilde{\varepsilon}_m(\omega) + \varepsilon_d} = -\tilde{k}_x^2\frac{\varepsilon_d}{\tilde{\varepsilon}_m(\omega)}\,, \tag{1.164}$$

$$\tilde{k}_z^m(\omega)^2 = -\frac{\omega^2}{c^2}\frac{\tilde{\varepsilon}_m(\omega)^2}{\tilde{\varepsilon}_m(\omega) + \varepsilon_d} = -\tilde{k}_x^2\frac{\tilde{\varepsilon}_m(\omega)}{\varepsilon_d}\,, \tag{1.165}$$

$$\tilde{E}_z^m = -\tilde{E}_x^d\sqrt{\frac{\varepsilon_d}{\tilde{\varepsilon}_m(\omega)}}\,, \tag{1.166}$$

$$\tilde{E}_z^d = -\tilde{E}_x^d\sqrt{\frac{1}{\tilde{\varepsilon}_m(\omega)\varepsilon_d}}\,. \tag{1.167}$$

The conditions for the SPP solution in Eqs. (1.156, 1.157, 1.158) can be satisfied if:

$$\tilde{\varepsilon}_m(\omega) + \varepsilon_d \approx \mathrm{Re}[\tilde{\varepsilon}_m(\omega) + \varepsilon_d] < 0\,, \tag{1.168}$$

$$\tilde{\varepsilon}_m(\omega)\varepsilon_d \approx \mathrm{Re}[\tilde{\varepsilon}_m(\omega)\varepsilon_d] < 0\,. \tag{1.169}$$

Since dielectrics have a positive (and real) ε_d, this means that $\tilde{\varepsilon}_m(\omega)$ must be almost real and negative. This condition is largely fulfilled in several metals in the visible and near-infrared (near-IR) parts of the spectrum for which $\tilde{\varepsilon}_m(\omega)$ has a large negative real part, see Fig. 1.5.

1.3.1 *The Surface Plasmon-Polariton Wavevector*

Equation (1.163) represents the *dispersion relation* of the surface plasmon polariton mode (SPP-mode), i.e. the relationship between the frequency ω and the in-plane wavevector $\tilde{k}_x$ for SPPs propagating along the interface between a metal and a dielectric.

We now consider a simple interface between a dielectric and an ideal free-electron metal with a dielectric constant given by Eq. (1.122). The dispersion relation becomes

$$\hat{k}_x = \hat{\omega}\sqrt{\frac{\varepsilon_d(\hat{\omega}^2 - 1)}{(\varepsilon_d + 1)\hat{\omega}^2 - 1}}, \tag{1.170}$$

where $\hat{k}_x = \tilde{k}_x/(\frac{\omega_p}{c})$ is the normalized wavevector and $\hat{\omega} = \omega/\omega_p$ is the normalized frequency. This relation can be inverted yielding two solutions:

$$\hat{\omega}^2 = \frac{1}{2} + \frac{\varepsilon_d + 1}{2\varepsilon_d}\hat{k}_x^2 + \sqrt{\left(\frac{1}{2} + \frac{\varepsilon_d + 1}{2\varepsilon_d}\hat{k}_x^2\right)^2 - \frac{\hat{k}_x^2}{\varepsilon_d}}, \tag{1.171}$$

$$\hat{\omega}^2 = \frac{1}{2} + \frac{\varepsilon_d + 1}{2\varepsilon_d}\hat{k}_x^2 - \sqrt{\left(\frac{1}{2} + \frac{\varepsilon_d + 1}{2\varepsilon_d}\hat{k}_x^2\right)^2 - \frac{\hat{k}_x^2}{\varepsilon_d}}, \tag{1.172}$$

which are shown in Fig. 1.7.

The solution in the upper branch correspond to Eq. (1.171) and it is characterized by $\omega > \omega_p$: in this case the dielectric constant of the metal is positive and both $\tilde{k}_z^d(\omega)$ and $\tilde{k}_z^m(\omega)$ are imaginary, thus describing a propagating electromagnetic wave in the z direction.

A SPP-mode corresponds instead to Eq. (1.172) and lies entirely below the *light cone*, i.e. the dispersion law of electromagnetic waves in a vacuum $\omega = ck$. For $\omega \ll \omega_P$ the SPP dispersion curve tends to the light line $\omega = ck\sqrt{\varepsilon_d}$ and therefore the surface mode has the features of an ordinary electromagnetic wave.

In the opposite limit of infinitely large wavenumbers ($\omega \gg \omega_p$), the SPP dispersion curve approaches the surface plasmon frequency

$$\omega_{\text{SPP}} = \frac{\omega_p}{\sqrt{1 + \varepsilon_d}}. \tag{1.173}$$

Within the model of an ideal free electron plasma, the spectral range between the surface plasmon frequency and the plasma

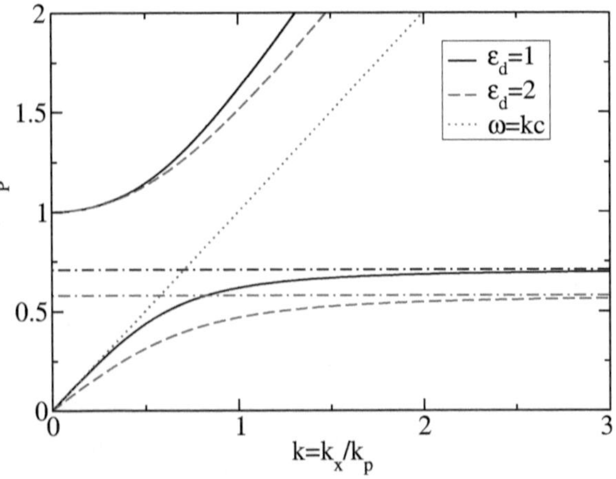

Figure 1.7 Normalized SPP dispersion relation for dielectric/ideal-free-electron-plasma interface for $\varepsilon_d = 1$ and $\varepsilon_d = 2$. The straight dotted line represents the dispersion law of electromagnetic waves in a vacuum $\omega = ck$.

frequency represents a *gap* where no propagating electromagnetic modes exist. Anyway, in real metals, the contribution from interband transitions modifies the SPP dispersion curve excluding an unlimited growth of the wavenumber at $\omega \to \omega_{\text{SPP}}$ [7].

As we can observe in Fig. 1.7 the magnitude of the SPP momentum $\tilde{k}_x$, is always larger than that of a photon in the dielectric medium bounding the metal, $\sqrt{\varepsilon_d}\omega/c$: this means that the SPP mode has a non-radiative or bound nature and cannot be excited by incident plane-wave light. In the context of photonics, light and SPP modes have to be coupled with different techniques (prism coupling, near-field excitation, diffraction grating…) [4, 7, 12, 14].

From the surface plasmon–polariton wavevector (Eq. (1.163)) we can then obtain the wavelength of the surface plasmon–polariton λ_{SPP} which represents the period of the surface charge density oscillation and of the associated field distribution of the mode (see Fig. 1.6). In particular, the SPP wavelength can be found from the complex dispersion relation Eq. (1.163) by taking the real part. Writing the complex relative permittivity of the metal as $\tilde{\varepsilon}_m(\omega) = -|\varepsilon'_m| + i\varepsilon''_m$ we can write the complex SPP wavevector as $\tilde{k}_x = k'_x + ik''_x$

and consider its real part given by:

$$k'_x \approx \frac{\omega}{c}\sqrt{\frac{\varepsilon_d|\varepsilon'_m|}{|\varepsilon'_m| - \varepsilon_d}}, \tag{1.174}$$

which is valid under the assumption that $|\varepsilon'_m| \gg |\varepsilon''_m|$ and $|\varepsilon'_m| > \varepsilon_d$. Thus the SPP wavelength is:

$$\lambda_{\mathrm{SPP}} = \frac{2\pi}{k'_x} = \lambda_0\sqrt{\frac{1}{\varepsilon_d} - \frac{1}{|\varepsilon'_m|}}, \tag{1.175}$$

i.e. always slightly less than, the free-space wavelength, λ_0. Again, the fact that $\lambda_{\mathrm{SPP}} < \lambda_0$ reflects the bound nature of SPP modes on a planar surface. The length λ_{SPP} is very important to realize nanophotonics structures of the right dimensions [7, 12, 14].

1.3.1.1 Propagation length

The propagation length of a SPP mode, δ_{SPP}, is the distance over which the field falls to $1/e$ of its initial value, and it is given by $\delta_{\mathrm{SPP}} = 1/k''_x$ where k''_x is the imaginary part of the SPP wavevector Eq. (1.163):

$$k''_x \approx \frac{\omega}{c}\frac{\varepsilon''_m}{2|\varepsilon'_m|^2}\left(\frac{|\varepsilon'_m|\varepsilon_d}{|\varepsilon'_m| - \varepsilon_d}\right)^{3/2} \tag{1.176}$$

which is valid under the assumption that $|\varepsilon'_m| \gg |\varepsilon''_m|$ and $|\varepsilon'_m| > \varepsilon_d$. The propagation length is thus:

$$\delta_{\mathrm{SPP}} = \frac{\lambda_0}{2\pi}\frac{2|\varepsilon'_m|^2}{\varepsilon''_m}\left(\frac{1}{\varepsilon_d} - \frac{1}{|\varepsilon'_m|}\right)^{3/2}. \tag{1.177}$$

From Eq. (1.177) we can observe that for a long propagation length we require a large negative real part of the relative permittivity of the metal, ε'_m, and a small imaginary part, ε''_m so that the absorptive nature of metals in the visible and near-IR places considerable restrictions on the construction of photonic components or circuits based on SPPs.

In addition, we see that the SPP propagation length is significantly greater than the SPP wavelength. This allows surface plasmon-polaritons to be manipulated by wavelength scale gratings and other periodic surface structures [7, 12, 14].

1.3.1.2 Field-penetration depths

The penetration of the fields into the dielectric and the metal can be found by considering Eq. (1.164) and Eq. (1.165), respectively. The electric field falls to $1/e$ of its value at distances (from the interface) when:

$$\delta_d = \frac{1}{\mathrm{Re}\left[\tilde{k}_z^d(\omega)\right]} \approx \frac{\lambda_0}{2\pi}\frac{\sqrt{|\varepsilon'_m| - \varepsilon_d}}{\varepsilon_d}, \tag{1.178}$$

$$\delta_m = \frac{1}{\mathrm{Re}\left[\tilde{k}_z^m(\omega)\right]} \approx \frac{\lambda_0}{2\pi}\frac{\sqrt{|\varepsilon'_m| - \varepsilon_d}}{|\varepsilon'_m|}, \tag{1.179}$$

where we have assumed once more that $|\varepsilon'_m| \gg |\varepsilon''_m|$ and $|\varepsilon'_m| > \varepsilon_d$.

For $|\varepsilon'_m| > \varepsilon_d$ the penetration depth into the dielectric is larger than the one in the metal. For a silver/vacuum interfaces at visible wavelengths we have $\delta_m \approx$ 20–30 nm and $\delta_d \approx$ 300–500 nm.

For a simple interface between a dielectric and an ideal free-electron metal with dielectric constant given by Eq. (1.122) we can introduce the normalized penetration depths:

$$\hat{\delta}_d = \delta_d \frac{\omega_p}{c} = \frac{\sqrt{1-(1+\varepsilon_d)\hat{\omega}^2}}{\varepsilon_d \hat{\omega}}, \tag{1.180}$$

$$\hat{\delta}_m = \delta_m \frac{\omega_p}{c} = \frac{\sqrt{1-(1+\varepsilon_d)\hat{\omega}^2}}{(\hat{\omega}^2 - 1)}, \tag{1.181}$$

which are defined for $\omega < \omega_{\mathrm{SPP}}$ and shown in Fig. 1.8.

For a wide range of frequencies and ε_d, we can see that $\hat{\delta}_m = 1$ and thus:

$$\delta_m \approx \frac{c}{\omega_p} = \frac{\lambda_P}{2\pi}, \tag{1.182}$$

where λ_p is the wavelength associated with the plasma frequency. For $\hbar\omega_p = 10$ eV, we have $\delta_m \approx 19.7$ nm. The penetration depth into the metal is a measure of the required thickness of metal films that allow coupling to freely propagating light and of the sizes needed to control SPPs for nanophotonics applications [7, 12, 14].

1.4 Localized Surface Plasmon Resonances

Until now we have analyzed the electron plasma excitations inside metals due to the interaction of radiation with a bulk (volume

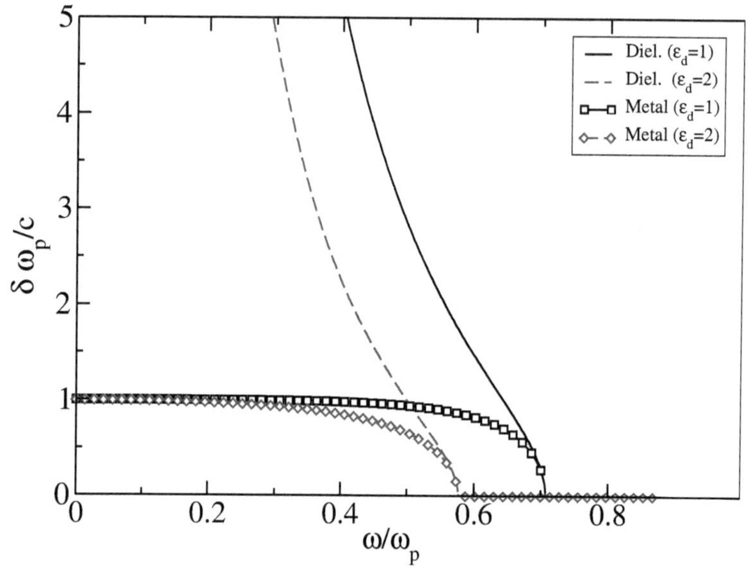

Figure 1.8 Normalized field penetrations for the dielectric/ideal free-electron plasma interface for $\varepsilon_d = 1$ and $\varepsilon_d = 2$.

plasmon) or with a surface (surface plasmon polaritons). Another important class of plasmons exists which arises naturally from the scattering of a small, sub-wavelength conductive nanoparticle in an oscillating electromagnetic field.

For isolated metal nanoparticles (MNPs), whose sizes fall in the range of the penetration depth of an electromagnetic field into the metal (e.g., 20 nm for silver in the optical spectral range), an external electromagnetic field can penetrate into them and shift the free conduction electrons with respect to the nuclei and the core electrons. The surface charges of opposite sign thus appearing on the opposite sides of the particle produce a restoring local field within the particle. The coherently shifted electrons in the MNP generate a resonance, whose characteristics are mainly defined by the electron density and the geometry of the MNP. This resonance is called *localized surface plasmon resonance* (LSPR) and leads to field amplification both inside and in the near-field zone outside the MNP. LSPRs are non-propagating excitations of the conduction electrons of metallic nanostructures coupled to the electromagnetic field which, in contrast to propagating SPPs, can be excited by direct light illumination. Spectrally, the LSPRs are found somewhere in the visible to near-infrared spectral region, depending on the particle shape, the environment and the metal dielectric constant. If the MNP is excited at resonance, the amplitude of the induced electromagnetic field can exceed the exciting fields by orders of magnitude (optical near field enhancement). Finally, we note that MNPs usually have more than a single oscillation mode. The different modes differ in their charge- and field- distribution. For the lowest LSPR, the distributions are dominated by a dipolar character while the higher energetic modes can be described with multipolar charge distributions.

1.4.1 *Nanosphere in the Quasi-Static Approximation*

If we consider a MNP with size much smaller than the wavelength of the excited light, the phase of the oscillating exciting field can be considered constant over the particle volume. This is the *quasi-static approximation* and it allows to compute the spatial field distribution solving the simplified problem of a particle in an uniform electrostatic field. The harmonic time dependence can then be added to the solution once the field distributions are known. In this section we are describing the lowest-order approximation of the full scattering problem, which will be analyzed in Sec. 1.6. The results presented in this section can be used to adequately describe the optical properties of MNPs of dimensions below 100 nm.

To simplify the analytical treatment, we consider a homogeneous, isotropic sphere of radius R located at the origin in a uniform, static electric field along the z direction ($\mathbf{E}_0 = E_0\widehat{\mathbf{z}}$). The dielectric response of the MNP is described by the complex dielectric function $\tilde{\varepsilon}_p(\omega)$, while the surrounding environment is isotropic and non-absorbing with a real dielectric constant ε_{out}. The geometrical configuration is reported in Fig. 1.9.

This is thus an electrostatic problem, which requires the solution of the the *Laplace equation* for the electrostatic potential $\tilde{\Phi}$ (in the Fourier domain):

$$\nabla^2\tilde{\Phi}(\mathbf{r}) = 0\,, \tag{1.183}$$

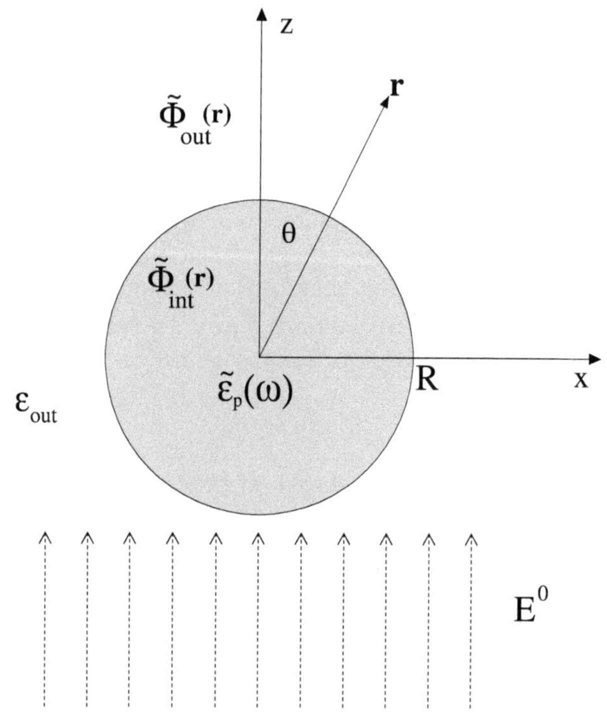

Figure 1.9 Geometry of the nanosphere electrostatic problem: a nanosphere of radius R and dielectric constant $\tilde{\varepsilon}_p(\omega)$ is located in an external environment of dielectric constant ε_{out} and excited by an uniform electric field in the z direction. $\tilde{\Phi}_{int}$ is the electrostatic potential internal to the sphere and $\tilde{\Phi}_{out}$ is the electrostatic potential outside it.

which is valid inside and outside the sphere, while at the sphere surface the boundary conditions have to be satisfied. The time-dependent potential can be obtained from $\Phi(\mathbf{r}, t) = \mathrm{Re}\left[\tilde{\Phi}(\mathbf{r})e^{-\mathrm{i}\omega t}\right]$.

It can be seen that this problem has an azimuthal symmetry and the general solution is of the form [1]:

$$\tilde{\Phi}(r, \theta) = \sum_{l=0}^{\infty} \left[A_l r^l + B_l r^{-l(l+1)}\right] P_l(\cos\theta)\,, \qquad (1.184)$$

where $P_l(\cos\theta)$ are the Legendre polynomials of order l [$P_0(x) = 1$, $P_1(x) = x$, $P_2(x) = 1/2(3x^2 - 1), \ldots$] and θ the angle between the position vector $\mathbf{r}$ and the z-axis. As there are no charges at the origin, the potential must be finite there and the solution for the potentials $\tilde{\Phi}_{int}$ inside and $\tilde{\Phi}_{out}$ outside the sphere can be written as:

$$\tilde{\Phi}_{int}(r,\theta) = \sum_{l=0}^{\infty} A_l r^l P_l(\cos\theta)\,, \tag{1.185}$$

$$\tilde{\Phi}_{out}(r,\theta) = \sum_{l=0}^{\infty} \left[B_l r^l + C_l r^{-(l+1)}\right] P_l(\cos\theta)\,. \tag{1.186}$$

The coefficients A_l , B_l and C_l can be obtained considering the boundary conditions at $r \to \infty$ and at the sphere surface ($r = R$). The condition that $\tilde{\Phi}_{out} \to -E_0 z = -E_0 r\cos\theta$ as $r \to \infty$ requires that $B_{l=1} = -E_0$ and $B_l = 0$ for $l \neq 1$. The other coefficients A_l and C_l are defined by the boundary conditions at $r = R$. From Eq. (1.52) we have:

$$-\frac{1}{R}\frac{\partial\tilde{\Phi}_{int}}{\partial\theta}\bigg|_{r=R} = -\frac{1}{R}\frac{\partial\tilde{\Phi}_{out}}{\partial\theta}\bigg|_{r=R}\,, \tag{1.187}$$

$$-\tilde{\varepsilon}_p(\omega)\frac{\partial\tilde{\Phi}_{int}}{\partial r}\bigg|_{r=R} = -\varepsilon_{out}\frac{\partial\tilde{\Phi}_{out}}{\partial r}\bigg|_{r=R}\,. \tag{1.188}$$

From these two conditions we obtain that $A_l = C_l = 0$ for $l \neq 1$, and calculating of the remaining coefficients A_l and C_l we obtain the final expression for the potentials:

$$\tilde{\Phi}_{int}(r,\theta) = -\frac{3\varepsilon_{out}}{\tilde{\varepsilon}_p(\omega) + 2\varepsilon_{out}} E_0 r\cos\theta\,, \tag{1.189}$$

$$\tilde{\Phi}_{out}(r,\theta) = -E_0 r\cos\theta + R^3\frac{\tilde{\varepsilon}_p(\omega) - \varepsilon_{out}}{\tilde{\varepsilon}_p(\omega) + 2\varepsilon_{out}} E_0 \frac{\cos\theta}{r^2}\,. \tag{1.190}$$

Eq. (1.190) indicates that $\tilde{\Phi}_{out}$ is given the superposition of the external applied potential and that of a dipole located at the particle center. We can simplify $\tilde{\Phi}_{out}$ by introducing the dipole moment $\tilde{\boldsymbol{\mu}}_{sph}$ of the sphere as

$$\tilde{\Phi}_{out}(\mathbf{r}) = -\mathbf{E}_0\cdot\mathbf{r} + \frac{1}{\varepsilon_{out}}\frac{\tilde{\boldsymbol{\mu}}_{sph}\cdot\mathbf{r}}{r^3}\,, \tag{1.191}$$

with

$$\tilde{\boldsymbol{\mu}}_{sph} = \varepsilon_{out} R^3 \frac{\tilde{\varepsilon}_p(\omega) - \varepsilon_{out}}{\tilde{\varepsilon}_p(\omega) + 2\varepsilon_{out}} \mathbf{E}_0\,. \tag{1.192}$$

The external applied field induces a dipole moment inside the sphere of a magnitude proportional to $\mathbf{E}_0$. Thus we can introduce the *effective polarizability of the sphere* $\tilde{\alpha}_{sph}(\omega)$, defined via $\tilde{\boldsymbol{\mu}}_{sph} = \varepsilon_{out}\tilde{\alpha}_{sph}(\omega)\mathbf{E}_0$,

$$\tilde{\alpha}_{sph}(\omega) = R^3 \frac{\tilde{\varepsilon}_p(\omega) - \varepsilon_{out}}{\tilde{\varepsilon}_p(\omega) + 2\varepsilon_{out}}. \tag{1.193}$$

Equation (1.193) represents the complex polarizability of a small sphere of sub-wavelength diameter in the quasi-static approximation. Note that it shows the same functional form as the *Clausius-Mossotti* polarizability [1].

The distribution of the electric field $\tilde{\mathbf{E}} = -\nabla\tilde{\Phi}$ can be directly evaluated from the potentials in Eq. (1.189) and Eq. (1.190):

$$\tilde{\mathbf{E}}_{int} = \frac{3\varepsilon_{out}}{\tilde{\varepsilon}_p(\omega) + 2\varepsilon_{out}} \mathbf{E}_0 \,, \tag{1.194}$$

$$\tilde{\mathbf{E}}_{out} = \mathbf{E}_0 + \frac{1}{\varepsilon_{out}} \frac{3\hat{\mathbf{r}}\left(\hat{\mathbf{r}} \cdot \tilde{\boldsymbol{\mu}}_{sph}\right) - \tilde{\boldsymbol{\mu}}_{sph}}{r^3} \,. \tag{1.195}$$

The second term of Eq. (1.195) is the *dipolar electrostatic field*, for a dipole $\tilde{\boldsymbol{\mu}}_{sph}$ located at the origin and it is shown as a vector field in Fig. 1.10a.

Figure 1.10b shows the total electric field inside and outside the dielectric sphere with $\varepsilon_{out} = 1$ and $\tilde{\varepsilon}_p(\omega) = 4$. We can see that the field lines "curve" near the sphere with dielectric constant larger than the environment and that the field inside the sphere is smaller than outside: $E_{int}/E_0 = 1/2$, see Eq. (1.194).

It is evident that the polarizability $\tilde{\alpha}_{sph}(\omega)$ experiences a resonant enhancement when $\left|\tilde{\varepsilon}_p(\omega) + 2\varepsilon_{out}\right|$ is a minimum or vanishes. For the case of small or slowly-varying $\mathrm{Im}\left[\tilde{\varepsilon}_p(\omega)\right]$ around the resonance, the resonance condition simplifies to:

$$\mathrm{Re}\left[\tilde{\varepsilon}_p(\omega)\right] \approx -2\varepsilon_{out} \,. \tag{1.196}$$

This condition is well known with the name of *Fröhlich condition* and the mode which satisfies it, is named *dipole surface plasmon.* Condition (1.196) requires a negative dielectric constant which can be obtained for metals. Using the ideal free-electron Drude dielectric constant (see Eq. (1.122)) we obtain the condition:

$$\omega_{\mathrm{LSP}} = \frac{\omega_p}{\sqrt{1 + 2\varepsilon_{out}}} \tag{1.197}$$

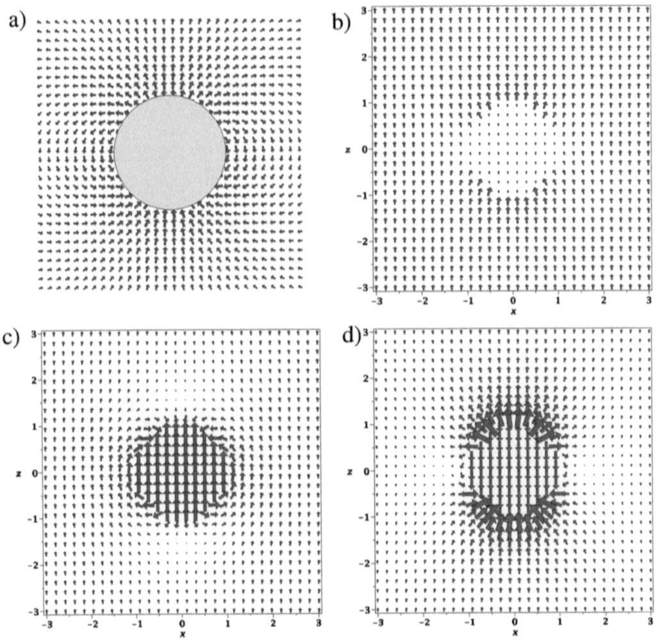

Figure 1.10 (a) Dipolar field outside the nanosphere for $\tilde{\mu}_{sph} = 1$. (b) Total electric field for $\tilde{\varepsilon}_p(\omega) = 4$, (c) Total electric field for $\tilde{\varepsilon}_p(\omega) = -1$, (d) Total electric field for $\tilde{\varepsilon}_p(\omega) = -3$. In all cases $\varepsilon_{out} = 1$ and the length of the vectors is in a logarithmic scale.

which can be compared with the one for SPP, see Eq. (1.173): ω_{LSP} is thus always lower than the one for SPPs.

Figure 1.10c shows the total electric field for a metal sphere with $\tilde{\varepsilon}_p(\omega) = -1$, or $\omega = \omega_p/\sqrt{2} > \omega_{\text{LSP}}$ in the ideal free-electron model. In this case $\tilde{\alpha}_{sph} = -2R^3$, i.e. it is negative and larger than the previous case of a dielectric sphere in Figure 1.10b ($\tilde{\alpha}_{sph} = 1/2R^3$): the dipole moment of the sphere is thus in *opposite* direction with respect to the external field. Figure 1.10c clearly shows that the total electric field "turns around" the sphere and that the field inside the sphere is 3 times the external field, see Eq. (1.194). Decreasing further the $\tilde{\varepsilon}_p(\omega)$ towards the Fröhlich condition, will magnify these effects.

For $\tilde{\varepsilon}_p(\omega) = -3$ we have that $\omega = \omega_p/2$ is less than ω_{LSP}: in this case we see another interesting effect. The field inside the sphere is now negative (−3 times the external field) and more interestingly the sphere polarizability is now again positive and much larger ($\tilde{\alpha}_{sph} = 4R^3$): it strongly enhances (9 times) the electric field at the surface.

Thus in metal nanoparticles a large *field enhancement* is present at their surface.

We can define the *perpendicular field enhancement factor* along the z axis at a distance d from the surface, i.e for $\mathbf{r} = (R + d)\widehat{\mathbf{z}}$, as:

$$A_{\perp}(\omega, d) = \left|\frac{\mathbf{E}_{out}((R+d)\widehat{\mathbf{z}})}{E_0}\right| = \left|1 + \frac{2\tilde{\alpha}_{sph}(\omega)}{(R+d)^3}\right| \tag{1.198}$$

$$= \left|1 + 2\frac{\tilde{\varepsilon}_p(\omega) - \varepsilon_{out}}{\tilde{\varepsilon}_p(\omega) + 2\varepsilon_{out}}\frac{R^3}{(R+d)^3}\right|, \tag{1.199}$$

and the *parallel field enhancement factor* along the x (or y) axis at a distance d from the surface, i.e for $\mathbf{r} = (R + d)\widehat{\mathbf{x}}$ as:

$$A_{\parallel}(\omega, d) = \left|\frac{\mathbf{E}_{out}((R+d)\widehat{\mathbf{x}})}{E_0}\right| = \left|1 - \frac{\tilde{\alpha}_{sph}(\omega)}{(R+d)^3}\right| \tag{1.200}$$

$$= \left|1 - \frac{\tilde{\varepsilon}_p(\omega) - \varepsilon_{out}}{\tilde{\varepsilon}_p(\omega) + 2\varepsilon_{out}}\frac{R^3}{(R+d)^3}\right|. \tag{1.201}$$

Then, considering the field at the surface ($d = 0$) and the ideal free-electron Drude dielectric constant for the nanoparticle (see Eq. (1.122)), we obtain:

$$A_{\perp}(\hat{\omega}, 0) = 3\left|\frac{(\hat{\omega}^2 - 1)}{(1 + 2\varepsilon_{out})\hat{\omega}^2 - 1}\right| \tag{1.202}$$

$$A_{\parallel}(\hat{\omega}, 0) = 3\left|\frac{\varepsilon_{out}\hat{\omega}^2}{(1 + 2\varepsilon_{out})\hat{\omega}^2 - 1}\right| \tag{1.203}$$

with $\hat{\omega} = \omega/\omega_p$. These field enhancement factors are reported in Fig. 1.11. For low frequency the $A_{\perp}$ approaches 3, whereas $A_{\parallel}$ vanishes: in fact at small frequencies the dielectric constant of the nanoparticle approaches $-\infty$. Both field enhancements diverge when $\omega = \omega_{\text{LSP}}$, see Eq. (1.197). For $\omega = \omega_p$, $A_{\perp}$ vanishes while $A_{\parallel} = 3/2$: in fact in this case the dielectric constant is zero. For high frequency and $\varepsilon_{out} = 1$ both field enhancements approach 1, as the dielectric constant of the sphere approaches 1 (i.e. no sphere is present).

Equation (1.196) further expresses the strong dependence of the resonance frequency on the dielectric environment: the resonance

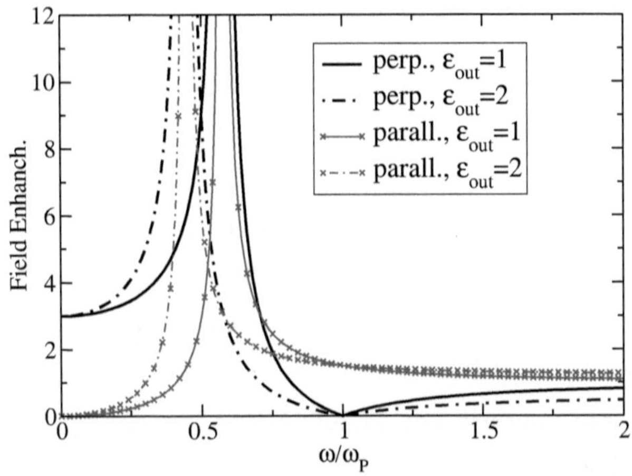

Figure 1.11 Field enhancements at the surface of a ideal free-electron sphere (perpendicular and parallel directions), for two different external dielectric constants.

red-shifts as ε_{out} increases. Metal nanoparticles are thus ideal platforms for the optical sensing of refractive index changes of the environment (see Chapter 6).

For a small sphere with $R \ll \lambda$, its representation as an ideal dipole is valid in the quasi-static regime, i.e. allowing for time-varying fields but neglecting spatial retardation effects over the particle volume. Under plane-wave excitation with $\mathbf{E}_0(\mathbf{r}, t) = \mathrm{Re}[\mathbf{E}_0 \exp(\mathrm{i}\mathbf{k} \cdot \mathbf{r} - \mathrm{i}\omega t)]$ and with $k \ll R$, this quasi-static representation is still valid, and the incident field will induce an oscillating dipole moment $\boldsymbol{\mu}(t) = \mathrm{Re}\left[\tilde{\boldsymbol{\mu}}_{sph} e^{-\mathrm{i}\omega t}\right]$, with $\tilde{\boldsymbol{\mu}}_{sph}$ given by the electrostatic result in Eq. (1.192). An oscillating dipole is however a *radiating dipole*.

1.5 Dipole Radiation

When an electronic charge is oscillating, it emits electromagnetic radiation [1]. It this section we will consider the electromagnetic field emitted by a point-like oscillating dipole, which also describes quite well the field emitted by molecules, which are much smaller then the electromagnetic wavelength.

Let's consider a point-like dipole $\tilde{\mu}_0$ located at $\mathbf{r}_0 = 0$ and oscillating with frequency ω. In the most general case $\tilde{\mu}_0$ is a complex vector. It has a polarization field given by:

$$\mathbf{P}(\mathbf{r}, t) = \mathrm{Re}\left[\tilde{\mathbf{P}}(\mathbf{r})\,\mathrm{e}^{-\mathrm{i}\omega t}\right] = \mathrm{Re}\left[\tilde{\mu}\,\mathrm{e}^{-\mathrm{i}\omega t}\delta(\mathbf{r})\right]. \tag{1.204}$$

This point-like dipole can be thought to generate *external source terms,* namely a charge density given by Eq. (1.8) and a current density given by Eq. (1.9). Using the Fourier transformation we have:

$$\tilde{\rho}_e(\mathbf{r}) = -\nabla \cdot \tilde{\mathbf{P}}(\mathbf{r}) = -\tilde{\mu}_0 \cdot \nabla\delta(\mathbf{r}), \tag{1.205}$$

$$\tilde{\mathbf{J}}_e(\mathbf{r}) = -\mathrm{i}\omega\tilde{\mathbf{P}}(\mathbf{r}) = -\mathrm{i}\omega\tilde{\mu}_0\delta(\mathbf{r}). \tag{1.206}$$

We can now use Eqs. (1.205, 1.206) into Eqs. (1.104, 1.106) obtaining:

$$\begin{aligned}\tilde{\phi}(\mathbf{r}) &= -4\pi\,\tilde{\mu}_0 \cdot \nabla g(\mathbf{r}) \\ &= -\tilde{\mu}_0 \cdot \hat{\mathbf{r}}\left(\mathrm{i}k - \frac{1}{r}\right)\frac{\mathrm{e}^{ikr}}{r},\end{aligned} \tag{1.207}$$

$$\tilde{\mathbf{A}}(\mathbf{r}) = -\mathrm{i}\frac{4\pi\,\omega}{c}\tilde{\mu}_0 g(\mathbf{r}) = -\mathrm{i}k\tilde{\mu}_0\frac{\mathrm{e}^{ikr}}{r}, \tag{1.208}$$

where $r = ||\mathbf{r}||$ and $g(\mathbf{r})$ is the scalar Green's function (see Eq. (1.99)). The electric and magnetic fields in the free-space can be directly obtained from Eqs. (1.107, 1.110):

$$\tilde{\mathbf{H}}(\mathbf{r}) = -\mathrm{i}4\pi k\nabla g(\mathbf{r}) \times \tilde{\mu}_0 \tag{1.209}$$

$$= k^2(\hat{\mathbf{r}} \times \tilde{\mu}_0)\left(\frac{1}{r} - \frac{1}{\mathrm{i}kr^2}\right)\mathrm{e}^{ikr}, \tag{1.210}$$

$$\tilde{\mathbf{E}}(\mathbf{r}) = 4\pi\left(\nabla\nabla + \frac{\omega^2}{c^2}\mathbb{I}\right)g(\mathbf{r}) \cdot \tilde{\mu}_0 \tag{1.211}$$

$$\begin{aligned}&= k^2\left((\hat{\mathbf{r}} \times \tilde{\mu}) \times \hat{\mathbf{r}}\right)\frac{\mathrm{e}^{ikr}}{r} \\ &\quad + \left[3(\tilde{\mu}_0 \cdot \hat{\mathbf{r}})\hat{\mathbf{r}} - \tilde{\mu}_0\right]\left(\frac{1}{r^3} - \frac{ik}{r^2}\right)\mathrm{e}^{ikr}.\end{aligned} \tag{1.212}$$

where $\mathbb{I}$ is the identity dyadic. From Eq. (1.211) it is possible to define the *free-space dyadic Green's function for the electric field* :

$$\tilde{\mathbb{G}}^0(\mathbf{r}, \mathbf{r}') = 4\pi\left(\nabla\nabla + \frac{\omega^2}{c^2}\mathbb{I}\right)g(\mathbf{r} - \mathbf{r}') = \left(\nabla\nabla + \frac{\omega^2}{c^2}\mathbb{I}\right)\frac{\mathrm{e}^{ik||\mathbf{r}-\mathbf{r}'||}}{||\mathbf{r} - \mathbf{r}'||}, \tag{1.213}$$

so that:

$$\tilde{\mathbf{E}}(\mathbf{r}) = \int \tilde{\mathbb{G}}^0(\mathbf{r}, \mathbf{r}') \cdot \tilde{\boldsymbol{P}}(\mathbf{r}')\mathrm{d}^3\mathbf{r}' \tag{1.214}$$

$$= \tilde{\mathbb{G}}^0(\mathbf{r}, \mathbf{r}_0) \cdot \tilde{\boldsymbol{\mu}}_0 \,. \tag{1.215}$$

Thus the free-space dyadic Green's function relates the electric field in the free-space at point $\mathbf{r}$ with the source dipole at point $\mathbf{r}_0$. The magnetic field is instead easier derived from Eq. (1.210).

In Fig. 1.12 we report the real part of electric field $\mathrm{Re}\left[\tilde{\mathbf{E}}(\mathbf{r})\right]$ and the magnetic field $\mathrm{Re}\left[\tilde{\mathbf{H}}(\mathbf{r})\right]$ generated by an unit dipole in the z direction and with $k = 1$. The plots refer to the xz plane and we report the x and the z components of the electric field (all other components are zero in this plane). For the electric field we also plot the norm of the field.

These plots clearly show that the electric field resembles a static dipole field near the center, whereas it assumes a propagating spherical wave at larger distances. In fact in the *near-field zone*, i.e. for $kr \ll 1$, we obtain:

$$\tilde{\mathbf{H}}(\mathbf{r}) \approx \mathrm{i}k(\widehat{\mathbf{r}} \times \tilde{\boldsymbol{\mu}}_0)\frac{1}{r^2}\,, \tag{1.216}$$

$$\tilde{\mathbf{E}}(\mathbf{r}) \approx \left[3(\tilde{\boldsymbol{\mu}}_0 \cdot \widehat{\mathbf{r}})\widehat{\mathbf{r}} - \tilde{\boldsymbol{\mu}}_0\right]\frac{1}{r^3}\,. \tag{1.217}$$

Equation (1.217) coincides with the electrostatic result for the field of an electrostatic dipole, see Eq. (1.195). We can also see that within the near field, the fields are predominantly electric in nature, since the magnitude of the magnetic field is about a factor $kr \ll 1$ smaller than that of the electric field.

In the *far-field zone* (defined by $kr \gg 1$), the fields have the well-known *spherical wave* form:

$$\tilde{\mathbf{H}}(\mathbf{r}) \approx k^2(\widehat{\mathbf{r}} \times \tilde{\boldsymbol{\mu}}_0)\frac{\mathrm{e}^{\mathrm{i}kr}}{r}\,, \tag{1.218}$$

$$\tilde{\mathbf{E}}(\mathbf{r}) \approx k^2\left((\widehat{\mathbf{r}} \times \tilde{\boldsymbol{\mu}}_0) \times \widehat{\mathbf{r}}\right)\frac{\mathrm{e}^{\mathrm{i}kr}}{r} = k^2(\mathbb{I} - \widehat{\mathbf{r}}\widehat{\mathbf{r}}) \cdot \tilde{\boldsymbol{\mu}}_0\frac{\mathrm{e}^{\mathrm{i}kr}}{r}\,. \tag{1.219}$$

In the far-field the electric and the magnetic fields have the *same* magnitude but perpendicular directions (and normal to $\widehat{\mathbf{r}}$). If the dipole $\tilde{\boldsymbol{\mu}}_0$ is real and along the z axis then the fields are zero in the z direction while the largest contributions are in the x, y directions, see Fig. 1.12.

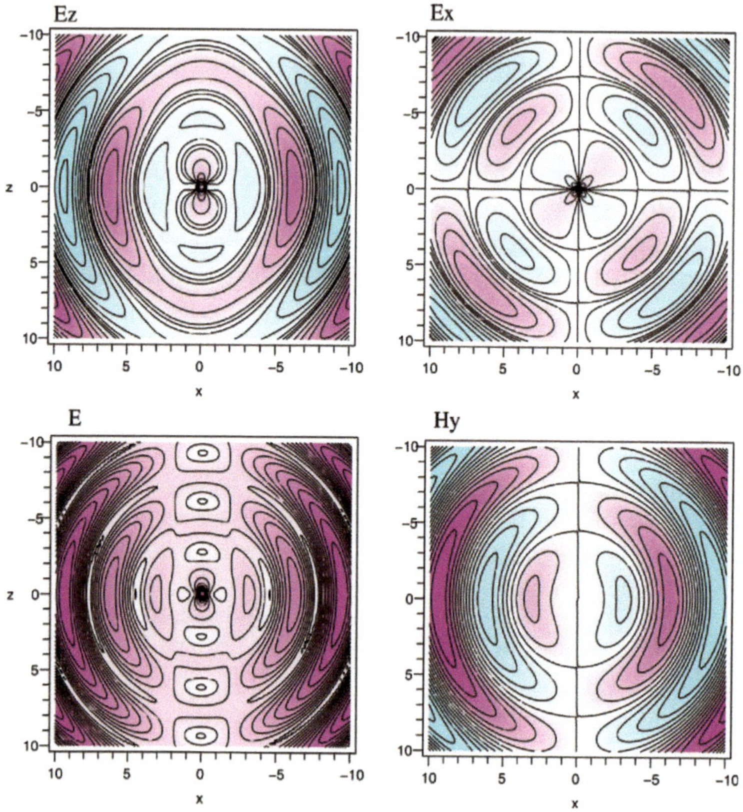

Figure 1.12 Electromagnetic field (plane xz) generated by a unit dipole in the z direction and $k = 1$: electric (z and x non-zero component and the field norm $E = \sqrt{|E_x|^2 + |E_z|^2}$) and magnetic ($y$ non-zero component) field. For visualization purpose, the intensity of the field is multiplied by r^2. Colors: pink (positive), white (null), cyan (negative).

Equation (1.219) also defines the dyadic Green's function in the far-field:

$$\tilde{\mathbb{G}}^0(\mathbf{r}, \mathbf{r}_0) \to k^2(\mathbb{I} - \widehat{\mathbf{r}\mathbf{r}})\frac{\mathrm{e}^{\mathrm{i}kr}}{r} \exp\left(-\mathrm{i}k\widehat{\mathbf{r}} \cdot \mathbf{r}_0\right), \tag{1.220}$$

which is valid in the limit $r = ||\mathbf{r}|| \gg ||\mathbf{r}_0||$.

From the electromagnetic fields, the Poynting vector can be obtained. It can be shown that the real part of the complex Poynting vector is (for a real dipole) [1]:

$$\begin{aligned}\mathrm{Re}\left[\tilde{\mathbf{S}}(\mathbf{r})\right] &= \frac{c}{8\pi}\mathrm{Re}\left[\tilde{\mathbf{E}}(\mathbf{r}) \times \tilde{\mathbf{H}}^*(\mathbf{r})\right] \\ &= \frac{ck^4}{8\pi}\left|\left|\boldsymbol{\mu}_0\right|\right|^2 \sin^2(\theta)\frac{\widehat{\mathbf{r}}}{r^2}, \end{aligned} \tag{1.221}$$

where θ is the angle between $\widehat{\mathbf{r}}$ and $\boldsymbol{\mu}_0$. This is easily obtained in the far-field from Eqs. (1.218, 1.219) where the complex Poynting vector is actually real, but Eq. (1.221) holds also in the near-field region. Note that Eq. (1.221) cannot be obtained from Eqs. (1.216, 1.217), as higher order terms are required to obtain the correct $1/r^2$ dependence. Equation (1.221) means that the power radiated by a dipole is mostly confined in a plane perpendicular to the dipole.

In the far-field region we further have:

$$\mathrm{Re}\left[\tilde{\mathbf{S}}(\mathbf{r})\right] \approx \frac{c}{8\pi}\left|\left|\tilde{\mathbf{E}}(\mathbf{r})\right|\right|^2, \tag{1.222}$$

whereas in the near-field the norm of the electric field is:

$$\left|\left|\tilde{\mathbf{E}}(\mathbf{r})\right|\right|^2 \approx \left|\left|\boldsymbol{\mu}_0\right|\right|^2 \frac{1+3\cos^2(\theta)}{r^6}, \tag{1.223}$$

and thus it shows a less angular dependence than in the far-field and it is maximized at $\theta = 0$, in contrast to the far-field, where it is maximized for $\theta = 90°$.

The flux over a spherical surface of radius $R \to \infty$ of the Poynting vector is:

$$\oint_S \mathrm{Re}\left[\tilde{\mathbf{S}}\right] \cdot \widehat{\mathbf{n}}\mathrm{d}A = \frac{ck^4\left|\left|\boldsymbol{\mu}_0\right|\right|^2}{3} = \frac{\omega^4\left|\left|\boldsymbol{\mu}_0\right|\right|^2}{3c^3} = < P^0_{em} >_T . \tag{1.224}$$

The quantity $<P^0_{em}>_T$ represents the averaged *dipole emitted power* per time-period $T = 2\pi/\omega$, in the vacuum, and it rapidly increases with the frequency.

1.5.1 *Radiation Pattern*

An emitting dipole, as well as any electromagnetic antenna, can be characterized by its *radiation pattern* $D^0(\theta, \phi)$ (where the apex ^{0} indicates radiation in the vacuum), defined from:

$$\mathrm{Re}\left[\tilde{\mathbf{S}}(\mathbf{r})\right] = \frac{< P^0_{em} >_T}{4\pi r^2} D^0(\theta, \phi)\widehat{\mathbf{r}}. \tag{1.225}$$

The radiation pattern represents thus the normalized emitted power within an unit solid angle $d\Omega = \sin(\theta)d\theta d\phi$ and satisfies:

$$\frac{1}{4\pi}\int_{\Omega} D^0(\theta,\phi)\mathrm{d}\Omega = 1\,. \tag{1.226}$$

An "isotropic" antenna has $D(\theta,\phi) = 1$. For a dipole in the z direction we have, using Eqs. (1.221, 1.224):

$$D_z^0(\theta,\phi) = \frac{3}{2}\sin^2(\theta)\,. \tag{1.227}$$

In the case of a dipole aligned on the x-axis we have instead:

$$D_x^0(\theta,\phi) = \frac{3}{2}\left(1-\cos^2(\phi)\sin^2(\theta)\right)\,, \tag{1.228}$$

where θ is the angle with $\hat{\mathbf{z}}$ and ϕ is the angle with $\hat{\mathbf{x}}$.

We can also define the integral only over the ϕ coordinates as

$$\eta^0(\theta) = \frac{1}{2\pi}\int_0^{2\pi} D^0(\theta,\phi)\mathrm{d}\phi\,, \tag{1.229}$$

with the normalization condition:

$$2 = \int_0^{\pi} \eta^0(\theta)\sin(\theta)\mathrm{d}\theta\,. \tag{1.230}$$

For the dipoles along z and x we obtain:

$$\eta_z^0(\theta) = \frac{3}{2}\sin^2(\theta)\,, \tag{1.231}$$

$$\eta_x^0(\theta) = \frac{3}{4}(1+\cos^2(\theta))\,. \tag{1.232}$$

These definitions will be used later, in section Sec. 1.7.4.

1.5.2 *Dissipated Power*

The emitted power radiated far away from the dipole must be generated by an external force that keeps the dipole oscillating. In fact the following equation holds:

$$< P_{diss} >_T = -\frac{1}{2}\int \mathrm{Re}\left[\tilde{\mathbf{J}}_e^*(\mathbf{r})\cdot\tilde{\mathbf{E}}(\mathbf{r})\right]\mathrm{d}^3\mathbf{r}\,. \tag{1.233}$$

The external current $\tilde{\mathbf{J}}_e$ that generates the electric field is kept constant. Using Eq. (1.206) for a real dipole source at point $(\mathbf{r}_0)$ in the vacuum we obtain:

$$< P_{diss}^0 >_T = \frac{\omega}{2}\mathrm{Im}\left[\boldsymbol{\mu}_0^*\cdot\tilde{\mathbf{E}}(\mathbf{r}_0)\right] \tag{1.234}$$

$$= \frac{\omega}{2}\mathrm{Im}\left[\boldsymbol{\mu}_0^*\cdot\tilde{\mathbb{G}}^0(\mathbf{r}_0,\mathbf{r}_0)\cdot\boldsymbol{\mu}_0\right]\,. \tag{1.235}$$

The function $\mathbb{G}(\mathbf{r}, \mathbf{r}_0)$ seems to have a singularity at $\mathbf{r} = \mathbf{r}_0$, but this happens only for the real part. In fact from a Taylor expansion we have [4]:

$$\mathrm{Im}\left[\tilde{\mathbb{G}}^0(\mathbf{r}_0, \mathbf{r}_0)\right] \to \frac{2}{3}k^3\mathbb{I}\,, \tag{1.236}$$

$$\mathrm{Re}\left[\tilde{\mathbb{G}}^0(\mathbf{r}, \mathbf{r}_0)\right] \to -\frac{4\pi}{3}\mathbb{I}\delta(\mathbf{r} - \mathbf{r}_0)\,. \tag{1.237}$$

Thus we finally obtain:

$$< P^0_{diss} >_T = \frac{\omega k^3}{3}\,||\boldsymbol{\mu}_0||^2\,, \tag{1.238}$$

which coincides with Eq. (1.224). In fact the total energy must be conserved (see Eq. (1.44)): all the energy consumed by an external force to keep the current constant, is radiated away in the free-space.

1.5.3 *Classical Approach*

The expression for emitted or dissipated power in Eq. (1.224) or Eq. (1.238) can be also obtained by a simple classical model that can represent the physical system. In fact an *emitting* atom or a molecule can be modeled as a classical oscillator consisting of a fixed charge $(-e)$ located at point $\mathbf{r}_0$ and a charge $(+e)$ located at point $\mathbf{r}' = \mathbf{r}_0 + \widehat{\boldsymbol{\mu}}d(t)$, i.e. oscillating in the direction $\widehat{\boldsymbol{\mu}}$, see Fig. 1.13. The total dipole is $\boldsymbol{\mu}(t) = e\widehat{\boldsymbol{\mu}}d(t) = \widehat{\boldsymbol{\mu}}\mu(t)$. The equation of motion for the dipole $\mu(t)$ in the vacuum is:

$$\frac{m_e}{e}\frac{\mathrm{d}^2\mu(t)}{\mathrm{d}t^2} + \frac{m_e}{e}\gamma_r^0\frac{\mathrm{d}\mu(t)}{\mathrm{d}t} + \frac{m_e}{e}\omega_0^2\mu(t) = 0\,, \tag{1.239}$$

where γ_r^0 represents a *friction* term which dissipates energy, radiating it away: γ_r^0 is called the *spontaneous emission decay rate* in the vacuum. The pedex r indicate that is a *radiative* decay rate, to distinguish from other types of decay (see Sec. 1.7.3).

Assuming an $\exp(-\mathrm{i}\tilde{\omega}t)$ time-dependence (with $\tilde{\omega}$ complex), we obtain from Eq. (1.239) the condition:

$$-\tilde{\omega}^2 - \mathrm{i}\tilde{\omega}\gamma_r^0 + \omega_0^2 = 0\,. \tag{1.240}$$

Considering that γ_r^0 is of the order of $10^9 - 10^7 s^{-1}$ for allowed optical transitions, while $\hbar\omega_0$, the energy of an emitted photon, of the order

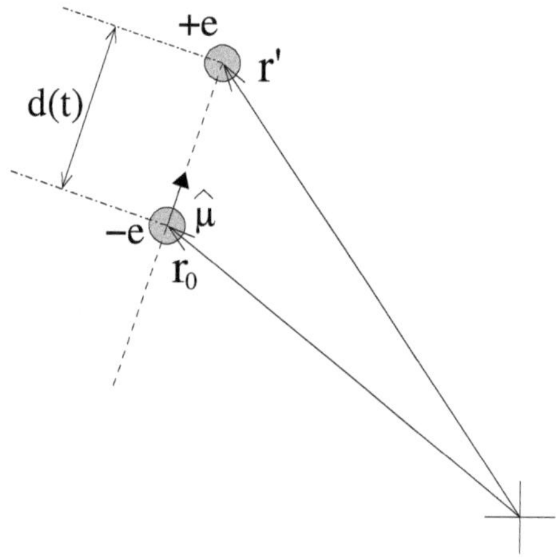

Figure 1.13 A radiating dipole can be represented as two classical point charges, $+e$ and $-e$, with an oscillating distance $d(t)$.

of 1–10 eV, we have that $\omega_0 \gg \gamma_r^0$. Thus the solution of Eq. (1.240) becomes:

$$\tilde{\omega} = \frac{i\gamma_r^0}{2} + \frac{1}{2}\sqrt{4\omega_0^2 - (\gamma_r^0)^2} \approx \omega_0 + i\gamma_r^0/2\,, \qquad (1.241)$$

and thus the solution of Eq. (1.239) can be written as a *damped harmonic oscillator*:

$$\mu(t) = ||\boldsymbol{\mu}_0|| \exp\left(-\frac{\gamma_r^0}{2}t\right)\cos(\omega_0 t)\,, \qquad (1.242)$$

where $||\boldsymbol{\mu}_0||$ is the maximum amplitude of the dipole (which needs to be determined by other constrains). Equation (1.242) is a product of a fast component which oscillates at optical frequencies and of a slow component which decays slowly. Thus we can set $t = t_{slow} + t'$ with

$$t_{slow} = \left(j + \frac{1}{2}\right)T\,, \qquad (1.243)$$

$$t' \in [-T/2, T/2]\,, \qquad (1.244)$$

and j in an integer that counts the periods $T = 2\pi/\omega_0$. Equation (1.242) thus becomes:

$$\mu(t) = ||\boldsymbol{\mu}_0|| \exp\left(-\frac{\gamma_r^0}{2}\left(j + \frac{1}{2}\right)T\right)\cos(\omega_0 t')\,, \qquad (1.245)$$

and the total dipole can be thus written in the Fourier space as:

$$\boldsymbol{\mu}[j] = \widehat{\boldsymbol{\mu}} \, ||\boldsymbol{\mu}_0|| \exp\left(-\frac{\gamma_r^0}{2}\left(j+\frac{1}{2}\right)T\right). \tag{1.246}$$

We now consider the fact that the total emitted energy is that of one emitted photon with energy $\hbar\omega_0$. Thus it must be that:

$$\hbar\omega_0 = \int_0^{+\infty} P_{em}^0(t)dt\,. \tag{1.247}$$

The instantaneous emitted power $P_{em}^0(t)$ is also a product of a fast component which oscillates at optical frequencies and a slow component which decays slowly. In the j-th period the dipole $\boldsymbol{\mu}[j]$ emits the power given by Eq. (1.224), i.e. the averaged emitted power is:

$$< P_{em}^0 >_{Tj} = \frac{\omega_0^4 \, ||\boldsymbol{\mu}_0||^2}{3c^3} \exp\left(-\gamma_r^0\left(j+\frac{1}{2}\right)T\right). \tag{1.248}$$

Equation (1.247) thus becomes:

$$\begin{aligned}\hbar\omega_0 &= \sum_{j=0}^{\infty}\int_{jT}^{(j+1)T} P_{em}^0(t)dt \approx \sum_{j=0}^{\infty} T < P_{em}^0 >_{Tj} \\ &\approx \frac{1}{\gamma_r^0}\frac{\omega_0^4\,||\boldsymbol{\mu}_0||^2}{3c^3}\,.\end{aligned} \tag{1.249}$$

Thus the classical approach predicts the following spontaneous emission decay rate:

$$\gamma_r^0 = \frac{\omega_0^3\,||\boldsymbol{\mu}_0||^2}{3\hbar c^3} = \frac{< P_{em}^0 >_T}{\hbar\omega_0}\,, \tag{1.250}$$

where the second identity is consistent with Eq. (1.224).

Another relation between the value of $||\boldsymbol{\mu}_0||$ and the photon energy can be found considering the total energy of the oscillator, which is the sum of the kinetic and potential energy:

$$E(t) = \frac{1}{2}\frac{m}{e^2}\left|\left|\frac{\mathrm{d}\boldsymbol{\mu}(t)}{\mathrm{d}t}\right|\right|^2 + \frac{1}{2}\frac{m\omega_0^2}{e^2}\,||\boldsymbol{\mu}(t)||^2\,. \tag{1.251}$$

At $t = 0$ the oscillator must contain the total energy which will be radiated, i.e.

$$\hbar\omega_0 = E(0) = \frac{1}{2}\frac{m\omega_0^2}{e^2}\,||\boldsymbol{\mu}_0||^2\,, \tag{1.252}$$

and thus

$$||\boldsymbol{\mu}_0|| = e\sqrt{\frac{2\hbar}{m_e\omega_0}}\,. \tag{1.253}$$

Using Eq. (1.253) in Eq. (1.250) we obtain:

$$\gamma_r^0 = \frac{2\omega_0^2 e^2}{3m_e c^3}\,, \tag{1.254}$$

which is often used to described the decay rate of atoms [4].

1.6 Scattering Theory

In Sec. 1.1.2 we considered an electromagnetic plane-wave incident to a semi-infinite medium with a different dielectric constant. In this section we will consider instead a plane-wave incident to a nanoparticle of finite-size.

Let's consider a nanoparticle occupying a volume region V described by the complex dielectric susceptibility $\tilde{\chi}_p(\mathbf{r},\omega)$. In absence of external sources, Eq. (1.21) can be rewritten as [15]:

$$-\nabla\times\nabla\times\tilde{\mathbf{E}}(\mathbf{r}) + \frac{\omega^2}{c^2}\tilde{\mathbf{E}}(\mathbf{r}) = -\mathrm{i}\frac{4\pi\omega}{c^2}\tilde{\mathbf{J}}_b(\mathbf{r},\omega)\,, \tag{1.255}$$

with

$$\tilde{\mathbf{J}}_b(\mathbf{r},\omega) = -\mathrm{i}\omega\tilde{\chi}_p(\mathbf{r},\omega)\,\tilde{\mathbf{E}}(\mathbf{r})\,, \tag{1.256}$$

where we used Eq. (1.14). If the current $\tilde{\mathbf{J}}_b$ is zero then Eq. (1.255) is the equation for the propagation in the free-space and the solution $\tilde{\mathbf{E}}^0(\mathbf{r}) = \tilde{\mathbf{E}}^{inc}(\mathbf{r})$ can be a single plane-wave being $\tilde{\mathbf{E}}^{inc}(\mathbf{r})$ the incident field.

Otherwise, using Eqs. (1.214, 1.12), the solution can be written in the form

$$\tilde{\mathbf{E}}^{tot}(\mathbf{r}) = \tilde{\mathbf{E}}^{inc}(\mathbf{r}) + \int \tilde{\mathbb{G}}^0(\mathbf{r},\mathbf{r}')\cdot\tilde{\chi}_p(\mathbf{r}',\omega)\,\tilde{\mathbf{E}}^{tot}(\mathbf{r}')\mathrm{d}^3\mathbf{r}', \tag{1.257}$$

which is called the *Lippmann-Schwinger equation*. The second term at the right-end side is the scattered field $\tilde{\mathbf{E}}_{sca}(\mathbf{r})$. To simplify the notation we introduce here a *super-operator* notation in which integration over spatial variable is implicit in the product. Equation (1.257) thus simplifies to:

$$\tilde{\mathbf{E}}^{tot} = \tilde{\mathbf{E}}^{inc} + \tilde{\mathbb{G}}^0\cdot\tilde{\chi}_p(\omega)\tilde{\mathbf{E}}^{tot}\,. \tag{1.258}$$

Note that Eq. (1.258) defines the total electric field in term of itself, and thus it can be solved by iterative methods or by defining a direct inversion of operators:

$$\tilde{\mathbf{E}}^{tot} = \left[\mathbb{1} - \tilde{\chi}_p(\omega)\tilde{\mathbb{G}}^0\right]^{-1} \cdot \tilde{\mathbf{E}}^{inc}, \qquad (1.259)$$

where $\mathbb{1}(\mathbf{r}, \mathbf{r}') = \mathbb{I}\delta(\mathbf{r} - \mathbf{r}')$. Otherwise one can introduce the *dyadic Green's function for the total electric field* $\tilde{\mathbb{G}}_p^{tot}(\mathbf{r}, \mathbf{r}', \omega)$ so that:

$$\tilde{\mathbf{E}}^{tot}(\mathbf{r}) = \tilde{\mathbb{G}}_p^{tot}(\mathbf{r}, \mathbf{r}_0, \omega) \cdot \tilde{\boldsymbol{\mu}}_0, \qquad (1.260)$$

thus relating the *total* electric field in a scattering environment at point $\mathbf{r}$ to a source dipole $\tilde{\boldsymbol{\mu}}_0$ at point $\mathbf{r}_0$. Note that $\tilde{\mathbb{G}}_p^{tot}$ depends on the nanoparticle characteristics and we explicitly indicate the ω dependence in $\tilde{\mathbb{G}}^{tot}$ because it contains the dielectric constant of the nanoparticle.

In this way Eq. (1.258) can be rewritten as:

$$\tilde{\mathbf{E}}^{tot} = \tilde{\mathbf{E}}^{inc} + \tilde{\mathbb{G}}_p^{tot}(\omega) \cdot \tilde{\chi}_p(\omega)\tilde{\mathbf{E}}^{inc}. \qquad (1.261)$$

Equation (1.261) defines the total electric field only in terms of the incident field $\tilde{\mathbf{E}}^{inc}$. Comparing Eq. (1.258) with Eq. (1.261) one immediately obtain that:

$$\tilde{\mathbb{G}}_p^{tot}(\omega) = \tilde{\mathbb{G}}^0 + \tilde{\mathbb{G}}^0 \cdot \tilde{\chi}_p(\omega)\tilde{\mathbb{G}}_p^{tot}(\omega), \qquad (1.262)$$

which is called the *Dyson equation* (see also Sec. 1.2.3). Note that Eq. (1.262) defines the total dyadic Green's function in term of itself, and thus it can be solved by iterative methods or by direct inversion:

$$\tilde{\mathbb{G}}_p^{tot}(\omega) = \left[\mathbb{1} - \tilde{\chi}_p(\omega)\tilde{\mathbb{G}}^0\right]^{-1} \cdot \tilde{\mathbb{G}}^0. \qquad (1.263)$$

Equation (1.262) can also be rewritten as:

$$\tilde{\mathbb{G}}_p^{tot}(\omega) = \tilde{\mathbb{G}}^0 + \tilde{\mathbb{G}}^0 \cdot \tilde{\mathbb{T}}_p(\omega) \cdot \tilde{\mathbb{G}}^0, \qquad (1.264)$$

where

$$\tilde{\mathbb{T}}_p(\omega) = \tilde{\chi}_p(\omega)\mathbb{I} + \tilde{\chi}_p(\omega)\tilde{\mathbb{G}}^0 \cdot \tilde{\mathbb{T}}_p(\omega), \qquad (1.265)$$

is the *scattering operator* for the nanoparticle, which is a tensor but *local* in the space (i.e. $\tilde{\mathbb{T}}_p(\omega) = \tilde{\mathbb{T}}_p(\mathbf{r}, \omega)$). Equation (1.264) is an explicit expression of the total dyadic Green's function for the total electric field. The explicit expression of the scattering operator is:

$$\tilde{\mathbb{T}}_p(\omega) = \left[\mathbb{1} - \tilde{\chi}_p(\omega)\tilde{\mathbb{G}}^0\right]^{-1} \tilde{\chi}_p(\omega). \qquad (1.266)$$

We can also introduce *dyadic Green's function for the scattered electric field* $\tilde{\mathbb{G}}_p^{scat}(\mathbf{r}, \mathbf{r}', \omega)$, so that:

$$\tilde{\mathbf{E}}^{sca}(\mathbf{r}) = \tilde{\mathbb{G}}_p^{scat}(\mathbf{r}, \mathbf{r}_0, \omega) \cdot \tilde{\boldsymbol{\mu}}_0 , \tag{1.267}$$

thus relating the scattered electric field at point $\mathbf{r}$ to a source dipole $\tilde{\boldsymbol{\mu}}_0$ at point $\mathbf{r}_0$. The total, free and the scattered dyadic Green's are simply related:

$$\tilde{\mathbb{G}}_p^{tot}(\mathbf{r}, \mathbf{r}_0, \omega) = \tilde{\mathbb{G}}^0(\mathbf{r}, \mathbf{r}_0) + \tilde{\mathbb{G}}_p^{scat}(\mathbf{r}, \mathbf{r}_0, \omega) . \tag{1.268}$$

Thus from Eq. (1.264) we have:

$$\tilde{\mathbb{G}}_p^{scat}(\omega) = \tilde{\mathbb{G}}^0 \cdot \tilde{\mathbb{T}}_p(\omega) \cdot \tilde{\mathbb{G}}^0 . \tag{1.269}$$

The scattering operator can be also defined to have the following property:

$$\tilde{\chi}_p(\omega)\tilde{\mathbf{E}}^{tot} = \tilde{\mathbb{T}}_p(\omega) \cdot \tilde{\mathbf{E}}^{inc} , \tag{1.270}$$

which is obtained from Eqs. (1.266, 1.259). Note that the quantity in Eq. (1.270) is simply the polarization $\tilde{\mathbf{P}}$: thus $\tilde{\mathbb{T}}_p(\omega)$ can be also called as an *effective polarizability* per unit-volume as it relates the *incident* external field to the polarization. In this way Eq. (1.258) becomes:

$$\tilde{\mathbf{E}}^{sca} = \tilde{\mathbf{E}}^{tot} - \tilde{\mathbf{E}}^{inc} = \tilde{\mathbb{G}}^0 \cdot \tilde{\mathbb{T}}_p(\omega) \cdot \tilde{\mathbf{E}}^{inc} . \tag{1.271}$$

It is interesting to compare Eq. (1.261) with Eq. (1.271): the first computes the scattered field from the *total* Green's function and the *simple susceptibility*, the second employs the simple *free* Green's function in the free-space but requires the *full* scattering operator.

Combining Eqs. (1.262, 1.265) we obtain a direct relation between the scattering operator and the total dyadic Green's function:

$$\tilde{\mathbb{T}}_p(\omega) = \tilde{\chi}_p(\omega)\mathbb{I} + \tilde{\chi}_p(\omega)\tilde{\mathbb{G}}_p^{tot}(\omega)\tilde{\chi}_p(\omega) . \tag{1.272}$$

1.6.1 *Scattering Cross-Section*

Let's assume that we have scatterer (e.g. a metal nanoparticle) inside the vacuum which is excited by an incident uniform plane-wave with wavevector $\mathbf{k}_{inc} = k\widehat{\mathbf{k}}^{inc}$. The electric and magnetic incident fields are (see Sec. 1.1.1):

$$\tilde{\mathbf{E}}^{inc}(\mathbf{r}) = E_0\widehat{\mathbf{e}}^{inc} \exp\left(ik\widehat{\mathbf{k}}^{inc} \cdot \mathbf{r}\right) , \tag{1.273}$$

$$\tilde{\mathbf{H}}^{inc}(\mathbf{r}) = E_0\widehat{\mathbf{k}}^{inc} \times \widehat{\mathbf{e}}^{inc} \exp\left(ik\widehat{\mathbf{k}}^{inc} \cdot \mathbf{r}\right), \tag{1.274}$$

with $\widehat{\mathbf{k}}^{inc} \cdot \widehat{\mathbf{e}}^{inc} = 0$.

The total (incident plus scattering) electromagnetic field can be written as [15]:

$$\tilde{\mathbf{E}}(\mathbf{r}) = \tilde{\mathbf{E}}^{inc}(\mathbf{r}) + \tilde{\mathbf{E}}^{sca}(\mathbf{r}), \tag{1.275}$$

$$\tilde{\mathbf{H}}(\mathbf{r}) = \tilde{\mathbf{H}}^{inc}(\mathbf{r}) + \tilde{\mathbf{H}}^{sca}(\mathbf{r}). \tag{1.276}$$

If we define a surface far away around the nanoparticle, the averaged power absorbed will be:

$$< P_{abs} >_T = -\oint_A \mathrm{Re}\left[\tilde{\mathbf{S}}(\mathbf{r})\right] \cdot \widehat{\mathbf{n}} \mathrm{d}A \quad \geq 0. \tag{1.277}$$

Note that all the sources of the field are outside the surface A, and thus $< P_{abs} >_T$ can be positive or zero (if the imaginary part of dielectric constant of the nanoparticle is zero).

Equation (1.277) can be rewritten as:

$$< P_{abs} >_T = < P_{inc} >_T - < P_{sca} >_T + < P_{ext} >_T, \tag{1.278}$$

with

$$< P_{inc} >_T = -\oint_A \mathrm{Re}\left[\tilde{\mathbf{S}}^{inc}(\mathbf{r})\right] \cdot \widehat{\mathbf{n}}\, \mathrm{d}A \quad = 0, \tag{1.279}$$

$$< P_{sca} >_T = +\oint_A \mathrm{Re}\left[\tilde{\mathbf{S}}^{sca}(\mathbf{r})\right] \cdot \widehat{\mathbf{n}}\, \mathrm{d}A \quad \geq 0, \tag{1.280}$$

$$< P_{ext} >_T = -\oint_A \mathrm{Re}\left[\tilde{\mathbf{S}}^{ext}(\mathbf{r})\right] \cdot \widehat{\mathbf{n}}\, \mathrm{d}A \quad \geq 0, \tag{1.281}$$

where

$$\tilde{\mathbf{S}}^{inc}(\mathbf{r}) = \frac{c}{8\pi}\tilde{\mathbf{E}}^{inc}(\mathbf{r}) \times \tilde{\mathbf{H}}^{inc}(\mathbf{r})^* = \frac{c}{8\pi}\widehat{\mathbf{k}}^{inc} E_0^2, \tag{1.282}$$

$$\tilde{\mathbf{S}}^{sca}(\mathbf{r}) = \frac{c}{8\pi}\tilde{\mathbf{E}}^{sca}(\mathbf{r}) \times \tilde{\mathbf{H}}^{sca}(\mathbf{r})^*, \tag{1.283}$$

$$\tilde{\mathbf{S}}^{ext}(\mathbf{r}) = \frac{c}{8\pi}\tilde{\mathbf{E}}^{inc}(\mathbf{r}) \times \tilde{\mathbf{H}}^{sca}(\mathbf{r})^* + \frac{c}{8\pi}\tilde{\mathbf{E}}^{sca}(\mathbf{r}) \times \tilde{\mathbf{H}}^{inc}(\mathbf{r})^*. \tag{1.284}$$

These equations have been simply obtained using Eqs. (1.275, 1.276) in Eq. (1.41).

Equation (1.279) states that $<P_{inc}>_T = 0$ as the surrounding medium is non-absorbing and the sources of the incident field are outside the surface A. The scattered power (see Eq. (1.280)) is instead generated by the metal nanoparticle which re-irradiates the power in different directions. Thus it is redefined with a positive sign. The averaged extinction power $< P_{ext} >_T$ can be thus redefined as:

$$< P_{ext} >_T = < P_{abs} >_T + < P_{sca} >_T \, , \tag{1.285}$$

and it originates from the interaction between the scattered and the incident field. Then it is possible to define the *scattering cross-section*, the *absorption cross-section* and the *extinction cross-section* as

$$C_{abs} = \frac{< P_{abs} >_T}{S_{inc}} \, , \quad C_{sca} = \frac{< P_{sca} >_T}{S_{inc}} \, ,$$
$$C_{ext} = \frac{< P_{ext} >_T}{S_{inc}} \, , \tag{1.286}$$

respectively. All the cross-sections have a dimension of an area.

The extinction cross-section for a sample of nanoparticles with concentration n, can be easily measured experimentally, as the light intensity is attenuated as [16]:

$$I(\omega) = I_0(\omega) \exp\left(-n\, d\, C_{ext}(\omega)\right), \tag{1.287}$$

where d is the optical path. Only for molecules or very small nanoparticles, where the scattering is negligible, the light intensity attenuation can give direct information on $C_{abs}(\omega) \approx C_{ext}(\omega)$.

1.6.2 *Optical Theorem*

In the far-field it is possible to show (see also Eq. (1.218) and Eq. (1.219)) that the scattering field can be written as [15]:

$$\tilde{\mathbf{E}}^{sca}(r\hat{\mathbf{r}}) \to \frac{e^{ikr}}{r} \tilde{\mathbb{F}}\left(\hat{\mathbf{r}}, \hat{\mathbf{k}}^{inc}\right) \cdot \hat{\mathbf{e}}^{inc} E_0 \, , \tag{1.288}$$

$$\tilde{\mathbf{H}}^{sca}(r\hat{\mathbf{r}}) \to \frac{e^{ikr}}{r} \hat{\mathbf{r}} \times \tilde{\mathbb{F}}\left(\hat{\mathbf{r}}, \hat{\mathbf{k}}^{inc}\right) \cdot \hat{\mathbf{e}}^{inc} E_0 \, , \tag{1.289}$$

where $\tilde{\mathbb{F}}\left(\hat{\mathbf{r}}, \hat{\mathbf{k}}^{inc}\right)$ is called the *scattering dyadic* and describes in which directions the incident electromagnetic wave is scattered.

Using these definitions into Eq. (1.284) it is possible to show that [15]:

$$C_{ext} = \frac{4\pi}{k}\mathrm{Im}\left[\widehat{\mathbf{e}}^{inc}\cdot\tilde{\mathbb{F}}\left(\widehat{\mathbf{k}}^{inc},\widehat{\mathbf{k}}^{inc}\right)\cdot\widehat{\mathbf{e}}^{inc}\right] \tag{1.290}$$

$$= C_{abs} + C_{sca}\,. \tag{1.291}$$

Equation (1.290) represents the *optical theorem*. This theorem is also know as the *extinction* or the *forward scattering* theorem, because it allows to compute the extinction cross section only in terms of the scattering dyad in the forward (i.e. the incident) direction.

1.6.3 *Rayleigh Scattering by a Nanosphere*

We now consider a simple application for the scattering formalism introduced above. We consider a dielectric or a metal nanosphere in the vacuum, centered at the $\mathbf{r}_0 = 0$ and excited by a a plane-wave with S_{inc} given by Eq. (1.282). If the sphere is very small ($R \ll \lambda$) we have the so called *Rayleigh scattering* and we can use the quasi-static solution of Sec. 1.4.1. In particular we can assume that sphere can be modeled by a dipole

$$\tilde{\boldsymbol{\mu}}_{sph} = \int_V \tilde{\mathbf{P}}(\mathbf{r})\mathrm{d}^3\mathbf{r} = \int_V \frac{\tilde{\chi}_p - 1}{4\pi}\tilde{\mathbf{E}}_{int}(\mathbf{r})\mathrm{d}^3\mathbf{r} \tag{1.292}$$

$$= \tilde{\alpha}_{sph}(\omega)\mathbf{E}_0 \tag{1.293}$$

located at the origin. Note that $\tilde{\alpha}_{sph}(\omega)$ is given Eq. (1.192) as the total electric field inside the nanosphere is uniform. As $<P_{sca}>_T$ in Eq. (1.280) coincides with the emitted power definition in Eq. (1.224), we obtain:

$$C_{sca}(\omega) = \frac{1}{S_{inc}}\frac{\omega^4}{3c^3}\left|\left|\tilde{\boldsymbol{\mu}}_{sph}\right|\right|^2 \tag{1.294}$$

$$= \frac{8\pi}{c}\frac{\omega^4}{3c^3}|\tilde{\alpha}_{sph}(\omega)|^2 \tag{1.295}$$

$$= \frac{8\pi k^4}{3}R^6\left|\frac{\tilde{\varepsilon}_p(\omega) - 1}{\tilde{\varepsilon}_p(\omega) + 2}\right|^2\,. \tag{1.296}$$

Concerning the absorption cross-section, it can be computed from the dissipated energy by the Joule effect (see Eq. (1.46)).

Considering that the field inside the sphere is actually uniform and it is given by Eq. (1.194), we obtain:

$$C_{abs}(\omega) = \frac{\omega}{cE_0^2}\int \mathrm{Im}\left[\tilde{\varepsilon}_p(\omega)\right] \left|\left|\tilde{\mathbf{E}}_{int}(\mathbf{r})\right|\right|^2 \mathrm{d}^3\mathbf{r} \tag{1.297}$$

$$= \frac{\omega}{c} V \mathrm{Im}\left[\tilde{\varepsilon}_p(\omega)\right] \left|\frac{3}{\tilde{\varepsilon}_p(\omega)+2}\right|^2 \tag{1.298}$$

$$= \frac{4\pi\omega}{c}\frac{3V}{4\pi}\mathrm{Im}\left[\frac{\tilde{\varepsilon}_p(\omega)-1}{\tilde{\varepsilon}_p(\omega)+2}\right] = \frac{4\pi\omega}{c}\mathrm{Im}\left[\tilde{\alpha}_{sph}(\omega)\right], \tag{1.299}$$

which is often rewritten as:

$$C_{abs}(\omega) = 9k\frac{4\pi R^3}{3}\frac{\mathrm{Im}\left[\tilde{\varepsilon}_p(\omega)\right]}{\left|\tilde{\varepsilon}_p(\omega)+2\right|^2}. \tag{1.300}$$

C_{sca} and C_{abs} have a different dependence from the sphere radius (R^6 and R^3, respectively): for small spheres the absorption will be larger than the scattering, while the opposite is true for large R. However for a large particle retardation effects cannot be neglect. Thus the above formulas are valid only for $R \ll \lambda$.

We now compute C_{ext} from Eq. (1.290). First we solve Eq. (1.265) in the limit of a nanosphere with $V \to 0$ localized in $\mathbf{r}_0 = 0$. Using Eq. (1.237) we obtain:

$$\tilde{\mathbb{T}}_p(\mathbf{r}_0, \omega) = \frac{\tilde{\chi}_p(\omega)}{1+\frac{4\pi}{3}\tilde{\chi}_p(\omega)}\mathbb{I} = \frac{3}{4\pi}\frac{\tilde{\varepsilon}_p(\omega)-1}{\tilde{\varepsilon}_p(\omega)+2}\mathbb{I} = \frac{\tilde{\alpha}_{sph}(\omega)}{V}\mathbb{I}, \tag{1.301}$$

which is consistent with the fact that the scattering operator $\tilde{\mathbb{T}}_p$ represents an effective polarizability per unit-volume. Then we use Eqs (1.220, 1.301) in Eq. (1.271) to obtain:

$$\tilde{\mathbf{E}}^{sca}(\mathbf{r}) = k^2(\mathbb{I} - \widehat{\mathbf{r}}\widehat{\mathbf{r}})\frac{\mathrm{e}^{ikr}}{r}\tilde{\alpha}_{sph}(\omega)\cdot\widehat{\mathbf{e}}^{inc}E_0. \tag{1.302}$$

Comparing the above expression with Eq. (1.288) we obtain:

$$\tilde{\mathbb{F}}\left(\widehat{\mathbf{r}}, \widehat{\mathbf{k}}^{inc}\right) = k^2(\mathbb{I} - \widehat{\mathbf{r}}\widehat{\mathbf{r}})\tilde{\alpha}_{sph}(\omega), \tag{1.303}$$

which is correctly independent from $\widehat{\mathbf{k}}^{inc}$ for an isotropic nanosphere. Using Eq. (1.303) in Eq. (1.290) and recalling that $\widehat{\mathbf{k}}^{inc}\cdot\widehat{\mathbf{e}}^{inc} = 0$, we have:

$$C_{ext}(\omega) = 4\pi k\,\mathrm{Im}\left[\tilde{\alpha}_{sph}(\omega)\right], \tag{1.304}$$

which coincides with C_{abs} in Eq. (1.299). It follows (see Eq. (1.291)) that no scattering is predicted, in contrast to Eq. (1.296). Moreover a dielectric nanoparticle without dissipation (i.e. with a null imaginary part of $\tilde{\alpha}_{sph}(\omega)$) is predicted to have a null extinction cross-section. The only conclusion is that the sphere polarizability of Eq. (1.193) is in conflict with the optical theorem, as it doesn't explicitly take into account the radiation of the particle. This inconsistency can be solved by the introduction of a radiative damping term.

1.6.4 *Radiative Damping*

In this section we will show how the sphere polarizability can be modified to satisfy the optical theorem [17].

We consider again Eq. (1.265) for a nanosphere located in $\mathbf{r}_0 = 0$ with volume $V = 4\pi R^3/3 \to 0$, but we use the limit of the Green's function for both the real and the imaginary part, see Eqs. (1.236, 1.237):

$$\tilde{\mathbb{T}}_p(\mathbf{r}, \omega) = \frac{\tilde{\chi}_P(\omega)}{1 + \frac{4\pi}{3}\tilde{\chi}_P(\omega) - \mathrm{i}V\frac{2}{3}k^3\tilde{\chi}_P(\omega)}\mathbb{I} = \frac{\tilde{\alpha}'_{sph}(\omega)}{V}\mathbb{I}, \qquad (1.305)$$

with

$$\tilde{\alpha}'_{sph}(\omega) = \tilde{\alpha}_{sph}(\omega)\left(1 - \mathrm{i}\frac{2}{3}k^3\tilde{\alpha}_{sph}(\omega)\right)^{-1}. \qquad (1.306)$$

The term $\mathrm{i}\frac{2}{3}k^3\tilde{\alpha}_{sph}(\omega)$ is named *radiative damping* [18] of the sphere polarizability. This imaginary term takes directly into account that the dipole induced in the (metallic or dielectric) sphere by the external incident field, re-irradiates the energy in the free-space. As this scattering process occurs in different directions, then, the net effect is that the incident field is attenuated in the propagation direction as it will be for a pure absorption process.

We now computed $C'_{ext}, C'_{sca}, C'_{abs}$ using this corrected polarizability. The total field inside the nanoparticle can be obtained from Eq. (1.257) using Eqs. (1.236, 1.237):

$$\tilde{\mathbf{E}}^{tot}(\mathbf{r}_0) = \frac{3}{\tilde{\varepsilon}_p(\omega) + 2}\left(1 - \mathrm{i}\frac{3V}{4\pi}\frac{2}{3}k^3\frac{\tilde{\varepsilon}_p(\omega) - 1}{\tilde{\varepsilon}_p(\omega) + 2}\right)^{-1}\tilde{\mathbf{E}}^{inc}(\mathbf{r}_0). \qquad (1.307)$$

Equation (1.307) replaces the quasi-static result of Eq. (1.194).

Correcting Eq. (1.299) with Eq. (1.307), the absorption cross section becomes:

$$C'_{abs}(\omega) = \frac{4\pi\omega}{c}\frac{3V}{4\pi}\mathrm{Im}\left[\frac{\tilde{\varepsilon}_p(\omega)-1}{\tilde{\varepsilon}_p(\omega)-2}\right] \times \left|1 - \mathrm{i}\frac{3V}{4\pi}\frac{2}{3}k^3\frac{\tilde{\varepsilon}_p(\omega)-1}{\tilde{\varepsilon}_p(\omega)+2}\right|^{-2} \quad (1.308)$$

$$= 4\pi k\frac{\mathrm{Im}\left[\tilde{\alpha}_{sph}(\omega)\right]}{\left|1-\mathrm{i}\frac{2}{3}k^3\tilde{\alpha}_{sph}(\omega)\right|^2}. \quad (1.309)$$

The extinction cross section is, from the definition in Eq. (1.290):

$$C'_{ext}(\omega) = 4\pi k\mathrm{Im}\left[\tilde{\alpha}'_{sph}(\omega)\right] = C'_{abs}(\omega) + \frac{8\pi}{3}k^4|\tilde{\alpha}'_{sph}(\omega)|^2, \quad (1.310)$$

where we use that for a complex of the form $\tilde{x} = \tilde{a}/(1 - iq\tilde{a})$ its imaginary part can be rewritten as $\mathrm{Im}[x] = \mathrm{Im}[a]/|1 - iq\tilde{a}|^2 + q|\tilde{x}|^2$. The second term on the right-end side is simply the C'_{sca} (see Eq. (1.295), with $\tilde{\alpha}'_{sph}(\omega)$ polarizability). Thus the polarizability $\tilde{\alpha}'_{sph}(\omega)$ satisfies the optical theorem.

For additional details, see Sec. 2.4.3.2.

1.6.5 *Mie Scattering*

The Rayleigh theory of scattering and absorption of radiation by a small sphere predicts a resonant field enhancement due to a resonance of the polarizability if the Frölich condition (1.196) is satisfied. At resonance condition, the nanoparticle acts as an oscillating dipole, resonantly absorbing and scattering electromagnetic fields. This simple theory of dipole resonance is only valid for vanishingly small particles; in practice it provides a reasonably good approximation for MNPs with dimensions below 100 nm excited with visible or near-infrared light. For larger particles whose sizes are such that phase-changes of the driving field inside them cannot be neglected, the quasi-static approximation can largely fail and a full electrodynamic approach is required.

The formal solution for the case of a sphere is available since the beginning of XX century. In 1908 Gustav Mie developed a complete theory of the scattering and absorption of electromagnetic radiation by a sphere, in order to understand the colors of colloidal gold particles in solution [19–21]. According to the *Mie theory,* the different eigenmodes of the spherical particles are dipolar or multipolar in character and their excitation strength can be determined by the expansion of the internal and scattered fields into a set of normal modes described by vector spherical harmonics. The quasi-static results valid for sub-wavelength spheres are then recovered by retaining only the first term in the power series expansion of the absorption and scattering coeffcients.

By calculating the analytical solution of the exact electromagnetic problem, the extinction and scattering cross sections can be expressed as a series of multipolar contributions [15, 20–22]:

$$C_{ext} = \frac{2\pi}{k^2}\sum_{n=1}^{\infty}(2n+1)\mathrm{Re}\left[-\tilde{A}_n - \tilde{B}_n\right], \qquad (1.311)$$

$$C_{sca} = \frac{2\pi}{k^2}\sum_{n=1}^{\infty}(2n+1)\left(|\tilde{A}_n|^2 + |\tilde{B}_n|^2\right). \qquad (1.312)$$

The *Mie scattering coefficients* $\tilde{A}_n$ and $\tilde{B}_n$ are used to expand the magnetic and the electric field in spherical waves and the multipolar mode is defined by n ($n = 1$ for dipole, $n = 2$ for quadrupole, $n = 3$ for octupole, and so on). The Mie coefficients are complex numbers and are functions of the radius of the sphere (R) and of $k = 2\pi/\lambda = \omega/c$ and $\tilde{k}_p = \sqrt{\tilde{\varepsilon}_p(\omega)}2\pi/\lambda$, the wavevector outside and inside the nanoparticle, respectively:

$$\tilde{A}_n = \frac{j_n(kR)\psi_n'(\tilde{k}_pR) - j_n(\tilde{k}_pR)\psi_n'(kR)}{j_n(\tilde{k}_pR)\tilde{\xi}_n'(kR) - \tilde{h}_n(kR)\psi_n'(\tilde{k}_pR)}, \qquad (1.313)$$

$$\tilde{B}_n = \frac{j_n(kR)\psi_n'(\tilde{k}_pR) - \tilde{\varepsilon}_p j_n(\tilde{k}_pR)\psi_n'(kR)}{\tilde{\varepsilon}_p j_n(\tilde{k}_pR)\tilde{\xi}_n'(kR) - \tilde{h}_n(kR)\psi_n'(\tilde{k}_pR)}. \qquad (1.314)$$

Here $j_n(x)$ is the spherical Bessel function, $\tilde{h}_n(x) = \tilde{h}_n^{(1)}(x) = j_n(x)+\mathrm{i}y_n(x)$ is the first spherical Hankel function (complex valued). We also used the Riccati-Bessel functions $\psi_n(x) \equiv xj_n(x)$, $\tilde{\xi}_n(x) \equiv x\tilde{h}_n(x)$ and their derivatives $\psi_n'(x) \equiv \frac{\mathrm{d}}{\mathrm{d}x}\psi_n(x)$, $\tilde{\xi}_n'(x) \equiv \frac{\mathrm{d}}{\mathrm{d}x}\tilde{\xi}_n(x)$. We note that often in literature the Mie coefficients are defined as $\tilde{a}_n \to -\tilde{B}_n$ and $\tilde{b}_n \to -\tilde{A}_n$.

The absorption cross section can be computed as $C_{abs} = C_{ext} - C_{sca}$: the Mie theory solves the exact electromagnetic problem, thus the optical theorem is satisfied. If the sphere dielectric constant is real then simple relations hold [15, 22]:

$$-\mathrm{Re}\left[\tilde{A}_n\right] = \left|\tilde{A}_n\right|^2 \quad ; \quad -\mathrm{Re}\left[\tilde{B}_n\right] = \left|\tilde{B}_n\right|^2, \qquad (1.315)$$

and thus the absorption and scattering coefficients vanish. The absorption scattering coefficient can be also computed directly from the dissipation inside the sphere of radius R (see Eq. (1.297)), yielding the same result [23].

The first Mie coefficients ($n = 1$, the dipolar ones), for vanishing kR (i.e. small nanoparticle and/or long wavelength, Rayleigh approximation) are (by Taylor expansion):

$$\tilde{A}_1 \approx \frac{\mathrm{i}(\tilde{\varepsilon}_p - 1)}{45}(kR)^5 + O\left((kR)^7\right), \tag{1.316}$$

$$\begin{aligned}\tilde{B}_1 \approx & \frac{2\mathrm{i}(\tilde{\varepsilon}_p - 1)}{3(\tilde{\varepsilon}_p(\omega)_p + 2)}(kR)^3 + \frac{2\mathrm{i}(-3\tilde{\varepsilon}_p + 2 + \tilde{\varepsilon}_p^2)}{5(\tilde{\varepsilon}_p + 2)^2}(kR)^5 \\ & - \frac{4(\tilde{\varepsilon}_p - 1)^2}{9(\tilde{\varepsilon}_p + 2)^2}(kR)^6 + O\left((kR)^7\right).\end{aligned} \tag{1.317}$$

We note that the third term in Eq. (1.317) hasn't the unit imaginary number prefactor, while the second term is negligible with respect the first one as well as the coefficient $\tilde{A}_1$. Thus Eq. (1.317) can be rewritten as:

$$\tilde{B}_1 \approx \frac{2\mathrm{i}k^3}{3}\tilde{\alpha}_{sph}(\omega) - \frac{4k^6}{9}\tilde{\alpha}_{sph}(\omega)^2 \tag{1.318}$$

$$= \frac{2\mathrm{i}k^3}{3}\tilde{\alpha}_{sph}(\omega)\left(1 + \mathrm{i}\frac{2k^3}{3}\tilde{\alpha}_{sph}(\omega)\right), \tag{1.319}$$

where we used $\tilde{\alpha}_{sph}(\omega)$ as defined in Eq. (1.193).

Substituting Eq. (1.318) in Eq. (1.311) we obtain [22]:

$$\begin{aligned}C_{ext}(\omega) &\approx -\frac{2\pi}{k^2}3\mathrm{Re}\left[\tilde{B}_1\right] \\ &= 4\pi k\mathrm{Im}\left[\tilde{\alpha}_{sph}(\omega)\right] + \frac{8\pi k^4}{3}\mathrm{Re}\left[\tilde{\alpha}_{sph}(\omega)^2\right],\end{aligned} \tag{1.320}$$

$$C_{sca}(\omega) \approx \frac{2\pi}{k^2}3|\tilde{B}_1|^2 \approx \frac{8\pi k^4}{3}|\tilde{\alpha}_{sph}(\omega)|^2, \tag{1.321}$$

so that the optical theorem is satisfied, for vanishing small $\mathrm{Im}\left[\tilde{\varepsilon}_p(\omega)\right]$. Note also that Eq. (1.321) recovers Eq. (1.295).

1.7 Dipole Radiation in Scattering Environments

So far we have considered surfaces and nanoparticles under the illumination of a plane-wave and a dipole radiation in the free-space. In molecular plasmonics instead we have to deal with the combination of the two previous situations, i.e. electromagnetic fields emitted by molecules near surfaces and nanoparticles. As emitting molecules can be considered as point-like radiating dipoles (their dimensions are in fact much smaller than electromagnetic wavelengths), then in this section we will discuss the dipole radiation near dielectric/metallic surfaces/nanoparticles (or any scattering environment). In the vacuum the electromagnetic field is radiated away from the oscillator. When instead a scattering environment is present close to the emitting dipole, these systems can reflect or scatter the emitted radiation generating an electric field $\mathbf{E}^{sca}$ at the dipole position.

Let's consider an emitting point-like dipole located at $\mathbf{r}_0$. Then according to Sec. 1.6 and Eq. (1.215) the total electric field at $\mathbf{r}_0$ will be:

$$\tilde{\mathbf{E}}^{tot}(\mathbf{r}_0) = \tilde{\mathbb{G}}^0(\mathbf{r}_0, \mathbf{r}_0) \cdot \boldsymbol{\mu}_0 + \tilde{\mathbf{E}}^{sca}(\mathbf{r}_0), \qquad (1.322)$$

where $\tilde{\mathbf{E}}^{sca}(\mathbf{r}_0)$ is the field scattered back by the other objects to the dipole position. Note that in Eq. (1.322) we are *not* considering any external incident plane-wave: the "incident" field is in this case a pure dipolar radiation. Using Eq. (1.322) in Eq. (1.234) we have [4]:

$$\frac{< P^{tot}_{diss} >_T}{< P^0_{diss} >_T} = 1 + \frac{3}{2k^3\,||\boldsymbol{\mu}_0||}\mathrm{Im}\left[\widehat{\boldsymbol{\mu}} \cdot \tilde{\mathbf{E}}^{sca}(\mathbf{r}_0)\right], \qquad (1.323)$$

where we used Eq. (1.238) and where $\widehat{\boldsymbol{\mu}}$ is the versor of $\boldsymbol{\mu}_0$. Equation (1.323) clearly shows that the averaged power dissipated by the dipole is different from the free-space situation due to the presence of the scattering field.

1.7.1 *Forced Damped Oscillator*

We now consider how the classical model in Sec. 1.5.3 is modified in a scattering environment, which generates an electric field $\mathbf{E}^{sca}$ at the oscillating charge position $\mathbf{r}'$. Equation (1.239) becomes:

$$\frac{m_e}{e}\frac{\partial^2}{\partial t^2}\mu(t) + \frac{m_e}{e}\gamma_r^0\frac{\partial}{\partial t}\mu(t) + \frac{m_e}{e}\omega_0^2\mu(t) = e\widehat{\boldsymbol{\mu}} \cdot \mathbf{E}^{sca}(\mathbf{r}', t), \quad (1.324)$$

which represents a *forced damped oscillator*. If we assume that the $\mathbf{E}^{sca}(\mathbf{r}', t) \approx \mathbf{E}^{sca}(\mathbf{r}_0, t)$ and the same $\exp(-i\tilde{\omega}t)$ time-dependence (with $\tilde{\omega}$ complex) for all quantities, we obtain:

$$
\begin{aligned}
& -\tilde{\omega}^2 - \mathrm{i}\tilde{\omega}\left\{\gamma_r^0 + \frac{e^2}{m_e\,||\boldsymbol{\mu}_0||\,\tilde{\omega}}\mathrm{Im}\left[\widehat{\boldsymbol{\mu}}\cdot\tilde{\mathbf{E}}^{sca}(\mathbf{r}_0)\right]\right\} \\
& +\left\{\omega_0^2 + \frac{e^2}{m_e\,||\boldsymbol{\mu}_0||}\mathrm{Re}\left[\widehat{\boldsymbol{\mu}}\cdot\tilde{\mathbf{E}}^{sca}(\mathbf{r}_0)\right]\right\} = 0\,, \qquad (1.325)
\end{aligned}
$$

which resembles Eq. (1.240). If the scattered field is small (*weak coupling regime*) then we can write a solution as:

$$
\tilde{\omega} \approx \omega_{tot} + \mathrm{i}\frac{\gamma^{tot}}{2}\,, \qquad (1.326)
$$

with

$$
\begin{aligned}
\gamma^{tot} &= \gamma_r^0 + \frac{e^2}{m_e\,||\boldsymbol{\mu}_0||\,\omega_0}\mathrm{Im}\left[\widehat{\boldsymbol{\mu}}\cdot\tilde{\mathbf{E}}^{sca}(\mathbf{r}_0)\right]\,, \qquad (1.327) \\
\omega_{tot} &= \omega_0 + \frac{1}{2}\frac{e^2}{m_e\,||\boldsymbol{\mu}_0||\,\omega_0^2}\mathrm{Re}\left[\widehat{\boldsymbol{\mu}}\cdot\tilde{\mathbf{E}}^{sca}(\mathbf{r}_0)\right]\,. \qquad (1.328)
\end{aligned}
$$

Equation (1.328) predicts also a shift in the frequency, but here we will focus only on the modification of the decay rate.

Equation (1.327) can be rewritten as:

$$
\begin{aligned}
\frac{\gamma^{tot}}{\gamma_r^0} &= 1 + \frac{e^2}{m_e\,||\boldsymbol{\mu}_0||\,\omega_0\gamma_r^0}\mathrm{Im}\left[\widehat{\boldsymbol{\mu}}\cdot\tilde{\mathbf{E}}^{sca}(\mathbf{r}_0)\right] \qquad (1.329) \\
&= 1 + \frac{3c^3}{2\,||\boldsymbol{\mu}_0||\,\omega_0^3}\mathrm{Im}\left[\widehat{\boldsymbol{\mu}}\cdot\tilde{\mathbf{E}}^{sca}(\mathbf{r}_0)\right]\,, \qquad (1.330)
\end{aligned}
$$

where in the last equation we used Eq. (1.254). Note that Eq. (1.330) concides with Eq. (1.323). The quantity γ^{tot} is named *total decay rate* and differs from the spontaneous emission decay rate γ_r^0, due to the presence of the scattering field.

The full solution of Eq. (1.325) is the same as in Eq. (1.246) with γ_r^0 replaced by γ^{tot}.

1.7.2 *Purcell's Effect*

An emitting atom or a molecule can be modeled as a two level quantum system: the ground state (with wavefunction Ψ_0 and energy E_0) and the excited state (with wavefunction Ψ_n and energy E_n). Despite the excited state is a stationary state of the quantum system, it is not so when quantum fluctuations of the vacuum are considered [24, 25], which are responsible for the *spontaneous decay*. Thus an excited atom/molecule will spontaneously emit a photon of energy $\hbar\omega = E_n - E_0$ with an intrinsic decay time γ_0. From *quantum electrodynamics* it can be shown that [24, 25]:

$$\gamma^{tot} = \frac{2}{\hbar}\boldsymbol{\mu}_{n0} \cdot \mathrm{Im}\left[\tilde{\mathbb{G}}^{tot}(\mathbf{r}_0, \mathbf{r}_0, \omega)\right] \cdot \boldsymbol{\mu}_{n0} , \tag{1.331}$$

where

$$\boldsymbol{\mu}_{n0} = \langle \Psi_n | (-e) \sum_i \hat{\mathbf{r}}_i | \Psi_0 \rangle , \tag{1.332}$$

is the *transition dipole moment* (see Chapter 4), i sums over all the electrons and it is assumed that the atom/molecule is located in $\mathbf{r}_0$.

For isotropic environment, it is possible to introduce the *photon local density of states (LDOS)*:

$$\rho^{tot}(\mathbf{r}, \omega) = \frac{1}{2\pi^2\omega}\mathrm{Tr}\left[\mathrm{Im}\left[\tilde{\mathbb{G}}^{tot}(\mathbf{r}, \mathbf{r}, \omega)\right]\right] , \tag{1.333}$$

so that Eq. (1.331) becomes:

$$\gamma^{tot} = \frac{4\pi^2\omega}{3\hbar}\rho^{tot}(\mathbf{r}_0, \omega) \, ||\boldsymbol{\mu}_{n0}||^2 . \tag{1.334}$$

In the vacuum, using Eq. (1.236) we have the well known result:

$$\gamma^0 = \gamma_r^0 = \frac{4\omega^3}{3\hbar c^3} \, ||\boldsymbol{\mu}_{n0}||^2 , \tag{1.335}$$

$$\rho^0(\mathbf{r}, \omega) = \frac{\omega^2}{\pi^2 c^3} . \tag{1.336}$$

For further discussion about the relation between the quantum and the classical description, see Chapter 5.

In a scattering environment, we can use Eq. (1.268) and Eq. (1.335) to have:

$$\frac{\gamma^{tot}}{\gamma_r^0} = 1 + \frac{3}{2k^3}\widehat{\boldsymbol{\mu}} \cdot \mathrm{Im}\left[\tilde{\mathbb{G}}^{scat}(\mathbf{r}_0, \mathbf{r}_0, \omega)\right] \cdot \widehat{\boldsymbol{\mu}} , \tag{1.337}$$

where $\widehat{\boldsymbol{\mu}}$ is the direction of $\boldsymbol{\mu}_{n0}$. Note that Eq. (1.337) equals Eq. (1.323) and Eq. (1.330). This modification of the lifetime of a molecule is known as the *Purcell's effect* [26]. Purcell's prediction has been verified in different experimental settings such as close to planar interfaces [27, 28], in cavities [29] and photonic crystals [30], and close to near-field optical probes [31]. Excited-state lifetimes of single molecules have been measured as a function of their separation from laser-irradiated metal boundaries and satisfactory agreement with theory has been achieved [28, 31, 32].

1.7.3 *Fluorescence Lifetime and Quantum Yield*

So far we have modeled an emitting molecule as an oscillating dipole which radiates *all* the energy. However, in the case of fluorescence emission, we need also to consider intrinsic losses due to internal non-radiative processes [16]. In the vacuum (or at infinite distance from any scattering object) the emitting molecule has an *intrinsic total decay rate*

$$\gamma^0 = \gamma_r^0 + \gamma_{nr}^0 , \tag{1.338}$$

and an *intrinsic fluorescence quantum yield* given by:

$$q^0 = \frac{\gamma_r^0}{\gamma^0} = \frac{\gamma_r^0}{\gamma_r^0 + \gamma_{nr}^0} , \tag{1.339}$$

where γ_{nr}^0 is the *intrinsic non-radiative decay rate*. For a discussion about the non-radiative processes in organic molecules, see Chapter 4. The fluorescence quantum yield measures the efficiency of the fluorescence process and it is defined as the ratio between the number of photons emitted and the number of photons absorbed. When a molecule is an excellent emitter, then $\gamma_{nr}^0 \approx 0$, $q^0 \approx 1$ and $\gamma^0 \approx \gamma_r^0$.

The following relations also hold:

$$\frac{\gamma_{nr}^0}{\gamma^0} = 1 - q^0 , \tag{1.340}$$

$$1 + \frac{\gamma_{nr}^0}{\gamma_r^0} = \frac{1}{q^0} . \tag{1.341}$$

In a scattering environment the above equations are modified. First of all we note that the total non-radiative decay rate will become:

$$\gamma_{nr}^{tot} = \gamma_{nr}^0 + \gamma_{nr}^{env} , \tag{1.342}$$

where non-radiative decay rate due to the environment is (see Eq. (1.46)):

$$\gamma_{nr}^{env} = \frac{1}{\hbar\omega}\frac{1}{2}\int_{V_s} \mathrm{d}^3\mathbf{r}\,\sigma(\mathbf{r})\left|\left|\tilde{\mathbf{E}}^{tot}(\mathbf{r})\right|\right|^2 . \tag{1.343}$$

Here the integral is limited to the volume V_s of the scattering objects with conductibility σ. If the scattering environment is non dissipative, then the non-radiative decay rate is not modified.

The total decay rate in Eq. (1.331) can be decomposed as:

$$\gamma^{tot} = \gamma_r^{tot} + \gamma_{nr}^{env} + \gamma_{nr}^{0}, \tag{1.344}$$

and Eq. (1.337) can be rewritten as:

$$\frac{\gamma^{tot}}{\gamma_r^0} = \frac{1}{q^0} + \frac{3}{2k^3}\hat{\boldsymbol{\mu}} \cdot \mathrm{Im}\left[\tilde{\mathbb{G}}^{scat}(\mathbf{r}_0, \mathbf{r}_0, \omega)\right] \cdot \hat{\boldsymbol{\mu}}, \tag{1.345}$$

where we used Eq. (1.341). If $\gamma_{nr}^0 = 0$, we have that $q^0 = 1$ and thus Eq. (1.345) coincides with Eq. (1.337). In Eq. (1.344) γ_r^{tot} is the total (e.g. emitting dipole plus nanoparticle) radiative decay and can be calculated from the flux of the total Poynting vector (see Chapter 5 for more details).

The quantum yield of an emitting molecule coupled to the environment is thus [33]:

$$q^{tot} = \frac{\gamma_r^{tot}}{\gamma^{tot}} = \frac{\gamma_r^{tot}/\gamma_r^0}{(\gamma_r^{tot} + \gamma_{nr}^{env})/\gamma_r^0 + 1/q^0 - 1}. \tag{1.346}$$

As both the radiative and the non-radiative decay rates are modified by the environment, q^{tot} can be smaller or larger than the intrinsic quantum yield, as it will be shown in the next sections.

Equation (1.345) considers the total decay rate in a scattering environment normalized to the radiative decay rate in the vacuum (γ_r^0). It is also possible to normalize to the total decay rate in the vacuum (γ^0):

$$\begin{aligned}\hat{\gamma} = \frac{\gamma^{tot}}{\gamma^0} &= \frac{\gamma_r^{tot} + \gamma_{nr}^{env} + \gamma_{nr}^0}{\gamma^0} = q^0 \frac{\gamma_r^{tot} + \gamma_{nr}^{env} + \gamma_{nr}^0 + \gamma_r^0 - \gamma_r^0}{\gamma_r^0} \\ &= 1 + q^0 \frac{\gamma_r^{tot} - \gamma_r^0 + \gamma_{nr}^{env}}{\gamma_r^0} = 1 - q^0 + q^0 \frac{\gamma_r^{tot} + \gamma_{nr}^{env}}{\gamma_r^0}.\end{aligned} \tag{1.347}$$

In similar way we have:

$$\hat{\gamma}_r = \frac{\gamma_r^{tot}}{\gamma^0} = q^0 \frac{\gamma_r^{tot}}{\gamma_r^0}, \tag{1.348}$$

$$\hat{\gamma}_{nr} = \frac{\gamma_{nr}^{tot}}{\gamma^0} = \hat{\gamma} - \hat{\gamma}_r = 1 - q^0 + q^0 \frac{\gamma_{nr}^{env}}{\gamma_r^0}. \tag{1.349}$$

In molecular fluorescence experiments the total emission intensity is measured, after a short impulse excitation. The total emission intensity (I_{em}) originates from a two-step procedure which first involves a molecular absorption process (with a absorption scattering cross-section σ_{abs}) and then an emission probability (i.e. the quantum yield q) [33, 34], i.e. $I_{em} = kN\sigma_{abs}q$ where N is the number of molecules and k is an instrument constant. Thus the ratio between the measured emission intensity with and without a scattering environment (e.g. a metal nanoparticle) is:

$$\frac{I_{em}^{tot}}{I_{em}^{0}} = \frac{\sigma_{abs}^{env}}{\sigma_{abs}^{0}} \frac{q^{tot}}{q^{0}} . \tag{1.350}$$

In Eq. (1.350) I_{em}^{tot} is the total emission intensity of the *coupled* molecule-metal system and it is measured after that the excitation signal has ended, while σ_{abs}^{env} is the molecular scattering cross section near the metal. The excitation contribution $\sigma_{abs}^{env}/\sigma_{abs}^{0}$ is simply proportional to the square of the field enhancement factor A, see Eq. (1.199) or (1.201), as the molecular absorption is proportional to the square of incident field (for details see Chapter 4). As the metal nanoparticle enhances the electric field in a region close to the surface, the N surrounding molecular emitters will feel a larger exciting field and thus they will be excited with an increased rate. This is often the dominant term for the total *metal enhanced fluorescence* (MEF) effect (see Chapter 7). For further details see Chapter 5.

The second factor ($\frac{q^{tot}}{q^{0}}$) is instead more complex and it is related to both the radiative and non-radiative decay rates modification due to the presence of a scattering environment. We will now consider two of the most interesting examples of molecular emission near metallic systems.

1.7.4 *Fluorescence Emission Near a Metal Surface*

Let's consider an emitting dipole at a distance d from a metal surface with dielectric constant $\tilde{\varepsilon}_m(\omega)$. Without loss of generality we can consider two orientations, parallel and perpendicular to the surface, see Fig. 1.14.

Chance, Prock and Silbey [28] computed explicit expressions for the normalized decay rates near a metal surface, expanding the dipolar field in plane-waves and considering for each of them the reflection coefficients from the metal substrate:

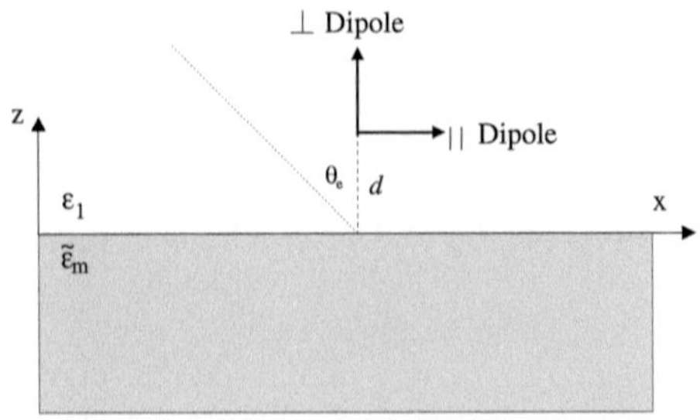

Figure 1.14 An emitting dipole (parallel or perpendicular orientation) in a medium of dielectric constant ε_1 at distance d from a metal substrate of dielectric constant $\tilde{\varepsilon}_m(\omega)$. In the *parallel* configuration the dipole is parallel to the substrate (i.e. aligned with the x-axis); in the *perpendicular* configuration the dipole is perpendicular to the substrate (i.e. aligned with the z-axis).

$$\hat{\gamma}(\perp) = 1 - q^0 + \frac{3}{2}q^0 \,\mathrm{Re}\left[\int_0^\infty du\,(1 + \tilde{r}_\| \mathrm{e}^{2\mathrm{i}l_1\hat{d}})\frac{u^3}{l_1}\right] \tag{1.351}$$

$$= 1 + \frac{3}{2}q^0 \,\mathrm{Re}\left[\int_0^\infty \mathrm{d}u\, \tilde{r}_\| \mathrm{e}^{2\mathrm{i}l_1\hat{d}}\frac{u^3}{l_1}\right], \tag{1.352}$$

$$\hat{\gamma}(\|) = 1 - q^0 + \frac{3}{4}q^0 \,\mathrm{Re}\left[\int_0^\infty du\,(1 + \tilde{r}_\perp \mathrm{e}^{2\mathrm{i}l_1\hat{d}})\frac{u}{l_1}\right.$$

$$\left. +(1-u^2)(1 - \tilde{r}_\| \mathrm{e}^{2\mathrm{i}l_1\hat{d}})\frac{u}{l_1}\right] \tag{1.353}$$

$$= 1 + \frac{3}{4}q^0 \,\mathrm{Re}\left[\int_0^\infty du\,(\tilde{r}_\perp - (1-u^2)\tilde{r}_\|)\mathrm{e}^{2\mathrm{i}l_1\hat{d}}\frac{u}{l_1}\right], \tag{1.354}$$

where (following the definitions in Ref. [35]):

$$\tilde{r}^\| = \frac{\tilde{\varepsilon}_m(\omega)l_1 - \varepsilon_1\tilde{l}_m}{\tilde{\varepsilon}_m(\omega)l_1 + \varepsilon_1\tilde{l}_m}, \tag{1.355}$$

$$\tilde{r}^\perp = \frac{l_1 - \tilde{l}_m}{l_1 + \tilde{l}_m}, \tag{1.356}$$

$$l_1 = \sqrt{1-u^2}, \tag{1.357}$$

$$\tilde{l}_m = \sqrt{\tilde{\varepsilon}_m(\omega)/\varepsilon_1 - u^2}, \tag{1.358}$$

$$\hat{d} = k_1 d = \frac{\omega n_1 d}{c}. \tag{1.359}$$

For the reflection coefficients ($\tilde{r}^\|$,$\tilde{r}^\perp$) see Sec. 1.1.2; l_1 and $\tilde{l}_m$ are normalized wavevectors and correspond to k_z^r/k_1, $\tilde{k}_z^t/k_1$, respectively, being $k_1 = n_1\omega/c$ the wavevector in the free medium 1; $\hat{d}$ is the normalized distance. Note that Eq. (1.352) is obtained from Eq. (1.351) considering that $\mathrm{Re}\left[\int_0^\infty du\, u^3/l_1\right] = 2/3$, and Eq. (1.354) from Eq. (1.353) considering that $\mathrm{Re}\left[\int_0^\infty du\, u/l_1\right] = 1$.

In all expressions (1.351–1.354) we have:

$$u = \sin(\theta_e) = \frac{k_x^r}{k_1}, \qquad (1.360)$$

where θ_e is the emission angle with respect to the surface normal and $(k_x^r)^2 + (k_z^r)^2 = k_1^2$. Thus for the light coming out above the metal surface, indicated with the symbol $(\uparrow)$, which contributes to the radiative rates, u can only vary between 0 and 1. Non-radiative rates include integrals with $u>1$: in this case k_z^r will have a complex value, thus describing evanescent waves propagating inside the metal (see Sec. 1.1.2).

It can be shown that the *apparent radiative decay rates* are [28]:

$$\hat{\gamma}_\uparrow(\perp) = \frac{q^0}{2}\int_0^1 \eta_\perp(u)\frac{u}{l_1}\mathrm{d}u, \qquad (1.361)$$

$$\hat{\gamma}_\uparrow(\parallel) = \frac{q^0}{2}\int_0^1 \eta_\parallel(u)\frac{u}{l_1}\mathrm{d}u, \qquad (1.362)$$

where the ϕ-averaged radiation pattern (see Sec. 1.5.1 with $u = \sin(\theta_e)$) are:

$$\eta_\perp(u) = \left(\frac{3}{2}u^2\right)\left(1 + 2\mathrm{Re}\left[\tilde{r}_\parallel e^{2il_1\hat{d}}\right] + |\tilde{r}_\parallel|^2\right), \qquad (1.363)$$

$$\eta_\parallel(u) = \frac{3}{4}\left(2 - u^2 + 2\mathrm{Re}\left[(\tilde{r}_\perp - (1-u^2)\tilde{r}_\parallel)e^{2il_1\hat{d}}\right]\right.$$
$$\left. + ((1-u^2)|\tilde{r}_\parallel|^2 + |\tilde{r}_\perp|^2)\right). \qquad (1.364)$$

These rates are named "apparent" because correspond only to what can be measured, collecting the light coming out above the metal substrate. Without the metal substrate all the reflection coefficients vanish and we have:

$$\eta_\perp(u) \to \eta_\perp^0(u) = \frac{3}{2}u^2, \qquad (1.365)$$

$$\eta_\parallel(u) \to \eta_\parallel^0(u) = \frac{3}{4}(2 - u^2), \qquad (1.366)$$

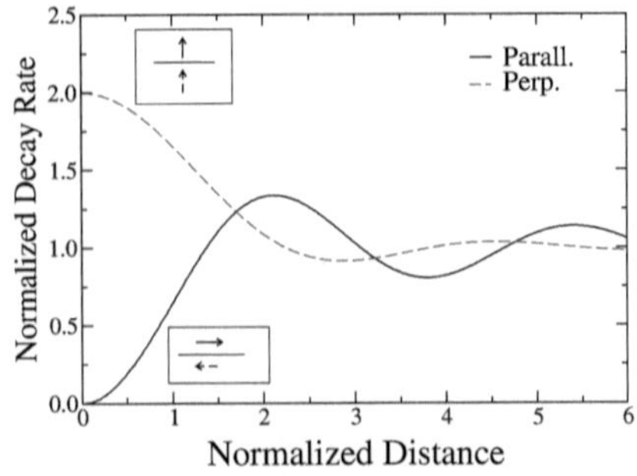

Figure 1.15 Normalized decay rate ($\hat{\gamma}$) vs normalized distance ($\hat{d}$) for a parallel and perpendicular dipole with $q^0 = 1$ on a perfect conductor (see Eq. (1.367) and Eq. (1.368)). Insets show the dipoles from the image theory.

so that $\hat{\gamma}_{\uparrow}(\perp) = \hat{\gamma}_{\uparrow}(\parallel) = q^0/2$, because the apparent rates consider only one half of the space (the one above the metal substrate). In fact the normalization condition in Eq. (1.230) becomes (for the upper half space) $\int_0^1 \mathrm{d}u\, u/l_1\eta^0(u) = 1$.

For a *perfect reflecting metal* (i.e. $\tilde{\varepsilon}_m(\omega) \rightarrow \infty$) we have $\tilde{r}^{\parallel} = -\tilde{r}^{\perp} = 1$ and expressions Eq. (1.352) and Eq. (1.354) simplify to:

$$\hat{\gamma}(\perp) = \hat{\gamma}_{\uparrow}(\perp) = 1 - \frac{3}{2}q^0\, \mathrm{Re}\left[\left(\frac{2\mathrm{i}}{(2\hat{d})^3} + \frac{2}{(2\hat{d})^2}\right)\mathrm{e}^{2\mathrm{i}\hat{d}}\right], \quad (1.367)$$

$$\hat{\gamma}(\parallel) = \hat{\gamma}_{\uparrow}(\parallel) = 1 - \frac{3}{4}q^0\, \mathrm{Re}\left[\left(\frac{2\mathrm{i}}{(2\hat{d})^3} + \frac{2}{(2\hat{d})^2} - \frac{2\mathrm{i}}{(2\hat{d})}\right)\mathrm{e}^{2\mathrm{i}\hat{d}}\right]. \quad (1.368)$$

In this case the electromagnetic waves are perfectly reflected and no dissipation occurs inside the metal (i.e. the decay rates are purely radiative, for $q^0 = 1$). Plots of (1.367) and Eq. (1.368) are shown in Fig. 1.15 (for $q^0 = 1$). For the parallel orientation the decay rate vanishes for small distances [36]: in fact, following the *image theory* [1] for a perfect reflecting surface, the emitting dipole is canceled out by an equivalent image dipole in the metal substrate (see the inset at the bottom). On the other hand, for a perpendicular orientation, the decay rate is twice the value in the vacuum, as the image dipole sums up with the emitting dipole (see the inset at the top). At larger distances the radiative rate oscillates due to radiative interference effects, reaching the vacuum value for $\hat{d} \rightarrow \infty$.

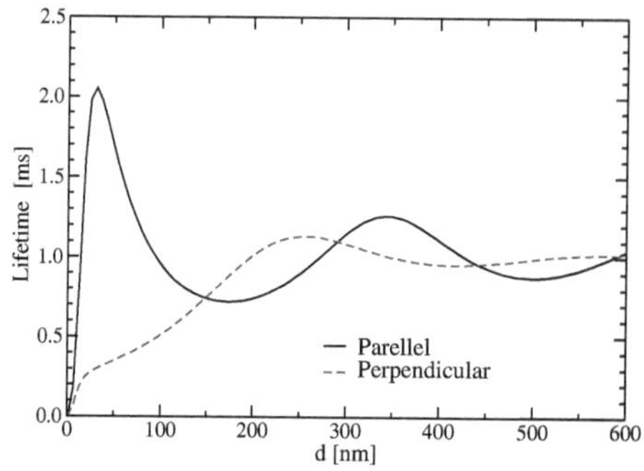

Figure 1.16 Total lifetime for a dipole at distance d above a silver surface at $\lambda = 612$ nm. The dipole is oriented parallel (black solid line) or perpendicular (dashed line) to the surface. The intrinsic quantum yield is $q^0 = 1$ and the intrinsic lifetime is $\tau_0 = 1$ ms.

As a real example we now consider an ideal emitter with $q^0 = 1$, $\gamma^0 = 1\ \mathrm{ms}^{-1}$, an emission wavelength $\lambda = 612$ nm and located in the vacuum ($n_1 = 1.0$) above a silver substrate with $n_m + i\kappa_m|_{\lambda=612\,\mathrm{nm}} = 0.131+3.88i$ (or $\tilde{\varepsilon}_m(\omega) = -15.037+1.016i$). Figure 1.16 shows how the total lifetime $\tau^{tot} = 1/\gamma^{tot}$ changes as a function of the distance d from the metal substrate. At small distances the lifetime goes to zero (the decay rate diverges) also for the perpendicular orientation. This is due to non-radiative effects (i.e. emission quenching due to the dissipation into the metal, not included in Eq. (1.367) and Eq. (1.368)). At higher distances the dissipation into the metal rapidly decreases, and the oscillations resemble the ones in Fig. 1.15.

The integrand of Eq. (1.351) and Eq. (1.353) is often indicated as *dissipated power* [36, 37] $f(u)$ and it is reported in Fig. 1.17 for one distance (e.g. d = 61 nm). The big peak at $u \approx 1.035$ correspond to the SPP. In fact from Eq. (1.163) we have:

$$u = \frac{\tilde{k}_x c}{\omega} = \sqrt{\frac{(1)\tilde{\varepsilon}_m(\omega)}{\tilde{\varepsilon}_m(\omega) + (1)}} \approx 1.035\,. \tag{1.369}$$

In Fig. 1.18 we report the ϕ-averaged radiation pattern (see Eq. (1.363) and Eq. (1.364)) for dipole with $d = 61$ nm as well as the η^0 in the free-space. For the perpendicular dipole, no emission is present in the forward direction ($\theta_e = 0$) both above the metal and in the vacuum. Reflection at the substrate also cancels emission parallel to the plane ($\theta_e = 90°$); this is also true for the parallel

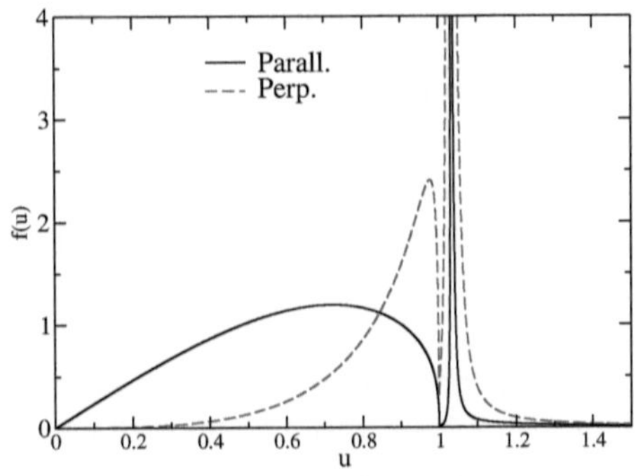

Figure 1.17 Integrand of Eq. (1.351) and Eq. (1.353) for a dipole 61 nm above a silver substrate, with $\lambda = 612$ nm.

dipole orientation, where the emission is strongly enhanced in the forward direction while no emission is present parallel to the plane ($\theta_e = 90°$). Note that for the considered distance ($d = 61$ nm) the radiative emission is similar for the parallel and the perpendicular directions.

The analysis presented in this section has been generalized to quantum electrodynamics [38] and to describe multilayer structure above the metal [28, 35–37, 39, 40].

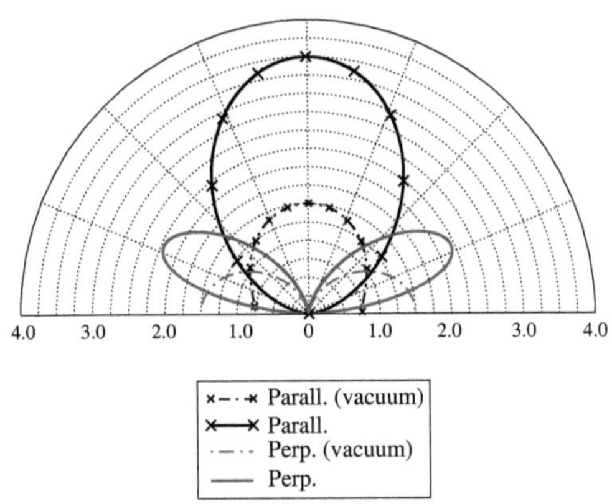

Figure 1.18 Polar plot of ϕ-averaged radiation pattern ($\eta(\theta)$) for a parallel and perpendicular direction 61 nm above a silver substrate at $\lambda = 612$ nm. Dashed lines represent $\eta(\theta)$ for a dipole in the vacuum.

1.7.5 *Fluorescence Emission Near a Metal Nanosphere*

We conclude this first chapter with the most important case in molecular plasmonics: an emitting dipole near a metallic nanosphere. For this problem the exact electrodynamic solution can be obtained, as shown by Ruppin [41], Chew [42] and by Kim *et al.* [43].

The emitter is described as a classical dipole positioned at a distance d from the surface of a sphere with radius R; we consider a parallel (i.e. tangential) or perpendicular (i.e. radial) orientation (see Fig. 1.19). The sphere has a complex dielectric constant $\tilde{\varepsilon}_p(\omega)$ and it is in the vacuum. Generalization with a sphere embedded in an external medium with a real dielectric constant ε_{out} can be directly obtained by considering the sphere with a relative dielectric constant $\tilde{\varepsilon}_p(\omega)/\varepsilon_{out}$.

The normalized (with respect the vacuum radiative rate) total and radiative decay rates in presence of the sphere are:

$$\frac{\gamma^{tot}(\perp)}{\gamma_r^0} = \frac{1}{q^0} + \frac{3}{2}\sum_{n=1}^{\infty}(2n+1)\,n\,(n+1)\,\mathrm{Re}\left[\tilde{B}_n\left(\frac{\tilde{h}_n(kr)}{kr}\right)^2\right], \tag{1.370}$$

$$\frac{\gamma_r^{tot}(\perp)}{\gamma_r^0} = \frac{3}{2}\sum_{n=1}^{\infty}(2n+1)\,n\,(n+1)\,\frac{\left|j_n(kr)+\tilde{B}_n\tilde{h}_n(kr)\right|^2}{(kr)^2}, \tag{1.371}$$

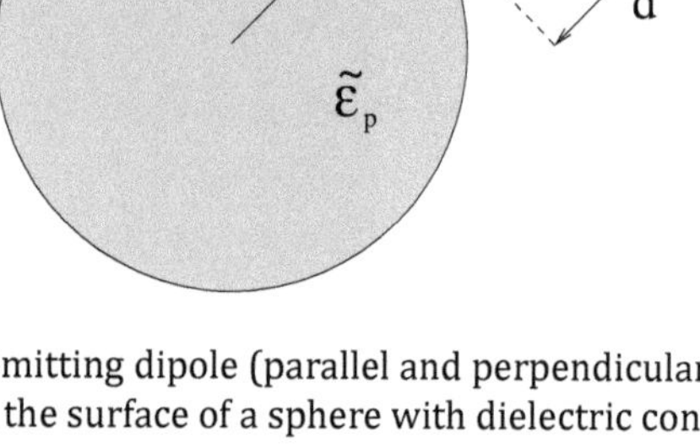

Figure 1.19 An emitting dipole (parallel and perpendicular orientation) at a distance d from the surface of a sphere with dielectric constant $\tilde{\varepsilon}_p(\omega)$.

$$\frac{\gamma^{tot}(\parallel)}{\gamma_r^0} = \frac{1}{q^0} + \frac{3}{2}\sum_{n=1}^{\infty}\left(n+\frac{1}{2}\right)\mathrm{Re}\left[\tilde{B}_n\left(\frac{\tilde{\zeta}_n'(kr)}{kr}\right)^2 + \tilde{A}_n\left(\tilde{h}_n(kr)\right)^2\right], \tag{1.372}$$

$$\frac{\gamma_r^{tot}(\parallel)}{\gamma_r^0} = \frac{3}{2}\sum_{n=1}^{\infty}\left(n+\frac{1}{2}\right) \times \left[\left|j_n(kr)+\tilde{A}_n\tilde{h}_n(kr)\right|^2 + \frac{\left|\psi_n'(kr)+\tilde{B}_n\tilde{\zeta}_n'(kr)\right|^2}{(kr)^2}\right], \tag{1.373}$$

where $r = R + d$ is the total distance of the dipole position from the center of the sphere. For the definition of the special functions as well as for the Mie scattering coefficients ($\tilde{A}_n$, $\tilde{B}_n$) which contain the information about the dimension of the sphere, see Sec. 1.6.5.

The expressions for the radiative rates, Eqs. (1.371, 1.373) have been obtained considering the flux of the Poynting vector of the total field (the dipolar source plus the field scattered from the sphere) over a spherical surface with radius $r \to \infty$. The expressions for the total decay rates have been obtained from Eq. (1.345).

If $\tilde{\varepsilon}_p(\omega) = 1$, i.e. if the sphere is not present, Eqs. (1.370, 1.372) correctly equal to $1/q^0$, as all Mie coefficient vanishes. Eqs. (1.371, 1.373) also equal $1/q^0$, using sum rules of the spherical Bessel functions [42].

If the imaginary part of $\tilde{\varepsilon}_p(\omega)$ is not zero, e.g. if the sphere is metallic, the dissipation inside the sphere has to be considered. The non-radiative decay rate (see Eq. (1.343) has been obtained by Ruppin [41], computing the total electric field inside the sphere. These expressions are quite complicated and it is easier to obtain the non-radiative rate as a difference between the total and the radiative rates. It can be shown that [44]

$$\frac{\gamma_{nr}^{env}(\perp)}{\gamma_r^0} = \frac{3}{2}\sum_{n=1}^{\infty}(2n+1)\,n\,(n+1)\left[-\mathrm{Re}\left[\tilde{B}_n\right] - \left|\tilde{B}_n\right|^2\right]\left|\frac{\tilde{h}_n(kr)}{kr}\right|^2, \tag{1.374}$$

$$\frac{\gamma_{nr}^{env}(\parallel)}{\gamma_r^0} = \frac{3}{2}\sum_{n=1}^{\infty}\left(n+\frac{1}{2}\right)\left\{\left[-\mathrm{Re}\left[\tilde{B}_n\right] - \left|\tilde{B}_n\right|^2\right]\left|\frac{\tilde{\zeta}_n'(kr)}{kr}\right|^2 + \left[-\mathrm{Re}\left[\tilde{A}_n\right] - \left|\tilde{A}_n\right|^2\right]\left|\tilde{h}_n(kr)\right|^2\right\}. \tag{1.375}$$

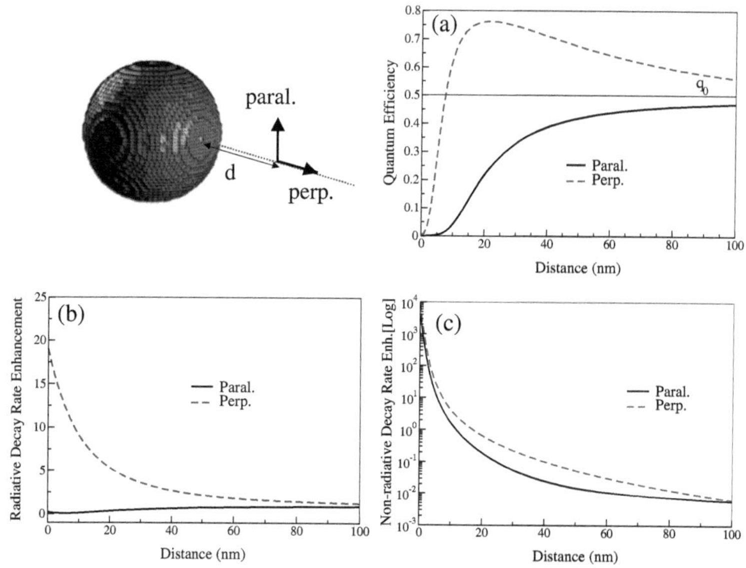

Figure 1.20 (a) Quantum efficiency (q^{tot}), (b) enhancement of the total radiative decay rate $\gamma_r^{tot}/\gamma_r^0$ (c) enhancement of environment non-radiative decay rate $\gamma_r^{env}/\gamma_r^0$ of a dipole (with intrinsic quantum efficiency $q_0 = 0.5$) emitting parallel (solid lines) and perpendicular (dashed lines) to a 100 nm-diameter Ag nanosphere at a distance (d) from the surface (see inset). The curves are obtained from the exact electrodynamic theory using a free software available online [45]. The external medium refractive index is 1, and the exciting wavelength was fixed to 612 nm.

If the sphere dielectric constant is real, then Eq. (1.315) holds and the non-radiative rate γ_{nr}^{env} correctly vanishes.

As an example of application we consider a dipole oscillating parallel or perpendicular to a 100 nm-diameter silver nanosphere (see inset of Fig. 1.20). We use an intrinsic quantum efficiency $q^0 = 0.5$ and an emitting wavelength of 612 nm.

Figure 1.20c shows that the non-radiative decay rate enhancements (i.e. $\gamma_{nr}^{env}/\gamma_r^0$) due to the environment (see Eq. (1.374) and Eq. (1.375)) are very high close to the nanosphere but then they decay very rapidly away from the surface.

The total radiative decay rate enhancements (see Eq. (1.370) and Eq. (1.372)) have a different behavior (see Fig. 1.20b). The perpendicular one is strongly enhanced, while the parallel is always very low: this behavior resembles the one in Fig. 1.15 and thus can be explained with an image dipole effect. The resulting total quantum yield (q^{tot}) is shown in Fig. 1.20a. While for the parallel dipole the quantum yield is always lower than the one in the vacuum, for the perpendicular one, it is larger, for distances greater than 10 nm. Thus the *presence of the metal nanoparticle increases the quantum yield of the molecule.* This is one of main effects used for molecular plasmonics applications.

1.7.5.1 The Gersten-Nitzan model

In the limit of $k \to 0$, only the dipolar term (with $n = 1$) counts for the radiative rates, and Eq. (1.371) and Eq. (1.373) simplify to:

$$\frac{\gamma_r^{tot}(\perp)}{\gamma_r^0} \approx 9\frac{\left|j_1(kr) + \tilde{B}_1\tilde{h}_1(kr)\right|^2}{(kr)^2} \tag{1.376}$$

$$\approx \left|1 + \frac{2(\tilde{\varepsilon}_p - 1)}{\tilde{\varepsilon}_p + 2}\left(\frac{R}{R+d}\right)^3\right|^2, \tag{1.377}$$

$$\frac{\gamma_r^{tot}(\|)}{\gamma_r^0} \approx \frac{9}{4}\frac{\left|\psi_1'(kr) + \tilde{B}_1\tilde{\zeta}_1'(kr)\right|^2}{(kr)^2} \tag{1.378}$$

$$\approx \left|1 - \frac{\tilde{\varepsilon}_p - 1}{\tilde{\varepsilon}_p + 2}\left(\frac{R}{R+d}\right)^3\right|^2, \tag{1.379}$$

where we used Eq. (1.317). Eqs. (1.377, 1.379) accurately describe the radiative decay rates for metal nanoparticles used in applications (R of the order of 10 nm), despite their quasi-static limit derivation.

It is interesting to note that Eq. (1.377) and Eq. (1.379) coincide exactly with the square of Eq. (1.199) and Eq. (1.201), respectively, which describe the field-enhancements of a dielectric sphere in the quasi-static limit. This is a manifestation of the so called *optical reciprocity theorem* and it can be shown that it holds for arbitrary geometry [46] (see also Sec. 5.3.3).

For the total decay rate, instead, it is not correct to include only the dipolar term: in fact the total decay rate also contains dissipation into the metal which requires the inclusion of high-order terms [47].

The formulas for the radiative and non-radiative decays rate in the quasi-static approximation were also derived by Gersten and Nitzan (GN) [48] who extended the quasi-static treatment to *spheroids* based on an expansion in terms of an orthogonal set of eigenfunctions, so that shape-induced shifts of radiative and non-radiative decay rates can be described. The accuracy of the GN decay rates versus the exact electrodynamic theory has been described in literature [47, 49, 50] for spherical nanoparticles, while no exact analytic solution exists for spheroids [51].

Exact and/or approximated expressions of the decay rate of an emitting dipole near dielectric/metallic body of complex shapes/structures can be found in literature [52–54].

1.7.5.2 The Carminati model

We can also take the limit of Eq. (1.370) and Eq. (1.372) as $R \to 0$ to obtain:

$$\frac{\gamma^{tot}(\perp)}{\gamma_r^0} = 1 + 6k^3 \mathrm{Im}\left[\tilde{\alpha}_{sph} e^{2ikd}\left(-\frac{1}{(kd)^4} - \frac{2i}{(kd)^5} + \frac{1}{(kd)^6}\right)\right], \tag{1.380}$$

$$\frac{\gamma^{tot}(\|)}{\gamma_r^0} = 1 + \frac{3}{2}k^3 \times \mathrm{Im}\left[\tilde{\alpha}_{sph} e^{2ikd}\left(+\frac{1}{(kd)^2} + \frac{2i}{(kd)^3} - \frac{3}{(kd)^4} - \frac{2i}{(kd)^5} + \frac{1}{(kd)^6}\right)\right]. \tag{1.381}$$

These equations have been also obtained by Carminati *et al.* [17] and directly show a d^{-6} dependence at large nanoparticle-molecule distances. In the Carminati model the metal nanosphere is considered as a *point-like* (located in $\mathbf{r}_p$) polarizable entity with polarizability $\tilde{\alpha}'_{sph}$ (i.e. due to radiative damping, see Sec. 1.6.4). Thus from Eqs. (1.269, 1.301, 1.337) we simply have:

$$\frac{\gamma^{tot}}{\gamma_r^0} = 1 + \frac{3}{2k^3}\hat{\boldsymbol{\mu}} \cdot \mathrm{Im}\left[\tilde{\alpha}_{sph} \tilde{\mathbb{G}}^0(\mathbf{r}_0, \mathbf{r}_p) \cdot \tilde{\mathbb{G}}^0(\mathbf{r}_p, \mathbf{r}_0)\right] \cdot \hat{\boldsymbol{\mu}} \tag{1.382}$$

where $\hat{\boldsymbol{\mu}}$ is the direction of the dipole. In real situation, the Carminati model (also known as the *dipole approximation* for the nanosphere) can be applied only for nanoparticle-molecule distances 2–3 times the nanoparticle radius. Otherwise the total decay rate is largely underestimated. Higher multipole moments are needed for a correct description [34, 47], such as in the GN model.

Acknowledgments

This work was partially funded by the ERC Starting Grant FP7 Project DEDOM, Grant Agreement No. 207441.

References

1. Jackson, J. D. (ed.) (1999) *Classical Electrodynamics*, 3rd edn, John Wiley & Sons, New York.
2. Hecht, E. (ed.) (1998) *Optics* 3rd edn, Addison-Wesley, San Francisco.
3. Smith, D. R., Pendry, J. B., and Wiltshire, M. C. K. (2004) Metamaterials and negative refractive index, *Science*, **305**, 788–792.
4. Novotny, L., and Hecht, B. (2006) *Principles of Nano-Optics*, Cambridge University Press, Cambridge.
5. Wooten, F. (1972) *Optical properties of solids*, Academic Press, New York.
6. Drude, P. (1900). Zur elektronentheorie der metalle, *Ann. Phys. (Berlin)*, **306**, 3, 566–613.
7. Maier, S. A. (ed.) (2007) *Plasmonics: Fundamentals and Applications*, Springer Science + Business Media LLC, United Kingdom.
8. Kittel, C. (ed.) (1986) *Introduction to Solid State Physics*, 6th edn, John Wiley & Sons, New York.
9. Palik, E. D. (ed.) (1985) *Handbook of Optical Constants of Solids*, Academic Press, New York.
10. Johnson, P. B., and Christy, R. W. (1972) Optical constants of the noble metals, *Phys. Rev. B*, **6**, 4370–4379.
11. Martin, R. M. (ed.) (2004) *Electronic Structure: Basic Theory and Practical Methods*, Cambridge University Press, Cambridge.
12. Brongersma, M. L., and Kik, P. G. (eds) (2007) *Surface Plasmon Nanophotonics*, Springer.
13. Pitarke, J. M., Silkin, V. M., Chulkov, E. V., and Echenique, J. M. (2007) Theory of surface plasmon and surface-plasmon polaritons, *Rep. Prog. Phys.*, **70**, 1–87.

14. Barnes, W. L, Dereux, A., and Ebbesen, T. W., (2003) Surface plasmon subwavelength optics, *Nature*, **424**, 824–830.
15. Mishchenko, M. I., Travis, L. D., and Lacis, A. A. (2002) *Scattering, Absorption, and Emission of Light by Small Particles*, Cambridge University Press, Cambridge.
16. Lakowicz, J. R. (ed.) (2006) *Principles of Fluorescence Spectroscopy*, 3rd edn, Springer.
17. Carminati, R., Greffet, J.-J., Henkel, C., and, Vigoureux, J. M. (2006) Radiative and non-radiative decay of a single molecule close to a metallic nanoparticle, *Opt. Commun.*, **261**, 368–375.
18. Wokaun, A., Gordon, J. P., and Liao, P. F. (1982) Radiation damping in surface-enhanced Raman scattering, *Phys. Rev. Lett.*, **48**, 957–960.
19. Mie, G. (1908) Beiträge zur Optik trb̈er Medien, speaiell kolloidaler Metallösungen, *Ann. Phys. (Berlin)*, **25**, 377–445.
20. Bohren, C. F., and Huffman, D. R. (eds) (1983) *Absorption and Scattering by Small Particles*, Wiley, New York.
21. Kerker, M. (ed.) (1969) *The Scattering of Light and Other Electromagnetic Radiation* Academic Press, New York.
22. Tsang, L, Kong, J. A., and Ding, K.-H. (eds) (2000) *Scattering of Electromagnetic Waves: Theories and Applications*, John Wiley & Sons.
23. Karam, M. A., and Fung, A. K. (1993) Electromagnetic energy absorbed within a Mie sphere, *J. Electr. Waves. Appl.*, **7**, 1379–1387.
24. Milonni, P. W. (ed.) (1994) *The quantum vacuum: an introduction to quantum electrodynamics*, Academic Press.
25. Craig, D. P., Thirunamachandran, T. (eds) (1998) *Molecular Quantum Electrodynamics*, Dover Publications.
26. Purcell, E. M. (1946) Resonance absorption by nuclear magnetic moments in a solid, *Phys. Rev.*, **69**, 37–38.
27. Drexhage, K. H. (1974) Interaction of light with monomolecular dye layers, *Prog. Opt.*, **12**, 163–232.
28. Chance, R. R., Prock, A., and Silbey, R. (1978) Molecular fluorescence and energy transfer near interfaces, *Adv. Chem. Phys.*, **37**, 1–65.
29. Kleppner, D. (1981) Inhibited Spontaneous Emission, *Phys. Rev. Lett.*, **47**, 233–236.
30. Lodahl, P., van Driel, F., Nikolaev, I. S., Irman, A., Overgaag, K., Vanmaekelbergh, D., and Vos, W. L. (2004) Controlling the dynamics of spontaneous emission from quantum dots by photonic crystals, *Nature*, **430**, 654–657.

31. Bian, R. X., Dunn, R. C., Xie S. X., and Leung, P. T. (1995) Single molecule emission characteristics in near-field microscopy, *Phys. Rev. Lett.*, **75**, 4772–4775.

32. Buchler, B. C., Kalkbrenner, T., Hettich, C., and Sandoghdar, V. (2005) Measuring the quantum efficiency of the optical emission of single radiating dipoles using a scanning mirror, *Phys. Rev. Lett.*, **95**, 063003/1-4.

33. Bharadwaj, P., and Novotny, L. (2007) Spectral dependence of single molecule fluorescence enhancement *Opt. Express*, **15**, 14266.

34. Anger, P., Bharadwaj, P., and Novotny, L. (2006) Enhancement and quenching of single-molecule fluorescence, *Phys. Rev. Lett.*, **96**, 113002–113006.

35. Ford, G. W., and Weber, W. H. (1984) Electromagnetic interactions of molecules with metal surfaces, *Phys. Rep.*, **113**(4), 195–287.

36. Barnes, W. L. (1998) Fluorescence near interfaces: The role of photonic mode density, *J. Mod. Optics*, **45**, 661–699.

37. Barnes, W. L. (1999) Electromagnetic crystals for surface plasmon polaritons and the exctraction of light from emissive devices *J. Lightwave Tech.*, **17**, 2170–2182.

38. Wylie, J. M., and Sipe, J. E. (1984) Quantum electrodynamics near an interface, *Phys. Rev. A*, **84**, 1185–1193.

39. Lukosz, W. (1980) Theory of optical-environment-dependent spontaneous-emission rates for emitters in thin layers *Phys. Rev. B.*, **22**, 3030–3038.

40. Celebi, K, Heidel, T. D., and Baldo, M. A. (2007) Simplified calculation of dipole energy transport in a multilayer stack using dyadic Green's functions, *Opt. Express*, **15**, 1762–1772.

41. Ruppin, R. (1982) Decay of an excited molecule near a small metal sphere, *J. Chem. Phys.*, **76**, 1681–1684.

42. Chew, H. (1987) Transition rates of atoms near spherical surfaces, *J. Chem. Phys.*, **87**, 1355–1360.

43. Kim, Y. S., Leung, P. T., and George, T. F. (1988) Classical decay rates for molecules in the presence of a spherical surface: A complete treatment, *Surf. Sci.*, **195**, 1–14.

44. Colas des Francs, G., Bouhelier, A., Finot, E., Weeber, J. C., Dereux, A., Girard, C, and Dujardin, E. (2008) Fluorescence relaxation in the near-field of a mesoscopic metallic particle: distance dependence and role of plasmon modes, *Opt. Express*, **16**, 17654–17666.

45. Mertens, H., and Polman, A., software available at http://erbium.amolf.nl/.

46. Gersten, J. (2005) Theory of fluorophore-metallic surface interaction, in *Topics in Fluorescence Spectroscopy, Volume 8: Radiative Decay Engineering* (ed. by Geddes, C. D., and Lakowicz, J. R., Springer), New-York.

47. Moroz, A. (2010) Non-radiative decay of a dipole emitter close to a metallic nanoparticle: Importance of higher-order multipole contributions, *Opt. Commun.*, **283**, 2277–2287.

48. Gersten, J., and Nitzan, A. (1981) Spectroscopic properties of molecules interacting with small dielectric particles, *J. Chem. Phys.*, **75**, 1139–1152.

49. Mertens, H., Koenderink, A. F., and Polman, A. (2007) Plasmon-enhanced luminescence near noble-metal nanospheres: Comparison of exact theory and an improved Gersten and Nitzan model, *Phys. Rev. B*, **76**, 115123/1-12.

50. Mertens, H., and Polman, A. (2009) Strong luminescence quantum-efficiency enhancement near prolate metal nanoparticles: Dipolar versus high-order modes, *J. Appl. Phys.*, **105**, 044302/1-8.

51. Klimov, V. V., Ducloy, M., and Letokhov, V. S. Spontaneous emission of an atom places near a prolate nanospheroid, *Eur. Phys. J. D.*, **20**, 133–148.

52. Klimov, V. V, Ducloy, M., and Letokhov, V. S., (2001) Spontaneous emission of an atom in the presence of nanobodies, *Quant. Electr.*, **31**, 569–586.

53. Moroz, A. (2005) A recursive transfer-matrix solution for a dipole radiating inside and outside a stratified sphere, *Ann. Phys. (NY)*, **315**, 352–418.

54. Chang, D. E, Sorensen, A. S., Hemmer, P. R., and Lukin, M. D. (2007) Strong coupling of single emitters to surface plasmons, *Phys. Rev. B.*, **76**, 035420/1-26.

Part I

Theory and Computational Methods

Chapter 2

Computational Approaches for Plasmonics

M. A. Yurkin[a,b]

[a] *Institute of Chemical Kinetics and Combustion SB RAS, Institutskaya 3 Novosibirsk, 630090, Russia*
[b] *Novosibirsk State University, Pirogova 2 Novosibirsk, 630090, Russia*
yurkin@gmail.com

In this chapter the problem of elastic light scattering, i.e. interaction of electromagnetic waves with finite objects, is discussed. A detailed overview of one of the widely used methods for plasmonics, the *discrete dipole approximation* (DDA), is presented. This includes the theory of the DDA, practical recommendations for using available computer codes, and discussion of the DDA accuracy.

2.1 Introduction

Although the field of plasmonics has experienced a rapid growth in recent decades, it started long before that. In particular, the analytical solution for light scattering by a sphere was developed by Mie [1] to explain the color of colloidal gold. Since then a number of methods to solve Maxwell's equations has been developed [2] mainly motivated by other fields, such as astrophysics

Handbook of Molecular Plasmonics
Edited by Fabio Della Sala and Stefania D'Agostino

ISBN 978-981-4303-20-0 (Hardcover), 978-981-4303-21-7 (eBook)
www.panstanford.com

and atmospheric remote sensing [3]. Nowadays, an opposite process is taking place—more and more methods of numerical electromagnetics are applied to nanoparticles.

In this chapter we mainly discuss the problem of *elastic light scattering*, i.e. interaction of electromagnetic waves with finite objects. However, problems involving infinite scatterers, such as surfaces or periodic objects, are also addressed. Other physical processes, such as two-photon or Raman scattering, can be simulated using the near-field distribution obtained during the solution of the light scattering problem. Moreover, this chapter is limited to non-magnetic scatterers.

The first goal of this chapter is to review existing methods to simulate light scattering by plasmonic systems, including the applicability of bulk electric permittivity to nanoparticles. The second goal is to provide a detailed overview of one of the widely used methods, the DDA. This includes the theory of the DDA, practical recommendations for using available computer codes, and discussion of the DDA accuracy.

2.2 Overview of Different Methods

2.2.1 *Finite Scatterers in Homogeneous Medium*

The *Mie theory* [1] and the *T-matrix method* [4] are very efficient for (multilayered) spheres and axisymmetric particles (with moderate aspect ratios), respectively. Several methods, applicable to particles of arbitrary shapes, have been used in plasmonic simulations: the *boundary element method (BEM)* [5, 6], the DDA [7–9], the *finite-difference time-domain method (FDTD)* [10, 11], the *finite element method (FEM)* [12, 13], the *finite integration technique (FIT)* [14] and the *null-field method with discrete sources (NFM-DS)* [15, 16]. There is also quasi-static approximation for spheroids [12], but it is not discussed here.

The BEM, the DDA, the FEM, and the NFM-DS solve the Maxwell's equations in the *frequency domain*. The BEM and the DDA solve by discretization the corresponding surface- and volume-integral equation, respectively. In the FEM the differential form of Maxwell's equations is solved by volume discretization. The NFM-DS is based on expansion of the electromagnetic field as a combination of spherical wave functions with different centers (so-called, discrete sources), which amplitudes are obtained from boundary conditions at the particle surface.

The FDTD and the FIT solve the *time-domain* Maxwell's equations in the original and modified form, respectively. These two methods, as well as the FEM, need to discretize not only the particle but also some space around it.

There are a few reviews of computational methods in plasmonics [12, 17–19], but none of them aims at full comparison of the methods. General properties of the methods allow one to analyze their scaling properties, e.g. how the simulation time and memory requirements scale with volume of the particle [17]. It does not help, however, in answering the main question: "Given a scattering problem *and required accuracy*, which method is the fastest?" For instance, the variation of refractive index influences computational resources both directly and indirectly through deteriorating accuracy, requiring one to refine discretization.

A systematic comparison of the methods should include the simulation of the several (the more—the better) test problems by these methods running on the same hardware. Such comparisons were performed for dielectric particles, see, e.g. Refs. [20–22], but they are not relevant for the plasmonics. On contrary, in plasmonics such comparisons are very rare. We can cite three examples, which both considered a single specific scattering problem, making it hard to generalize the conclusions. In particular, the FDTD and the FEM were compared for computation of near-field around 50-nm silver cube interacting with 600-nm plane-wave [12]. Accuracy of the FEM was worse than that of the FDTD but still satisfactory. The FEM simulation required 4 hours on a single 3.4 GHz processor, while FDTD—8 hours on 256 double-core 2.6 GHz processors. Another comparison [14] addressed the DDA and the FIT (the latter implemented in the commercial software) for simulation of refractive index sensitivity of rhombic hybrid Au–Ag nanostructure array. Both methods obtained the same value of sensitivity, but the DDA was faster (not specified how much).

The third example [19] is the most systematic one. The DDA, the FEM, and the FDTD were used for the calculation of scattering spectrum of 80-nm gold sphere with several discretizations levels for each method. Unfortunately, the accuracy and simulation times are discussed separately, so it is impossible to say which time corresponds to which accuracy (discretization level). General trend is that DDA is faster but less accurate than the FEM. Simulation time of the FDTD is comparable to that of the FEM, but its accuracy is the worst of three methods. Summarizing all three examples, truly systematic comparisons in these fields will definitely benefit the community.

Finally, the basic configuration for all methods is a finite particle *in vacuum*. However, a scattering problem of a finite particle in homogeneous non-absorbing dielectric medium (with a real refractive index n_0) is equivalent to the basic one after adjusting the wavelength $\lambda \to \lambda/n_0$ and all refractive indices $\tilde{n} \to \tilde{n}/n_0$, due to the corresponding invariance of the frequency-domain Maxwell's equations without sources.

2.2.2 *Periodic Scatterers*

Configurations, obtained by 1D or 2D periodic repetition of a nanoparticle, are common in plasmonics. And most methods can, after some modifications, simulate such configurations, discretizing only one instance of the nanoparticle. Corresponding modifications for the FDTD, the FIT, and the FEM consist in replacing absorbing boundary condition on the outer boundary of the computational domain by a periodic one, see, e.g. Ref. [11].

Modifications required for the BEM and the DDA are more involved. The core entity in these two methods is the Green's tensor, which describes interaction between two discretization elements (e.g. see Eq. (2.6) below for the DDA). To account for particle periodicity, the Green's tensor should be modified to include contributions from all periodic copies of the same discretization element, which involves infinite sums. For the BEM such approach is discussed in Ref. [23], for the DDA in Refs. [24, 25], including implementation in the freely-available computer codes (see Sec. 2.5.4).

2.2.3 *Scatterers Near an Infinite Plane Surface*

Another common configuration in plasmonics is a particle upon or near semi-infinite substrate. This substrate may have no plasmonic properties by itself, e.g. glass, but it still modifies the light scattering properties of the nanoparticle. The FDTD, the FIT, and the FEM can naturally handle such problems, because the effect of the interface is automatically accounted by the part of the substrate falling into the computational domain, when appropriate absorbing boundary conditions are used [11, 13].

The BEM and the DDA can solve this problem discretizing only the particle itself, if the Green's tensor is adjusted to account for the substrate. A resulting Green's tensor is expressed in terms of *Sommerfeld integrals*, accurate evaluation of which is challenging by itself [26]. This has been implemented into the DDA by a number of different researchers [27–30]. Moreover, a simpler way to calculate the Green's tensor in the presence of substrate has been recently proposed by Mackowski [31]. The NFM-DS can also simulate such configurations, taking account of the substrate by additional boundary conditions [16].

Alternative way to approach this scattering problem by the DDA in their standard form is a brute-force discretization of a large block of substrate together with the particle [32, 33]. Finite size of this block does introduce certain artifacts in the computed results. However, they can be diminished by using either the Gaussian beam with size smaller than that of the block [34] or by using empirical compensation formula [32], which also uses result of simulations for truncated substrate without the particle. Similar ideas should, in principle, be also applicable to the BEM.

Finally, periodic scatterers and particles on a substrate, discussed above, are the simplest examples of the scattering problem in a complex background. Real applications may present more elaborate configurations, like multi-layered substrate or periodic array in a film. However, this presents no qualitative difference. In particular, using a proper Green's tensor for the complex background [35], the standard DDA or BEM can be used, discretizing only the scatterer itself.

2.3 Electric Permittivity

There is a consensus that Maxwell's equations work fine for nanoparticles with size down to at least 1 nm. In other words, a good fit of experimental data can be obtained using a rigorous simulation method and proper data for the complex electric permittivity $\tilde{\varepsilon}$ or equivalently the complex refractive index $\tilde{n}$ [36]. This makes choosing a particular value of $\tilde{\varepsilon}$ an important practical question, which can be divided into two parts:

(1) How to choose a correct value of $\tilde{\varepsilon}$ for bulk material?
(2) How the bulk values should be adjusted for small nanoparticles?

First question is a consequence of existence of several sources of data for each material with sometimes significant differences. In particular, for gold there are two widely used sources: by Johnson and Christy [37] and by Palik [38]; however, several other options are also available [7, 36]. Although Khlebtsov [36] provided a prescription based on his experience, choosing the best set of $\tilde{\varepsilon}$ values is still ambiguous. To reliably choose one option over the others one should use a precise experimental data, in which $\tilde{\varepsilon}$ is the most important uncertainty. This strict requirement can, in principle, be complied by single-particle experiments, see, e.g. Ref. [18].

For other plasmonic materials the situation is similar to that of gold. In particular, both sources [37, 38] contain also data for silver and copper. Moreover, handbook by Palik [38] contains data for much more materials with a certain critique justifying the choice of particular values. Some of the known sources for many materials can be found in an online database [39].

Apart from tabulated experimental data, there exist analytical models for the refractive indices. The most widely used is the *Drude free-electron model* [37] (see Sec. 1.2.1):

$$\tilde{\varepsilon}(\omega) = 1 - \frac{\omega_p^2}{\omega\left(\omega + \mathrm{i}/\tau\right)}, \tag{2.1}$$

where ω_p is plasmon frequency and τ is the mean relaxation time. While this model is generally accurate for low frequencies, it is usually not so for $\omega \sim \omega_p$. Therefore, it is not recommended to use this model in the simulations. The only exception is time-domain methods (e.g. the FDTD), in which only an analytical model for $\tilde{\varepsilon}(\omega)$ can be used, not a set of values. But even then it is recommended to use more elaborate multi-parametric models to better approximate the experimental values of $\tilde{\varepsilon}$ [11].

Second question arises because particles may be comparable or smaller than electron mean free path. Thus, $\tilde{\varepsilon}$, determined mostly by free electrons, is effected by reflections from surface. The common way to include this effect is [36]:

$$\tilde{\varepsilon}(\omega, a) = \tilde{\varepsilon}_b(\omega) + \Delta\tilde{\varepsilon}_{\mathrm{SD}}(\omega, a), \tag{2.2}$$

$$\begin{aligned}\Delta\tilde{\varepsilon}_{\mathrm{SD}}(\omega, a) &= \tilde{\varepsilon}_p^{\mathrm{Dr}}(\omega, a) - \tilde{\varepsilon}_b^{\mathrm{Dr}}(\omega) \\ &= \left[1 - \frac{\omega_p^2}{\omega\,(\omega + \mathrm{i}/\tau_b + \mathrm{i}/\tau_s)}\right] - \left[1 - \frac{\omega_p^2}{\omega\,(\omega + \mathrm{i}/\tau_b)}\right] \\ &= \frac{\omega_p^2}{\omega\,(\omega + \mathrm{i}/\tau_b)} - \frac{\omega_p^2}{\omega\,(\omega + \mathrm{i}/\tau_b + \mathrm{i}/\tau_s)},\end{aligned} \tag{2.3}$$

where $\Delta\tilde{\varepsilon}_{\mathrm{SD}}$ is the *surface damping correction*, $\tilde{\varepsilon}_b(\omega)$ is bulk permittivity, a is the size of the nanoparticle, $\tilde{\varepsilon}_b^{\mathrm{Dr}}(\omega)$ and $\tilde{\varepsilon}_p^{\mathrm{Dr}}(\omega, a)$ are bulk and small-particle Drude models respectively, τ_b and τ_s are bulk and *surface-induced damping time* respectively. Surface damping is expressed as [36]:

$$\frac{1}{\tau_s} = \frac{A\nu_{\mathrm{F}}}{L_{eff}}, \tag{2.4}$$

where ν_{F} is the Fermi velocity, L_{eff} is the effective electron mean free path, and A is a dimensionless parameter determined by the details of scattering of electrons by the particle surface (which is often simply set equal to 1). Although Eq. (2.4) was originally derived by simple physical, or even geometrical, considerations, later works using ab initio quantum analysis lead to the same result with only difference in value of constant A (reviewed in Ref. [36]).

The only remaining question is dependence of L_{eff} on particle size. For a sphere with radius a, $L_{eff} = a$ and $L_{eff} = 4a/3$ for isotropic and diffuse scattering respectively. For particles of other shapes the following empirical formula can be used $L_{eff} = 4V/S$ [36], where V and S are volume and surface area of the particle, respectively.

Refractive indices of metals, obtained from value of $\tilde{\varepsilon}$ discussed above, vary from moderate values (close to 1 in complex plane) to values with large $\mathrm{Im}[\tilde{n}] = \kappa$ and $|\tilde{n}|$. Although real applications may employ different domains of $\tilde{n}$, in the remainder of this chapter "plasmonic refractive index" denotes the large values of $|\tilde{n}|$. The main motivation for that is that such values of $\tilde{n}$ are the most problematic for the DDA simulation method (see Sec. 2.4.2.1).

2.4 Theory of the DDA

This section is largely based on review [40]. However, it is here restructured and updated with special emphasis given to plasmonic applications.

2.4.1 *General Framework*

We assume $\exp(-i\omega t)$ time dependence of all fields and that the scatterer is non-magnetic. For simplicity the ω dependence of all quantities is omitted. Quantities indicated with a $\tilde{}$ are complex numbers. Also the electric permittivity $\tilde{\varepsilon}$ is assumed isotropic to simplify the derivations; however, extension to arbitrary dielectric tensors is straightforward. Rigorous derivation of the DDA starts with the *integral equation* governing the electric field inside the finite dielectric scatterer [41, 42]:

$$\tilde{\mathbf{E}}(\mathbf{r}) = \tilde{\mathbf{E}}^{inc}(\mathbf{r}) + \int_{V\setminus V_0} \mathrm{d}^3\mathbf{r}' \tilde{\mathbb{G}}^0(\mathbf{r},\mathbf{r}')\,\tilde{\chi}(\mathbf{r}')\tilde{\mathbf{E}}(\mathbf{r}') + \tilde{\mathbf{M}}(V_0,\mathbf{r}) - \mathbb{L}(\partial V_0,\mathbf{r})\tilde{\chi}(\mathbf{r})\tilde{\mathbf{E}}(\mathbf{r}), \tag{2.5}$$

where $\tilde{\mathbf{E}}^{inc}(\mathbf{r})$ and $\tilde{\mathbf{E}}(\mathbf{r})$ are the incident and total electric field at location $\mathbf{r}$, and $\tilde{\chi}(\mathbf{r}) = (\tilde{\varepsilon}(\mathbf{r}) - 1)/4\pi$ is the susceptibility of the medium at point $\mathbf{r}$. V is the volume of the particle, i.e. the volume that contains all points where the susceptibility is not zero. V_0 is a small volume around $\mathbf{r}$, $\tilde{\mathbb{G}}^0(\mathbf{r},\mathbf{r}')$ is the *free-space Green's tensor*:

$$\tilde{\mathbb{G}}^0(\mathbf{r},\mathbf{r}') = \frac{\exp(ikR)}{R}\left[k^2\left(\mathbb{I} - \frac{\mathbf{RR}}{R^2}\right) - \frac{1 - ikR}{R^2}\left(\mathbb{I} - 3\frac{\mathbf{RR}}{R^2}\right)\right], \tag{2.6}$$

where $\mathbb{I}$ is the identity tensor, $k = \omega/c$ is the free-space wavevector, $\mathbf{R} = \mathbf{r} - \mathbf{r}'$, $R = ||\mathbf{R}||$, and $\mathbf{RR}$ is a tensor defined as $(\mathbf{RR})_{\mu\nu} = R_\mu R_\nu$ (μ and ν are Cartesian components). Note that Eq. (2.6) is obtained directly from Eq. (1.213).

$\tilde{\mathbf{M}}$ is the following integral associated with the finiteness of the exclusion volume V_0:

$$\tilde{\mathbf{M}}(V_0, \mathbf{r}) = \int_{V_0} \mathrm{d}^3\mathbf{r}' \left[\tilde{\mathbb{G}}^0(\mathbf{r}, \mathbf{r}')\,\tilde{\chi}(\mathbf{r}')\tilde{\mathbf{E}}(\mathbf{r}') - \mathbb{G}^{st}(\mathbf{r}, \mathbf{r}')\,\tilde{\chi}(\mathbf{r})\tilde{\mathbf{E}}(\mathbf{r})\right], \tag{2.7}$$

where $\mathbb{G}^{st}(\mathbf{r}, \mathbf{r}')$ is the static limit ($k \to 0$) of $\tilde{\mathbb{G}}^0(\mathbf{r}, \mathbf{r}')$:

$$\mathbb{G}^{st}(\mathbf{r}, \mathbf{r}') = -\frac{1}{R^3}\left(\mathbb{I} - 3\frac{\mathbf{RR}}{R^2}\right). \tag{2.8}$$

$\mathbb{L}$ is the so-called *self-term tensor*:

$$\mathbb{L}(\partial V_0, \mathbf{r}) = -\int_{\partial V_0} \mathrm{d}^2\mathbf{r}' \frac{\hat{\mathbf{n}}'\mathbf{R}}{R^3}, \tag{2.9}$$

where $\hat{\mathbf{n}}'$ is an external normal to the surface ∂V_0 at point $\mathbf{r}'$. $\mathbb{L}$ is always a real, symmetric tensor with trace equal to 4π [43], which does not depend on the size of the volume V_0. On the contrary, $\tilde{\mathbf{M}}$ does depend on the size of the volume; moreover it approaches zero when the size of the volume decreases [42] (if both $\tilde{\chi}(\mathbf{r})$ and $\tilde{\mathbf{E}}(\mathbf{r})$ are continuous inside V_0).

A large variety of methods to solve Eq. (2.5) is thoroughly discussed in Ref. [40]. This chapter is devoted to the mainstream DDA, compatible with the fast Fourier transform (FFT) acceleration (Sec. 2.4.2). It is obtained by discretization of Eq. (2.5) on a regular cubical grid [27] by dividing the scatterer into N cubical subvolumes (dipoles) V_i ($i = 1, \ldots, N$). Size of each dipole is d, and its volume $V_d = d^3$. Setting $V_0 = V_i$ and $\mathbf{r} = \mathbf{r}_i$, the center of cube V_i, Eq. (2.5) can be rewritten as:

$$\tilde{\mathbf{E}}_i = \tilde{\mathbf{E}}_i^{inc} + \sum_{j \neq i}^{N} \int_{V_j} \mathrm{d}^3\mathbf{r}'\tilde{\mathbb{G}}^0(\mathbf{r}_i, \mathbf{r}')\tilde{\chi}(\mathbf{r}')\tilde{\mathbf{E}}(\mathbf{r}') + \tilde{\mathbf{M}}(V_i, \mathbf{r}_i) - \mathbb{L}_i\tilde{\chi}_i\tilde{\mathbf{E}}_i, \tag{2.10}$$

where $\tilde{\mathbf{E}}_i = \tilde{\mathbf{E}}(\mathbf{r}_i)$, $\tilde{\mathbf{E}}_i^{inc} = \tilde{\mathbf{E}}^{inc}(\mathbf{r}_i)$, $\tilde{\chi}_i = \tilde{\chi}(\mathbf{r}_i)$, $\mathbb{L}_i = \mathbb{L}(\partial V_i, \mathbf{r}_i)$. Moreover, for this specific (cubic) geometry of V_i and $\mathbf{r}_i$, $\mathbb{L}_i$ is calculated analytically yielding [25]:

$$\mathbb{L}_i = \frac{4\pi}{3}\mathbb{I}. \tag{2.11}$$

The set of Eq. (2.10) (for all i) is exact except for possible errors in replacing V by a set of cubical cells V_i (see Sec. 2.4.3.4). The principal approximating assumptions are:

$$\int_{V_j} d^3\mathbf{r}'\tilde{\mathbb{G}}^0(\mathbf{r}_i,\mathbf{r}')\tilde{\chi}(\mathbf{r}')\tilde{\mathbf{E}}(\mathbf{r}') = V_d\,\tilde{\mathbb{G}}^0_{ij}\,\tilde{\chi}(\mathbf{r}_j)\,\tilde{\mathbf{E}}(\mathbf{r}_j)\,, \qquad (2.12)$$

$$\tilde{\mathbf{M}}(V_i,\mathbf{r}_i) = \tilde{\mathbb{M}}_i\,\tilde{\chi}(\mathbf{r}_i)\,\tilde{\mathbf{E}}(\mathbf{r}_i)\,. \qquad (2.13)$$

They state that corresponding integrals linearly depend upon the values of $\tilde{\chi}$ and $\tilde{\mathbf{E}}$ at point $\mathbf{r}_i$ and allow one to rewrite Eq. (2.10) as:

$$\tilde{\mathbf{E}}_i = \mathbf{E}_i^{inc} + \sum_{j\neq i}^{N} \tilde{\mathbb{G}}^0_{ij} V_j \tilde{\chi}_j \tilde{\mathbf{E}}_j + \left(\tilde{\mathbb{M}}_i - \mathbb{L}_i\right)\tilde{\chi}_i \tilde{\mathbf{E}}_i\,. \qquad (2.14)$$

The particular expressions for $\tilde{\mathbb{M}}_i$ and $\tilde{\mathbb{G}}^0_{ij}$ determine the quality of the approximation and are discussed in Sec. 2.4.3.2 and Sec. 2.4.3.1. The most widely used method to justify assumptions in Eq. (2.12) and Eq. (2.13) is assuming $\tilde{\mathbf{E}}(\mathbf{r})$ and $\tilde{\chi}(\mathbf{r})$ constant inside each subvolume:

$$\tilde{\mathbf{E}}(\mathbf{r}) = \tilde{\mathbf{E}}_i\,,\ \ \tilde{\chi}(\mathbf{r}) = \tilde{\chi}_i \ \ \text{for}\ \ \mathbf{r}\in V_i\,, \qquad (2.15)$$

implying:

$$\tilde{\mathbb{M}}_i = \int_{V_i} d^3\mathbf{r}'\left(\tilde{\mathbb{G}}^0(\mathbf{r}_i,\mathbf{r}') - \mathbb{G}^{st}(\mathbf{r}_i,\mathbf{r}')\right)\,, \qquad (2.16)$$

$$\tilde{\mathbb{G}}^0_{ij} = \frac{1}{V_d}\int_{V_j} d^3\mathbf{r}'\tilde{\mathbb{G}}^0(\mathbf{r}_i,\mathbf{r}')\,. \qquad (2.17)$$

Such formulation is equivalent to the method of moments [41] applied to Eq. (2.5), using unit pulse and delta-function (point-matching) as basis and testing functions respectively.

Two equivalent forms of Eq. (2.14) are [41, 44]:

$$\tilde{\mathbf{E}}_i^{inc} = \tilde{\mathbf{E}}_i^{exc} - \sum_{j\neq i}\tilde{\mathbb{G}}^0_{ij}\,\tilde{\alpha}_j\,\tilde{\mathbf{E}}_j^{exc}\,, \qquad (2.18)$$

$$\tilde{\mathbf{E}}_i^{inc} = \tilde{\alpha}_i^{-1}\tilde{\mathbf{P}}_i - \sum_{j\neq i}\tilde{\mathbb{G}}^0_{ij}\tilde{\mathbf{P}}_j\,. \qquad (2.19)$$

They seek for the *exciting electric fields* (excluding the field by the dipole itself)

$$\tilde{\mathbf{E}}_i^{exc} = \left(\mathbb{I} + \left(\mathbb{L}_i - \tilde{\mathbb{M}}_i\right)\tilde{\chi}_i\right)\tilde{\mathbf{E}}_i\,, \qquad (2.20)$$

and the *dipole polarizations*

$$\tilde{\mathbf{P}}_i = \tilde{\alpha}_i \tilde{\mathbf{E}}_i^{exc} = V_d \tilde{\chi}_i \tilde{\mathbf{E}}_i \,, \tag{2.21}$$

respectively. DDA implementations are mostly related to the dipole polarization of Eq. (2.19) and Eq. (2.21).

The dipole *polarizability tensor* $\tilde{\alpha}_i$ is defined as:

$$\tilde{\alpha}_i = V_d \tilde{\chi}_i \left(\mathbb{I} + \left(\mathbb{L}_i - \tilde{\mathbb{M}}_i\right) \tilde{\chi}\right)^{-1} = \tilde{\alpha}_i^{\mathrm{CM}} \left(\mathbb{I} - \tilde{\mathbb{M}}_i \tilde{\alpha}_i^{\mathrm{CM}} / V_d\right)^{-1} \,, \tag{2.22}$$

where we used Eq. (2.11) and $\tilde{\alpha}_i^{CM}$ is the *Clausius–Mossotti (CM) polarizability* obtained assuming $\tilde{\mathbb{M}}_i = \mathbb{O}$:

$$\tilde{\alpha}_i^{\mathrm{CM}} = \mathbb{I} V_d \frac{3}{4\pi} \frac{\tilde{\varepsilon}_i - 1}{\tilde{\varepsilon}_i + 2} \,, \tag{2.23}$$

with $\tilde{\varepsilon}_i = \tilde{\varepsilon}(\mathbf{r}_i)$. Note that Eq. (2.21) follows Eq. (1.270) and, for a sphere, Eq. (2.23) is the effective polarizability of Eq. (1.192).

Solution of system of linear equations in Eq. (2.14), or one of its equivalent forms, constitutes the most challenging computational part of the DDA, discussed in Sec. 2.4.2. All measurable quantities of interest can be computed from the obtained internal (total) fields or dipole polarizations.

Electric fields near or far from scatterers (called *near-fields* and *scattered-fields*, respectively) are obtained directly from Eq. (2.5) for a point $\mathbf{r}$ outside the scatterer, where $\tilde{\chi}(\mathbf{r}) = 0$. After discretization of the scatterer volume it becomes:

$$\tilde{\mathbf{E}}(\mathbf{r}) = \tilde{\mathbf{E}}^{inc}(\mathbf{r}) + \sum_i^N \int_{V_i} \mathrm{d}^3\mathbf{r}' \tilde{\mathbb{G}}^0\left(\mathbf{r}, \mathbf{r}'\right) \tilde{\chi}(\mathbf{r}') \tilde{\mathbf{E}}(\mathbf{r}') \,, \tag{2.24}$$

similar to Eq. (2.10). To make Eq. (2.24) ready for practical use, approximations similar to Eq. (2.12) should be used

$$\int_{V_i} \mathrm{d}^3\mathbf{r}' \tilde{\mathbb{G}}^0\left(\mathbf{r}, \mathbf{r}'\right) \tilde{\chi}(\mathbf{r}') \tilde{\mathbf{E}}(\mathbf{r}') = \tilde{\mathbb{G}}_i^0(\mathbf{r}) \tilde{\mathbf{P}}_i \,. \tag{2.25}$$

The simplest possible expression for $\tilde{\mathbb{G}}_i^0(\mathbf{r})$ is:

$$\tilde{\mathbb{G}}_i^0(\mathbf{r}) = \tilde{\mathbb{G}}^0(\mathbf{r}, \mathbf{r}_i) \,. \tag{2.26}$$

For $\mathbf{r} \to \infty$ it is more convenient to use the *scattering amplitude* $\tilde{\mathbf{F}}$ instead of the total electric field $\tilde{\mathbf{E}}$ (see, e.g. Ref. [45]):

$$\tilde{\mathbf{E}}(\mathbf{r}) - \tilde{\mathbf{E}}^{inc}(\mathbf{r}) = \tilde{\mathbf{E}}^{sca}(r\hat{\mathbf{r}}) = \frac{\exp(ikr)}{-ikr} \tilde{\mathbf{F}}(\hat{\mathbf{r}}) \,, \tag{2.27}$$

where $\widehat{\mathbf{r}} = \mathbf{r}/r$ is the unit vector in the scattering direction. In particular, Eq. (2.25) and Eq. (1.220) imply:

$$\tilde{\mathbf{F}}(\widehat{\mathbf{r}}) = -\mathrm{i}k^3 \left(\mathbb{I} - \widehat{\mathbf{r}}\widehat{\mathbf{r}}\right) \sum_i^N \tilde{\mathbf{P}}_i \exp\left(-\mathrm{i}k\mathbf{r}_i \cdot \widehat{\mathbf{r}}\right) . \tag{2.28}$$

The DDA is perfectly applicable to any incident field, provided only that its values can be calculated for all $\mathbf{r}_i$. However, for simplicity we further assume a plane incident wave with unit amplitude:

$$\tilde{\mathbf{E}}^{inc}(\mathbf{r}) = \widehat{\mathbf{e}}^{inc} \exp\left(\mathrm{i}k\widehat{\mathbf{k}}^{inc} \cdot \mathbf{r}\right) , \tag{2.29}$$

where $\widehat{\mathbf{k}}^{inc}$ is the incident direction. The amplitude and *Mueller scattering matrices* for direction $\widehat{\mathbf{r}}$ are determined from $\tilde{\mathbf{F}}(\widehat{\mathbf{r}})$ calculated for two orthogonal incident polarizations [46]. Note that $\tilde{\mathbf{F}}(\widehat{\mathbf{r}}) = -\mathrm{i}k\tilde{\mathbb{F}}\left(\widehat{\mathbf{r}}, \widehat{\mathbf{k}}^{inc}\right) \cdot \widehat{\mathbf{e}}^{inc}$, following the definitions of Sec. 1.6.2.

Integral scattering quantities, such as *scattering cross-section* C_{sca} and *extinction cross-section* C_{ext} (see Sec. 1.6.1) can also be obtained from $\tilde{\mathbf{F}}(\widehat{\mathbf{r}})$ [46, 47]:

$$C_{sca} = \frac{1}{k^2} \int_A \mathrm{d}A \left|\left|\mathbf{F}(\widehat{\mathbf{n}})\right|\right|^2 , \tag{2.30}$$

$$\begin{aligned} C_{ext} &= 4\pi k \sum_i^N \int_{V_i} \mathrm{d}^3\mathbf{r}' \mathrm{Im}\left[\tilde{\chi}(\mathbf{r}')\tilde{\mathbf{E}}(\mathbf{r}') \cdot \tilde{\mathbf{E}}^{inc*}(\mathbf{r}')\right] \\ &= \frac{4\pi}{k^2} \mathrm{Re}\left[\left(\widehat{\mathbf{e}}^{inc}\right)^* \cdot \tilde{\mathbf{F}}(\widehat{\mathbf{k}}^{inc})\right] . \end{aligned} \tag{2.31}$$

Absorption cross section C_{abs} is derived directly from the internal (total) fields [47]:

$$C_{abs} = 4\pi k \sum_i^N \int_{V_i} \mathrm{d}^3\mathbf{r}' \mathrm{Im}\left[\tilde{\chi}(\mathbf{r}')\right] \left|\left|\tilde{\mathbf{E}}(\mathbf{r}')\right|\right|^2 , \tag{2.32}$$

which is most commonly approximated using the notion of point dipoles [48],

$$C_{abs} = 4\pi k \sum_i^N \left[\mathrm{Im}\left[\tilde{\mathbf{P}}_i \cdot \tilde{\mathbf{E}}_i^{exc*}\right] - (2/3)\, k^3 \left|\left|\tilde{\mathbf{P}}_i\right|\right|^2\right] , \tag{2.33}$$

to enable practical evaluation. Other alternatives are discussed in Sec. 2.4.3.3. Radiation forces and torques can also be calculated [49–51].

2.4.2 *Numerical Scheme*

2.4.2.1 Iterative methods to solve the linear system

There are two general types of methods to solve linear systems of equations $\tilde{\mathbb{A}}\tilde{\mathbf{x}} = \tilde{\mathbf{y}}$, where $\tilde{\mathbf{x}}$ is an unknown vector and $\tilde{\mathbb{A}}$ and $\tilde{\mathbf{y}}$ are known complex matrix and complex vector, respectively: *direct* and *iterative* [52]. For a general $n \times n$ matrix (in the DDA $n = 3N$) computation time of direct inversion (e.g. *LU* decomposition) is $O(n^3)$ and storage requirements $O(n^2)$, while computation time for one iteration is $O(n^2)$ [53]. Generally, iterative methods converge in $O(n)$ iterations, and sometimes they do not converge at all. However, in many cases satisfactory accuracy can be obtained after a much smaller number of iterations. In these cases iterative methods can provide significant increases in speed, especially for large n. Moreover, most iterative methods access the matrix $\tilde{\mathbb{A}}$ only through matrix-vector multiplication (sometimes also with the transposed matrix), which allows the construction of special faster routines for calculation of these products (see Sec. 2.4.2.2). Throughout DDA history, mostly iterative methods were employed. At first, they were used to accelerate computations [54], but they also allowed larger numbers of dipoles to be simulated [55], since storage of the entire matrix is prohibitive for direct methods. The most widely used iterative methods in the DDA are *Krylov-space methods*, also known as *conjugate-gradient (CG)* family of methods [53]. A number of studies were devoted to comparison of different iterative methods in DDA simulations [56–60], but it is still hard to identify the most efficient one. In practical applications *bi-CG-stabilized (Bi-CGSTAB)* and *quasi-minimal-residual (QMR)* iterative solvers [53] are most commonly used, as implemented in production codes (see Sec. 2.5.4). However, less tested alternatives of these methods were also proposed [61–63], designed for better convergence in the finite-precision arithmetic. An important advantage of the QMR is that complex-symmetric property of the DDA interaction matrix can be used to halve the number of matrix-vector multiplications [64] (and thus computational time). Assume that the used formulation for $\tilde{\mathbb{G}}^0_{ij}$ retains the symmetry properties of $\tilde{\mathbb{G}}^0(\mathbf{r}_i, \mathbf{r}_j)$, i.e.

$$\tilde{\mathbb{G}}^0_{ij} = \tilde{\mathbb{G}}^0_{ji}\,. \tag{2.34}$$

In particular, this is true for all formulations described in Sec. 2.4.3.1. Then for any of the equivalent systems of DDA equations (Eqs. 2.14, 2.18, 2.19) $\tilde{\mathbb{A}}$ is complex-symmetric if and only if $\tilde{\alpha}_i$ is complex-symmetric for all i. The latter is true for all cases except chiral and non-diagonal anisotropic absorbing materials. Moreover, when $\tilde{\alpha}_i$ is complex-symmetric, it can be decomposed as $\tilde{\alpha}_i = \tilde{\beta}_i^T \tilde{\beta}_i$, leading to the following form of DDA equations [65]:

$$\tilde{\mathbb{A}}\tilde{\mathbf{x}} = \tilde{\mathbf{y}} \quad ; \quad \tilde{\mathbb{A}}_{ij} = \mathbb{I}\delta_{ij} - \tilde{\beta}_i \tilde{\mathbb{G}}^0_{ij} \tilde{\beta}_j^T \tag{2.35}$$

$$\tilde{\mathbf{x}}_i = \tilde{\beta}_i \tilde{\mathbf{E}}_i^{exc} = \left(\tilde{\beta}_i^T\right)^{-1} \tilde{\mathbf{P}}_i \quad ; \quad \tilde{\mathbf{y}}_i = \tilde{\beta}_i \tilde{\mathbf{E}}_i^{inc},$$

which is an intermediate one between Eq. (2.18) and Eq. (2.19).

An important part of the iterative solver is preconditioning, which effectively decreases the condition number of the matrix $\tilde{\mathbb{A}}$ and, therefore, speeds up convergence. Although there are a large variety of preconditioners in the literature [53], most of them apply to general dense or sparse matrices. However, any effective preconditioner for the DDA should not significantly modify the block-Toeplitz structure of $\tilde{\mathbb{A}}$ (see Sec. 2.4.2.2). That is why only the simplest Jacobi preconditioner (transforming $\tilde{\mathbb{A}}$ into a matrix with unit diagonal) has been used. For instance, Eq. (2.35) corresponds to Jacobi-preconditioned $\tilde{\mathbb{A}}$ [65]. Number of iterations required for a particular scattering problem is hard to predict a priori; the best estimate is usually provided by empirical data. Still, there are general theoretical results. Rahola [66] showed that the spectrum of the integral scattering operator for any homogeneous scatterer is a line in the complex plane going from 1 to $|\tilde{n}|^2$, except for a small amount of points, corresponding to resonances for the specific size and shape, e.g. so-called Mie resonances for spheres. Based on this he derived an estimate for the optimal (best) reduction factor for any Krylov-space iterative method (norm of the residual is multiplied by this factor every iteration),

$$\gamma = \left|\frac{\tilde{n} - 1}{\tilde{n} + 1}\right|, \tag{2.36}$$

which is especially accurate for particles much smaller than λ, when no above-described resonances are present. Since $\tilde{\mathbb{A}}$ is a discretization of the integral operator, their spectra are similar (see also [56]). Hence, Eq. (2.36) applies to $\tilde{\mathbb{A}}$ as well, although not exactly.

An important consequence of this analysis is that the spectrum of $\tilde{\mathbb{A}}$, and thus convergence should not depend significantly on the discretization. This fact was also confirmed empirically [56, 67, 68]. Budko *et al.* [69] derived optimal value of γ when using the general overrelaxation iterative method. This estimate of γ is similar but always greater than Eq. (2.36), since this stationary iterative method also constructs the solution in the Krylov subspace but not in the most optimal way.

To estimate the required number of iterations N_{iter} one can combine Eq. (2.36) with commonly used value of 10^{-5} for convergence threshold of the iterative solver:

$$N_{iter} \approx \frac{\ln\left(10^{-5}\right)}{\ln\left(\gamma\right)} = \frac{5\ln(10)}{\ln\left|(\tilde{n}+1)\,/\,(\tilde{n}-1)\right|}. \tag{2.37}$$

Applicability of Eq. (2.37) is illustrated by comparison with simulation data [67] in Fig. 2.1. Shown data is for nanospheres discretized with 128 dipoles per diameter; however, a particular value of this parameter is not important, as discussed above. More details about these data are given in Sec. 2.6. The estimate describes well the general behavior of N_{iter} versus λ—it is accurate within the factor of two and lies in between the two tested DDA formulations.

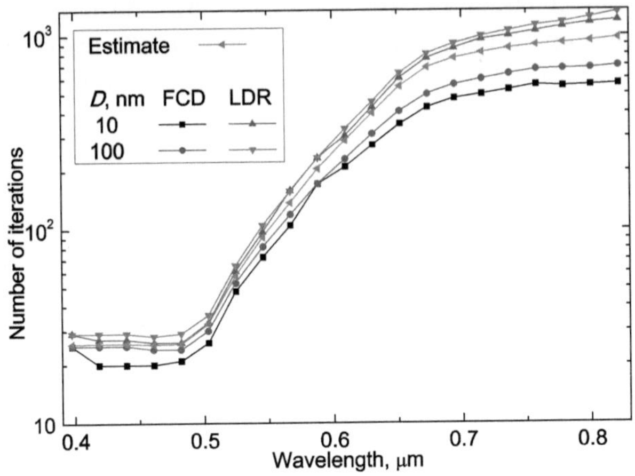

Figure 2.1 Number of iterations for DDA simulations of gold nanospheres (in vacuum) in logarithmic scale. Real data is for two diameters D and two DDA formulations (see Sec. 2.4.3); estimate is based on Eq. (2.37).

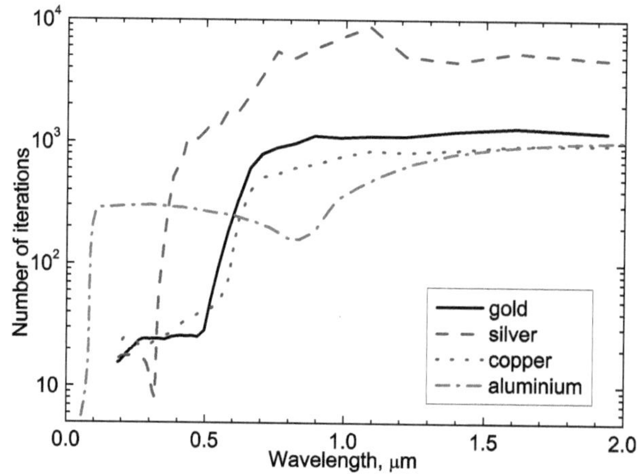

Figure 2.2 Estimated number of iterations, based on Eq. (2.37), for DDA simulations of metallic nanoparticles (in vacuum) in logarithmic scale.

Equation (2.37) is further used to estimate N_{iter} in a wider wavelength range for metals commonly used in plasmonics. The result is shown in Fig. 2.2, using reference refractive index of gold, silver, and copper from [37] and of aluminium from [70]. The general features are small N_{iter} for small λ (< 50–500 nm depending on the metal), much larger and almost constant N_{iter} for large λ (>1−1.5 μm), and rapid change somewhere in between. To understand such behavior let us perform an asymptotic expansion of Eq. (2.37) for large $|\tilde{n}|$:

$$N_{iter} \approx \frac{5\ln(10)}{\eta + O(\eta^2)} \approx \frac{5\ln(10)}{2}\frac{1+|\tilde{n}|^2}{\mathrm{Re}[\tilde{n}]}, \quad \eta = \frac{2\mathrm{Re}[\tilde{n}]}{1+|\tilde{n}|^2}, \tag{2.38}$$

and apply it to the Drude model (see Eq. (2.1)). In particular, for most metals in the near-infrared the Drude model is accurate and the following limit is satisfied:

$$1/\tau \ll \omega \ll \omega_p \Rightarrow N_{iter}(\omega) \approx 5\omega_p\tau\ln(10)\,. \tag{2.39}$$

This allows one to quickly estimate N_{iter} for any other metal in the near-infrared using its Drude parameters. Unfortunately, the Drude model is usually inaccurate for $\omega \approx \omega_p$ and, hence, does not provide an accurate estimate of the wavelength position of the transitions in Fig. 2.2. If a nanoparticle is located in dielectric medium with refractive index $\tilde{n}_0$, the level of the plateau in Fig. 2.2 is $\tilde{n}_0$ times smaller, which follows from Eq. (2.38). Budko and Samokhin [71] generalized Rahola's results to arbitrary inhomogeneous and anisotropic scatterers. Moreover, they described a convex region in the complex plane containing the whole spectrum of the integral scattering operator, including resonances. This region depends only on the values of $\tilde{n}$ inside the scatterer but not on its size and shape. For instance, the size can be comparable or larger than λ. However, size and shape do influence the particular distribution of eigenvalues inside this region; hence, the geometry of the region can be used only to deduce the upper bound of N_{iter}. In particular, for $\tilde{n}$ with zero or small imaginary part this region comes close to the origin, which potentially leads to large N_{iter}. The latter was shown in DDA simulations for real $\tilde{n}$ and large size [59]. Moreover, Ayranci *et al.* [72] observed the decrease of N_{iter} with increasing $\mathrm{Im}[\tilde{n}]$ for sizes comparable with λ. However, for particles much smaller than the wavelength Eq. (2.36) is valid, which has no special behavior for real $\tilde{n}$.

2.4.2.2 Block-Toeplitz structure and FFT acceleration

A square matrix $\mathbb{Q}$ of size $N \times N$ is called *Toeplitz matrix* of order N if $\mathbb{Q}_{ij} = B_{i-j}$, i.e. matrix elements on any line parallel to the main diagonal are the same [52].

In a block-Toeplitz (BT) matrix (of order N_1) elements B_i are not numbers, but square matrices of size $f \times f$ themselves:

$$\mathbb{Q} = \begin{bmatrix} \mathbb{B}_0 & \mathbb{B}_1 & \dots & \mathbb{B}_{N_1-1} \\ \mathbb{B}_{-1} & \mathbb{B}_0 & \ddots & \vdots \\ \vdots & \ddots & \ddots & \mathbb{B}_1 \\ \mathbb{B}_{-N_1+1} & \dots & \mathbb{B}_{-1} & \mathbb{B}_0 \end{bmatrix} \quad (2.40)$$

In a *2-level BT matrix* (of order N_1, N_2) the components $\mathbb{B}_i$ are BT matrices themselves (of order N_2 and size $f \times f$). Proceeding recursively a *M-level BT* (MBT) matrix of order $(N_1, N_2, \dots, N_3)$ and size $f \times f$ for any number of levels can be defined.

Consider a rectangular lattice $n_x \times n_y \times n_z$, numbered in the following way:

$$i = n_y n_z (n_x - 1) i_x + n_z (n_y - 1) i_y + n_z i_z, \quad (2.41)$$

where $i_\mu \in \{1, \ldots, n_\mu\}$ indicates the position of the element along the axes. Let us also define the vector index $\mathbf{i} = (i_x, i_y, i_z)$. Then $\tilde{\mathbb{G}}^0_{ij}$, defined by Eq. (2.17), satisfies the following:

$$\tilde{\mathbb{G}}^0_{ij} = \tilde{\mathbb{G}}'_{\mathbf{i}-\mathbf{j}} \tag{2.42}$$

Analogously to Eq. (2.34) it is further assumed that Eq. (2.42) is satisfied for all used formulations for $\tilde{\mathbb{G}}^0_{ij}$. First, Eq. (2.42) can be used to greatly reduce the storage requirements of iterative methods by use of indirect addressing from $O(N^2)$ to $O(N)$. Second, together with Eq. (2.34) it defines a symmetric 3-level BT matrix $\tilde{\mathbb{G}}^0$ (orders of subsequent levels are n_x, n_y, n_z) whose smallest blocks are 3×3 matrices (tensors) $\tilde{\mathbb{G}}^0_{ij}$. The BT structure can be used for acceleration of direct methods to solve the linear system, which was applied to the DDA by Flatau *et al.* [73, 74]. However, these methods are not discussed here in details due to two major limitations. First, their computational complexity is still larger than that of iterative methods (see below). Second, the whole interaction matrix $\tilde{\mathbb{A}}$ (not only $\tilde{\mathbb{G}}^0$) has to be BT, which is true if and only if polarizabilities of all dipoles in the rectangular grid are the same or, equivalently, the scatterer is homogeneous and rectangular.

The major practical application of the BT structure is acceleration of matrix-vector product, which is the computational bottleneck of any Krylov-subspace iterative method. Goodman *et al.* [75] showed that multiplication of 3 level BT matrix $\mathbb{G}$ by a vector can be transformed into a discrete convolution:

$$\tilde{\mathbf{y}}_{\mathbf{i}} = \sum_{j=1}^{N} \tilde{\mathbb{G}}^0_{ij}\tilde{\mathbf{x}}_j = \sum_{\mathbf{j}=(1,1,1)}^{(n_x,n_y,n_z)} \tilde{\mathbb{G}}'_{\mathbf{i}-\mathbf{j}}\tilde{\mathbf{x}}_{\mathbf{j}} = \sum_{\mathbf{j}=(1,1,1)}^{(2n_x,2n_y,2n_z)} \tilde{\mathbb{G}}'_{\mathbf{i}-\mathbf{j}}\tilde{\mathbf{x}}'_{\mathbf{j}}, \tag{2.43}$$

where $\tilde{\mathbb{G}}'_{\mathbf{i}-\mathbf{j}}$ is defined by Eq. (2.42) for $|i_\mu| \le |n_\mu|$ (and $\mathbb{G}'_{\mathbf{0}} = \mathbb{0}$) and

$$\tilde{\mathbf{x}}'_{\mathbf{j}} = \begin{cases} \tilde{\mathbf{x}}_{\mathbf{j}}, & \forall\mu\ 1 \le j_\mu \le n_\mu \\ 0, & \text{otherwise.} \end{cases} \tag{2.44}$$

Both $\mathbb{G}'$ and $\tilde{\mathbf{x}}'$ are then regarded as periodic in each dimension μ with period $2n_\mu$. A discrete convolution can be transformed by FFT to an element-wise product of two vectors, which is easily computed in $O(N)$ operations. It requires evaluation of a direct and inverse FFT for each matrix-vector product. Each of them is a 3D FFT of order $2n_x \times 2n_y \times 2n_z$ with $O(N\ln(N))$ complexity. This operation is done for each of the 3 Cartesian components of $\tilde{\mathbf{x}}'$, while FFT of 6 independent tensor components of $\mathbb{G}'$ is done only once during the initialization of the iterative solver.

Alternative methods of FFT acceleration were proposed [73, 76], but the resulting complexity is only slightly different from the one discussed above. It is important to note that for any form of DDA equations [Eqs. 2.14, 2.18, 2.19, 2.35] product of interaction matrix by vector is reduced to product of $\tilde{\mathbb{G}}^0$ by vector through a few $O(N)$ operations, such as element-wise product of two vectors. Therefore, in all these cases the complexity of one iteration of an iterative solver is $O(N\ln(N))$.

As noted above, both the BT structure and the FFT-acceleration require a cubical lattice of dipoles that completely fills a rectangular box. This has two drawbacks: (1) additional void dipoles should be introduced to build up the grid up to the full box enclosing the particle; (2) position and size of the dipoles cannot be chosen arbitrarily to better describe the shape of the scatterer. These drawbacks are minor for solid and relatively smooth particles, which is usually the case in plasmonics. However, they may become critical for highly porous particles or clusters of particles, where the monomer has a size comparable to a single dipole. Another method to accelerate the DDA computations without the above limitations is the fast multipole method (FMM). It was originally proposed by Greengard and Rokhlin [77] for efficient evaluation of the potential and force fields in N-body simulations. The FMM is based on truncated potential expansions [78] and is principally an approximate method, contrary to FFT-based acceleration. A careful tuning of approximation parameters is important to reach both satisfactory accuracy and low computational complexity. The FMM can be single- or multilevel; the latter is based on hierarchical grouping of particles [79]. The FMM was widely applied in connection with surface-discretization methods [80, 81]. For them the single- and multi-level FMM has complexity of $O(N^{1.5})$ [80] and $O(N\ln^2(N))$ [82], respectively.

The FMM naturally fits the DDA, since the matrix-vector multiplication is actually computing the total field on each single dipole due to all other dipoles [79]. And it was actually implemented in the framework of the DDA [56, 78, 83]. However, rigorous error analysis (required for the rigorous complexity estimate) is still lacking for the FMM–DDA. Moreover, the multi-level FMM has poor parallel performance, making single-level one more applicable to supercomputers with a large number of cores [80]. Hence, for such hardware performance of the FMM is generally much worse than that of FFT-based methods. Since the FMM is not implemented in any production DDA code, it is not discussed further.

2.4.2.3 Orientation averaging and repeated calculations

In many plasmonic applications one is interested in optical properties of an ensemble of randomly oriented particles. When the concentration of particles is small, multiple scattering is negligible and the optical properties are obtained by averaging single-particle scattering over different particle orientations. Orientation averaging can be described as the integral over the Euler's orientation angles (including a probability distribution function if necessary), which is brought down to a sum by appropriate quadrature. The problem therefore consists in calculation of some scattering property for a set of different orientations of the same particle. The easiest way is to calculate it by solving sequentially and independently each problem from the set. However, the large size of this set calls for some means of reducing the calculations.

Singham *et al.* [84] noted that the set of problems described above is physically equivalent to a fixed orientation of the particle and different incident and scattering directions. Fixed scatterer geometry has two immediate advantages. First, $\tilde{\mathbb{A}}$ is kept constant (although minor variation is possible, see Sec. 2.4.3.2) and need to be computed only once. Second, any scattering quantity (for any angle) is quickly obtained after the linear system is solved for two incident polarizations. Hence, integration over one Euler angle is relatively fast. This finalizes the current state of the art implemented in the production DDA codes. In the following several interesting ideas proposed in the literature are explained, which have not (yet) found a widespread use.

If inverse or *LU* decomposition of $\tilde{\mathbb{A}}$ is available, then scattering problems for each different orientation can be solved in $O(N^2)$

operations [84, 85]. Moreover, analytical averaging over orientations can be performed in $O(N^2)$ operations [84, 86, 87], similarly to the T-matrix method [4]. However, this is still slower than FFT-accelerated iterative methods, when $N_{iter} \ll N$ (which is usually the case).

Another option is to first compute the *T-matrix* of the particle. The T-matrix formalism is based on the multipole expansion truncated at some order n_0, which usually equals to several times size parameter x [88, 89]. The order (number of rows) of the T-matrix equals $2n_0(n_0+2)$. The straightforward way to evaluate the T-matrix based on the DDA is to solve for every incident spherical wave (i.e. for each row of the T-matrix) independently [90] with possible optimization for the repeated calculation discussed above. Using iterative techniques, computation time is $n_0^2 N\left[O(N_{iter}\ln(N)) + O(n_0^2)\right]$, where the first term in the sum is the time for solving the linear system, and the second one is for the computation of the T-matrix components themselves. A new method to obtain the T-matrix from the DDA interaction matrix was proposed by Mackowski [90], requiring two summations with computational time $O(n_0^2 N\ln(N))$ and $O(n_0^4 N)$. For particles larger than λ $O(n_0^4 N)$ usually dominates the computational time of both straightforward and Mackowski's methods, thus decreasing the difference between the two. However, for metallic nanoparticles n_0 can be rather small, while N_{iter} – up to few thousands (see Sec. 2.4.2.1). In this case Mackowski's method to calculate T-matrix will be much faster than the straightforward one, and may prove a faster alternative to calculate orientation-averaged properties. Another class of possible improvements is optimizing multiple runs of iterative solvers with different $\tilde{\mathbf{y}}$. First option is a "heavy" preconditioner, e.g. incomplete factorization preconditioner [53], which has large initialization cost (computed only once) but better convergence rate. Second option is block-iterative methods (see, e.g. [91–93]), which solve the linear system for many $\tilde{\mathbf{y}}$ simultaneously. The main limitation of the latter is significantly larger memory requirements. However, none of these two options have even been tested in the DDA. Okada *et al.* [94] proposed the third option – interpolating the internal field from already simulated orientations to obtain a "good guess" for the initial vector in the simulation for another orientation. Tests showed that N_{iter} can be decreased several times for a dense grid or orientation angles.

Additionally to performing the DDA simulations for a number of orientations the final accuracy is also affected by the used quadrature (except, when analytical averaging is performed). In the production DDA codes the Simpson rule [95] and the Romberg integration [65] are used, while Okada [96] recently advocated the use of quasi-Monte-Carlo techniques. It is not clear what the best method is, and the answer may depend on a particular problem.

Finally, there are many plasmonic applications requiring DDA simulations for multiple particle sizes (averaging over size distribution) or wavelengths (calculation of a spectrum). Matrix $\tilde{\mathbb{A}}$ depends both on size divided by λ, e.g. x, and $\tilde{n}$. In plasmonics $\tilde{n}$ usually strongly depends on λ. Therefore, simulations for different sizes (or λ) are largely independent. The only known general way to optimize calculations is reusing initial fields from previous simulations to compute a good guess for the iterative solver [97], similarly as was noted above for orientation averaging. However, other improvements are possible in a special case of particles much smaller than the wavelength. Then the interaction matrix for Eq. (2.14) is size-invariant and its part, matrix $\tilde{\mathbb{G}}^0$, is real symmetric. This makes size averaging much easier and simplifies spectrum calculation. First, spectral representation formalism (see [98] for an overview) allows one to reduce the calculation of the whole spectrum to diagonalization of $\tilde{\mathbb{G}}^0$. Second, multiple runs of the iterative solver varying only the refractive index can be significantly optimized using shift-invariance properties of the CG-type iterative methods [99]. Unfortunately, it is hard to specify a size limit for the underlying assumptions to be valid. Usually, size is required to be at least 10 times smaller than all used values [100] of λ. But even stricter requirements may apply for large $|\tilde{n}|$.

2.4.3 *Existing Formulations*

2.4.3.1 Interaction term

Three formulations for the interaction term are known. The simplest is the interaction of point dipoles:

$$\tilde{\mathbb{G}}^0_{ij} = \tilde{\mathbb{G}}^0(\mathbf{r}_i, \mathbf{r}_j). \tag{2.45}$$

The *integrated Green's tensor* (IGT, [101]) is based on a direct numerical integration of Eq. (2.17). This integration takes a lot of time; however, it can be reduced by performing exact calculation only for nearby dipoles, when the difference between Eq. (2.17) and Eq. (2.45) is the largest [65].

The *filtered coupled dipoles* (FCD, [102]) is based on the sampling theory applied to evaluation of the integrals in Eq. (2.5). The electric field and the susceptibility are sampled:

$$\tilde{\chi}(\mathbf{r}')\tilde{\mathbf{E}}(\mathbf{r}') \approx V_d \sum_i h_r(\mathbf{r}' - \mathbf{r}_i)\tilde{\chi}(\mathbf{r}_i)\tilde{\mathbf{E}}(\mathbf{r}_i), \tag{2.46}$$

where h_r is the impulse response function of an *antialiasing filter*:

$$h_r(R) = \frac{\sin(k_F R) - k_F R\cos(k_F R)}{2\pi^2 R^3}, \tag{2.47}$$

and $k_F = \pi/d$ is the wavenumber corresponding to the grid. Equation (2.5) is then transformed to Eq. (2.14) with the so-called *filtered free-space Green's tensor*, defined as:

$$\begin{aligned}\mathbb{G}^F_{ij} &= \int_{R^3/V_0} \mathrm{d}^3\mathbf{r}'\tilde{\mathbb{G}}^0(\mathbf{r}_i, \mathbf{r}')h_r(\mathbf{r}' - \mathbf{r}_j) \\ &= \mathbb{I}\left(k^2 g_F(R) + \frac{g_F'(R)}{R} + \frac{4\pi}{3}h_r(R)\right) \\ &\quad + \frac{\mathbf{RR}}{R^2}\left(g_F''(R) - \frac{g_F'(R)}{R}\right),\end{aligned} \tag{2.48}$$

where the symbol $'$ indicates the derivative; the integral is evaluated analytically [102], taking V_0 to be infinitesimally small, and g_F is the *filtered scalar Green's function*:

$$\begin{aligned}g_F(R) = \frac{1}{\pi R}\Big\{ &\sin(kR)\left[\pi\mathrm{i} + \mathrm{Ci}\left((k_F - k)R\right) - \mathrm{Ci}\left((k_F + k)R\right)\right] \\ &+ \cos(kR)\left[\mathrm{Si}\left((k_F + k)R\right) + \mathrm{Si}\left((k_F - k)R\right)\right]\Big\}.\end{aligned} \tag{2.49}$$

To apply this formulation k_F must be larger than k, i.e. $kd < \pi$. But use of the DDA with larger dipoles has little sense anyway. Additional computational time for using the FCD is comparable to a single iteration of the iterative solver, which is negligible in most cases.

The default $\tilde{\mathbb{G}}^0_{ij}$, given by Eq. (2.45), is used in almost all applications, while others are considered in only a few papers. However, the effect due to different $\tilde{\mathbb{G}}^0_{ij}$ can be much larger than that due to different $\tilde{\alpha}_i$, because the whole matrix $\tilde{\mathbb{A}}$ instead of only its diagonal is affected. For instance, the IGT is known to perform very good for small scatterers with large and almost real refractive indices [101], while the FCD significantly improves the accuracy of the DDA for large $|\tilde{n}|$ [67, 98, 102, 103] (see also Sec. 2.6). Moreover, both the IGT and the FCD may decrease N_{iter} several times in certain cases [98, 101, 103], cutting down the simulation time proportionally. For the FCD this was explained by closer resemblance between spectra of the interaction matrix and the integral scattering operator (see Sec. 2.4.2.1) than that when using CM [98]. It was also proved theoretically that convergence of the IGT with refining discretization for shapes exactly described by a set of cubes is quadratic in kd, contrary to linear convergence of all other DDA formulations except the FCD [104].

Fortunately, all three formulations for the interaction term are implemented in the production codes. So it is recommended to try different options for a particular application, especially when dealing with plasmonic refractive indices.

2.4.3.2 Polarizability prescription

The simplest formula for polarizability (CM) was defined above by Eq. (2.23) and used in the original DDA formulation [54]. All others are equivalent to using non-zero $\tilde{\mathbb{M}}_i$ in Eq. (2.22). The *radiative reaction* correction is based on the radiative damping (see Sec. 1.6.4) of a finite dipole [105]:

$$\tilde{\mathbb{M}}^{\mathrm{RR}} = \frac{2}{3}\mathrm{i}\,(kd)^3\,\mathbb{I}\,. \tag{2.50}$$

It was proposed [48] to satisfy the optical theorem, in particular, to make C_{abs} defined by Eq. (2.33) exactly equal zero for purely real $\tilde{n}$. RR is also included in many other formulae discussed below.

Two other formulations are based on approximate evaluation of Eq. (2.16) replacing the cubical dipole by an equi-volume sphere with radius $a_d = d\,(3/4\pi)^{1/3}$. Obtained result, advocated by

Lakhtakia and others [106, 107], is denoted as LAK,

$$\tilde{\mathbb{M}}^{\mathrm{LAK}} = \frac{8\pi}{3}\mathbb{I}\left[(1 - \mathrm{i}ka_d)\exp(\mathrm{i}ka_d) - 1\right], \tag{2.51}$$

while its series expansion up to the third order of *kd* was proposed under the name *digitized Green's function* (DGF) [45, 106]:

$$\tilde{\mathbb{M}}^{\mathrm{DGF}} = \mathbb{I}\left(b_1^{\mathrm{DGF}}(kd)^2 + \frac{2}{3}\mathrm{i}(kd)^3\right), \tag{2.52}$$

$$b_1^{\mathrm{DGF}} = (4\pi/3)^{1/3} \approx 1.6111992\,. \tag{2.53}$$

Peltoniemi [108] improved the LAK formulation by relaxing the assumption in Eq. (2.15). Instead of assuming $\tilde{\mathbf{E}}(\mathbf{r})$ constant inside a dipole it is expanded in Taylor series around the dipole center up to the third order of *kr*. Then second derivative of $\tilde{\mathbf{E}}(\mathbf{r})$ is expressed in terms of $\tilde{\mathbf{E}}$ itself with help of the Maxwell equations, leading to:

$$\tilde{\mathbb{M}}^{\mathrm{PEL}} = \mathbb{I}\left[b_1^{\mathrm{DGF}}\left(1 + \frac{1}{10}\tilde{n}^2\right)(kd)^2 + \frac{2}{3}\mathrm{i}(kd)^3\right], \tag{2.54}$$

where the assumption of a spherical dipole is also used and $\tilde{n}$ is the refractive index of the point considered.

Another class of polarizability formulations is based not on the integral equation in Eq. (2.5) but on the notion of a set of point dipoles. Draine and Goodman [100] found an optimal $O((kd)^2)$ correction to the CM polarizability in the sense that an infinite lattice of point dipoles with such polarizability would lead to the same propagation of a plane-wave as in a homogeneous medium with a given refractive index. This polarizability was called the *lattice dispersion relation* (LDR):

$$\tilde{\mathbb{M}}^{\mathrm{LDR}} = \mathbb{I}\left[\left(b_1^{\mathrm{LDR}} + b_2^{\mathrm{LDR}}\tilde{n}^2 + b_3^{\mathrm{LDR}}\tilde{n}^2 S\right)(kd)^2 + \frac{2}{3}\mathrm{i}(kd)^3\right], \tag{2.55}$$

$$b_1^{\mathrm{LDR}} \approx 1.8915316,\ b_2^{\mathrm{LDR}} \approx -0.1648469,$$
$$b_3^{\mathrm{LDR}} \approx 1.7700004, \tag{2.56}$$

$$S = \sum_{\mu=x,y,z}\left(\hat{e}_\mu^{inc}\hat{k}_\mu^{inc}\right)^2. \tag{2.57}$$

The LDR prescription can be averaged over all possible incident polarizations [100], resulting in:

$$S = \frac{1}{2}\left(1 - \sum_{\mu=x,y,z}\left(\hat{k}_\mu^{inc}\right)^4\right). \tag{2.58}$$

Later a minor flaw in the LDR derivation was found and corrected [109]. This *corrected LDR* (CLDR) is independent on the incident polarization but leads to a diagonal polarizability tensor instead of scalar:

$$\tilde{\mathbb{M}}^{\mathrm{CLDR}} = \delta_{\mu\nu}\left[\left(b_1^{\mathrm{LDR}} + b_2^{\mathrm{LDR}}\tilde{n}^2 + b_3^{\mathrm{LDR}}\tilde{n}^2\left(\hat{k}_\mu^{inc}\right)^2\right)(kd)^2 + \frac{2}{3}\mathrm{i}\,(kd)^3\right], \tag{2.59}$$

where $\delta_{\mu\nu}$ is the Kronecker symbol.

By construction the LDR is expected to be especially inaccurate near the particle surface. To alleviate this problem Rahmani, Chaumet, and Bryant [110] (RCB) proposed to determine polarizability based on the solution of the electrostatic problem for the same scatterer, described by a tensor $\tilde{\mathbb{D}}(\mathbf{r})$:

$$\tilde{\mathbf{E}}(\mathbf{r}) = \tilde{\mathbb{D}}(\mathbf{r})\tilde{\mathbf{E}}^{inc}. \tag{2.60}$$

Such simple relation is possible only in the static limit, when $\tilde{\mathbf{E}}^{inc}$ does not depend on $\mathbf{r}$. Using Eq. (2.21) and Eq. (2.60) in Eq. (2.19), the latter is exactly satisfied for any $\tilde{\mathbf{E}}^{inc}$ if and only if the following polarizability is used

$$\tilde{\alpha}_i^{\mathrm{RCB}} = V_d\tilde{\chi}_i\tilde{\mathbb{D}}_i\left(\mathbb{I} + V_d\sum_{j\neq i}\mathbb{G}_{ij}^{st}\tilde{\chi}_j\tilde{\mathbb{D}}_j\right)^{-1}, \tag{2.61}$$

where $\tilde{\mathbb{D}}_i = \tilde{\mathbb{D}}(\mathbf{r}_i)$ and $\mathbb{G}_{ij}^{st} = \mathbb{G}_{ij}^{st}(\mathbf{r}_i, \mathbf{r}_j)$. Although the RCB polarizability looks completely different from the CM, they differ significantly only for dipoles closer than $2d$ to the interface [111]. Any $\tilde{\mathbf{M}}$-based polarizability corrections can be further applied to the RCB, replacing $\tilde{\alpha}_i^{\mathrm{CM}}$ by $\tilde{\alpha}_i^{\mathrm{RCB}}$ in Eq. (2.22). For instance, the RR correction was used in the original manuscript [110] and Collinge and Draine [111] applied the CLDR correction leading to the *surface-corrected LDR.*

The RCB has been applied only to simple shapes so far, for which $\tilde{\mathbb{D}}$ is constant (independent of $\mathbf{r}$): spheres, ellipsoids, infinite slabs and cylinders. Application to other shapes is connected with two problems. First, it requires preliminary solution of the electrostatic problem for the same shape, which is not trivial. Second, when $\tilde{\mathbb{D}}$ is not constant the RCB polarizability is generally not symmetric, which is physically impossible in the static case [112]. However, it is not clear whether this apparent contradiction affects the accuracy of the final results.

Two other polarizability formulations are consequences of formulations for the interaction term (see Sec. 2.4.3.1). In the framework of the FCD [102], the filtered Green's tensor (see Eq. (2.48)) does not have a singularity for zero argument anymore. This can be used for direct evaluation of polarizability [98]:

$$\tilde{\mathbb{M}}^{\mathrm{FCD}} = V_d \lim_{R\to 0} \mathbb{G}^F(\mathbf{R}) = \frac{4}{3}\mathbb{I}(kd)^2 + \frac{2}{3}\left(\mathrm{i} + \frac{1}{\pi}\ln\frac{\pi - kd}{\pi + kd}\right)\mathbb{I}(kd)^3 . \tag{2.62}$$

In the framework of the IGT Eq. (2.16) is evaluated numerically to a high accuracy using a *Weyl expansion* of the Green's tensor [101]. However, it is much easier to obtain the series expansion of this result up to the third order of kd. Proceeding similar to the derivation of the DGF but keeping the cubical integration volume one can obtain:

$$\tilde{\mathbb{M}}^{\mathrm{IGT}} = \mathbb{I}\left(b_1^{\mathrm{IGT}}(kd)^2 + \frac{2}{3}\mathrm{i}(kd)^3\right), \tag{2.63}$$

$$b_1^{\mathrm{IGT}} = \frac{4}{3}\int_0^1\int_0^1\int_0^1 \frac{\mathrm{d}x\mathrm{d}y\mathrm{d}z}{\sqrt{x^2+y^2+z^2}} = \frac{4}{3}\left[\ln\left(5+3\sqrt{3}\right) - \frac{\ln(2)}{2} - \frac{\pi}{4}\right] \approx 1.586718 . \tag{2.64}$$

Finally, Dungey and Bohren [113], using results by Doyle [114], proposed to use the electric dipole coefficient $a_1 = -B_1$ (see Sec. 1.6.5) from the Mie theory computed for the inscribed sphere with diameter d and electric permittivity $\tilde{\varepsilon}_s$ determined by the *Maxwell-Garnett effective medium theory* [46]:

$$\frac{\pi}{6}\frac{\tilde{\varepsilon}_s - 1}{\tilde{\varepsilon}_s + 2} = \frac{\tilde{\varepsilon} - 1}{\tilde{\varepsilon} + 2}, \tag{2.65}$$

where $\pi/6$ is the volume filling factor. Other effective medium theories may also be used [60]. The dipole polarizability is defined as [114]

$$\tilde{\mathbb{Q}}^{Mie} = \mathrm{i}\frac{3}{2k^3}a_1\mathbb{I}, \tag{2.66}$$

which recovers the usual polarizability of the sphere (see Sec. 1.6.5). This formulation is denoted as the a_1-term method [115]. It is expected to be suitable for simulation of clusters of small spheres, where each of them can be modeled by a single dipole.

A large part of the described polarizability prescriptions is implemented in the production codes, namely: CM, RR, LDR, CLDR, DGF, LAK, FCD, IGT. In plasmonic applications dipoles are usually very small, i.e. $kd \ll 1$. This implies that $O((kd)^2)$ differences between the formulations are minor, making the choice of the best prescription not that important. There is, however, an exception. Two prescriptions from the above list, LDR and CLDR, contain terms proportional $(\tilde{n}kd)^2$, which may be significant for not-very-small particles and large $\tilde{n}$ typical in the near-IR. Although LDR is still the most widely used polarizability prescription, its applicability to such $\tilde{n}$ is very debatable, especially for particles larger than λ [98].

2.4.3.3 Calculating measurable quantities

The most widely used way to calculate measurable quantities is using Eqs. 2.26, 2.28, 2.30, 2.31, and 2.33. The major advantage of these equations is that they are exact for the set of point dipoles. Hence, the optical theorem (a consequence of energy conservation), expressed, e.g., as:

$$C_{ext} = C_{abs} + C_{sca}\,, \tag{2.67}$$

is a direct implication of the main DDA equations (e.g. Eq. (2.19)). Therefore, the inaccuracy of Eq. (2.67) is comparable with the final residual of the iterative solver (assuming Eq. (2.30) is evaluated to a high accuracy), which is usually much smaller than the error of the DDA itself (due to other approximations). In particular, Eq. (2.67) can be used for an accurate evaluation of C_{sca} without a laborious integration implied by Eq. (2.30), even when it is up to 100 times smaller than both C_{ext} and C_{abs}.

Other expressions for C_{abs} have also been proposed. Originally, Purcell and Pennypacker [54] used Eq. (2.33) without the second term, but that works satisfactory only in combination with the CM polarizability. Otherwise, physical artifacts occur, such as non-zero C_{abs} for purely real $\tilde{n}$. A more advanced formula was proposed [101] based on radiation correction of a *finite* dipole instead of a *point* dipole:

$$C_{abs} = 4\pi k \sum_i \operatorname{Im}\left[\tilde{\mathbf{P}}_i \cdot \tilde{\mathbf{E}}_i^*\right], \tag{2.68}$$

which is also a direct approximation to Eq. (2.32). There is no difference between Eq. (2.33) and Eq. (2.68) for the CM, the RR, and the FCD polarizability formulations and also for real refractive indices [40]. Otherwise, the optical theorem is no more exactly satisfied. To alleviate this problem a corresponding correction to C_{ext} was proposed [65]. However, it is not yet clear whether using Eq. (2.68) improves the accuracy of the DDA.

Another possibility is to improve Eq. (2.26) using advanced formulations for the interaction term (Sec. 2.4.3.1). In particular, the following expressions naturally follow from the IGT and the FCD:

$$\tilde{\mathbb{G}}_i^{\mathrm{IGT}}(\mathbf{r}) = \frac{1}{V_d} \int_{V_i} \mathrm{d}^3\mathbf{r}' \tilde{\mathbb{G}}^0(\mathbf{r}_i, \mathbf{r}'), \tag{2.69}$$

$$\tilde{\mathbb{G}}_i^F(\mathbf{r}) = \mathbb{G}^F(\mathbf{r}, \mathbf{r}_i). \tag{2.70}$$

These expressions may improve the accuracy of the near-field calculation but, unfortunately, this has never been tested. The production codes allow one to calculate the near-field only using the simplest Eq. (2.26).

The importance of choosing the right expression for $\tilde{\mathbb{G}}_i(\mathbf{r})$ diminishes with increasing r. In particular, it is easy to show that:

$$\tilde{\mathbb{G}}_i^{\mathrm{IGT}}(\mathbf{r}) \overset{r\to\infty}{\longrightarrow} \eta(kd, \hat{\mathbf{r}})\tilde{\mathbb{G}}^0(\mathbf{r}, \mathbf{r}_i) \tag{2.71}$$

$$\tilde{\mathbb{G}}_i^F(\mathbf{r}) \overset{r\to\infty}{\longrightarrow} \tilde{\mathbb{G}}^0(\mathbf{r}, \mathbf{r}_i) \tag{2.72}$$

where

$$\eta(x, \hat{\mathbf{r}}) = \prod_{\mu=x,y,z} \frac{\sin(\hat{r}_\mu x/2)}{\hat{r}_\mu x/2} = 1 - (1/24)x^2 + O(x^4). \tag{2.73}$$

In other words, the FCD does not change the expression for $\tilde{\mathbf{F}}(\hat{\mathbf{r}})$ given by Eq. (2.28), while the IGT adds the multiplicative factor $\eta(kd, \hat{\mathbf{r}})$, which only slightly differs from unity.

Although all of the described improvements, except those to calculate the near-field, are implemented in the production codes, the default one seems satisfactory. However, further research is required to evaluate possible virtues of the IGT.

2.4.3.4 Decreasing shape errors

The above discussion of different DDA formulations implicitly assumed that particle shape can be exactly described by a set of cubes. Hence, all errors were due to the discretization of the integral equation and related simplifications. In this section so-called shape errors [68] are discussed, which are caused by violation of the above assumption by the real particle shape. First, it is important to note that none of the existing improvements is implemented in the production codes, partly because of the existing internal data structure capable of handling only several different values for polarizabilities of all dipoles [65]. Hence, the main goal of this section is to put forward promising ideas for further development.

A standard way to improve description of the particle shape in numerical solution of integral equations is adaptive discretization, using smaller dipoles near the particle surface. Application of this idea to the DDA is discussed in Ref. [40], but it is incompatible with the FFT acceleration (Sec. 2.4.2.2). Therefore, the only practically viable option is to keep the regular grid of cubical dipoles, but adjust the properties of the boundary dipoles.

Evans and Stephens [50] proposed to modify the susceptibility of the boundary dipole using the Lorentz-Lorenz mixing rule:

$$\frac{\tilde{\chi}_i^e}{4\pi\tilde{\chi}_i^e + 3} = f\frac{\tilde{\chi}_i}{4\pi\tilde{\chi}_i + 3}, \tag{2.74}$$

where the interface between the particle and vacuum is considered, $\tilde{\chi}_i^e$ is the effective susceptibility, and f is the volume fraction of the subvolume actually occupied by scatterer.

A more advanced averaging, called the *weighted discretization* (WD), was proposed by Piller [116]. It modifies both the susceptibility and the self-term of the boundary dipoles. The particle surface, crossing the subvolume V_i, is assumed linear and divides the subvolume into two parts: the *principal volume* V_i^p that contains the center and a *secondary volume* V_i^s with susceptibilities $\tilde{\chi}_i^p$, $\tilde{\chi}_i^s$ and electric fields $\tilde{\mathbf{E}}_i^p \equiv \tilde{\mathbf{E}}_i$, $\tilde{\mathbf{E}}_i^s$, respectively. Electric fields are considered constant inside each part and related to each other via a boundary condition tensor $\tilde{\mathbb{T}}_i$:

$$\tilde{\mathbf{E}}_i^s = \tilde{\mathbb{T}}_i\tilde{\mathbf{E}}_i. \tag{2.75}$$

Then the total polarization of the subvolume can be evaluated as follows:

$$\tilde{\mathbf{P}}_i = \int_{V_i} \mathrm{d}^3\mathbf{r}' \tilde{\chi}(\mathbf{r}')\tilde{\mathbf{E}}(\mathbf{r}') = V_i^p \tilde{\chi}_i^p \tilde{\mathbf{E}}_i^p + V_i^s \tilde{\chi}_i^s \tilde{\mathbf{E}}_i^s = V_d \tilde{\mathbb{X}}_i^e \tilde{\mathbf{E}}_i, \qquad (2.76)$$

with an effective (averaged) susceptibility:

$$\tilde{\mathbb{X}}_i^e = \left(V_i^p \tilde{\chi}_i^p \mathbb{I} + V_i^s \tilde{\chi}_i^s \tilde{\mathbb{T}}_i\right) / V_d \,. \qquad (2.77)$$

The susceptibility of the boundary subvolume is replaced by an effective one. The effective self-term is evaluated directly starting from Eq. (2.7), considering $\tilde{\chi}$ and $\tilde{\mathbf{E}}$ constant inside each part:

$$\tilde{\mathbb{M}}_i^e \tilde{\mathbb{X}}_i^e = \int_{V_i^p} \mathrm{d}^3\mathbf{r}' \left(\tilde{\mathbb{G}}^0(\mathbf{r}_i, \mathbf{r}') - \mathbb{G}^{st}(\mathbf{r}_i, \mathbf{r}')\right) \tilde{\chi}_i^p$$
$$+ \int_{V_i^s} \mathrm{d}^3\mathbf{r}' \left(\tilde{\mathbb{G}}^0(\mathbf{r}_i, \mathbf{r}') - \mathbb{G}^{st}(\mathbf{r}_i, \mathbf{r}')\right) \tilde{\chi}_i^s \tilde{\mathbb{T}}_i \,. \qquad (2.78)$$

Piller [116] evaluated the integrals in Eq. (2.78) numerically. The polarizability is obtained from Eq. (2.22) using $\tilde{\mathbb{X}}_i^e$ and $\tilde{\mathbb{M}}_i^e \tilde{\mathbb{X}}_i^e$, while Eq. (2.19) remains unchanged. Hence, the WD does not modify the general numerical scheme. A significant improvement of DDA accuracy due to the WD was shown both by theoretical analysis [104] and in sample simulations [103, 116].

Two other improvements are not aimed exclusively at shape errors, but they affect only dipoles close to the boundary. The first one is the RCB formulation discussed in Sec. 2.4.3.2. Second one is the spectral filtering of $\tilde{\chi}$ that was proposed in combination with the FCD [102] to smooth the change of $\tilde{\chi}$ at the particle boundary by removing high-frequency components.

2.5 Practical Aspects of DDA Simulations

2.5.1 *General Applicability*

The principal advantage of the DDA is that it is completely flexible regarding the geometry of the scatterer, being limited only by the need to use the dipole size d small compared to both any structural length in the scatterer and λ. For particles with size comparable to or larger than λ the following rule of thumb is commonly used [40]:

$$d = \lambda/10\,|\tilde{n}|\,. \tag{2.79}$$

It implies that the number of dipoles N increase with both size and $\tilde{n}$. The expected accuracy of cross sections is then several percents, if $\tilde{n}$ belongs to the well-tested range, approximately described as

$$|\tilde{n} - 1| < 2. \tag{2.80}$$

For larger $\tilde{n}$ the accuracy of the simulation with default dipole size deteriorates, and smaller, hence more dipoles must be used to improve it. However, there is no strict limitation on $\tilde{n}$ - accurate results can be obtained for a wide range of $\tilde{n}$, given enough computational resources. In this respect state-of-the-art DDA formulations (Sec. 2.4.3) can be of great help.

For particles smaller than λ, e.g. nanoparticles, Eq. (2.79) is definitely not relevant. In this case the main requirement for d is that it should be small enough to resolve fine details of particle shape. For instance, for a compact shape (like sphere or cube) it is recommended to use at least 10 dipoles along the smallest dimension, no matter how small the particle is. Therefore, for nanoparticles N is almost independent of size. However, N does depend on $\tilde{n}$, which is critical due to the wide range of $\tilde{n}$ used in plasmonic applications. In more details these issues are discussed in Sec. 2.6.

2.5.2 *System Requirements*

The major price paid for versatility of the DDA is its large computational costs, even for simple scatterers. The most important factor is the number of dipoles N, which determines both memory requirements M and computational time of one iteration t_{iter}. More precisely, the relevant quantity is the number of dipoles composing the enclosing rectangular box (see Sec. 2.4.2.2), which is typically one-two times larger than N (for compact particles). System requirements largely vary depending on the particular DDA code, arithmetic precision, and iterative solver. The following are the guiding values:

$$M \approx 1000N \text{ bytes}, \tag{2.81}$$

$$t_{iter} \approx 3 \times 10^{-7} N \ln(N)\, s. \tag{2.82}$$

The latter is an estimate when using a single core of a modern processor. More accurate estimates can be found in the manuals of particular codes (see Sec. 2.5.4).

Required memory is the major factor limiting N when a DDA simulation is run on a single PC. This limitation can be alleviated by parallelization of a single simulation among a cluster of computers, which is implemented in some of the DDA codes. Corresponding parallel efficiency is generally close to unity [65, 117], which means that M is effectively distributed among several computers and t_{iter} is decreased proportionally to the total number of processor cores. However, for efficient parallelization the number of cores must not be greater than the maximum number of dipoles along a coordinate axis.

Total time t of DDA simulation can be decomposed into the following parts:

$$t \approx t_{init} + N_{or}\left(N_{iter}t_{iter} + t_{quan}\right), \tag{2.83}$$

where t_{init} and t_{quan} are time of initialization and calculation of measurable quantities, respectively, and N_{or} is the number of distinct particle orientations (see Sec. 2.4.2.3). Generally, t_{init} is comparable to t_{iter} and t_{quan} is even smaller. However, the latter is proportional to the number of considered scattering angles or near-field probe points and, thus, may become large in certain conditions. Existing theory concerning N_{iter} is discussed in Sec. 2.4.2.1, here we only note that it is also sensitive to the used arithmetic precision. Using single instead of double precision halves M, but increases round-off errors, which in turn accumulate and increase N_{iter}, especially when the latter is already large.

2.5.3 *Free Parameters*

Many parameters of the DDA codes are directly determined by a scattering problem to be simulated. However, there are also some free parameters that can be tuned. In the following they are listed in the order of decreasing importance:

- the level of discretization (i.e. size or number of dipoles),
- DDA formulation,
- the number of orientations for orientation averaging,
- iterative solver and convergence threshold,
- the volume correction.

The level of discretization is a result of compromise between accuracy and required computational resources. In many cases the required accuracy is specified *a priori*, but finding an appropriate d (or N) is not an easy task. Existing benchmark studies are discussed in Sec. 2.6, but their coverage of the field of plasmonics is relatively poor. Therefore, whenever one plans to run more that a single DDA simulation for a class of similar particles, it is recommended first to perform an accuracy study. For that one should choose a single test particle and perform DDA simulations with different d around the values typically used for similar simulations in the literature. The estimate of d required for a particular accuracy can be obtained from a variation of results with decreasing d. Moreover, the estimation can be made much more rigorous by using an extrapolation technique, as proposed by Yurkin *et al.* [68] and applied in Refs. [67, 98, 118] (see also Sec. 2.6). Seeking a careful compromise is important due to a steep dependence of N (and t) on required accuracy. In particular, two-fold decrease of simulation error usually requires eight-fold increase of N [68, 104].

Different DDA formulations are discussed in Sec. 2.4.3 together with some practical recommendations. The number of orientations required for accurate orientation-averaged results is also hard to estimate *a priori*, since this issue is rarely discussed in details [96, 119]. However, contrary to the DDA simulation itself, it is relatively easy to estimate the accuracy of the orientation averaging during the process and increase the number of orientations, if needed. In other words, orientation averaging can be performed in semi-automatic adaptive regime [65, 96]. Moreover, the dependence of t on accuracy of orientation averaging is not that steep as discussed above for DDA simulation for fixed orientation. Therefore, it may be acceptable to use "large enough" number of orientations.

Different iterative solvers are discussed in Sec. 2.4.2.1 and there are several of them, which work fine in most cases. The convergence threshold is usually chosen small enough to introduce negligible errors, because in this case the extra accuracy is especially cheap. Convergence of the iterative solver is usually linear in logarithmic scale, so, e.g. ten-fold increase of the typical threshold value of 10^{-5} decreases t by only about 20%. Nevertheless, such acceleration can be useful for very long simulations.

Discretization of a particle in the framework of the DDA is usually performed by testing the dipole centers for belonging to the particle volume. Then the total volume of the set of dipoles differs from volume of the particle. The *volume correction*, implemented and used by default in some DDA codes, changes the size of each dipole (scales the whole set of dipoles) to remove this difference. This is believed to increase the accuracy of DDA, especially for small scatterers, although the effect is not always beneficial [44]. However, in plasmonic applications N is usually large (to get satisfactory accuracy), which decrease the effect of volume correction.

To conclude, although DDA is one of the most conceptually simple and easy-to-use methods to simulate light scattering, performing accurate and fast DDA simulations is still, to some extent, an art. It requires one to consider open questions that still remain, because the plasmonics is a relatively new application domain of the DDA. Hence, to perform reliable and efficient DDA simulations one should take a critical attitude towards results and be ready to experiment with free parameters, instead of hoping for "black box" to produce correct results by itself. Although the DDA codes usually have some built-in empirics to set up the free parameters without any input from user, these empirics may be completely inappropriate in some cases. Finally, in the art of DDA simulations an advice from more experienced users can be invaluable. Apart from traditional means, such advice can also be obtained through an internet discussion group [120].

2.5.4 *Available Codes*

To our knowledge, there exist four freely-available DDA codes: DDSCAT [44], ADDA [65], OpenDDA [117], and DDA-SI toolbox [27]. More DDA codes exist, and some of them are discussed in Ref. [119], but these are not freely available to the community. OpenDDA and DDA-SI toolbox have been released only recently and not yet well-tested by the community, but they do have attractive features. OpenDDA contains highly-optimized computational kernel which can run efficiently both on multicore processors using OpenMP [121] and on computer clusters using MPI (message passing interface [122]). This is probably the fastest DDA code when using the standard formulation of the DDA. DDA-SI toolbox can rigorously handle particles near the surface (see Sec. 2.2.3) but, unfortunately, do not employ FFT acceleration.

In the following two mature codes are compared, based on their current versions: DDSCAT 7.2 [123] and ADDA 1.1 [124]. However, one should refer to the corresponding manuals [95, 125] for more details. Both codes are fast, accurate and feature-rich, so only the differences are pointed out.

- **DDA formulations.** DDSCAT can rigorously handle 1D and 2D periodic scatterers (see Sec. 2.2.2). ADDA implements state-of-the-art DDA formulations, in particular, the IGT and the FCD (see Sec. 2.4.3.1).
- **Portability.** DDSCAT is written in Fortran 90, while ADDA is written in C99. Both are highly portable and can be compiled on a wide variety of platforms. Additionally DDSCAT developers provide compiled binaries for 32-bit Windows, while ADDA – both for 32-bit and 64-bit Windows.
- **Use of modern hardware.** DDSCAT uses OpenMP to effectively run on multi-core processors and can use MPI to parallelize orientation averaging. ADDA parallelizes a *single* DDA simulation using MPI. Hence, huge N (very large particles or very fine discretization) can be used when running on a large computer cluster. However, for a single multi-core processor the performance of MPI is generally worse than that of OpenMP. Also ADDA features GPU-acceleration, which potentially allows about 10 times acceleration using modern graphics cards.
- **Speed.** The speed of the codes on the same basic hardware can also differ due to differences in algorithms and optimizations. In 2007 Penttila *et al.* [119] compared, among other codes, DDSCAT 6.1 and ADDA 0.7a. ADDA was from

1.2 to 4 times faster than DDSCAT for simulation of light scattering by cubes and spheroids in fixed orientation. However, performance of both codes has been significantly improved since that comparison In particular, DDSCAT currently includes a fast FFT-based routine to calculate near-field [126].

- **Orientation averaging.** DDSCAT employs the Simpson rule as a quadrature for orientation averaging, while ADDA – the Romberg integration. The latter is potentially more accurate and can be used in adaptive regime. However it requires more careful consideration to be used efficiently, and is less flexible in the choice of number of orientations [119].
- **Spectrum calculation.** DDSCAT completely automates calculations for a spectrum of wavelengths in a convenient manner. ADDA can perform such automatic calculations only with the help of additional scripts.
- **User interface.** DDSCAT accepts most of the simulation parameters through a special file. ADDA accepts such parameters through a command line, which is more convenient for scripted parallel runs [119]. However, both codes also use large input files, e.g. to describe an arbitrary particle. Also ADDA features a built-in help system, which should be especially useful for new users.
- **History and development process.** First version of DDSCAT was publicly released in 1993; hence, it is an extremely well-tested and reliable code. In particular, it was used in several hundreds of journal papers. Development of ADDA started in 1990, but the first public release was in 2006. It is also well-tested, exemplified by its use in more than 80 journal papers. Currently ADDA features an open development process with several developers from different countries.

2.6 Accuracy of the DDA

First, it should be stressed that although the name of DDA contains "approximation" due to historical reasons, it is a numerically exact method in the sense that accuracy can be made however good given enough computational resources. The only limitation on the achievable accuracy is enforced by the machine precision. The proof of DDA numerical exactness was provided by the rigorous convergence analysis [104] based on the derivation of the DDA from the Maxwell's equations using no physical approximations (see Sec. 2.4.1). It was also confirmed by many numerical convergence studies, see, e.g. Refs. [72, 104], also for very large $|\tilde{n}|$ [67, 98, 127].

Accuracy of the DDA in general is discussed in review [40] as well as in more recent benchmark study [72]. However, they are mostly limited to the moderate range of $\tilde{n}$ (see Eq. (2.80)), for which the rule-of-thumb is commonly used (see Sec. 2.5.1). Use of the DDA in plasmonics is mainly application-driven with, unfortunately, little attention paid to error analysis of the DDA itself. Moreover, even in the recent plasmonic papers the DDA is sometimes (e.g. Ref. [12]) wrongly regarded as an approximate method (see the discussion above). This misconception may be caused by two reasons. First, the DDA accuracy for plasmonic $\tilde{n}$ is indeed much worse than for moderate $\tilde{n}$ in the sense that corresponding computational resources needed to achieve the same accuracy are larger. Hence, for some plasmonic applications good DDA accuracy cannot be achieved on, e.g. standard desktop PC. Second, the DDA accuracy is often judged by comparison with experiment, which is also largely affected by uncertainties in the proper values of $\tilde{n}$ (see Sec. 2.3) and in experimental conditions, e.g. in particle shape [128, 129]. So in the following existing literature data on DDA accuracy for metallic nanoparticles is reviewed with particular emphasis on gold. However, the results for other metals are expected to be similar with corresponding changes in λ, because of similar variation ranges of $\tilde{n}$ (see Section 2.4.2.1).

The accuracy of the DDA for gold nanoparticles is usually quoted as "good enough if a large enough number of dipoles is used", i.e. the discussion is qualitative with no error measurements available [130–133]. In a couple of papers DDA results are plotted together with exact reference results in the same graph for spheres [131] and spheroids [132]. These graphs show that errors of extinction efficiency at particular wavelengths can be as large as 50%. Moreover, a reliable identification of small side peaks in the spectrum is hampered by DDA errors [131]. Similar results have been obtained for silver nanospheres [134].

Although DDA simulations for nanoparticles mostly focus on extinction efficiency, as this is usually measured experimentally, several researchers have studied its constituents—absorption and scattering efficiencies—separately [132, 135, 136]. This should result in a better understanding of DDA errors, especially their size dependence. Moreover, absorption efficiency is relevant to practical applications involving optical heating of nanoparticles.

Recently a systematic error analysis of the DDA for gold nanoparticles was performed [67]. In the following this study is considered in details. Its important features were using a wide range of discretizations and relatively new FCD formulation, additionally to the standard LDR (see Sec. 2.4.3.2). The following particles were considered: two spheres (diameters $D = 10$ and 100 nm), two cubes (edge sizes $D = 10$ and 100 nm), and a rod (cylinder with hemispherical caps, diameter 20 nm, total length 90 nm). The incident light propagated along the z-axis and was polarized (E-field) along the x-axis. Cubes were oriented with edges along the coordinate axes, and the rod was oriented with symmetry axis along the x-axis, emphasizing the longitudinal plasmon resonance. All particles were considered in vacuum using the wavelength range [0.398,0.822] μm. The latter is informative for gold, since it contains both moderate and plasmonic $\tilde{n}$, see, e.g. Fig. 2.2. Values of $\tilde{n}$ were taken from Johnson and Christy [37] without size correction (see Sec. 2.3).

Absorption and scattering efficiencies (Q_{abs}, Q_{sca}) were calculated using ADDA 0.79 varying the discretization level, characterized by number of dipoles N_x along the x-axis. Reference results were obtained using Mie theory [46], T-matrix method [137], and extrapolation technique combined with the DDA [68] for the spheres, the rod, and the cubes respectively. Some of the results [67] for Q_{abs} and Q_{sca} are presented in Fig. 2.3 and Fig. 2.4, respectively. Since the accuracy of these two quantities weakly depends on the size [67] (see also Fig. 2.5), only results for the smaller sphere and cube are shown.

The immediate conclusion is that errors can be huge (100% and more) when using moderate N_x. On the other hand, such a

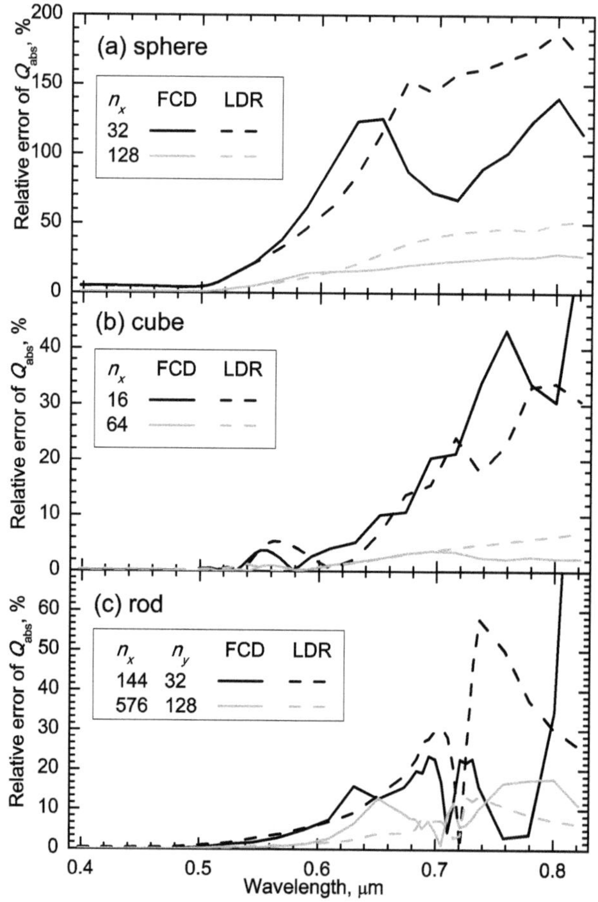

Figure 2.3 Relative errors of absorption efficiency Q_{abs} computed using the FCD and LDR formulations of the DDA, varying the number of dipoles, for gold (a) 10-nm sphere, (b) 10-nm cube, and (c) 20×90-nm rod.

reasonable task as obtaining Q_{abs} of a sphere (or a rod) in the near-IR with at least 10% accuracy requires N_x up to 256 ($N \sim 10^7$), which is still very challenging for a desktop PC, considering large N_{iter} (see Fig. 2.2) and potential need for orientational averaging. Fortunately, the errors for the cube are an order of magnitude smaller than for the sphere and the rod for the same dipole size.

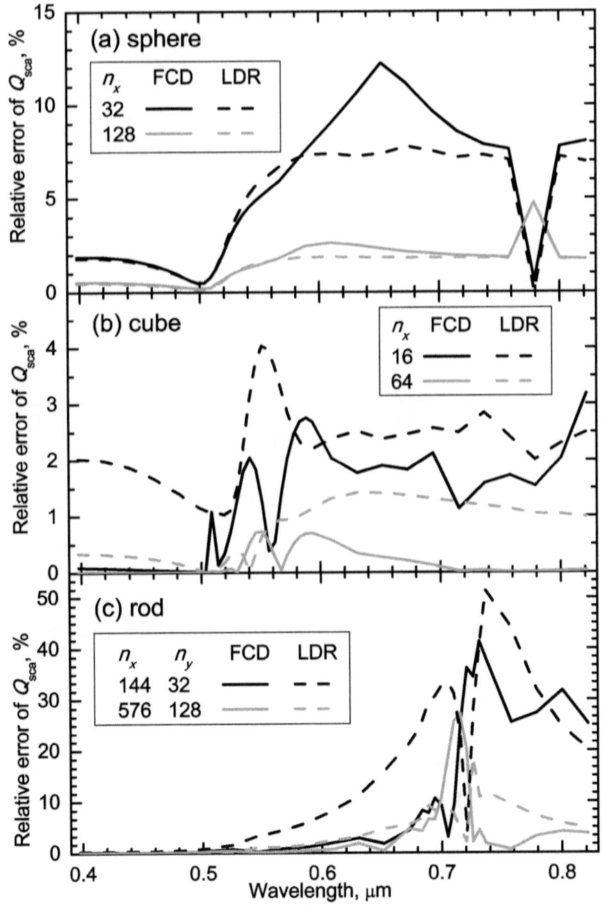

Figure 2.4 Same as Fig. 2.3 but for scattering efficiency Q_{sca}.

Supposedly, the situation is similar for other shapes that can be exactly described by a set of cubical dipoles due to absence of shape errors. Therefore, for other shapes the accuracy of DDA is expected to be significantly improved by the WD (see Sec. 2.4.3.4) or similar ideas, which calls for their implementation in the production DDA codes.

The relative errors of Q_{sca} are about 10 times less than that of Q_{abs} for the spheres and the cubes, but only 2 times for the rod. The reasons for such difference are still unclear. However, it

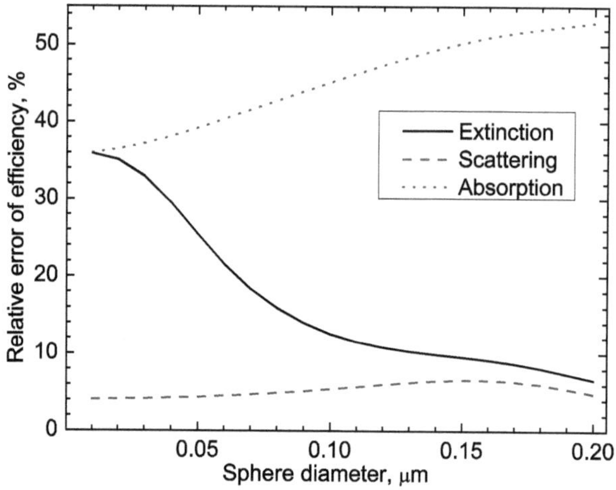

Figure 2.5 Relative errors of Q_{ext}, Q_{sca}, and Q_{abs} of a gold sphere computed using the FCD formulation of the DDA as a function of sphere diameter ($\lambda = 0.694$ μm, $N_x = 64$).

implies that accuracy of Q_{ext} has strong size dependence due to very different size dependences of Q_{abs} and Q_{sca}, which are $O(D/\lambda)$ and $O((D/\lambda)^4)$, respectively, for $D < \lambda$ [46]. In other words, accuracy of Q_{ext} is similar to that of Q_{abs} for smaller and to that of Q_{sca} for larger particles, which is illustrated in Fig. 2.5 for a specific value of λ [67]. The boundary value of D for the spheres is 50–200 nm depending on λ [67]. Therefore, accuracy of Q_{ext} can be unusually poor for sufficiently small nanoparticles, as compared to earlier studies using larger particles (e.g. 300 nm spheres, [131]).

In certain plasmonic applications the only quantity of interest is the position and the amplitude of the spectral peak. For these values the DDA accuracy is much better than that discussed above. Using moderate discretization (32 dipoles per shortest particle dimension), the position is determined with accuracy better than 10 nm and the amplitude—better than 3% (except the FCD results for the rod [67]). Comparing the two DDA formulations, neither the FCD nor the LDR can claim conclusive superiority in terms of accuracy, although the FCD is generally more accurate. However, the FCD is recommended over the LDR for general use, since this accelerates the convergence of the iterative solver up to two times (see Fig. 2.1).

To conclude, the DDA accuracy in the field of plasmonics varies in a wide range depending on particle, wavelength, measurable quantity to be simulated, and parameters of the particular DDA code. It is hard to give general recipes, because universally accurate DDA results may be too computationally expensive. Therefore, one should exercise caution when performing such simulations, as discussed in Sec. 2.5.3. Moreover, the DDA accuracy is expected to become significantly better in certain cases, if promising new DDA formulations are implemented in the production codes.

Acknowledgments

I thank Alfons Hoekstra for introducing me to a beautiful world of the DDA and for many fruitful discussions we had on this subject. I also thank David de Kanter for his invaluable contribution to the study of DDA accuracy when applied to gold nanoparticles. I acknowledge support of the program of the Russian Government "Research and educational personnel of innovative Russia" (contracts 8752 and 8804), by grant from the Russian Government 11.G34.31.0034, and of the Russian Foundation of Basic Research (grant 12-04-00737-a).

References

1. Mie, G. (1908) Beitrage zur optik truber medien, speziell kolloidaler metallosungen, *Ann. Phys. (NY)*, **330**, 377–445.
2. Mishchenko, M. I., Travis, L. D., and Lacis, A. A. (2002) *Scattering, Absorption, and Emission of Light by Small Particles*, Cambridge University Press, Cambridge.
3. Mishchenko, M. I., Hovenier, J. W., and Travis, L. D., (2000) *Light Scattering by Nonspherical Particles, Theory, Measurements, and Applications*, Academic Press, New York.
4. Mishchenko, M. I., Travis, L. D., and Mackowski, D. W. (1996) T-matrix computations of light scattering by nonspherical particles: A review, *J. Quant. Spectrosc. Radiat. Transfer*, **55**, 535–575.

5. Myroshnychenko, V., Carbó-Argibay, E., Pastoriza-Santos, I., Pérez-Juste, J., Liz-Marzán, L. M., and Garcia de Abajo, F. J. (2008) Modeling the optical response of highly faceted metal nanoparticles with a fully 3D boundary element method, *Adv. Mater.*, **20**, 4288–4293.
6. Kern, A. M., and Martin, O. J. F. (2009) Surface integral formulation for 3D simulations of plasmonic and high permittivity nanostructures, *J. Opt. Soc. Am. A*, **26**, 732–740.
7. Ungureanu, C., Rayavarapu, R. G., Manohar, S., and van Leeuwen, T. G. (2009) Discrete dipole approximation simulations of gold nanorod optical properties: Choice of input parameters and comparison with experiment, *J. Appl. Phys.*, **105**, 102032/1-7.
8. Yang, W. H., Schatz, G. C., and Vanduyne, R. P. (1995) Discrete dipole approximation for calculating extinction and Raman intensities for small particles with arbitrary shapes, *J. Chem. Phys.*, **103**, 869–875.
9. Kelly, K. L., Coronado, E., Zhao, L., and Schatz, G. C. (2003) The optical properties of metal nanoparticles: the influence of size, shape, and dielectric environment, *J. Phys. Chem. B*, **107**, 668–677.
10. Hao, F., Nehl, C. L., Hafner, J. H., and Nordlander, P. (2007) Plasmon resonances of a gold nanostar, *Nano Lett.*, **7**, 729–732.
11. Taflove, A., and Hagness, S. C. (2005) *Advances in Computational Electrodynamics: the Finite-Difference Time-Domain Method*, 3rd ed., Artech House, Boston.
12. Zhao, J., Pinchuk, A. O., McMahon, J. M., Li, S., Ausman, L. K., Atkinson, A. L., and Schatz, G. C. (2008) Methods for describing the electromagnetic properties of silver and gold nanoparticles, *Accounts Chem. Res.*, **41**, 1710–1720.
13. Monk, P. (2003) *Finite Element Methods for Maxwell's Equations*, Oxford University Press, New York.
14. Zhu, S., Zhou, W., Park, G., and Li, E. (2010) Numerical design methods of nanostructure array for nanobiosensing, *Plasmonics*, **5**, 267–271.
15. Hellmers, J., Riefler, N., Wriedt, T., and Eremin, Y. A. (2008) Light scattering simulation for the characterization of sintered silver nanoparticles, *J. Quant. Spectrosc. Radiat. Transfer*, **109**, 1363–1373.
16. Wriedt, T. (2007) Review of the null-field method with discrete sources, *J. Quant. Spectrosc. Radiat. Transfer*, **106**, 535–545.
17. Myroshnychenko, V., Rodriguez-Fernandez, J., Pastoriza-Santos, I., Funston, A. M., Novo, C., Mulvaney, P., Liz-Marzán, L. M., and Garcia de Abajo, F. J. (2008) Modelling the optical response of gold nanoparticles, *Chem. Soc. Rev.*, **37**, 1792–1805.

18. Khlebtsov, N. G., and Dykman, L. A. (2010) Optical properties and biomedical applications of plasmonic nanoparticles, *J. Quant. Spectrosc. Radiat. Transfer*, **111**, 1–35.

19. Parsons, J., Burrows, C. P., Sambles, J. R., and Barnes, W. L. (2010) A comparison of techniques used to simulate the scattering of electromagnetic radiation by metallic nanostructures, *J. Mod. Optics*, **57**, 356–365.

20. Comberg, U., and Wriedt, T. (1999) Comparison of scattering calculations for aggregated particles based on different models, *J. Quant. Spectrosc. Radiat. Transfer*, **63**, 149–162.

21. Wriedt, T., and Comberg, U. (1998) Comparison of computational scattering methods, *J. Quant. Spectrosc. Radiat. Transfer*, **60**, 411–423.

22. Gilev, K. V., Eremina, E., Yurkin, M. A., and Maltsev, V. P. (2010) Comparison of the discrete dipole approximation and the discrete source method for simulation of light scattering by red blood cells, *Opt. Express*, **18**, 5681–5690.

23. Gallinet, B., Kern, A. M., and Martin, O. J. F. (2010) Accurate and versatile modeling of electromagnetic scattering on periodic nanostructures with a surface integral approach, *J. Opt. Soc. Am. A*, **27**, 2261–2271.

24. Chaumet, P. C., Rahmani, A., and Bryant, G. W. (2003) Generalization of the coupled dipole method to periodic structures, *Phys. Rev. B*, **67**, 165404/1-5.

25. Draine, B. T., and Flatau, P. J. (2008) Discrete-dipole approximation for periodic targets: theory and tests, *J. Opt. Soc. Am. A*, **25**, 2693–2703.

26. Paulus, M., Gay-Balmaz, P., and Martin, O. J. F. (2000) Accurate and efficient computation of the Green's tensor for stratified media, *Phys. Rev. E*, **62**, 5797–5807.

27. Loke, V. L. Y., and Menguc, M. P. (2010) Surface waves and atomic force microscope probe-particle near-field coupling: discrete dipole approximation with surface interaction, *J. Opt. Soc. Am. A*, **27**, 2293–2303.

28. Schmehl, R., Nebeker, B. M., and Hirleman, E. D. (1997) Discrete-dipole approximation for scattering by features on surfaces by means of a two-dimensional fast Fourier transform technique, *J. Opt. Soc. Am. A*, **14**, 3026–3036.

29. Wijers, C. (1986) Rayleigh scattering from single-site polysylane adsorbed on silicon: Theory, *Surf. Sci.*, **168**, 816–822.

30. Taubenblatt, M. A., and Tran, T. K. (1993) Calculation of light-scattering from particles and structures on a surface by the coupled-dipole method, *J. Opt. Soc. Am. A*, **10**, 912–919.

31. Mackowski, D. W. (2010) A generalization of image theory to predict the interaction of multipole fields with plane surfaces, *J. Quant. Spectrosc. Radiat. Transfer*, **111**, 802–809.
32. D'Agostino, S., Pompa, P. P., Chiuri, R., Phaneuf, R. J., Britti, D. G., Rinaldi, R., Cingolani, R., and Della Sala, F. (2009) Enhanced fluorescence by metal nanospheres on metal substrates, *Opt. Lett.*, **34**, 2381–2383.
33. Parviainen, H., and Lumme, K. (2008) Scattering from rough thin films: discrete-dipole-approximation simulations, *J. Opt. Soc. Am. A*, **25**, 90–97.
34. Mackowski, D. W. (2007) Direct simulation of scattering and absorption by particle deposits, in *Proceedings of the 10th Conference on Electromagnetic and Light Scattering* (ed. Videen, G., Mishchenko, M. I., and Menguc, M. P.) Bodrum, Turkey, pp. 113–116.
35. Martin, O. J. F. (2004) Efficient scattering calculations in complex backgrounds, *AEU-Int. J. Electron. Commun.*, **58**, 93–99.
36. Khlebtsov, N. G. (2008) Optics and biophotonics of nanoparticles with a plasmon resonance, *Quantum Electron.*, **38**, 504–529.
37. Johnson, P. B., and Christy, R. W. (1972) Optical constants of the noble metals, *Phys. Rev. B*, **6**, 4370–4379.
38. Palik, E. D., (ed.) (1985) *Handbook of Optical Constants of Solids*, Academic Press, San Diego.
39. http://refractiveindex.info/
40. Yurkin, M. A., and Hoekstra, A. G. (2007) The discrete dipole approximation: an overview and recent developments, *J. Quant. Spectrosc. Radiat. Transfer*, **106**, 558–589.
41. Kahnert, F. M. (2003) Numerical methods in electromagnetic scattering theory, *J. Quant. Spectrosc. Radiat. Transfer*, **79**, 775–824.
42. Lakhtakia, A. (1992) Strong and weak forms of the method of moments and the coupled dipole method for scattering of time-harmonic electromagnetic fields, *Int. J. Mod. Phys. C*, **3**, 583–603.
43. Yanghjian, A. D. (1980) Electric dyadic Green's function in the source region, *IEEE Proc.*, **68**, 248–263.
44. Draine, B. T., and Flatau, P. J. (1994) Discrete-dipole approximation for scattering calculations, *J. Opt. Soc. Am. A*, **11**, 1491–1499.
45. Goedecke, G. H., and O'Brien, S. G. (1988) Scattering by irregular inhomogeneous particles via the digitized Green's function algorithm, *Appl. Optics*, **27**, 2431–2438.
46. Bohren, C. F., and Huffman, D. R. (1983) *Absorption and Scattering of Light by Small Particles*, Wiley, New York.

47. Hage, J. I., Greenberg, J. M., and Wang, R. T. (1991) Scattering from arbitrarily shaped particles: theory and experiment, *Appl. Optics*, **30**, 1141–1152.

48. Draine, B. T. (1988) The discrete dipole approximation and its application to interstellar graphite grains, *Astrophys. J.*, **333**, 848–872.

49. Hoekstra, A. G., Frijlink, M., Waters, L. B. F. M., and Sloot, P. M. A. (2001) Radiation forces in the discrete-dipole approximation, *J. Opt. Soc. Am. A*, **18**, 1944–1953.

50. Draine, B. T., and Weingartner, J. C. (1996) Radiative torques on interstellar grains. I. Superthermal spin-up, *Astrophys. J.*, **470**, 551–565.

51. Chaumet, P. C., Rahmani, A., Sentenac, A., and Bryant, G. W. (2005) Efficient computation of optical forces with the coupled dipole method, *Phys. Rev. E*, **72**, 046708/1-6.

52. Press, W. H., Flannery, B. P., Teukolsky, S. A., and Vetterling, W. T. (1990) *Numerical Recipes in C. The Art of Scientific Computing*, Cambridge University Press, New York.

53. Barrett, R., Berry, M., Chan, T. F., Demmel, J., Donato, J., Dongarra, J., Eijkhout, V., Pozo, R., Romine, C., and van der Vorst, H. A. (1994) *Templates for the Solution of Linear Systems: Building Blocks for Iterative Methods*, 2nd ed., SIAM, Philadelphia.

54. Purcell, E. M., and Pennypacker, C. R. (1973) Scattering and adsorption of light by nonspherical dielectric grains, *Astrophys. J.*, **186**, 705–714.

55. Hoekstra, A. G., Grimminck, M. D., and Sloot, P. M. A. (1998) Large scale simulations of elastic light scattering by a fast discrete dipole approximation, *Int. J. Mod. Phys. C*, **9**, 87–102.

56. Rahola, J. (1996) Solution of dense systems of linear equations in the discrete-dipole approximation, *SIAM J. Sci. Comput.*, **17**, 78–89.

57. Flatau, P. J. (1997) Improvements in the discrete-dipole approximation method of computing scattering and absorption, *Opt. Lett.*, **22**, 1205–1207.

58. Fan, Z. H., Wang, D. X., Chen, R. S., and Yung, E. K. N. (2006) The application of iterative solvers in discrete dipole approximation method for computing electromagnetic scattering, *Microw. Opt. Techn. Lett.*, **48**, 1741–1746.

59. Yurkin, M. A., Maltsev, V. P., and Hoekstra, A. G. (2007) The discrete dipole approximation for simulation of light scattering by particles much larger than the wavelength, *J. Quant. Spectrosc. Radiat. Transfer*, **106**, 546–557.

60. Lumme, K., and Rahola, J. (1994) Light-scattering by porous dust particles in the discrete-dipole approximation, *Astrophys. J.*, **425**, 653–667.
61. Freund, R. W., and Nachtigal, N. M. (1994) An implementation of the QMR method based on coupled 2-term recurrences, *SIAM J. Sci. Comput.*, **15**, 313–337.
62. Garcia, M. D., Florez, E., Suarez, A., Gonzalez, L., and Montero, G. (2005) New implementation of QMR-type algorithms, *Comp. Struct.*, **83**, 2414–2422.
63. Gutknecht, M. H., and Strakos, Z. (2000) Accuracy of two three-term and three two-term recurrences for Krylov space solvers, *SIAM J. Matrix Anal. Appl.*, **22**, 213–229.
64. Freund, R. W. (1992) Conjugate gradient-type methods for linear systems with complex symmetrical coefficient matrices, *SIAM J. Sci. Stat. Comp.*, **13**, 425–448.
65. Yurkin, M. A. and Hoekstra, A. G. (2011) The discrete-dipole-approximation code ADDA: capabilities and known limitations, *J. Quant. Spectrosc. Radiat. Transfer*, **112**, 2234–2247.
66. Rahola, J. (2000) On the eigenvalues of the volume integral operator of electromagnetic scattering, *SIAM J. Sci. Comput.*, **21**, 1740–1754.
67. Yurkin, M. A., de Kanter, D., and Hoekstra, A. G. (2010) Accuracy of the discrete dipole approximation for simulation of optical properties of gold nanoparticles, *J. Nanophotonics*, **4**, 041585/1–15.
68. Yurkin, M. A., Maltsev, V. P., and Hoekstra, A. G. (2006) Convergence of the discrete dipole approximation. II. An extrapolation technique to increase the accuracy, *J. Opt. Soc. Am. A*, **23**, 2592–2601.
69. Budko, N. V., Samokhin, A. B., and Samokhin, A. A. (2005) A generalized overrelaxation method for solving singular volume integral equations in low-frequency scattering problems, *Diff. Equat.*, **41**, 1262–1266.
70. Rakic, A. D. (1995) Algorithm for the determination of intrinsic optical constants of metal films: application to aluminum, *Appl. Optics*, **34**, 4755–4767.
71. Budko, N. V., and Samokhin, A. B. (2006) Spectrum of the volume integral operator of electromagnetic scattering, *SIAM J. Sci. Comput.*, **28**, 682–700.
72. Ayranci, I., Vaillon, R., and Selcuk, N. (2007) Performance of discrete dipole approximation for prediction of amplitude and phase of electromagnetic scattering by particles, *J. Quant. Spectrosc. Radiat. Transfer*, **103**, 83–101.

73. Flatau, P. J. (2004) Fast solvers for one dimensional light scattering in the discrete dipole approximation, *Opt. Express*, **12**, 3149–3155.
74. Flatau, P. J., Stephens, G. L., and Draine, B. T. (1990) Light-scattering by rectangular solids in the discrete-dipole approximation – a new algorithm exploiting the block-Toeplitz structure, *J. Opt. Soc. Am. A*, **7**, 593–600.
75. Goodman, J. J., Draine, B. T., and Flatau, P. J. (1991) Application of fast-Fourier-transform techniques to the discrete-dipole approximation, *Opt. Lett.*, **16**, 1198–1200.
76. Barrowes, B. E., Teixeira, F. L., and Kong, J. A. (2001) Fast algorithm for matrix-vector multiply of asymmetric multilevel block-Toeplitz matrices in 3-D scattering, *Microw. Opt. Techn. Lett.*, **31**, 28–32.
77. Greengard, L., and Rokhlin, V. (1987) A fast algorithm for particle simulations, *J. Comput. Phys.*, **73**, 325–348.
78. Rahola, J. (1996) Diagonal forms of the translation operators in the fast multipole algorithm for scattering problems, *BIT*, **36**, 333–358.
79. Hoekstra, A. G., and Sloot, P. M. A. (1994) New computational techniques to simulate light-scattering from arbitrary particles, *Part. Part. Sys. Charact.*, **11**, 189–193.
80. Taboada, J. M., Landesa, L., Obelleiro, F., Rodriguez, J. L., Bertolo, J. M., Araujo, M. G., Mourino, J. C., and Gomez, A. (2009) High scalability FMM-FFT electromagnetic solver for supercomputer systems, *IEEE Antennas Propag. Mag.*, **51**, 20–28.
81. Dembart, B., and Yip, E. (1998) The accuracy of fast multipole methods for Maxwell's equations, *IEEE Comp. Sci. Engin.*, **5**, 48–56.
82. Darve, E. (2000) The fast multipole method I: error analysis and asymptotic complexity, *SIAM J. Numer. Anal.*, **38**, 98–128.
83. Koc, S., and Chew, W. C. (2001) Multilevel fast multipole algorithm for the discrete dipole approximation, *J. Electromagnet. Wave.*, **15**, 1447–1468.
84. Singham, M. K., Singham, S. B., and Salzman, G. C. (1986) The scattering matrix for randomly oriented particles, *J. Chem. Phys.*, **85**, 3807–3815.
85. Lou, W., and Charalampopoulos, T. T. (1994) On the electromagnetic scattering and absorption of agglomerated small spherical particles, *J. Phys. D. Appl. Phys.*, **27**, 2258–2270.
86. McClain, W. M., and Ghoul, W. A. (1986) Elastic light scattering by randomly oriented macromolecules: Computation of the complete set of observables, *J. Chem. Phys.*, **84**, 6609–6622.

87. Khlebtsov, N. G. (2001) Orientational averaging of integrated cross sections in the discrete dipole method, *Opt. Spectrosc.*, **90**, 408–415.
88. Mackowski, D. W., and Mishchenko, M. I. (1996) Calculation of the T matrix and the scattering matrix for ensembles of spheres, *J. Opt. Soc. Am. A*, **13**, 2266–2278.
89. Mishchenko, M. I. (1993) Light-scattering by size shape distributions of randomly oriented axially-symmetrical particles of a size comparable to a wavelength, *Appl. Opt.*, **32**, 4652–4666.
90. Mackowski, D. W. (2002) Discrete dipole moment method for calculation of the T matrix for nonspherical particles, *J. Opt. Soc. Am. A*, **19**, 881–893.
91. Simoncini, V. (1997) A stabilized QMR version of block BICG, *SIAM J. Matrix Anal. Appl.*, **18**, 419–434.
92. Freund, R. W., and Malhotra, M. (1997) A block QMR algorithm for non-Hermitian linear systems with multiple right-hand sides, *Linear Algebra Appl.*, **254**, 119–157.
93. Boyse, W. E., and Seidl, A. A. (1996) A block QMR method for computing multiple simultaneous solutions to complex symmetric systems, *SIAM J. Sci. Comput.*, **17**, 263–274.
94. Okada, Y., Mann, I., Sano, I., and Mukai, S. (2008) Acceleration of the iterative solver in the discrete dipole approximation: Application to the orientation variation of irregularly shaped particles, *J. Quant. Spectrosc. Radiat. Transfer*, **109**, 1461–1473.
95. Draine, B. T., and Flatau, P. J. (2012) User guide for the discrete dipole approximation code DDSCAT 7.2. http://arXiv.org/abs/1202.3424.
96. Okada, Y. (2008) Efficient numerical orientation averaging of light scattering properties with a quasi-Monte-Carlo method, *J. Quant. Spectrosc. Radiat. Transfer*, **109**, 1719–1742.
97. Muinonen, K., and Zubko, E. (2006) Optimizing the discrete-dipole approximation for sequences of scatterers with identical shapes but differing sizes or refractive indices, *J. Quant. Spectrosc. Radiat. Transfer*, **100**, 288–294.
98. Yurkin, M. A., Min, M., and Hoekstra, A. G. (2010) Application of the discrete dipole approximation to very large refractive indices: Filtered coupled dipoles revived, *Phys. Rev. E*, **82**, 036703/1–12.
99. Budko, N. V., and Remis, R. F. (2004) Electromagnetic inversion using a reduced-order three-dimensional homogeneous model, *Inverse Probl.*, **20**, S17–S26.

100. Draine, B. T., and Goodman, J. J. (1993) Beyond Clausius–Mossotti – wave-propagation on a polarizable point lattice and the discrete dipole approximation, *Astrophys. J.*, **405**, 685–697.

101. Chaumet, P. C., Sentenac, A., and Rahmani, A. (2004) Coupled dipole method for scatterers with large permittivity, *Phys. Rev. E*, **70**, 036606/1–6.

102. Piller, N. B., and Martin, O. J. F. (1998) Increasing the performance of the coupled-dipole approximation: A spectral approach, *IEEE Trans. Antennas Propag.*, **46**, 1126–1137.

103. Piller, N. B. (1999) Coupled-dipole approximation for high permittivity materials, *Opt. Commun.*, **160**, 10–14.

104. Yurkin, M. A., Maltsev, V. P., and Hoekstra, A. G. (2006) Convergence of the discrete dipole approximation. I. Theoretical analysis, *J. Opt. Soc. Am. A*, **23**, 2578–2591.

105. Jackson, J. D. (1975) *Classical Electrodynamics*, 2nd edn, Wiley, New York.

106. Hage, J. I., and Greenberg, J. M. (1990) A model for the optical properties of porous grains, *Astrophys. J.*, **361**, 251–259.

107. Lakhtakia, A., and Mulholland, G. W. (1993) On 2 numerical techniques for light-scattering by dielectric agglomerated structures, *J. Res. Nat. Inst. Stand. Technol.*, **98**, 699–716.

108. Peltoniemi, J. I. (1996) Variational volume integral equation method for electromagnetic scattering by irregular grains, *J. Quant. Spectrosc. Radiat. Transfer*, **55**, 637–647.

109. Gutkowicz-Krusin, D., and Draine, B. T. (2004) Propagation of electromagnetic waves on a rectangular lattice of polarizable points. http://arxiv.org/abs/astro-ph/0403082.

110. Rahmani, A., Chaumet, P. C., and Bryant, G. W. (2002) Coupled dipole method with an exact long-wavelength limit and improved accuracy at finite frequencies, *Opt. Lett.*, **27**, 2118–2120.

111. Collinge, M. J., and Draine, B. T. (2004) Discrete-dipole approximation with polarizabilities that account for both finite wavelength and target geometry, *J. Opt. Soc. Am. A*, **21**, 2023–2028.

112. Rahmani, A., Chaumet, P. C., and Bryant, G. W. (2004) On the importance of local-field corrections for polarizable particles on a finite lattice: Application to the discrete dipole approximation, *Astrophys. J.*, **607**, 873–878.

113. Dungey, C. E., and Bohren, C. F. (1991) Light-scattering by nonspherical particles: a refinement to the coupled-dipole method, *J. Opt. Soc. Am. A*, **8**, 81–87.

114. Doyle, W. T. (1989) Optical properties of a suspension of metal spheres, *Phys. Rev. B*, **39**, 9852–9858.
115. Okamoto, H. (1995) Light scattering by clusters: the a1-term method, *Opt. Rev.*, **2**, 407–412.
116. Piller, N. B. (1997) Influence of the edge meshes on the accuracy of the coupled-dipole approximation, *Opt. Lett.*, **22**, 1674–1676.
117. McDonald, J., Golden, A., and Jennings, S. G. (2009) OpenDDA: a novel high-performance computational framework for the discrete dipole approximation, *Int. J. High Perf. Comp. Appl.*, **23**, 42–61.
118. Yurkin, M. A., Hoekstra, A. G., Brock, R. S., and Lu, J. Q. (2007) Systematic comparison of the discrete dipole approximation and the finite difference time domain method for large dielectric scatterers, *Opt. Express*, **15**, 17902–17911.
119. Penttila, A., Zubko, E., Lumme, K., Muinonen, K., Yurkin, M. A., Draine, B. T., Rahola, J., Hoekstra, A. G., and Shkuratov, Y. (2007) Comparison between discrete dipole implementations and exact techniques, *J. Quant. Spectrosc. Radiat. Transfer*, **106**, 417–436.
120. http://groups.google.com/group/adda-discuss.
121. http://openmp.org/.
122. http://www.mpi-forum.org/.
123. http://www.astro.princeton.edu/~draine/DDSCAT.html.
124. http://code.google.com/p/a-dda/.
125. Yurkin, M. A., and Hoekstra, A. G. (2012) User manual for the discrete dipole approximation code ADDA v.1.1. http://a-dda.googlecode.com/svn/tags/rel_1.1/doc/manual.pdf.
126. Flatau, P. J. and Draine, B. T. (2012) Fast near field calculations in the discrete dipole approximation for regular rectilinear grids, *Opt. Express*, **20**, 1247–1252.
127. Yurkin, M. A. and Kahnert, M. (2013) Light scattering by a cube: accuracy limits of the discrete dipole approximation and the T-matrix method, *J. Quant. Spectrosc. Radiat. Transfer*, http://dx.doi.org/10.1016/j.jqsrt.2012.10.001.
128. Pecharroman, C., Perez-Juste, J., Mata-Osoro, G., Liz-Marzán, L. M., and Mulvaney, P. (2008) Redshift of surface plasmon modes of small gold rods due to their atomic roughness and end-cap geometry, *Phys. Rev. B*, **77**, 035418/1–7.
129. Hao, E., Schatz, G., and Hupp, J. (2004) Synthesis and optical properties of anisotropic metal nanoparticles, *J. Fluoresc.*, **14**, 331–341.

130. Kooij, E. S., and Poelsema, B. (2006) Shape and size effects in the optical properties of metallic nanorods, *Phys. Chem. Chem. Phys.*, **8**, 3349–3357.

131. Shuford, K. L., Ratner, M. A., and Schatz, G. C. (2005) Multipolar excitation in triangular nanoprisms, *J. Chem. Phys.*, **123**, 114713/ 1–9.

132. Lee, K., and El-Sayed, M. A. (2005) Dependence of the enhanced optical scattering efficiency relative to that of absorption for gold metal nanorods on aspect ratio, size, end-cap shape, and medium refractive index, *J. Phys. Chem. B*, **109**, 20331–20338.

133. Yang, P., Portals, H., and Pileni, M. (2009) Identification of multipolar surface plasmon resonances in triangular silver nanoprisms with very high aspect ratios using the DDA method, *J. Phys. Chem. C*, **113**, 11597–11604.

134. Jensen, T., Kelly, L., Lazarides, A., and Schatz, G. C. (1999) Electrodynamics of noble metal nanoparticles and nanoparticle clusters, *J. Clust. Sci.*, **10**, 295–317.

135. Sosa, I. O., Noguez, C., and Barrera, R. G. (2003) Optical properties of metal nanoparticles with arbitrary shapes, *J. Phys. Chem. B*, **107**, 6269–6275.

136. Jain, P. K., Lee, K. S., El-Sayed, I. H., and El-Sayed, M. A. (2006) Calculated absorption and scattering properties of gold nanoparticles of different size, shape, and composition: applications in biological imaging and biomedicine, *J. Phys. Chem. B*, **110**, 7238–7248.

137. Mishchenko, M. I., and Travis, L. D. (1998) Capabilities and limitations of a current FORTRAN implementation of the T-matrix method for randomly oriented, rotationally symmetric scatterers, *J. Quant. Spectrosc. Radiat. Transfer*, **60**, 309–324.

Chapter 3

Size and Shape Dependence of Localized Surface Plasmon Resonances

S. D'Agostino
Dipartimento di Fisica, Università degli Studi di Pavia,
Via Bassi 6, I-27100 Pavia, Italy
National Nanotechnology Laboratory, Istituto Nanoscienze-CNR,
Via Arnesano 16, I-73100 Lecce, Italy
stefania.dagostino@unipv.it

In this chapter the influence of the structural parameters on the optical behavior of silver nanoparticles is analyzed. The absorption and scattering spectra are obtained for particles with different size and shape in the framework of the discrete dipole approximation. Radially symmetric nanoparticles, as well as finite-number faces nanoparticles or multi-tips objects are investigated under the excitation of uniform fields impacting with different polarizations and propagation directions. The optical responses can be assigned to the excitation of localized surface plasmon resonances of different order. The presented results can be used to interpret experimental measurements and/or to develop new high-performance substrates for molecular plasmonics applications.

Handbook of Molecular Plasmonics
Edited by Fabio Della Sala and Stefania D'Agostino

ISBN 978-981-4303-20-0 (Hardcover), 978-981-4303-21-7 (eBook)
www.panstanford.com

3.1 Introduction

Plasmonic resonances in metallic nanoparticles can be controlled by optimizing the nanoparticle topology, dimensions, and composition [1–3].

Among metallic particles used in plasmonics, silver nanoparticles are widely studied due to the particular optical, spectroscopic and catalytic properties of silver [4–8]. They have been largely used in catalysis [9, 10], biological labeling [11, 12], photonics [13–15] and surface-enhanced spectroscopies [16, 17]. Moreover a rich literature is now available for the synthesis of Ag nanoparticles [18, 19] (see also Chapter 10).

In this chapter, absorption and scattering efficiencies spectra will be presented for silver nanoparticles (NPs) with different shapes and dimensions. All the spectra are calculated in the discrete dipole approximation framework (see Chapter 2), with the Palik complex dielectric function $\tilde{\varepsilon}(\omega)$ [9] and in the lattice dispersion relation (LDR) prescription for the polarizability (see Sec. 2.4.3.2). For dimensions of the NPs smaller than the mean free path of the conduction electrons, the surface damping correction $\Delta\tilde{\varepsilon}_{\mathrm{SD}}$ is added to the Palik dielectric function (see Sec. 2.3), as several works have shown that the dielectric constant is strongly dependent on the size and the shape of the nanoparticle [21–23].

For both absorption and scattering, we will consider the *efficiency* as the ratio of the physical cross section on the effective geometrical one

$$Q_{abs} = C_{abs}/\left(\pi a_{eff}^2\right) \quad , \quad Q_{sca} = C_{sca}/\left(\pi a_{eff}^2\right) , \qquad (3.1)$$

where a_{eff} is the radius of a sphere with the same volume of the nanoparticle. The *extinction cross-section* and the *extinction efficiency* are defined as $C_{ext} = C_{abs} + C_{sca}$ and $Q_{ext} = Q_{abs} + Q_{sca}$, respectively.

Concerning the discretization accuracy of DDA simulations, the major part of the simulations is done with an unitary interdipolar distance ($d_{int} = 1$ nm) and this is assumed to be the default condition (unless otherwise specified). Anyway for some not cubically-shaped NPs (with curved surfaces) $d_{int} = 0.5$ nm has been used.

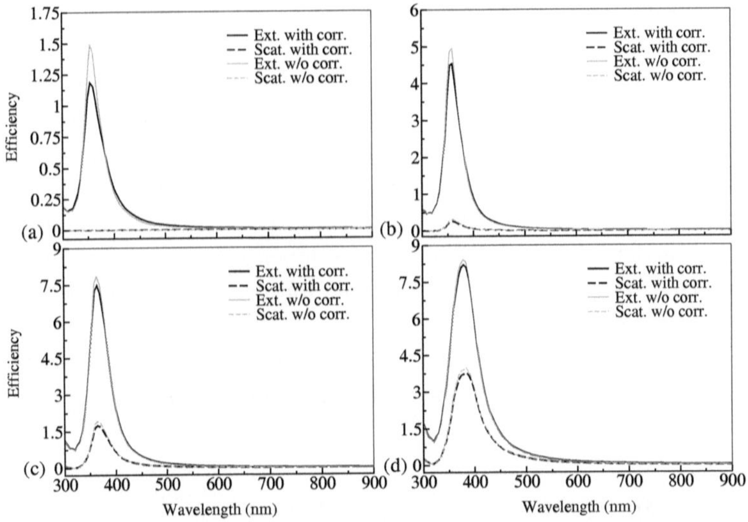

Figure 3.1 Extinction (solid lines) and scattering (dashed lines) efficiencies of four Ag spheres with small diameters: $d = 10$ nm (a), $d = 30$ nm (b), $d = 50$ nm (c) and $d = 70$ nm (d). The dielectric function of silver was corrected by including for each radius the corresponding surface damping correction (black lines). Spectra obtained without this correction (gray lines) are also reported. The interdipolar distance is $d_{int} = 0.5$ nm.

3.2 Size Dependence

3.2.1 *Small Nanoparticles and Surface Damping*

At small diameters ($d \leq 30$ nm) the absorption/extinction spectra are dominated by the dipole resonance ($l = 1$) and the incident electric field causes the conduction electrons to oscillate coherently along the E_0–polarization direction. Moreover the scattering, which is a radiative process and requires coupling between electric and magnetic fields, is negligible. Thus the extinction efficiency is very close to the absorption efficiency, as it can observed in the spectra reported in Fig. 3.1.

In all the spectra shown in Fig. 3.1 we can observe a feature that is inherent to the material properties and independent from the particle geometry: at about 320 nm the efficiencies present a minimum which is due to the vanishing dielectric function of silver in both real and imaginary parts.

Moreover, in considering the effects of the size in the optical response of a metallic nanoparticle, we must put in evidence that in the case of particles with dimensions comparable or smaller than the mean free path of its oscillating electrons (i.e. for gold and silver particles of radius $a < 10$ nm) the dielectric function of the particles becomes strongly size-dependent and the additional surface damping must be considered for a correct treatment of the resonances intensity.

For large diameters ($d \geq 50$ nm) the scattering contribution increases and the role of the surface-damping correction decreases, see Fig. 3.1c,d.

3.2.2 *Large Nanoparticles and Higher-Order Modes*

For larger nanoparticles the excitation processes are more complex than the simple electronic cloud oscillating all over the nanoparticle volume parallel to the incident field (dipole or $l = 1$ plasmonic mode): higher-order modes must be included into the theoretical treatment to explain the extinction spectra.

By coming back to Sec. 1.4.1, we can generalize the quasi-static solution of a nanosphere for an external potential of the type:

$$\Phi_l(r, \theta, \phi) = -C_l r^l Y_{l,0}(\theta, \phi), \tag{3.2}$$

where $Y_{l,0}$ is a spherical harmonic of degree l and order $m = 0$, and C_l is a normalization coefficient. The first spherical harmonics are: $Y_{1,0} = \sqrt{3/(4\pi)} \cos(\theta)$, $Y_{2,0} = \sqrt{5/(16\pi)}(3\cos^2(\theta) - 1), \ldots$. Solving the Laplace equation with the appropriate boundary condition it can be shown that the electrostatic potential outside the nanosphere is [24]:

$$\Phi_l^{out}(r, \theta, \phi) = \Phi_l(r, \theta, \phi) + C_l \tilde{\alpha}_l(\omega) \frac{1}{r^{l+1}} Y_{l,0}(\theta, \phi), \tag{3.3}$$

where the effective *multipolar polarizability* of the sphere is:

$$\tilde{\alpha}_l(\omega) = R^{2l+1} \frac{\tilde{\varepsilon}_p(\omega) - \varepsilon_{out}}{\tilde{\varepsilon}_p(\omega) + \frac{l+1}{l}\varepsilon_{out}}. \tag{3.4}$$

For $l = 1$, Eq. (3.3) and Eq. (3.4) recover Eq. (1.190) and Eq. (1.193) of Sec. 1.4.1, respectively.

For large particles, it is not possible to assume that the external field from an incident plane-wave is uniform inside the nanosphere, due to the phase factor, which is neglected when $\lambda \gg R$. As any external field can be expanded in multipoles, then we have that multipoles higher than the dipole can be excited by an external incident plane-wave. The *quadrupole* term ($l = 2$) is often very important to describe the extinction and scattering spectra: in spherical nanoparticles, quadrupolar modes become non-negligible already at particle radii around 30 nm, and they become significant in the visible range only for particle radii larger than 90. Octupolar modes are relevant only for much bigger particles.

The quadrupolar polarizability is:

$$\tilde{\alpha}_Q(\omega) = R^5 \frac{\tilde{\varepsilon}_p(\omega) - \varepsilon_{out}}{\tilde{\varepsilon}_p(\omega) + \frac{3}{2}\varepsilon_{out}}, \tag{3.5}$$

and contains the factor 3/2 in the the denominator, while for the dipolar case (see Eq. (1.193)) the corresponding factor is 2. Thus the quadrupolar resonance occurs, using the free-electron Drude dielectric constant of Eq. (1.122), at:

$$\omega_{\mathrm{LSP,Q}} = \frac{\omega_p}{\sqrt{1 + \frac{3}{2}\varepsilon_{out}}}, \tag{3.6}$$

i.e. at *higher* energies than the dipolar one (see Eq. (1.197)). The quadrupolar mode thus appears as a small shoulder at the shorter wavelength side in the optical absorption spectrum, as shown in Fig. 3.2, where we report the extinction and the scattering efficiencies of silver spheres with diameter from $d = 90$ nm to $d = 180$ nm.

Here results are obtained in the discrete dipole approximation, so they should give the exact solution to the electromagnetic problem and spectra identical to the Mie theory. Anyway, like already shown in literature [1], the quasistatic approach is able to retain the most important features in the spectra for spheres of these sizes.

While all the spectra are still dominated by the dipolar plasmon mode, the contribution from the quadrupolar mode increases as nanoparticle size increases: the quadrupolar polarizability in fact increases with the fifth power of the particles size, against the third power for the dipolar one.

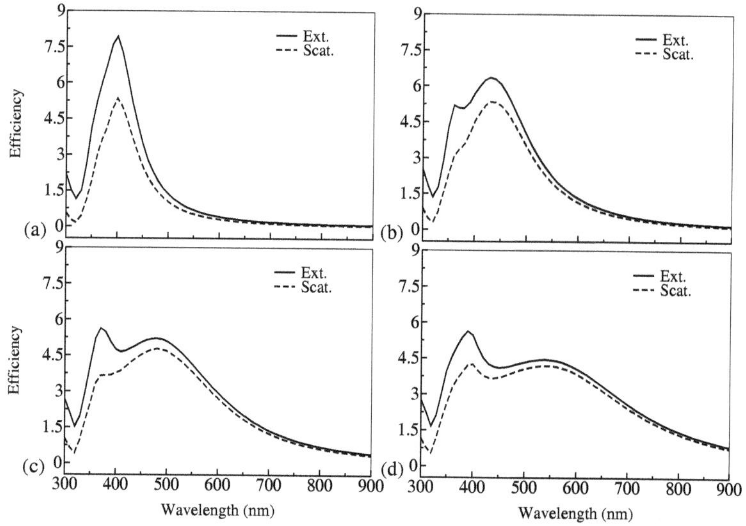

Figure 3.2 Extinction (solid lines) and scattering (dashed lines) efficiencies of four Ag spheres with large diameters: $d = 90$ nm (a), $d = 120$ nm (b), $d = 150$ nm (c) and $d = 180$ nm (d).

In general, when multipolar modes are present, the charge distributions generated by these modes give rise to additional electric fields that in turn cause a red-shift in the dipolar mode. In some cases (e.g. in small gold nanoparticles) the quadrupolar mode, being weak in intensity, is hidden by the interband transitions. In Ag nanoparticles instead, such overlap does not occur and therefore the quadrupolar mode is already evident at smaller sizes compared to gold.

For higher order modes, the resonance condition for the multipolar polarizability in Eq. (3.4) approaches to $\tilde{\varepsilon}_p(\omega) = -\varepsilon_{out}$, which is indeed the resonance condition of the planar surface plasmon polaritons of Sec. 1.3, i.e. plasmon resonance for a flat metal surface. This is because in multipolar modes fast charge oscillations practically cancel out the interaction between distant charges, therefore each small region on the surface of the metal behaves as in a planar bulk metal. These considerations however need to be taken with care, since at the large sizes at which multipolar modes are active, the optical behavior is additionally influenced by retardation and scattering.

In addition to multipolar modes, when the nanoparticles radius is larger than 30 nm, *retardation effects* influence the optical response. One of these retardation effects, the *energy-shifting effect*, arises because the particle size is not any more negligible with respect to the wavelength of the radiation and the wavelength of the incoming radiation cannot be considered as infinite, hence the field is not homogeneous everywhere inside the particle. More precisely, these effects appear already when the diameter of the particle is around 1/10 of the wavelength of the radiation in the medium surrounding the particle $\lambda_{out} = \lambda/\sqrt{\varepsilon_{out}}$. An elementary explanation of this effect can be sought by considering that when a dipole mode is excited in a spherical particle, the distance between opposite charges can be approximated to the nanoparticle diameter. Hence one side of the particle will feel any change occurring at the opposite side at the particle with a phase retardation equal to $2\pi \cdot 2R/\lambda_{out}$. The oscillation period of the dipole mode increases in order to take such retardation into account, which is equivalent to saying that the plasmon peak associated to a dipole mode shifts to the red as the particle size increases. When multipolar modes are involved, the distance between opposite charges on the surface of the particle is smaller than the particle diameter, and the phase retardation is smaller than in the dipolar mode. Indeed, the phase retardation in a spherical metal particle scales roughly as $2\pi \cdot 2R/(l \cdot \lambda_{out})$ with $l = 1, 2, 3, \ldots$ for dipole, quadrupole, octupole mode, and so on. Dipole modes therefore are influenced more by the particle size, due to retardation effects, than higher order modes.

Another retardation effect is due to *radiation scattering*, which becomes important already for particles radii bigger than 20 nm, and is the dominant effect for particles radii above 100 nm. In these larger particles the electrons are accelerated as a consequence of the electromagnetic field generated by the incident radiation. Therefore, they also start radiating in all directions, which causes them to lose energy, i.e. part of the energy of the plasmonic oscillations is converted into photons (see the radiative damping in 1.6.4). This leads to a broadening in the surface plasmon resonances and to a decrease in their intensities. Hence, as opposed to small particles, in large particles their interaction with the incoming light is dominated by radiative processes in which energy is absorbed by the incoming light and transformed again into photons.

3.3 Shape Dependence

The intensity, number, and width of the surface plasmon resonances are very sensitive to the nanoparticle morphology. For metal nanoparticles whose shape deviates from that of a sphere, exact analytic solutions of the extinction cross section are not available. Here we will describe, examples of absorption and scattering spectra obtained by discretizing the targets within the DDA framework.

3.3.1 *Radially Symmetric Nanoparticles*

3.3.1.1 Ellipsoids and spheroids

A particularly convenient case of particle whose shape deviates from that of a sphere, is that of an ellipsoid with semiaxes $a_1 \leq a_2 \leq a_3$ and $\frac{x^2}{a_1^2} + \frac{y^2}{a_2^2} + \frac{z^2}{a_3^2} = 1$, for which analytical solutions are known in the quasi-static limit [25].

From the solution of the scattering problem in this case, the polarizability turns out to be a diagonal tensor whose components along the three axes are:

$$\tilde{\alpha}_i = 4\pi a_1 a_2 a_3 \frac{\tilde{\varepsilon}_p(\omega) - \varepsilon_{out}}{3\varepsilon_{out} + 3L_i\left(\tilde{\varepsilon}_p(\omega) - \varepsilon_{out}\right)}, \tag{3.7}$$

where $\tilde{\varepsilon}_p(\omega)$ and ε_{out} are the dielectric functions of the nanoparticle and the external medium, respectively, and L_i is a geometrical factor which can be obtained by the integral:

$$L_i = \frac{a_1 a_2 a_3}{2} \int_0^\infty dq \frac{1}{\left(a_i^2 + q\right) f(q)}, \tag{3.8}$$

with $f(q) = \sqrt{\left(q + a_1^2\right)\left(q + a_2^2\right)\left(q + a_3^2\right)}$.

A particular class of ellipsoids are spheroids which are axially symmetric ellipsoidal particles in which two axes are equal: the two major ones for oblate spheroids ($a_1 = a_2$) and the two minor ones for prolate spheroids ($a_2 = a_3$).

In the case of spheroids, each of the two polarizabilities has a pole corresponding to the frequency vanishing the denominator (Eq. 3.7). These frequencies correspond to the resonant frequencies at which the conduction electrons move in phase along the major and minor axis, generating the longitudinal and transverse mode, respectively [2, 3, 26, 27]. Like it can be observed by comparing Eq. 3.7 with Eq. 1.193, the longitudinal resonance turns out to be more red-shifted, if compared to the resonance of a sphere with the same volume, as the aspect ratio is larger. Anyway Eq. (3.7) retains its validity only if the major axis is much smaller than the radiation wavelength.

In Fig. 3.3 the extinction efficiency is reported for a prolate spheroid excited with an electric field $\tilde{\mathbf{E}}^0$ polarized at $\pi/4$ respect to the y-axis. As we can observe, the dipolar resonance mode splits into one mode at small wavelengths (the *transverse mode*) and one mode at longer wavelengths (the *longitudinal mode*); this in agreement with Eq. (3.7) which for axially symmetric ellipsoidal particles provides two resonances.

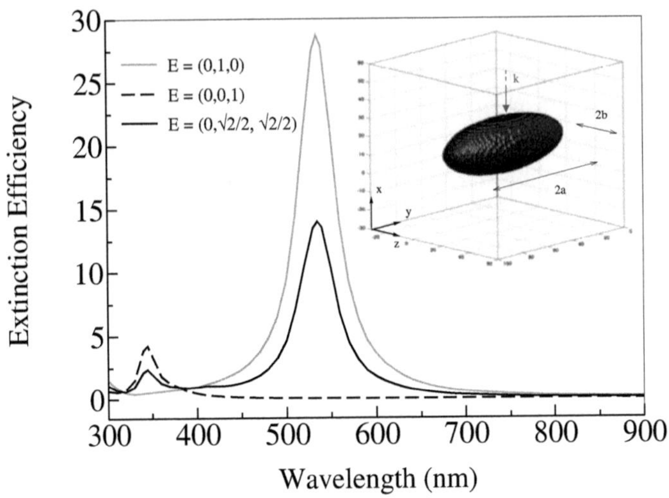

Figure 3.3 Extinction efficiencies of an Ag prolate spheroid with $a = 45$ nm and $b = 15$ nm. The three curves represent the responses of the nanoparticle for an impacting plane-wave with different polarization directions: along the minor axis $\tilde{\mathbf{E}}^0 = (0, 0, 1)$, along the major axis $\tilde{\mathbf{E}}^0 = (0, 1, 0)$ and with both the components $\tilde{\mathbf{E}}^0 = (0, \sqrt{2}/2, \sqrt{2}/2)$.

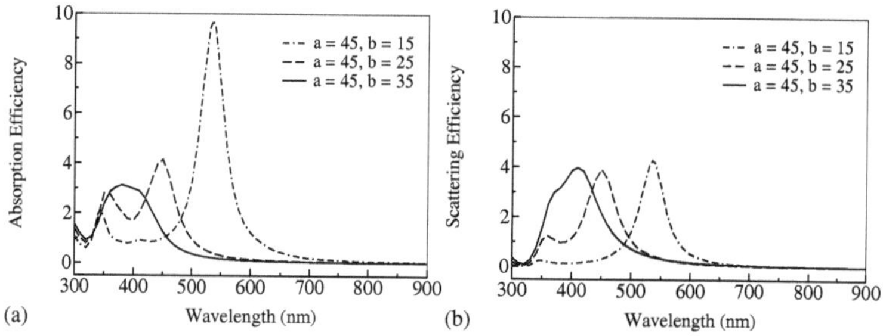

Figure 3.4 Absorption (a) and scattering spectra (b) of three Ag spheroids with different aspect ratio $f = a/b$. The incident field polarization is fixed at $\pi/4$ respect to the y-axis, $\tilde{\mathbf{E}}^0 = (0, \sqrt{2}/2, \sqrt{2}/2)$.

It is interesting to observe that the splitting between the resonances along the two axes is much more evident for larger aspect ratio (panel (a) in Fig. 3.4).

3.3.1.2 Cylinders and disks

Metallic nanorods and nanowires present the plasmon band of the metal splitted in two: the longitudinal plasmon band, deriving from charge oscillation along the long axis of the particle, and the transverse plasmon band, corresponding to the resonant motion of the conduction electrons along the short axis of the particle [1, 2, 28–30, 32]. The bigger the aspect ratio $s = h/d$ (length divided by width) the more red-shifted the longitudinal plasmon band [33]. In addition weaker bands, corresponding to higher-order modes can also be found [1].

Au nanorods (NRs) are particularly useful for biological applications since their near-field and far-field optical properties can be tuned controllably throughout the NIR water window by varying their geometry [34]. Metallic nanorods and nanowires can have enhanced electric fields at the tips compared to spheres, which would lead to increased SERS activity [1, 35–38]. Moreover it has been shown that Ag cylinders or nanorods exhibit extraordinary electrocatalytic activities towards the electro-reduction process, especially in the case of benzyl chloride [29].

Nanorods can be modeled as cylinders: in Fig. 3.5 the optical response of Ag cylinders excited with a plane-wave polarized

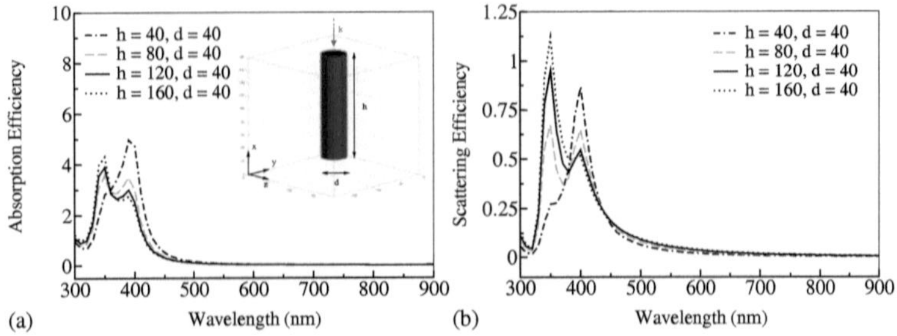

Figure 3.5 Absorption (a) and scattering spectra (b) of Ag cylinders with the same diameter d and different heights h. The incident field polarization is directed along the y-axis (the plane-wave propagates parallel to h).

parallel to the circular base (in-plane polarization) is reported, as a function of the height. This kind of ideal polarization could prove interesting also experimentally, being these particles alignable in several kinds of assemblies or patterns [30, 31]. For both the absorption and scattering spectra we can observe that also for a in-plane polarization of the incident field we have the appearance of two resonances with an opposite behavior at increasing aspect ratio: by increasing s, the intensity of Q_{abs} and Q_{sca} increases in the first peak and decreases in the second one. The fact that no shifts are present by changing h clarify the nature of the peaks: these are both resonances due to in-plane oscillations of electrons or transverse modes (dipolar and quadrupolar).

By moving from nanorods to nanodisks (see Fig. 3.6), for which $s < 1$ being $h < d$, a strong red-shift of the resonances appears in the spectra by increasing the dimensions of the base (d). This is due to retardation effects experienced by radiation inside a scatterer by enlarging its dimensions along the polarization direction. Moreover for a thin disk ($h = 10$ nm) only the dipolar resonance ($l = 1$) seems to be generated by an in-plane polarized incident field.

Of course, an average of the target orientations with respect to the incident field should produce richer spectra: an out-plane polarization should excite also longitudinal modes (along h direction) and this in agreement with the experimental measurements already done on solutions of this kind of nanoparticles [30, 39].

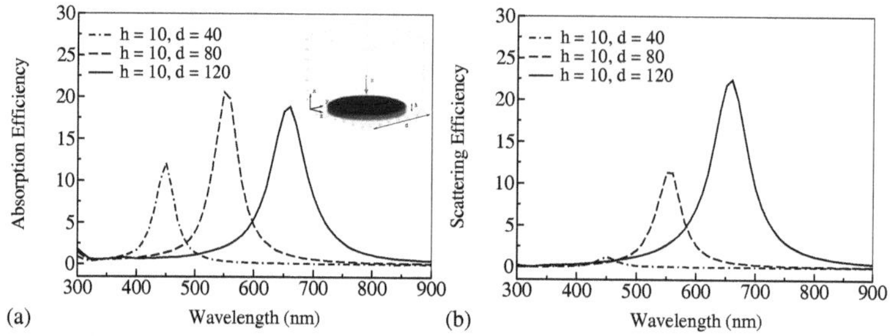

Figure 3.6 Absorption (a) and scattering spectra (b) of Ag disks with the same height h and different diameters d. The incident field polarization is directed along the y-axis (the plane-wave propagates parallel to h).

3.3.2 *Prisms and Finite-Number Facets Nanoparticles*

All the polyhedral particles show more than one resonance in contrast to sphere, where a single peak can be found.

Moreover, such kind of nanoparticles with a finite number of faces and sharper vertices respect to spherical objects, such as cubes [2, 3, 26, 27, 40–42], triangular prisms [1, 43, 44], tetrahedra [1, 2], pyramids [45] etc., exhibit tunable optical properties at visible and NIR wavelengths that depend on particle orientation with respect to the polarization direction. Such unique geometrical and optical characteristics make possible several applications in imaging and sensing so that polyhedral scatterers are widely investigated.

3.3.2.1 Cubes

Among polyhedra, the most studied target, for its spectra with a rich and articolated structure of peaks, is the cube [2, 3, 26, 27, 40, 41].

In Fig. 3.7 the absorption and scattering efficiencies are reported for three cubes with several dimensions (40 nm, 60 nm and 80 nm), excited with plane-waves polarized along one side direction (y-axis).

For wavelengths between 320 and 450 nm, the contribution to the extinction spectrum ($Q_{ext} = Q_{abs} + Q_{sca}$) comes from light absorption and scattering in the same manner, whereas for 450 nm $< \lambda_{exc} <$ 700 nm, it comes mainly from light scattering effects. The Q_{abs} spectra, as well as the scattering ones for λ_{exc} between 320 nm and 450 nm, show several resonances, inherent to the cubic

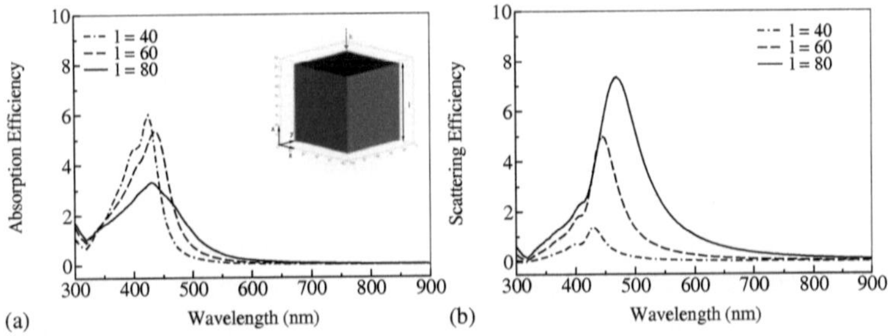

Figure 3.7 Absorption (a) and scattering spectra (b) of Ag cubes with several side dimensions *l*. The incident field polarization is directed along the *y*-axis.

geometry [46]. The major peaks correspond, for the three cubes in Fig. 3.7, to the dipolar resonances, whereas the peaks at smaller wavelengths are due to higher multipolar excitations.

More in general, for silver cubes the absorption spectra has been shown in literature [26] to have a rich structure of peaks. The presence of six planar faces and eight vertices makes more complex the charge distribution in a cube than in a sphere, thus different ways to distribute charge exist and a larger number of resonant modes appears, even in the quasi-static limit. At least six different modes are usually observed in literature [40], being the dipolar charge distribution the one that contributes more to the spectra and the best distinguishable (see Fig. 3.7). These six resonances were found by Fuchs [46] who calculated nine SPRs where only six of them account for more than the 96% of the spectrum.

For smaller cubes, the peaks are sharper. On the contrary, at increasing dimensions, the radiation damping effects, which red-shift the spectrum, make the peaks broader and reduce the intensity. For large cubes, the peaks at higher energies, due to high-multipolar contributions, become important, anyway the main features of the spectra due to the particular geometry, survive.

The rich structure of peaks can be better observed when the background medium is not vacuum ($n > 1$). For instance, when a silver nanocube is immersed in a medium with $n = 1.47$, it is possible to identify the six SPRs more easily than when $n = 1$ [3].

3.3.2.2 Triangular prisms

Another shape often found in literature for its interesting spectra is that of triangular nanoparticles or *nanoprisms*. For these nanoparticles, the quadrupole resonances seem to be much more intense than for spheroidal ones of equivalent dimensions, this being a surprising feature or peculiarity of such kind of targets [1].

Experimentally it is possible to use wet chemistry methods to make triangular prisms of silver that have atomically flat surfaces and do not aggregate [47] and plasmon excitation inside metallic prisms has been shown to play a critical role in the synthesis of these structures and highlight the power and versatility of plasmon-mediated synthetic strategies [43].

The optical responses of Ag triangular-shaped nanoprisms with several side and height dimensions are here reported in Fig. 3.8 and in Fig. 3.9. Here the incident field is assumed to propagate along the height direction and to be polarized along y-direction.

For small prisms, ($h = 10$ nm) the absorption and scattering spectra present only two peaks: these features could be attributed to the in-plane dipole resonance (longer wavelengths) and to the in-plane quadrupole resonance [44]. By increasing the side of the triangle, a clear red-shift of these two peaks appears (see Fig. 3.8) and this is also confirmed experimentally by ultraviolet–visible–near-infrared spectroscopy measurements [44]. This effect can be understood by considering the fact that by enlarging the area of

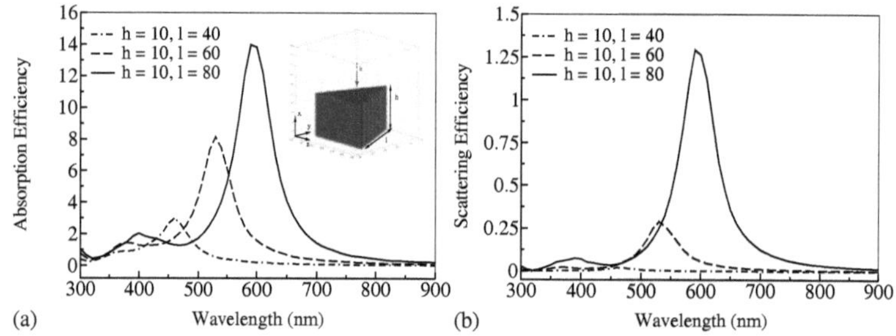

Figure 3.8 Absorption (a) and scattering spectra (b) of Ag triangular prisms with the same height h and several side dimensions l. The incident field polarization is directed along the y-axis (the plane-wave propagates parallel to h).

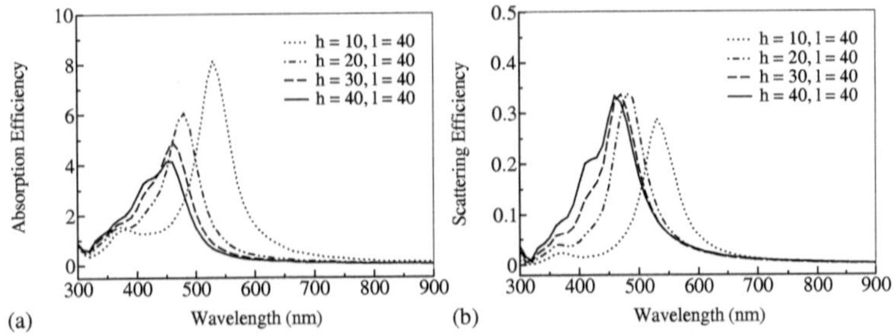

Figure 3.9 Absorption (a) and scattering spectra (b) of Ag triangular prisms with the same side dimension *l* and different heights *h*. The incident field polarization is directed along the *y*-axis (the plane-wave propagates parallel to *h*).

metal along the polarization plane, the amplitude of the resonant oscillation changes and with this also the restoring force acting on the oscillating electronic clouds.

Kelly *et al.* showed that two peaks also appear when Ag prisms are excited with the other two polarizations (short axis or long axis along the side) and that if for polarization along *h* we obtain two small resonances for small wavelengths, the two polarizations along the long axes of the prism (side and perpendicular bisector) closely resemble each other [1]. For both the in-plane and out-plane polarizations, the physical nature of the peaks can be clarified by the E-field enhancement contours maps calculated for the two major resonances of the spectra [1]: for the resonance at higher wavelengths most of the polarization inside the metal is parallel to the symmetry axis, this resembling a dipole resonance, while for smaller wavelengths, half the polarization points up and half down, as for a quadrupole resonance [1].

By increasing the path of radiation inside metal (10 nm $<$ h $<$ 40 nm), higher order modes can be excited and the spectra appear much more complicated (see Fig. 3.9). For thick prisms the resulting electronic oscillations lose their clear character (dipole or quadrupole).

Moreover, it should be noted how by moving from thin (h $=$ 10 nm) to thick prisms (h $=$ 40 nm) we have a blue-shift of the spectra (Fig. 3.9).

3.3.3 *Polyhedral Nanoparticles*

The location of the resonances of polyhedral nanoparticles with different shapes has been largely explored in literature by Noguez *et al.* [3, 27, 40, 41]. It is well known that particles with fewer faces and sharper vertices show resonances in a larger spectral range and that in the cases of truncated nanoparticles the main features of the spectra are blue-shifted, overlapping higher-order resonances, and thus, increasing their width. Here the major aspects of the optical response of these nanoparticles are summarized but readers are demanded to the original works for further details [3, 27, 40, 41].

3.3.3.1 Truncated cubes

In this subsection the optical behavior of different truncated cubes will be analyzed and compared to those of a cube and of a sphere of the same dimensions. By starting from cubes with a size length l, truncated cubes can be constructed by edging all the vertices of the cube by $l \times r$, where the number r, which satisfy the condition $0 < r \leq 1/2$, characterize the particular truncation. All the truncated cubes present thus six octagons and eight triangles [3, 40, 41].

In Fig. 3.10, the extinction efficiencies of truncated nanocubes with r from 1/8 to 1/2 (cuboctahedron) are shown in solid lines. Targets were generated by recurring Noguez's fortran routines available online [48]. The spectra for spherical (dotted line) and cubic (dashed line) NPs are also included for comparison. It is observed that even for the smallest truncation of $r = 1/8$, the spectrum is very sensitive to the morphology. The locations of the dipolar and quadrupolar resonances are now very close, such that, only one wide peak is observed around 420 nm, while the structure below 370 nm remains almost identical to the spectrum of the cube. The same trend is observed for larger truncations, and from Fig. 3.10 it can be observed that we can find that as r increases: (i) the main peak shifts to higher energies, (ii) the resonances at smaller wavelength are hidden due to the near primary resonance, and (iii) the main resonance becomes broader [41].

For comparison, we have included in Fig. 3.10 the spectrum of a silver nanosphere with 40 nm diameter. We can conclude that as

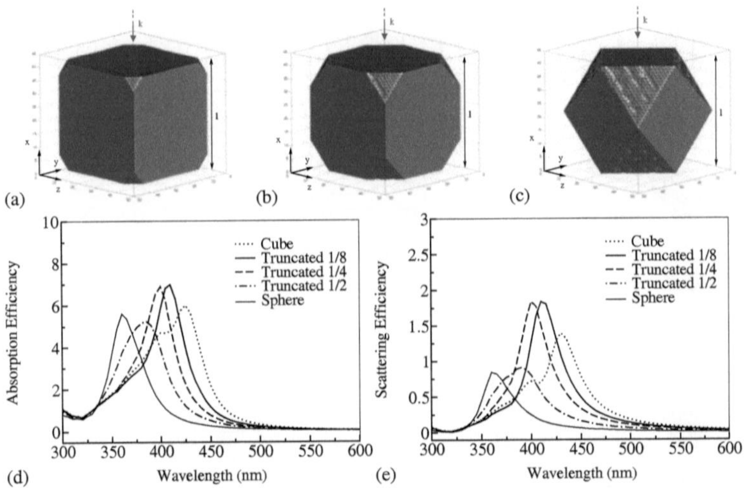

Figure 3.10 Absorption (d) and scattering spectra (e) of Ag truncated cubes with the same length of the cube side $l = 40$ nm and different truncation factor r: $r = 1/8$ (a), $r = 1/4$ (b) and $r = 1/2$ (c). The incident field polarization is directed along the y-axis.

the number of faces of the NP increases the energy range of the spectrum becomes smaller, the main resonance is blue-shifted and fewer resonances are observed. Therefore, with small modifications of the morphology, it is possible tune SPRs at different wavelengths.

3.3.3.2 Rounded, regular and marks decahedral nanoparticles

Another interesting morphology appearing for metal NPs by using several synthesis methods [49–53] is the decahedron or pentagonal bipyramid. The regular decahedron is made up with ten planar triangular faces which resemble two pentagons, as seen in Fig. 3.11, where three different orientations are reported ($1y$, $1x$ and $1z$). Being the decahedron an asymmetric particle, it presents an optical response strongly dependent by the orientation of the incident electromagnetic field.

In experiments, the most common shapes have truncated vertices and correspond to particles named *Marks decahedron* and the *round decahedron*.

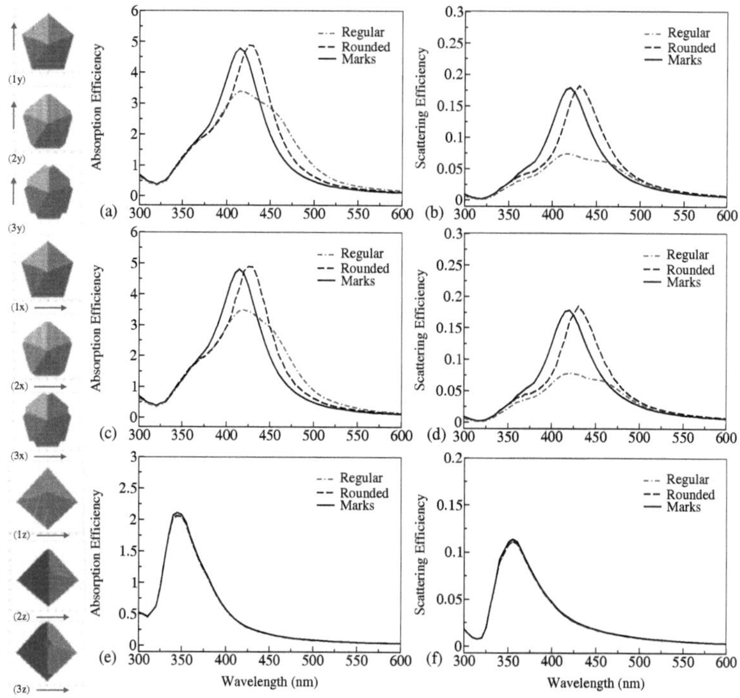

Figure 3.11 Absorption and scattering spectra of an Ag regular decahedron (panels 1x, 1y, and 1z), a rounded decahedron (panels 2x, 2y, and 2z) and a Marks decahedra (panels 3x, 3y, and 3z), all with a 40 nm-side. The polarization direction is assumed to be: parallel to the pentagonal motif and oriented along vertices $\tilde{\mathbf{E}}_y^0 = (0, 1, 0)$ in panels (a) and (b), parallel to the pentagonal motif and oriented along edges $\tilde{\mathbf{E}}_z^0 = (0, 0, 1)$ in panels (c) and (d), and perpendicular to the pantagonal motif $\tilde{\mathbf{E}}_x^0 = (1, 0, 0)$ in panels (e) and (f).

The first structure [54] contains extra $\{111\}$ facets and turns out to be quite stable. In particularly clean growth conditions (weak interactions with substrates), it results one of the predominant shapes for the size interval taken into account. An alternative way to describe the Marks decahedron is as a regular decahedron, which has truncations on its facets, as shown in Fig. 3.11 (3y, 3x and 3z).

The other type of decahedral particle, which is often observed, corresponds to the round pentagonal particle. An example of these particles is shown in Fig. 3.11 ($2y$, $2x$ and $2z$). This kind of particle, frequently formed when colloidal growth methods are used [50], can be seen as a truncated decahedron in which the truncation assumes a minimum possible value generating a contrast reduction in the borders. As already done in the works of González *et al.* [41], here we discuss the optical response of a regular decahedron, a Marks decahedra with a truncation of $r = 1/6$ and a rounded decahedron with a truncation of $r = 1/8$.

In Fig. 3.11, we show the three different orientations of the decahedrons with respect to the incident electromagnetic field: in panels (a-b) the electromagnetic field $\tilde{\mathbf{E}}^0$ is parallel to the pentagonal motif and $\tilde{\mathbf{E}}^0$ is along the vertices, in panels (c-d) $\tilde{\mathbf{E}}^0$ is also parallel but is along the edges, while in panels (e-f) $\tilde{\mathbf{E}}^0$ is perpendicular to the pentagonal motif. Targets were generated, like for truncated cubes, by recurring to free fortran routines [48].

In Fig. 3.11 we can note a large anisotropy of the spectra when the light incidence is such that the electric field is parallel and perpendicular to the pentagonal motif. When the electric field is parallel to the pentagon, the corresponding spectra are very wide with a FWHM (full width at half maximum) of about 90 nm, and a maximum at about 425 nm. Except near the maxima, where small differences are observed, the spectra for both parallel polarizations are almost identical. On the other hand, when the electric field is perpendicular to the pentagon, the spectrum shows a maximum at about 350 nm which is more than two times less intense and presents a FWHM of about 45 nm.

In Fig. 3.11 we observe for the perpendicular polarization, that the optical response of the regular decahedron does not change for small truncations, in both cases, the Marks and rounded decahedra. For both parallel polarizations, the spectra of the truncated decahedra show differences with respect to the regular ones. The observed effects are similar to those already seen in the case of truncated cubes (Fig. 3.10): as a result of the increment of the faces, the main resonance is blue-shifted and its FWHM decreases. Finally, as for more regularly shaped nanoparticles, also for such kind of NPs the spectra show a red-shift with increasing size as a consequence of the radiation effects.

3.3.4 *Multi-Tips Objects*

By starting from the maximum truncation of a regular decahedron which corresponds to a star decahedron ($r = 1/2$), in this section we will have a look at the optical behavior of multi-tips objects with the unique surface geometry consisting of several corners or thorns protruding from a central body.

Surface imperfections and roughness have a fundamental role in experiments involving field enhancement phenomena, like surface-enhanced Raman scattering (SERS) [55] and metal-enhanced fluorescence (MEF) [56] measurements. In fluorescence-based DNA detection systems, for example, the detection performance seems to be strongly dependent on the metallic substrate roughness, and in response to identical amounts of analyte, nanostructured Ag substrates provide a fluorescent sensor response a factor of 10 larger than the one achieved from planar Au substrates [56]. Moreover, rough particles show higher SERS performances, surface texturing or corrugation being responsible for significant shift and broadening of the localized surface plasmon resonances [55].

In light of this empirical evidence, in the past years roughness has passed from being an undesired artifact of the processes of synthesis to representing one of the most pursued tasks so that chemical etching methods have been developed to control roughening of the surface topography of nanoparticles [57] and complex multi-tips structures have been synthesized [58–62]. Many of these metallic scatterers have optical resonances in the near-infrared spectral window, which are able to penetrate deeply within biological tissues and do not cause photochemical damage differently from nanospheres [63]. Among these new-generation plasmonic objects, particular attention is reserved for nanourchins: nanoparticles which consist of numerous bumps or spines protruding from a central spherical core.

3.3.4.1 Stars

In Fig. 3.12 the absorption and scattering spectra for an Ag star (see the insets of Fig. 3.12), obtained by symmetrically edging the five vertices of a 40 nm-side regular decahedron (see Fig. 3.11) with

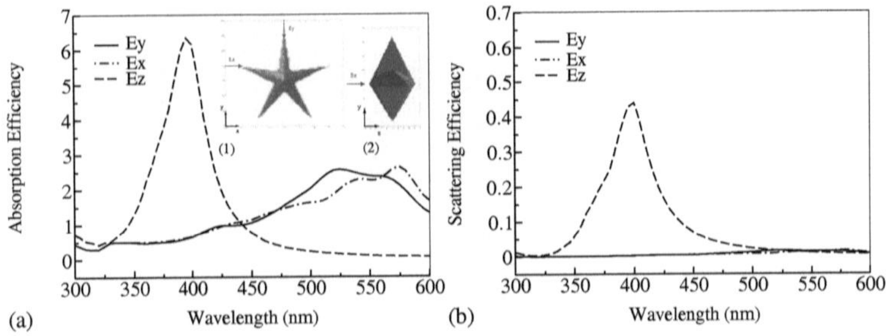

Figure 3.12 Absorption (a) and scattering spectra (b) of an Ag star 40 nm-size decahedron for the three incident field polarizations seen in Fig. 3.11. The insets (1) and (2) show respectively the top and the lateral views of the star.

the maximum truncation of $r = 1/2$, are reported. This target was produced by a free fortran routine [48].

The two Q_{abs} spectra obtained by exciting the nanoparticle with an incident field polarized parallel to the plane of the pentagonal motif result both broad and quite similar, while for the perpendicular polarization, the response of the star decahedron is completely different since it shows a sharp resonance at 400 nm with a FWHM of 50 nm. For the three polarizations, it should be noted, how the presence of the tips make this particle very peculiar and different from the other decahedron truncations (Fig. 3.11).

By comparing the spectra of a star (Fig. 3.12) with those of a regular decahedron (Fig. 3.11) we can observe that the main resonance is red-shifted to around 550 nm, and the spectra become very wide, due to the presence of several resonances. We can underline some similarities in the optical responses of a star and a cube: in both a large number of resonances make spectra quite broad. Moreover being characterized by sharp vertices, these two morphologies, can present very inhomogeneous charge distributions at the tips, this producing strong field enhancement phenomena [46].

3.3.4.2 Urchins-like nanoparticles

For urchins-like nanoparticles the effects of branches dimensions and of their mutual interactions on the optical response of Ag

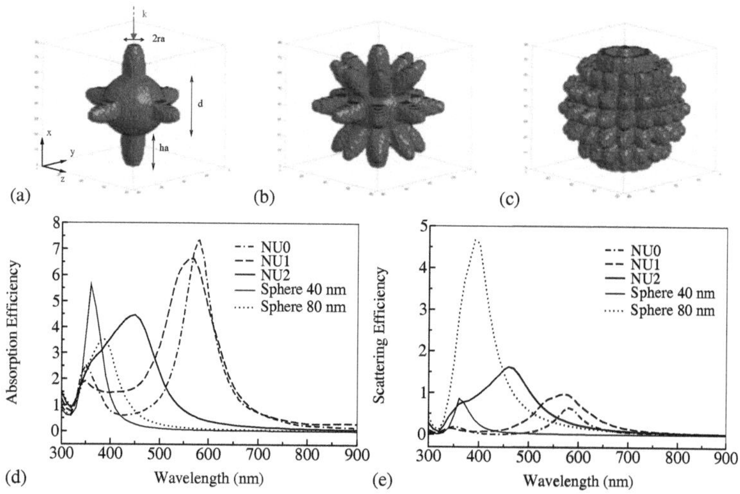

Figure 3.13 Absorption (a) and scattering spectra (b) of three Ag nanourchins with different coverages: NU0 (a), NU1 (b) and NU2 (c). The incident field polarization is directed along the y-axis.

nanourchins have been recently clarified [64]. The changes in the absorption and scattering spectra are analyzed by varying the number of aculei (or coverage), their width, and their height and are reported, respectively, in Fig. 3.13, Fig. 3.14 and Fig. 3.15.

The first Ag multi-branches structures presented in Fig. 3.13 are 40 nm Ag spherical cores with multi hemispheroidal Ag spines protruding from it, each with a semimajor (h_a) and a semiminor (r_a) axis, respectively, of 20 and 6 nm. The angular spacing between the spines is taken responding to the formula $\Delta\Theta = \pi/2^{(c+1)}$ with c = 0, 1, 2. The structures NU0 with $\Delta\Theta = \pi/2$, NU1 with $\Delta\Theta = \pi/4$ and NU2 with $\Delta\Theta = \pi/8$ are schematized respectively in the panels (a), (b) and (c) of Fig. 3.13.

These structures are excited with a plane-wave propagating in the x direction and polarized along the y direction. DDA tests done on the NU0 structure revealed no significant modifications of the spectra for a 3D rotation of the scatterer and this in agreement with the fact that multi-tips objects like these have a major probability of favorable alignment with the incident radiation, as compared to simpler shapes such as nanorods [65].

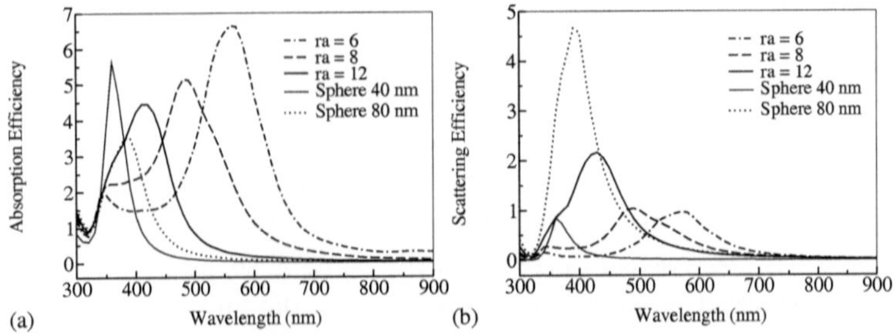

Figure 3.14 Absorption (a) and scattering spectra (b) of three Ag nanourchins with different aculei radii: $r_a = 6$ nm, $r_a = 8$ nm, and $r_a = 12$ nm. The incident field polarization is directed along the y-axis.

By going from a 40 nm sphere to a NU0 particle, the absorption efficiency peak increases, but, by moving to higher coverages, it decreases monotonically. Concerning the most important features of the spectra and their physical origin, the absorption of NU0 presents a small resonance at $\lambda = 350$ nm and a large one at $\lambda = 580$ nm. These two characteristic bands resemble the transverse surface plasmon resonance (SPR) and the longitudinal one usually appearing in the spectra of elongate particles [66, 67]. Here, as for ellipsoidal-shaped nanoparticles [7, 66, 67], the dipolar resonance mode is splitted into one mode at longer wavelengths (the longitudinal mode) and one mode at shorter wavelengths (the transverse mode). A detailed analysis done on the

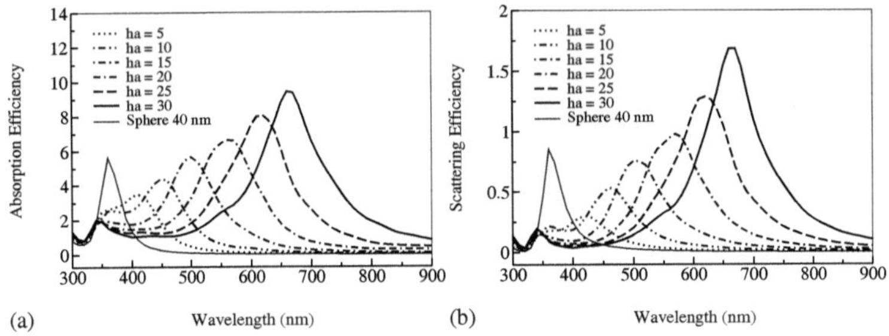

Figure 3.15 Absorption (a) and scattering spectra (b) of three Ag nanourchins with different aculei heights: $h_a = 5$ nm, $h_a = 10$ nm, $h_a = 15$ nm, $h_a = 20$ nm, $h_a = 25$ nm and $h_a = 30$ nm. The incident field polarization is directed along the y-axis.

absorption by separating the contribution of the branches dipoles from that deriving from the core [64], revealed how in the three cases the spines result to be much more efficient absorbers (per unit volume) than the core. The spines undergo a much more significant polarization than that inside the nanourchin, and this polarization affects the regions of the core near the bases of the spines. A progressive decrease of the spine absorption efficiencies at increasing coverage can be attributed to the mutual interactions between the spines and the mutual depolarization effects that they induce on the dipoles.

Concerning the spines radius effects, the spectra of structures with a $\Delta\Theta = \pi/4$ coverage (e.g., like the structure NU1, see Fig. 3.13b), a spine height of 20 nm, and a radius changing in the range $6 \div 12$ nm are reported in Fig. 3.14.

By looking at panel (a) of Fig. 3.14, we can observe a decrease of the absorption efficiencies by going from $r_a = 6$ nm to $r_a = 12$ nm, until reaching in the limit of $r_a \rightarrow \infty$, the spectra of a 80 nm diameter Ag sphere. Again, by increasing the radius of the spines, the interactions among these become larger so that phenomena of mutual depolarization can appear. In addition, the peaks blue-shift and become narrower at increasing r_a, but the intensity does not change substantially and this is in agreement with the idea that the dipole is mainly confined in the y−aligned spheroid. DDA tests performed on "equivalent" isolated Ag spheroids demonstrate that the peak blue-shifts at increasing r_a.

Finally, the absorption and scattering spectra of structures with a $\Delta\Theta = \pi/4$ coverage, $r_a = 6$ nm and different spine heights (h_a in the range $5 \div 30$ nm) are reported in Fig. 3.15. The plots clearly show that noninteracting high protusions produce huge enhancements in the absorption or in the scattering efficiency with respect to a 40 nm diameter sphere. By going from a 40 nm sphere to a $h_a =$ 30 nm NU1, for the resonant conditions, we obtain a 10% increase of the absorption. This can be justified by the fact that these kinds of nanourchins NUs have a large surface/volume ratio. In addition the spectra can be fully tuned over the entire visible region varying h_a: this is due to the fact that by enlarging the height of the spines, the spatial amplitude of the resonant oscillation changes. This result makes these objects surprisingly interesting for plasmonic applications if compared with smooth surface nanoparticles and this is in agreement with the early evidence now available [62, 63].

3.3.5 *Void Nanoparticles*

In this last section we will have a look at hollow nanoparticles with a particular attention focused on metal-dielectric *core-shell.*

It should be taken into account that the values of the absorption and scattering efficiencies, here reported, will result much higher than those related to the previous nanoparticles, being the efficiency given by the ratio of the physical cross section on the effective geometrical one. This kind of normalization to the effective geometrical cross section is useful to give us informations on the most useful or interesting shapes to be used in experiments for the same quantity of metal.

Moreover the dielectric function of the void metallic nanoparticles here presented, are corrected by including the surface damping term (see Sec. 2.3) into the bulk dielectric function.

3.3.5.1 Nanoshells

Among metallic nanoparticles which have recently attracted considerable attention, a particularly interesting structure is the metallic *nanoshell* which consists of a metallic layer grown over a solid dielectric core [34, 57, 68–76].

Metallic nanoshells can be very useful subwavelength optical components whose surface plasmons can be tuned across a large spectral range (from VIS to NIR spectral regions) by simply handling the dimensions of their core and shell layers [77]. These nanoparticles represent a promising system for enhancing light-matter interaction, since they support both sphere-like and cavity-like modes whose properties can be engineered and tuned as a function of geometrical parameters [78]. It has been shown that the tunability of nanoshell plasmons can be extended also into the FIR spectral region by recurring to the "nano-matryushka" geometry [79, 80]. Additionally, nanoshell plasmon resonances are very sensitive to the embedding dielectric medium [81].

The possibility offered by nanoshells to systematically control the optical frequency fields at the nanometer scale, represents a promising tool for applications such as strong-coupling interactions [78] and surface-enhanced spectroscopies [82–84]. This tunability has already enabled several applications, such as resonant photooxidation inhibitors [86], optical triggers for drug delivery implants [87, 88], environmental sensors [89], and Raman sensors [85]. In biology and medicine nanoshells result useful for whole-blood immunoassays [90], optically triggerable drug delivery [88], high-resolution bioimaging [91], and photothermal cancer therapy [92].

To have an intuitive understanding of the nanoshells tunability, we can recur to the plasmon hybridization theory [68, 69]. In this framework, the nanoshell resonances are viewed as due to the hybridization of the plasmon resonances of a nanosphere and a nanocavity. The plasmons of nanoshells can thus be understood as resulting from the interaction of the sphere and cavity plasmons. This interaction or hybridization gives origin to a splitting of the plasmon resonances into a lower energy symmetric (bonding) plasmon $|\omega_-\rangle$ and a higher energy antisymmetric (antibonding) plasmon $|\omega_+\rangle$. It has been shown that, for each $l-$order mode of the sphere, the frequencies of the deriving modes of the core-shell depend on the aspect ratio of the shell ($s = d_2/d_1$) (where d_1 and d_2 are respectively the outer and the inner diameters) and on the plasma frequency of the bulk metal ω_B [68, 71] according to the formula:

$$\omega_{l\pm}^2 = \frac{\omega_B^2}{2}\left[1 \pm \frac{1}{2l+1}\sqrt{1 + 4l\,(l+1)\left(\frac{d_2}{d_1}\right)^{2l+1}}\,\right]. \tag{3.9}$$

If the shell is thin, the interaction between sphere and cavity modes is strong and a large splitting of the plasmons appears. On the contrary, for a thick shell, the interaction becomes weak and a small splitting of the plasmons occurs. In the limit of vanishing aspect ratio the interaction between the cavity and sphere plasmons vanishes, and the shell plasmon resonances asynptotically approach the isolated sphere and the cavity plasmon resonances [78]. Moreover, to interpretate the feature in experimentally obtained extinction spectra, the different capability of these modes to couple to far-field radiation must be considered. The bonding plasmon has a larger dipole moment than the antibonding one. It assumes the semblance

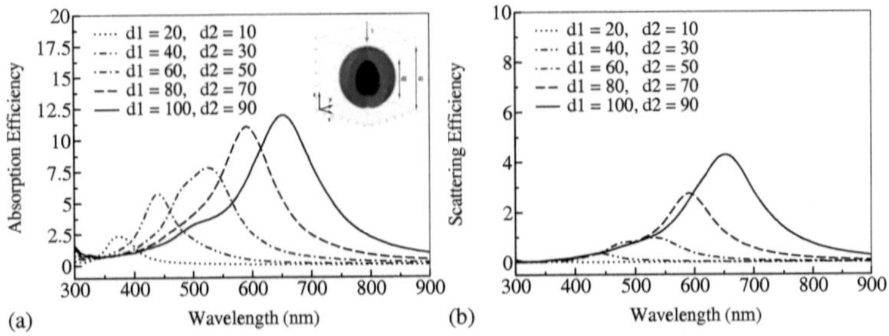

Figure 3.16 Absorption (a) and scattering spectra (b) of Ag nanoshells with several external (d_1) and internal (d_2) diameters, but with the same shell-thickness (5 nm). The incident field polarization is directed along the y-axis.

of a "bright" mode and, different from the antibonding one ("dark"), easily couples to incident electromagnetic radiation.

In Fig. 3.16 and Fig. 3.17 the spectra of Ag nanoshells (with an empty core) with external diameter d_1 and internal diameter d_2 are reported. In Fig. 3.16 the shell-thickness ($t = (d_1 - d_2)/2$) varies, while in Fig. 3.17 d_1 is fixed and d_2 is modified.

By comparing panels (a) and (b) in Fig. 3.16, we can observe that the absorption efficiencies for thin shells (small t) are much more interesting than the scattering ones. On the contrary, this relationship between the gain and the loss due to such kind of metallic NPs seem to invert for larger t (see Fig. 3.17).

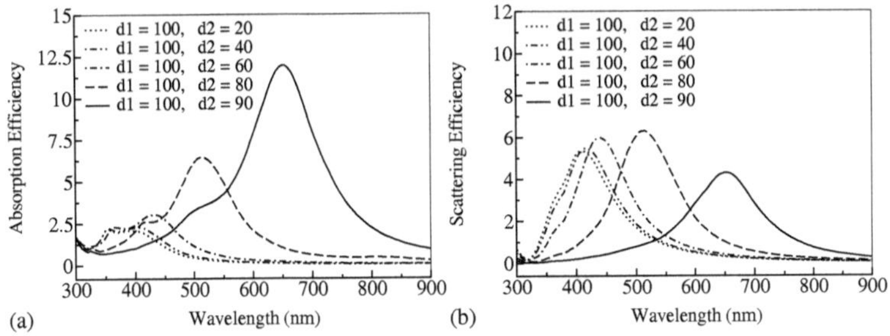

Figure 3.17 Absorption (a) and scattering spectra (b) of Ag nanoshells with the same external diameter (d_1) and different shell-thickness. The incident field polarization is directed along the y-axis.

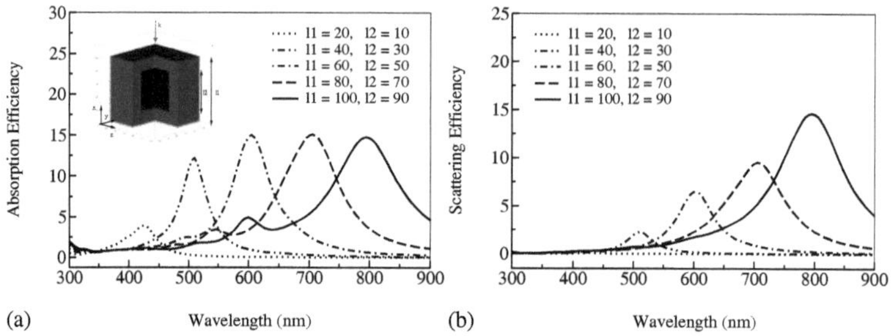

Figure 3.18 Absorption (a) and scattering spectra (b) of Ag void cubes with several external (l_1) and internal (l_2) side dimensions but with the same wall-thickness (5 nm). The incident field polarization is directed along the y-axis.

Moreover, by decreasing the shell thickness we can note (see Fig. 3.17) a red-shift of the plasmon peak and this in agreement with extinction spectroscopies measurements on Au core-shells [57, 73]. In the quasi-static limit, where the size of the nanoshell is much smaller than the wavelength of light, the plasmon resonance energies are determined by the aspect ratio [78, 93]. As the aspect ratio is increased, the plasmon resonance shifts to longer wavelengths [94, 95].

As previosly observed, the splitting of the plasmons appears only if the shell is thin enough. The two characteristic resonances which are the "fingerprints" of this kind of core-shell systems, can be identified in several measured extinction spectra [34, 57, 69, 70, 72–76].

For further details see Chapter 6.

3.3.5.2 Void cubes

Here we consider cubically shaped void nanoparticles (see inset of Fig. 3.18), with external side l_1 and internal side l_2: the thickness of the metallic "wall" is $t = (l_1 - l_2)/2$.

If by changing the effective volume and keeping t constant the spectra change only with a red-shift of the peaks (see Fig. 3.18), by changing t, we can observe important differences either in the profile of the curves or in the intensities of the resonances (see

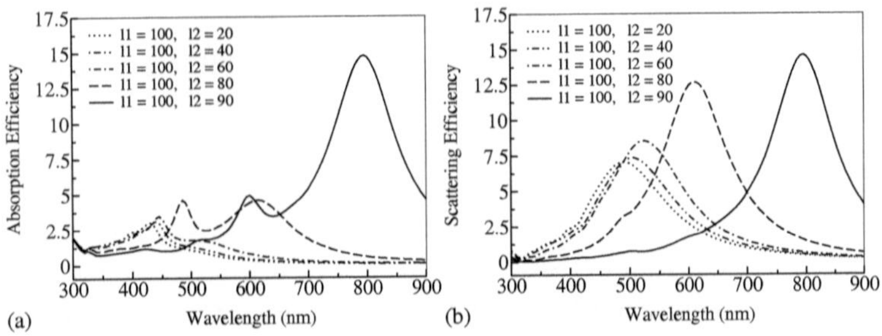

Figure 3.19 Absorption (a) and scattering spectra (b) of Ag void cubes with the same external side dimension (l_1) and different wall-thickness. The incident field polarization is directed along the y-axis.

Fig. 3.19). Moreover the singular property of void nanospheres to give origin to resonant states (bonding and anti-bonding) deriving from the hybridization of the resonances of the particle and the cavity structure, seems to appear also in the spectra of void-nanocubes (see Fig. 3.19).

For the same dimensions of void Ag spheres, void Ag cubes show more separate peaks covering wider spectral ranges.

For these, as well as for all the other shapes presented in this chapter, a direct comparison of the theoretical results with the experimental measurements of the optical properties of suspended nanoparticles would require an averaging procedure over the orientations with respect to the incident field polarization as well as a wide distribution of sizes and shapes. This averaging procedure might smooth out the main relevant features of the spectra associated to the size and shape of the nanoparticles.

3.4 Conclusions

In the last fifteen years we have assisted to a huge spread of nanomaterials in material science. In this panorama metallic nanoparticles (NPs) retain a fundamental role. The reason for this success lies in their interesting and often favorable optical behavior. The main characteristic of metallic NPs is the surface plasmon absorption which has 10^5–10^6 larger cross sections than ordinary molecular chromophores. Moreover their resonant frequencies can also be tuned from visible to near infrared acting on the NPs shape, size or environment. The spectral sensitivity, defined as the relative shift in resonance wavelength with respect to the refractive index change of surrounding materials, is dependent on the surface plasmon absorption position and width, which in turn depends on metal type and on particles shape. Silver is more sensible than gold to surface environment. Nanorods and nanoshells are better than spheres being the plasmon absorption bands more sharp, intense and tunable. The surface plasmons excitation in silver or gold NPs, find thus large applications in biology and medicine, due to sensitive spectral response to local environment and ease of monitoring the light signal due to the strong light absorption and scattering. In particular nanoparticles with significant SPR absorption (like triangular prism or truncated cubes) can be successfully used in the selective laser photothermal therapy of cancer or ablation of tumoral tissues. On the other hand, the strongly enhanced SPR scattering from nanoparticles, like large spheres or cubes, makes them useful as bright optical tags or labels for molecular specific biological imaging and detection using simple dark-field optical microscopy. Particles with sharp edge and cornes (like stars or nanourchins) are particularly interesting for surface field enhancement phenomena and applications (as strong-coupling) because fields near curved surface are intrinsically higher for higher curvatures. Moreover a variety of non-linear optical phenomena (such as second or third harmonic generation) can derive from the strong field enhancement caused by plasmon oscillation on metal particles surface.

Acknowledgments

This work was funded by the ERC Starting Grant FP7 Project DEDOM, Grant Agreement No. 207441 and by Fondazione Cariplo under Project no. 2010-0523. The author thanks Prof. Lucio Claudio Andreani for the openness and support which allowed work to be realized.

References

1. Kelly, K. L., Coronado, E., Zhao, L. L., Schatz, G. C. (2003) The optical properties of metal nanoparticles: The influence of size, shape, and dielectric environment, *J. Phys. Chem. B*, **107**, 668–677.
2. Sosa, I. O., Noguez, C., and Barrera, R. G. (2003) Optical properties of metal nanoparticles with arbitrary shapes, *J. Phys. Chem. B*, **107**, 6269–6275.
3. Noguez, C. (2007) Surface plasmons on metal nanoparticles: The influence of shape and physical environment, *J. Phys. Chem. C*, **111**, 3806–3819.
4. Pastoriza-Santos, I., and Liz-Marzán, L. M. (2002) Synthesis of silver nanoprisms in DMF, *Nano Lett.*, **2**, 903–905.
5. Jin, R. C., Cao, Y. W., Mirkin, C. A., Kelley, K. L., Schatz, G. C., and Zheng, J. G. (2001) Photoinduced conversion of silver nanospheres to nanoprisms, *Science*, **294**, 1901–1903.
6. Kreibig, U., and Vollmer, M. (eds) (1995) *Optical Properties of Metal Clusters*, Springer-Verlag, Berlin.
7. Billaud, P., Huntzinger, J.-R., Cottancin, E., Lerm, J., Pellarin, M., Arnaud, L., Broyer, M., Del Fatti, N., and Valle, F. (2007) Optical extinction spectroscopy of single silver nanoparticles, *Eur. Phys. J. D*, **43**, 271–274.
8. Jiang, Z.-J., Liu, C.-Y., and Sun, L.-W. (2005) Catalytic properties of silver nanoparticles supported on silica spheres, *J. Phys. Chem. B*, **109**, 1730–1735.
9. Schmid, G. (ed.) (1994) *Clusters & Colloidal: From Theory to Application*, VCH, Weinheim.
10. Xu, R., Wang, D., Zhang, J., and Li, Y. (2006) Shape-Dependent catalytic activity of silver nanoparticles for the oxidation of styrene, *Chem.-Asian J.*, **1**, 888–893.
11. Kim, J.-H., Kim, J.-S., Choi, H., Lee, S., Jun, B., Yu, K., Kuk, E., Kim, Y., Jeong, D. H., Cho, M., and Lee, Y. (2006) Nanoparticle probes with surface-enhanced Raman spectroscopic tags for cellular cancer targeting, *Anal. Chem.*, **78**, 6967–6973.
12. Malicka, J., Gryczynski, I., Fang, J., and Lakowicz, J. R. (2002) Photostability of Cy3 and Cy5-labeled DNA in the presence of metallic silver particles, *J. Fluoresc.*, **12**, 439–447.
13. Xia, Y., and Halas, N. J. (2005) Shape-controlled synthesis and surface plasmonic properties of metallic nanostructures, *MRS Bull.*, **30**, 338–344.

14. Barnes, W. L., Dereux, A., and Ebbese, T. W. (2003) Surface plasmon subwavelength optics, *Nature*, **424**, 824–830.
15. Krenn, J. R. (2003) Nanoparticle waveguides: Watching energy transfer, *Nat. Mater.*, **2**, 210–211.
16. Meyerab, M. W., and Smith, E. A. (2011) Optimization of silver nanoparticles for surface-enhanced Raman spectroscopy of structurally diverse analytes using visible and near-infrared excitation, *Analyst*, **136**, 3542–3549.
17. Chen, Y., Munechika, K., and Ginger, D. S. (2007) Dependence of fluorescence intensity on the spectral overlap between fluorophores and plasmon resonant single silver nanoparticles, *Nano Lett.*, **7**, 690–696.
18. Cobley, C. M., Skrabalak, S. E., Campbell, D. J., and Xia, Y. (2009) Shape-controlled synthesis of silver nanoparticles for plasmonic and sensing applications, *Plasmonics*, **4**, 171–179.
19. Evanoff, D. D., Jr., and Chumanov, G. (2005) Synthesis and optical properties of silver nanoparticles and arrays, *Chem. Phys. Chem.*, **6**, 1221–1231.
20. Palik, E. D. (ed.) (1985) *Handbook of Optical Constants of Solids*, Academic Press, New York.
21. Coronado, E. A., and Schatz, G. C. (2003) Surface plasmon broadening for arbitrary shape nanoparticles: A geometrical probability approach, *J. Chem. Phys.*, **119**(7), 3926–3934.
22. Liu, M., and Guyot-Sionnest, P. (2004) Synthesis and optical characterization of Au/Ag core/shell nanorods, *J. Phys. Chem. B*, **108**, 5882–5888.
23. Kreibig, U. (1974) Electronic properties of small silver particles: the optical constants and their temperature dependence, *J. Phys. F: Met. Phys.*, **4**, 999–1014.
24. Fuchs, R., and Claro, F. (1987) Multipolar response of small metallic spheres: Nonlocal theory, *Phys. Rev. B*, **35**, 3722–3727.
25. Maier, S. A. (2007) *Plasmonics: Fundamentals and Applications*, Springer Science & Business Media LLC, United Kingdom.
26. Noguez, C. (2005) Optical properties of isolated and supported metal nanoparticles, *Opt. Mater.*, **27**, 1204–1211.
27. Gonzalez, A. L., Reyes-Esqueda, J. A., and Noguez, C. (2008) Optical properties of elongated noble metal nanoparticles, *J. Phys. Chem. C*, **112**, 7356–7362.

28. Nikoobakht, B., and El-Sayed, M. A. (2003) Preparation and growth mechanism of gold nanorods (NRs) using seed-mediated growth method., *Chem. Mater.*, **15**, 1957–1962.

29. Maiyalagan, T. (2008) Synthesis, characterization and electrocatalytic activity of silver nanorods towards the reduction of benzyl chloride, *Applied Catal. A-Gen.*, **340**, 191–195.

30. Murphy, C. J., Sau, T. K., Gole, A. M., Orendorff, C. J., Gao, J., Gou, L., Hunyadi, S. E., and Li, T. (2005) Anisotropic metal nanoparticles: synthesis, assembly, and optical applications, *J. Phys. Chem. B*, **109**, 13857–13870.

31. Nicewarner-Pen, S. R., Freeman, R. G., Reiss, B. D., He, L., Pen, D. J., Walton, I. D., Cromer, R., Keating, C. D., and Natan, M. J. (2001) Submicrometer metallic barcodes, *Science*, **294**, 137–141.

32. Hornyak, G. L., Patrissi, C. J., and Martin, C. R. (1997) Fabrication, characterization, and optical properties of gold nanoparticle/porous alumina composites: The nonscattering Maxwell-Garnett limit, *J. Phys. Chem. B*, **101**, 1548–1555.

33. Jana, N. R., Gearheart, L., and Murphy, C. J. (2001) Wet chemical synthesis of high aspect ratio cylindrical gold nanorods, *J. Phys. Chem. B*, **105**, 4065–4067.

34. Bardhan, R., Grady, N. K., Cole, J. R., Joshi, A., and Halas, N. J. (2009) Fluorescence enhancement by Au nanostructures: Nanoshells and nanorods, *ACS Nano*, **3**, 744–752.

35. Imura, K., Nagahara, T., and Okamoto, H. (2004) Plasmon mode imaging of single gold nanorods, *J. Am. Chem. Soc.*, **126**, 12730–12731.

36. Tao, A., Kim, F., Hess, C., Goldberger, J., He, R., Sun, Y., Xia, Y., and Yang, P. (2003) Langmuir-Blodgett silver nanowire monolayers for molecular sensing using surface-enhanced Raman spectroscopy, *Nano Lett.*, **3**, 1229–1233.

37. Hanarp, P., Kall, M., and Sutherland, D. S. (2003) Optical properties of short range ordered arrays of nanometer gold disks prepared by colloidal lithography, *J. Phys. Chem. B*, **107**, 5768–5772.

38. Nikoobakht, B., and El-Sayed, M. A. (2003) Surface-enhanced Raman scattering studies on aggregated gold nanorods, *J. Phys. Chem. A*, **107**, 3372–3378.

39. Maillard, M., Giorgio, S., and Pileni, M.-P. (2003) Tuning the size of silver nanodisks with similar aspect ratios: Synthesis and optical properties, *J. Phys. Chem. B*, **107**, 2466–2470.

40. Gonzalez, A. L., and Noguez, C. (2007) Optical properties of silver nanoparticles, *Phys. Status Solidi C*, **4**, 4118–4126.
41. González, A. L., and Noguez, C. (2007) Influence of morphology on the optical properties of metal nanoparticles, *J. Comput. Theor. Nanosci.*, **4**, 231–238.
42. Sun, Y., and Xia Y. (2002) Shape-controlled synthesis of gold and silver nanoparticles, *Science*, **298**, 2176–2179.
43. Xue, C., Millstone, J. E., Li, S., and Mirkin, C. A. (2007) Plasmon-driven synthesis of triangular core-shell nanoprisms from gold seeds, *Angew. Chem.-Ger. Edit.*, **119**, 8588–8591.
44. Jin, R. C., Cao, Y. C., Hao, E. C., Metraux, G. S., Schatz, G. C., and Mirkin, C. A. (2003) Controlling anisotropic nanoparticle growth through plasmon excitation, *Nature*, **425**, 487–490.
45. Lee, J., Hasan, W., Stender, C. L., and Odom, T. W. (2008) Pyramids: A platform for designing multifunctional plasmonic particles, *Accounts Chem. Res.*, **41**, 1762–1771.
46. Fuchs, R. (1975) Theory of the optical properties of ionic crystal cubes, *Phys. Rev. B*, **11**, 1732–1740.
47. Jin, R., Cao, Y., Mirkin, C. A., Kelly, K. L., Schatz, G. C., and Zheng, J.-G. (2001) Photoinduced conversion of silver nanospheres to nanoprisms, *Science*, **294**, 1901–1903.
48. Noguez, C., Routines available at http://www.fisica.unam.mx/cecilia/codes.html.
49. Wang, Z. L. (2000) Transmission electron microscopy of shape-controlled nanocrystals and their assemblies, *J. Phys. Chem. B*, **104**, 1153–1175.
50. Yacamn, M. J., Ascencio, J. A., Liu, H. B., and Gardea-Torresdey, J. (2001) Structure shape and stability of nanometric sized particles, *J. Vac. Sci. Technol. B*, **19**, 1091–1103.
51. Kuo, C.-H., Chiang, T.-F., Chen, L.-J., and Huang, M. H. (2004) Synthesis of highly faceted pentagonal- and hexagonal-shaped gold nanoparticles with controlled sizes by sodium dodecyl sulfate, *Langmuir*, **20**, 7820–7824.
52. Wei, G., Zhou, H., Liu, Z., Song, Y., Wang, L., Sun, L., and Li, Z. (2005) One-step synthesis of silver nanoparticles, nanorods, and nanowires on the surface of DNA network, *J. Phys. Chem. B*, **109**, 8738–8743.
53. Nilius, N., Ernst, N., and Freund, H.-J. (2000) Photon emission spectroscopy of individual oxide-supported silver clusters in a scanning tunneling microscope, *Phys. Rev. Lett.*, **84**, 3994–3997.

54. Marks, L. D. (1994) Experimental studies of small particle structures, *Rep. Prog. Phys.*, **57**, 603–649.

55. Rodriguez-Fernandez, J., Funston, A. M., Perez-Juste, J., Alvarez-Puebla, R. A., Liz-Marzán, L. M., and Mulvaney, P. (2009) The effect of surface roughness on the plasmonic response of individual sub-micron gold spheres, *Phys. Chem. Chem. Phys.*, **11**, 5909–5914.

56. Peng, H.-I., Strohsahl, C. M., Leach, K. E., Krauss, T. D., and Miller, B. L. (2009) Label-free DNA detection on nanostructured Ag surfaces, *ACS Nano*, **3**, 2265–2273.

57. Wang, H., Goodrich, G. P., Tam, F., Oubre, C., Nordlander, P., and Halas, N. J. (2005) Controlled texturing modifies the surface topography and plasmonic properties of Au nanoshells, *J. Phys. Chem. B*, **109**, 11083–11087.

58. Chen, S. H., Wang, Z. L., Ballato, J., Foulger, S. H., and Carroll, D. L. (2003) Monopod, bipod, tripod, and tetrapod gold nanocrystals, *J. Am. Chem. Soc.*, **125**, 16186–16187.

59. Sau, T. K., and Murphy, C. J. (2004) Room temperature, high-yield synthesis of multiple shapes of gold nanoparticles in aqueous solution, *J. Am. Chem. Soc.*, **126**, 8648–8649.

60. Hao, E., Bailey, R. C., Schatz, G. C., Hupp, J. T., and Li, S. Y. (2004) Synthesis and optical properties of branched gold nanocrystals, *Nano Lett.*, **4**, 327–330.

61. Kuo, C.-H., and Huang, M. H. (2005) Synthesis of branched gold nanocrystals by a seeding growth approach, *Langmuir*, **21**, 2012–2016.

62. Bakr, O. M., Wunsch, B.-H., and Stellacci, F. (2006) High-yield synthesis of multi-branched urchin-like gold nanoparticles, *Chem. Mater.*, **18**, 3297–3301.

63. Hutter, E., Boridy, S., Labrecque, S., Lalancette, H., Kriz, J., Winnik, F. M., and Maysinger, D. (2010) Microglial response to gold nanoparticles, *ACS Nano*, **4**(5), 2595–2606.

64. D'Agostino, S., and Della Sala, F. (2011) Silver nanourchins in plasmonics: Theoretical investigation on the optical properties of the branches, *J. Phys. Chem. C*, **115**, 11934–11940.

65. Liberman, V., Rothschil, M., Bakr, O. M., and Stellacci, F. (2010) Optical limiting with complex plasmonic nanoparticles, *J. Opt.*, **12**, 065001/1-10.

66. Gonzlez, A. L., Reyes-Esqueda, J. A., and Noguez, C. (2008) Optical properties of elongated noble metal nanoparticles, *J. Phys. Chem. C*, **112**, 7356–7362.

67. Noguez, C. (2007) Surface plasmons on metal nanoparticles: The influence of shape and physical environment, *J. Phys. Chem. C*, **111**, 3806–3819.
68. Prodan, E., Radloff, C., Halas, N. J., and Nordlander, P. (2003) A hybridization model for the plasmon response of complex nanostructures, *Science*, **302**, 419–422.
69. Lal, S., Grady, N. K., Kundu, J., Levin, C. S., Lassiterde, J. B., and Halas, N. J. (2008) Tailoring plasmonic substrates for surface-enhanced spectroscopies, *Chem. Soc. Rev.*, **37**, 898–911.
70. Oldenburg, J., Jackson, J. B., Westcott, S. L., and Halas, N. J. (1999) Infrared extinction properties of gold nanoshells, *Appl. Phys. Lett.*, **75**, 2897–2899.
71. Steele, J. M., Grady, N. K., Nordlander, P., and Halas, N. J. (2007) Plasmon hybridization in complex nanostructures, in *Surface Plasmon Nanophotonics* (ed. Brongersma, M. L., and Kik, P. G.), Springer, The Netherlands, pp. 183–196.
72. Prodan, E., Nordlander, P., and Halas, N. J. (2003) Electronic structure and optical properties of gold nanoshells, *Nano Lett.*, **3**, 1411–1415.
73. Bardhan, R., Grady, N. K., Ali, T., and Halas, N. J. (2010) Metallic nanoshells with semiconductor cores: Optical characteristics modified by core medium properties, *ACS NANO* , **4**, 6169–6179.
74. Bardhan, R., Mukherjee, S., Mirin, N. A., Levit, S. D., Nordlander, P., and Halas, N. J. (2010) Nanosphere-in-a-nanoshell: A simple nanomatryushka, *J. Phys. Chem. C*, **114**, 7378–7383.
75. Lombardini, R., Acevedo, R., Halas, N. J., and Johnson, B. R. (2010) Plasmonic enhancement of Raman optical activity in molecules near metal nanoshells: Theoretical comparison of circular polarization methods, *J. Phys. Chem. C*, **114**, 7390–7400.
76. Levin, C. S., Hofmann, C., Ali, T. A., Kelly, A. T., Morosan, E., Nordlander, P., Whitmire, K. H., and Halas, N. J. (2009) Magnetic-plasmonic core-shell nanoparticles, *ACS Nano*, **3**, 1379–1388.
77. Oldenburg, S. J., Averitt, R. D., Westcott, S. L., Halas, N. J. (1998) Nanoengineering of optical resonances, *Chem. Phys. Lett.*, **288**, 243–247.
78. Alpeggiani, F., D'Agostino, S., and Andreani, L. C. (2012) Surface plasmons and strong light-matter coupling in metallic nanoshells, *Phys. Rev. B*, **86**, 035421-1/14.
79. Prodan, E., Radloff, C., Halas, N. J., Nordlander, P. (2003) A hybridization model for the plasmon response of complex nanostructures, *Science*, **302**, 419–422.

80. Radloff, C., and Halas, N. J. (2004) Plasmonic properties of concentric nanoshells, *Nano Lett.*, **4**, 1323–1327.

81. Love, J. C., Gates, B. D., Wolfe, B. D., Paul, K. E., and Whitesides, G. M. (2002) Fabrication and wetting properties of metallic half-shells with submicron diameters, *Nano Lett.*, **2**, 891–894.

82. Jackson, J. B., Westcott, S. L., Hirsch, L. R., West, J. L., and Halas, N. J. (2003) Controlling the surface-enhanced Raman effect via the nanoshell geometry, *Appl. Phys. Lett.*, **82**, 257–259.

83. Jackson, J. B., Halas, N. J. (2004) Surface-enhanced Raman scattering on tunable plasmonic nanoparticle substrates, *Proc. Natl. Acad. Sci. USA.*, **101**, 17930–17935.

84. Oldenburg, S. J., Westcott, S. L., Averitt, R. D., and Halas, N. J. (1999) Surface-enhanced Raman scattering in the near infrared using metal nanoshell substrates, *J. Chem. Phys.*, **111**, 4729–4736.

85. Jackson, J. B., Westcott, S. L., Hirsch, L. R., West, J. L., and Halas, N. J. (2003) Controlling the surface-enhanced Raman effect via the nanoshell geometry, *Appl. Phys. Lett.*, **82**, 257–259.

86. Hale, G. D., Jackson, J. B., Shmakova, O. E., Lee, T. R., and Halas, N. J. (2001) Enhancing the active lifetime of luminescent semiconducting polymers via doping with metal nanoshells, *Appl. Phys. Lett.*, **78**, 1502–1504.

87. Sershen, S., Westcott, S. L., West, J. L., and Halas, N. J. (2001) An opto-mechanical nanoshell-polymer composite, *Appl. Phys. B*, **73**, 379–381.

88. Sershen, S., Westcott, S. L., Halas, N. J., and West, J. L. (2000) Temperature-sensitive polymer-nanoshell composites for photothermally modulated drug delivery, *J. Biomed. Mater. Res.*, **51**, 293–298.

89. Sun, Y., and Xia, Y. (2002) Increased sensitivity of surface plasmon resonance of gold nanoshells compared to that of gold solid colloids in response to environmental changes, *Anal. Chem.*, **74**, 5297–5305.

90. Hirsch, L. R., Jackson, J. B., Lee, A., Halas, N. J., and West, J. L. (2003) A whole blood immunoassay using gold nanoshells, *J. Anal. Chem.*, **75**, 2377–2381.

91. Loo, C., Lin, A., Hirsch, L., Lee, M. H., Barton, J., Halas, N., West, J., and Drezek, R. (2004) Nanoshell-enabled photonics-based imaging and therapy of cancer, *Technol. Cancer Res. T.*, **3**, 33–40.

92. Hirsch, L. R., Stafford, R. J., Bankson, J. A., Sershen, S. R., Rivera, B., Price, R. E., Hazle, J. D., Halas, N. J., and West, J. L. (2003) Nanoshell-mediated near-infrared thermal therapy of tumors under magnetic resonance guidance, *Proc. Natl. Acad. Sci. USA*, **100**, 13549–13554.

93. Tam, F., Chen, A. L., Kundu, J., Wang, H., and Halas, N. J. (2007) Mesoscopic nanoshells: Geometry-dependent plasmon resonances beyond the quasistatic limit, *J. Chem. Phys.*, **127**, 204703–204708.
94. Averitt, R. D., Westcott, S. L., and Halas, N. J. (1999) The linear optical properties of gold nanoshells, *J. Opt. Soc. Am. B*, **16**, 1824–1832.
95. Averitt, R. D., Westcott, S. L., and Halas, N. J. (1999) The ultrafast optical properties of gold nanoshells, *J. Opt. Soc. Am.*, **16**, 1814–1824.

Chapter 4

Computational Molecular Photophysics

E. Fabiano
National Nanotechnology Laboratory, Istituto Nanoscienze-CNR, Via Arnesano 16, I-73100, Lecce, Italy
eduardo.fabiano@nano.cnr.it

In this chapter we present an introductory overview of the basic theoretical concepts of computational molecular photophysics. First, the nature and properties of electronic excitations are considered, with special attention to transition moments and vibrational contributions. Then, the main photophysical processes involving the electronic excited states are examined, focusing in particular on nonradiative deactivation phenomena. Finally, we present a brief review of computational methods commonly applied for the description of molecular excitations. Special emphasis is given to the configuration-interaction (CI) method and the time-dependent density functional theory (TD-DFT), discussing some technical details and outlining advantages and limitations.

4.1 Introduction

Recent years have witnessed a fast development of quantum-chemical methods for the calculation of molecular excited states.

Handbook of Molecular Plasmonics
Edited by Fabio Della Sala and Stefania D'Agostino

ISBN 978-981-4303-20-0 (Hardcover), 978-981-4303-21-7 (eBook)
www.panstanford.com

This makes now possible to perform computational studies of the excitations of realistic systems ranging from organic and organometallic dyes to biological systems. A detailed knowledge about the excited states of molecules is essential for the explanation and interpretation of molecular optical properties as well as for the description of photoinduced physical and chemical processes.

Excitation spectra arise from transitions between different quantum states of the system, corresponding to different nondegenerate solutions of the Schrödinger equation. In quantum chemistry it is common practice to treat the solution of the Schrödinger equation within the *Born–Oppenheimer approximation* [1] and separate the electronic and nuclear degrees of freedom. Consequently the excitation spectra are also separated into an electronic and a roto-vibrational spectrum. The former is studied mainly in optical (UV/vis) spectroscopy experiments and will constitute the main subject of this chapter; the latter, which can be investigated by infrared, microwave or Raman spectroscopy measurements, provides fine-structure corrections to the electronic spectrum.

Electronic spectra of molecular systems involve transitions between different electronic states of the molecule, with each state depending parametrically on the nuclear configuration. Therefore, the simulation and interpretation of electronic spectra is in general a hard task, especially for medium- and large-size systems. In the last years, a large number of methods was developed to deal with this difficult problem, including TD-DFT [2–4], configuration interaction (CI) techniques [3, 5], linear response [6–9] or equation-of-motion [10, 11] coupled cluster methods, and multireference approaches [12–16]. Nevertheless, the computation of excited electronic molecular states is still much more demanding than the corresponding ground-state problem and no reliable "black-box" methods are available at present in this context.

In this chapter we will briefly review some topics in the theory of excited states and in computational methods for the description of molecular electronic excitations, providing a general overview of the current state of the art. For more detailed information the reader is referred to more specialized publications [2–4, 12, 17–32]. In particular, we will consider electronic transitions and describe the basic photophysical processes of relevance in molecular systems. In the last part of the chapter we will focus on the description of the most relevant approaches for the simulation of molecular excited states, with special emphasis on the CI method and the TD-DFT.

4.2 Electronic Excitations

Molecular electronic excited states display a large diversity, reflecting the complexity of the molecular electronic structure and intramolecular interactions. A simple classification of the different electronic excitations is therefore desirable, in order to achieve a better understanding of the photophysical and optical processes, but is not straightforward. Moreover, the availability of a precise classification, would help in the correct selection of the most appropriate theoretical method for the description of selected excitations.

From a theoretical point of view, the electronic excited states are many-body quantum states. Therefore they have a rather complex structure and cannot be easily classified. Nevertheless, a first distinction of different excitations can be based on the spin multiplicity of the state (e.g., singlet, doublet, triplet, . . .). This classification is very important because, in general, in optical transitions, no change of the spin multiplicity is allowed, unless magnetic interactions are considered. Thus, different electronic excitations possessing different spins form separated excited-state manifolds, which do not mix easily. This concept is a central one in photophysics (see Sec. 4.3). Furthermore, the spin can have important consequences on the electronic distribution and energy of the states. For example, in the case of a triplet excited state, because of the parallel spin, the electrons tend to keep, on average, a larger separation than in a corresponding singlet state. Therefore, they feel a lower Coulomb repulsion and the energy of the triplet state is generally lower than that of the corresponding singlet state.

A different and more effective classification is based on the consideration of the single-particle character of the many-body

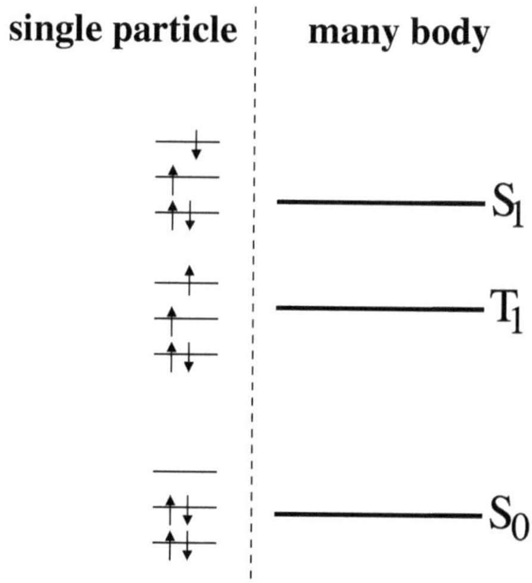

Figure 4.1 Pictorial representation of the many-body (right) and single-particle (left) representation of electronic excitations in molecular systems. The labels S and T denote singlet and triplet states, respectively.

excited states, in order to express the electronic excitations in terms of single-particle transitions. In practice, the many-body ground state is considered as a collection of single-particle molecular orbitals each with his occupation (in Hartree–Fock and DFT methods this corresponds to a single ground-state Slater determinant [33]), while the many-body excited states are described approximately by considering the promotion of one or more electrons from occupied to unoccupied single-particle molecular orbitals (in the CI method this corresponds to an excited or substituted Slater determinant [33]). This representation is depicted schematically in Fig. 4.1 and compared with the many-body description of excited states. Note that only the many-body description of the excited states is exact, while the single-particle representation is approximate, despite it holds very well for most low-lying one-photon excitations. However, the former provides little information on the characteristics of the electronic states, while the latter is intuitive and allows a detailed classification of different excitations.

Making use of the single-particle representation of the excited states we can classify the main molecular electronic excitations as follows:

- **Valence excitations**: These are excitations involving the promotion of a valence electron from a bonding or non-bonding orbital to a virtual orbital. According to the specific character of the orbitals involved (e.g., π, n, σ orbitals) the excitations can be further characterized as $\pi\pi^*$, $n\pi^*$, $\pi\sigma^*, \ldots$ The valence excitations are the most important for the optical and photophysical properties of organic molecules.
- **Core excitations**: These are excitations described by the promotion of a core electron to a virtual orbital. They are important especially for material science, as they provide detailed information on the local electronic environment around the nuclei [34–36] (e.g. assignment of oxidation states). The calculation of core excitations requires in general the use of specialized algorithms to deal with the extraction of high-lying roots [36, 37].
- **Rydberg excitations**: They involve single-particle excitations to virtual orbitals with large spatial extent. Because these orbitals resemble atomic orbitals with high principal quantum number, the excitation energies follow roughly a Rydberg series formula [38]. The calculation of Rydberg-type excitations requires special attention for the basis set and the asymptotic behavior of the potential, in order to correctly describe the highly delocalized virtual orbitals [39–43]
- **Local excitations**: These excitations involve single-particle orbitals displaying a significant overlap. Upon excitation the electron distribution of the molecule is thus barely changed.
- **Charge-transfer excitations**: These excitations are related with a net transfer of charge across the system [44, 45]. They are described in terms of molecular orbitals having small or no overlap. For this reason their theoretical description requires a proper inclusion of long-range and nonlocal interactions [46–49].

4.2.1 *Electronic Absorption Spectrum*

Upon absorption of a photon of energy $\hbar\omega$ a molecule can be promoted from the ground state $|\Psi_0\rangle$ of energy E_0 to a quantum electronic state $|\Psi_n\rangle$ of higher energy E_n if the relation

$$\hbar\omega = E_n - E_0 = \hbar\omega_{n0}\,, \tag{4.1}$$

is fulfilled. From the *Fermi golden rule* the transition probability per unit time is:

$$W_{0\rightarrow n} = \frac{2\pi}{\hbar}\sum_n \left|\langle\Psi_n|\,\hat{U}\,|\Psi_0\rangle\right|^2 \delta\left(\hbar\omega - (E_n - E_0)\right)\,, \tag{4.2}$$

where the operator $\hat{U}$ describes the interactions between the electrons of the molecule and the incident magnetic vector potential $\tilde{\mathbf{A}}\,(\mathbf{r},t) = \mathrm{Re}\left[\tilde{\mathbf{A}}^0 \exp(\mathrm{i}\mathbf{k}\cdot\mathbf{r} - \omega t)\right]$, is:

$$\hat{U} = \frac{e}{2m_e c}\sum_{i=1}^{N} \tilde{\mathbf{A}}^0 \cdot \hat{\mathbf{p}}_i \exp(\mathrm{i}\mathbf{k}\cdot\hat{\mathbf{r}}_i)\,. \tag{4.3}$$

Note that in Eq. (4.3) the position $\hat{\mathbf{r}}_i$ and momentum ($\hat{\mathbf{p}}_i$) of electrons are *operators*, while the magnetic vector potential is treated classically. This correspond to a quantum-mechanical treatment of the molecule associated with a classical description of the field.

As the molecular dimensions are much smaller than the wavelength, the exponential term can be neglected: this is known as the *dipole approximation*. Furthermore, we have:

$$\sum_{i=1}^{N}\langle\Psi_n|\,\hat{\mathbf{p}}_i\,|\Psi_0\rangle = \mathrm{i}\frac{m_e}{\hbar}(E_n - E_0)\boldsymbol{\mu}_{n0}\,, \tag{4.4}$$

where

$$\boldsymbol{\mu}_{n0} = \langle\Psi_n|\,\hat{\boldsymbol{\mu}}\,|\Psi_0\rangle = \langle\Psi_n|\sum_i(-e)\hat{\mathbf{r}}_i\,|\Psi_0\rangle \tag{4.5}$$

is the *transition dipole moment* and $\hat{\boldsymbol{\mu}}$ is *electric dipole operator.* Using Eq. (4.3) and Eq. (4.4) in Eq. (4.2) we have:

$$W_{0\rightarrow n} = \frac{2\pi}{\hbar}\left(\frac{\omega A_0}{2c}\right)^2 \left|\hat{\mathbf{a}}_0\cdot\boldsymbol{\mu}_{n0}\right|^2 \delta\left(\hbar\omega - (E_n - E_0)\right)\,, \tag{4.6}$$

where $\tilde{\mathbf{A}}^0 = A_0\hat{\mathbf{a}}_0$. Thus, the transition probability depends on the square of the scalar product between of the transition dipole moment and the $\widehat{\mathbf{a}}_0$. It is therefore orientational dependent, and this property can be exploited in spectroscopic experiments using polarized light sources. Note also that the matrix elements in Eq. (4.6) vanish for states of different spin. Non-zero transition probabilities can be instead found if the molecule displays a strong spin-orbit interaction [51], which causes the various spin states to mix, or if the magnetic dipole moment operator is considered. However, since the magnetic transition dipole moments are in general about five orders of magnitude smaller than the electric transition dipole moments, very small transition probabilities are often found for spin-changing (also called spin-forbidden) transitions.

Using (1.113) we have:

$$W_{0\rightarrow n} = S_{inc}\frac{4\pi^2}{\hbar c}\sum_n \left|\widehat{\mathbf{a}}_0 \cdot \boldsymbol{\mu}_{n0}\right|^2 \delta\left(\hbar\omega - (E_n - E_0)\right) . \quad (4.7)$$

Finally, the *absorption cross-section* is:

$$\sigma_{abs}(\omega) = \frac{\hbar\omega\, W_{0\rightarrow n}}{S_{inc}} = \frac{4\pi^2}{\hbar c}\sum_n \omega_{n0}\left|\widehat{\mathbf{a}}_0 \cdot \boldsymbol{\mu}_{n0}\right|^2 \delta\left(\omega - \omega_{n0}\right) , \quad (4.8)$$

where we used the relation $\delta(\omega) = \hbar\delta(\hbar\omega)$. Note that the quantum-mechanics absorption cross-section for a molecule is indicated with σ_{abs}, to distinguish it from the electromagnetic one for metal nanoparticles (C_{abs}).

For an isotropic incident radiation we have:

$$\sigma_{abs}(\omega) = \frac{2\pi^2}{m_e c}\sum_n f_{n0}\delta\left(\omega - \omega_{n0}\right) , \quad (4.9)$$

where the dimensionless *oscillator strength* is:

$$f_{n0} = \frac{2m_e}{3\hbar}\omega_{n0}\left|\boldsymbol{\mu}_{n0}\right|^2 . \quad (4.10)$$

The oscillator strength relates the strength of a molecular transition to the theoretical transition strength of a single electron in a harmonic potential (a simple model for a hydrogen atom). The oscillator strength can range from zero to about one. A strong transition will have a value of f close to one. Oscillator strengths greater than one result from the degeneracy of real electronic systems.

4.2.2 Vibrational Contributions

Within the *Born–Oppenheimer approximation*, excitations occur between two roto-vibrational levels of two electronic states. The excitation energy is thus:

$$\Delta E_{nm,\nu\lambda,jl} = E_{n,\nu,j} - E_{m,\lambda,l}, \tag{4.11}$$

with n, m; ν, λ; and j, l the electronic, vibrational and rotational quantum numbers respectively, and we consider a transition form the state m, λ, l to n, ν, j. Neglecting the rotational motion, which is rarely resolved in standard spectroscopy measurements, we are left with:

$$\Delta E_{nm,\nu\lambda} = E_n - E_m + \left(E_{n,\nu}^{vib} - E_{m,\lambda}^{vib}\right), \tag{4.12}$$

where E_n and E_m are now the energies of the electronic states n and m, and E^{vib} denotes the vibrational energy of a given vibrational state of a certain electronic state. Equation (4.12) can be further simplified considering that the transition only involves the vibrational ground-states (i.e. $\nu = \lambda = 0$), that is a good approximation for experiments at low-temperature. The excitation energy is then written as:

$$\Delta E_{nm} = E_n - E_m + \left(E_n^{\mathrm{ZPE}} - E_m^{\mathrm{ZPE}}\right), \tag{4.13}$$

where E_n^{ZPE} and E_m^{ZPE} are the *zero-point vibrational energies* (ZPE) of the electronic states n and m.

The proper calculation of the excitation energy requires thus the computation of the electronic energies of the two states of interest along with the calculation of all the corresponding vibrational modes. This procedure is relatively expensive from the computational point of view, because it requires the optimization of the excited-state geometries and the calculation of the Hessian matrix for the excited states. Moreover, the optimization of the geometry of excited states may be hindered by the presence of avoided crossings [52, 53] between different states, which makes impossible the use of standard optimization tools based on quasi-Newton algorithms [54]. For these reasons the excitation energies are often approximated by the sole electronic excitation energy (possibly also calculating both energies at the ground-state optimized geometry; i.e. vertical excitations are considered), neglecting vibrational contributions. This approximation introduces usually small errors of the order of 0.1 eV or less.

Concerning the transition probability, we saw in Eq. (4.6) that it is proportional to the square of the transition dipole

$$\boldsymbol{\mu}_{nm,\nu\lambda} = \langle \Psi_{n,\nu} | \hat{\boldsymbol{\mu}} | \Psi_{m,\lambda} \rangle . \qquad (4.14)$$

When the nuclear motion is taken into account the electric dipole moment operator must include nuclear contributions and it is written as:

$$\hat{\boldsymbol{\mu}} = (-e) \sum_{i=1}^{N} \hat{\mathbf{r}}_i + (e) \sum_{a=1}^{M} Z_a \hat{\mathbf{R}}_a , \qquad (4.15)$$

with $\hat{\mathbf{r}}_i$ and $\hat{\mathbf{R}}_a$ the coordinates of the i-th electron and of the a-th nucleus (with charge Z_a) respectively. The total wave function is partitioned, within the Born–Oppenheimer approximation, into an electronic an a nuclear part [55], i.e.:

$$\Psi_{n,\nu}(\underline{\mathbf{r}}, \underline{\mathbf{R}}) = \Phi_n(\underline{\mathbf{r}}, \underline{\mathbf{R}}) \Xi_\nu^{(n)}(\underline{\mathbf{R}}) , \qquad (4.16)$$

where $\underline{\mathbf{r}}$ and $\underline{\mathbf{R}}$ collect all the electron and nuclear coordinates, respectively. Neglecting the rotational and translational nuclear motion, the vibrational wave function $\Xi_\nu^{(n)}$ can be expressed, within the harmonic approximation, as the product of $3M - 6$ harmonic oscillators $\xi_{k,\nu_k}^{(n)}(q_k^{(n)})$, where $q_k^{(n)}$ is the k-th vibrational normal coordinate of the electronic state n. The transition dipole of Eq. (4.14) is thus obtained in terms of nuclear contributions of the form:

$$\begin{aligned} \boldsymbol{\mu}_{nm,\nu\lambda} &= \langle \prod_k \xi_{k,\nu_k}^{(n)}(q_k^{(n)}) \langle \Phi_n(\underline{\mathbf{r}}, \underline{\mathbf{R}}) | \hat{\boldsymbol{\mu}} | \Phi_m(\underline{\mathbf{r}}, \underline{\mathbf{R}}) \rangle \prod_h \xi_{h,\lambda_h}^{(m)}(q_h^{(m)}) \rangle \\ &= \langle \prod_k \xi_{k,\nu_k}^{(n)}(q_k^{(n)}) | \boldsymbol{\mu}_{nm}(\underline{\mathbf{R}}) | \prod_h \xi_{h,\lambda_h}^{(m)}(q_h^{(m)}) \rangle . \end{aligned} \qquad (4.17)$$

The quantity $\boldsymbol{\mu}_{nm}(\underline{\mathbf{R}})$ is the *electronic transition dipole* between the electronic states n and m, as a function of the normal nuclear coordinate $\underline{\mathbf{R}}$ (which can be represented also with the normal coordinated $q_k^{(n)}$ or $q_h^{(m)}$).

For a practical evaluation of integrals of the form of Eq. (4.17), two more issues must be still resolved. First of all, the normal coordinates of the state n must be transformed into the normal coordinates of the initial state m, in order to be able to carry on the integrals. This can be done by performing a *Duschinsky transformation* [56–61]. Secondly, in general no analytic form exists for the electronic transition dipole $\boldsymbol{\mu}_{nm}(\underline{\mathbf{R}})$. Collecting the normal coordinate for initial state m in $\underline{\mathbf{Q}}^{(m)}$, we expand the dipole into a Taylor series around the equilibrium geometry the initial state m (i.e. $\underline{\mathbf{Q}}_0^{(m)}$) to obtain:

$$\boldsymbol{\mu}_{nm}(\underline{\mathbf{Q}}^{(m)}) \approx \boldsymbol{\mu}_{nm}(\underline{\mathbf{Q}}_0^{(m)}) + \left.\frac{\partial \boldsymbol{\mu}_{nm}(\underline{\mathbf{Q}}^{(m)})}{\underline{\mathbf{Q}}^{(m)}}\right|_{\underline{\mathbf{Q}}_0^{(m)}} \cdot (\underline{\mathbf{Q}}^{(m)} - \underline{\mathbf{Q}}_0^{(m)}) + \cdots . \tag{4.18}$$

Truncating the expansion at zeroth-order yields the so called *Franck–Condon approximation* [58–60]:

$$\boldsymbol{\mu}_{nm,\nu\lambda} = \boldsymbol{\mu}_{nm}(\underline{\mathbf{Q}}_0^{(m)}) \left\langle \prod_k \xi_{k,\nu_k}^{(n)}(q_k^{(n)}) | \prod_h \xi_{h,\lambda_h}^{(m)}(q_h^{(m)}) \right\rangle . \tag{4.19}$$

Within this approximation the transition occurs between vibrational states with the highest overlap. Moreover, electric dipole forbidden transitions (i.e. when $\boldsymbol{\mu}_{nm}(\underline{\mathbf{Q}}_0^{(m)}) = 0$) cannot be described within this approximation. They can be treated however by considering the expansion in Eq. (4.18) to first order, which yields the *Franck–Condon–Herzberg–Teller approximation* [58, 61].

4.3 Photophysics

The term *photophysics* designates all the physical processes that occur upon interaction of electromagnetic radiation with matter. In the case of molecular photophysics this means all those phenomena which are connected and follow the absorption of photons from a molecular system. Closely related to molecular photophysics is also *photochemistry*, which is the branch of chemistry that studies photo-induced chemical processes [62].

A schematic summary of the most important photophysical processes occurring in molecules is given in Fig. 4.2 After the absorption (A) of one photon the molecule is promoted from its electronic ground state to an electronic excited state. Because for most molecules the ground state is a singlet state, we denoted

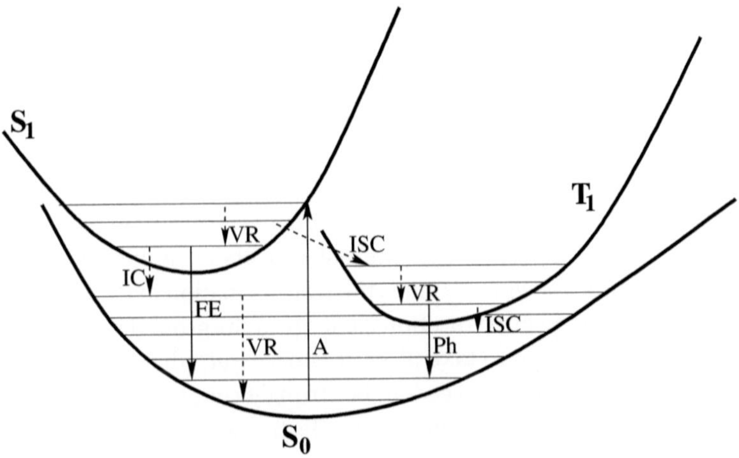

Figure 4.2 Pictorial representation of the possible photophysical processes occurring in molecules. Straight-line arrows denote radiative processes; dashed-line arrows denote non-radiative processes. *Abbreviation*: A, absorption; FE, fluorescence; Ph, phosphorescence; VR, vibrational relaxation; IC, internal conversion; ISC, intersystem crossing.

it as S_0 and the excitation occurs in the singlet manifold (from S_0 to S_1). After the excitation took place the molecule is in one vibrational state of the singlet excited state and possesses an excess energy corresponding to the adsorbed photon energy. The energy can then be dissipated through three different deactivation channels. On a time scale of about 0.1–1 fs a *vibrational relaxation* (VR) to a lower vibrational state of the same electronic state can occur. According to the *Kasha–Vavilov's rule* this process will lead very fast the molecule to the vibrational ground state of the electronic excited state. However, vibrational relaxation can only occur if the excess vibrational energy can be transmitted to an external environment, i.e. if the molecule interacts with some energy reservoir (solvent, other molecules in a gas/crystal), and cannot occur for isolated molecules.

On a time scale of 10^2–10^3 fs two more deactivation processes start to be competitive: *fluorescence emission* (FE) and *internal conversion* (IC) [62]. The fluorescence emission is simply the radiative process associated with the emission of one photon and brings the molecule into a vibrational state of its electronic ground state. Because the emission of the fluorescence occurs usually starting from the lowest vibrational state of the excited electronic state and ends in a excited vibrational state of the electronic ground state, the emitted photon has a lower frequency than the originally absorbed photon. The fluorescence emission is thus red-shifted with respect to the absorption frequency. This phenomenon is known as *Stokes' shift*. The internal conversion is instead a non-radiative process which implies the transformation of the electronic excitation energy into vibrational energy of the ground state. For the internal conversion to take place, it is necessary that the electronic and nuclear degrees of freedom be strongly coupled. This condition is in general only met when the two electronic states involved in the process are very close in energy (at conical intersections [63, 64]). Therefore, the internal conversion becomes non-negligible only in special conditions. Moreover, since the internal conversion process is reversible (can proceed from the electronic ground state to the electronic excited state) a fast vibrational relaxation in the electronic ground state is also a necessary condition to make the internal conversion an efficient deactivation channel.

So far we considered only photophysical processes occurring in the singlet manifold. However, alternatively to fluorescence and internal conversion, a different relaxation process, involving the energy transfer to a state of different spin, can occur. This phenomenon takes the name of *inter-system crossing* (ISC) [62, 65] and has a non-negligible rate probability only in presence of non-negligible intramolecular magnetic interactions (e.g., spin-orbit coupling [51]). The inter-system crossing probability is also enhanced when the singlet and triplet states have very close energies or intersect [66]. Nevertheless, it generally occurs on a time scale of 10^8–10^{12} fs, thus much slower than any photophysical process in the singlet manifold. After a triplet state is populated it usually undergoes a fast vibrational relaxation to its lowest vibrational state (on a time scale of about 0.1–1 fs). Then, the excitation energy can be dissipated either radiatively, through phosphorescence emission, or via a second non radiative inter-system crossing transition to the singlet ground state. *Phosphorescence* (Ph) is a spin-forbidden transition and has in general a low probability rate, which corresponds to a typical time scale of 10^8–10^{12} fs. Because the life time of triplet states is much longer than for the corresponding singlet states, they show in general an higher photo-induced reactivity and are often used in efficient photoinduced chemical reactions.

4.3.1 *Semiclassical Description of Internal Conversion and Inter-System Crossing*

Consider a molecule described by the total wave function $\Psi(\underline{\mathbf{r}}, \underline{\mathbf{R}}, t)$, which is solution of the *time-dependent Schrödinger equation*:

$$i\hbar\frac{\partial\Psi(\underline{\mathbf{r}}, \underline{\mathbf{R}}, t)}{\partial t} = \hat{H}\,\Psi(\underline{\mathbf{r}}, \underline{\mathbf{R}}, t) \tag{4.20}$$

where $\underline{\mathbf{r}}$ collects the coordinates of all electrons (including both spatial and spin coordinates), $\underline{\mathbf{R}}$ collects the coordinates of the nuclei, and $\mathcal{H}$ is the molecular Hamiltonian

$$\hat{H} = -\frac{\hbar^2}{2}\sum_a \frac{1}{M_a}\nabla^2_{\mathbf{R}_a} - \frac{\hbar^2}{2m}\sum_i \nabla^2_{\mathbf{r}_i} + V_{int}(\underline{\mathbf{r}}, \underline{\mathbf{R}}), \tag{4.21}$$

where M_a is the mass of nucleus a. The term V_{int} describes the electrostatic interaction between all electrons and all nuclei. To perform a separation of the electronic and nuclear degrees of freedom the wave function can be written using a *multi-configurational expansion*:

$$\Psi(\underline{\mathbf{r}}, \underline{\mathbf{R}}, t) = \sum_n \Phi_n(\underline{\mathbf{r}}, \underline{\mathbf{R}})\Xi_n(\underline{\mathbf{R}}, t), \tag{4.22}$$

where the electronic wave functions Φ_n will be assumed to be known and orthonormal, while the nuclear wave functions Ξ_i need to be neither orthogonal nor normalized.

Substituting Eq. (4.22) into Eq. (4.20), multiplying on the left by $\Phi^*_m(\underline{\mathbf{r}}, \underline{\mathbf{R}})$, and integrating over $\underline{\mathbf{r}}$ (i.e. N electronic coordinates) we obtain (expanding the nuclear laplacian):

$$\begin{aligned} i\hbar\frac{\partial\Xi_m(\underline{\mathbf{R}}, t)}{\partial t} &= -\frac{\hbar^2}{2}\sum_a \frac{1}{M_a}\nabla^2_{\mathbf{R}_a}\Xi_m(\mathbf{R}, t) + \sum_n H_{mn}(\underline{\mathbf{R}})\Xi_n(\underline{\mathbf{R}}, t) \\ &+\frac{\hbar^2}{2}\sum_{na}\frac{1}{M_a}L^a_{mn}(\underline{\mathbf{R}})\Xi_m(\underline{\mathbf{R}}, t) \\ &-\hbar^2\sum_{a,n\neq m}\frac{1}{M_a}\mathbf{D}^a_{mn}(\underline{\mathbf{R}})\cdot\nabla_{\mathbf{R}_a}\Xi_m(\underline{\mathbf{R}}, t), \end{aligned} \tag{4.23}$$

with

$$H_{mn}(\underline{\mathbf{R}}) = \int \Phi_m(\underline{\mathbf{r}}, \underline{\mathbf{R}}) \left[-\frac{\hbar^2}{2m} \sum_i \nabla^2_{\mathbf{r}_i} + V_{int}(\underline{\mathbf{r}}, \underline{\mathbf{R}}) \right] \Phi^*_n(\underline{\mathbf{r}}, \underline{\mathbf{R}}) \mathrm{d}^{3N}\underline{\mathbf{r}}, \tag{4.24}$$

$$L^a_{mn}(\underline{\mathbf{R}}) = -\int \Phi^*_m(\underline{\mathbf{r}}, \underline{\mathbf{R}}) \left[\nabla^2_{\mathbf{R}_a} \Phi_n(\underline{\mathbf{r}}, \underline{\mathbf{R}}) \right] \mathrm{d}^{3N}\underline{\mathbf{r}}, \tag{4.25}$$

$$\mathbf{D}^a_{mn}(\underline{\mathbf{R}}) = \int \Phi^*_m(\underline{\mathbf{r}}, \underline{\mathbf{R}}) \left[\nabla_{\mathbf{R}_a} \Phi_n(\underline{\mathbf{r}}, \underline{\mathbf{R}}) \right] \mathrm{d}^{3N}\underline{\mathbf{r}}. \tag{4.26}$$

The *classical limit* of Eq. (4.23) can be achieved by factoring the nuclear wave functions into amplitude and phase terms [67] as:

$$\Xi_n(\mathbf{R}, t) = A_n(\underline{\mathbf{R}}, t) e^{\mathrm{i}S_n(\underline{\mathbf{R}}, t)/\hbar}, \tag{4.27}$$

and then taking the limit $\hbar \rightarrow 0$. Substitution of Eq. (4.27) into Eq. (4.23) and reorganization of the different terms, yields two sets of coupled equations (dropping the $\underline{\mathbf{R}}, t$ dependence for simplicity):

$$\begin{aligned} \frac{\partial S_m}{\partial t} &+ \sum_a \frac{1}{2M_a} (\nabla_{\mathbf{R}_a} S_m)^2 + H_{mm} \\ &= \sum_a \frac{\hbar^2}{2M_a} \frac{\nabla^2_{\mathbf{R}_a} A_m}{A_m} - \sum_{na} \frac{\hbar^2}{2M_a} \frac{A_n}{A_m} L^a_{mn} e^{\mathrm{i}(S_n - S_m)/\hbar} \\ &+ \sum_{na} \frac{\hbar^2}{M_a} \frac{\mathbf{D}^a_{mn} \cdot (\nabla_{\mathbf{R}_a} A_n)}{A_m} e^{\mathrm{i}(S_n - S_m)/\hbar} \end{aligned} \tag{4.28}$$

$$\begin{aligned} \frac{\partial A_m}{\partial t} &+ \sum_a \frac{1}{M_a} \nabla_{\mathbf{R}_a} A_m \cdot \nabla_{\mathbf{R}_a} S_m + \sum_a \frac{1}{M_a} A_m \nabla^2_{\mathbf{R}_a} S_m \\ &+ \sum_{n \neq m} A_n \left[\sum_a \frac{1}{M_a} \mathbf{D}^a_{mn} \cdot \nabla_{\mathbf{R}_a} S_n + \frac{\mathrm{i}}{\hbar} H_{mn} \right] e^{\mathrm{i}(S_n - S_m)/\hbar} = 0. \end{aligned} \tag{4.29}$$

Equations (4.28) and (4.29) are completely equivalent to Eq. (4.23), but allowed the separation of diagonal H_{mm} terms [Eq. (4.28)] and off-diagonal H_{mn} coupling terms [Eq. (4.29)].

Performing the limit $\hbar \rightarrow 0$ for Eq. (4.28) gives:

$$\frac{\partial S_m(\underline{\mathbf{R}}, t)}{\partial t} + \sum_a \frac{1}{2M_a} (\nabla_{\mathbf{R}_a} S_m(\underline{\mathbf{R}}, t))^2 + H_{mm}(\underline{\mathbf{R}}) = 0. \tag{4.30}$$

We can consider Eq. (4.30) as a definition of the, up to now unknown, functions $S_m(\underline{\mathbf{R}}, t)$ and then recognize it as the *Hamilton–Jacobi equation* [68] for particles moving subject to the potential $H_{mm}(\underline{\mathbf{R}})$. Thus, the functions $S_m(\underline{\mathbf{R}}, t)$ can be identified with the *classical action* and the following relation holds:

$$\nabla_{\mathbf{R}_a} S_m(\underline{\mathbf{R}}, t) \leftrightarrow \mathbf{p}_{am} = M_a \mathbf{v}_{am}\,, \tag{4.31}$$

with $\mathbf{p}_{am}$ and $\mathbf{v}_{am}$ the classical momentum and the velocity of particle a on the potential energy surface defined by $H_{mm}(\underline{\mathbf{R}})$. Since the Hamilton–Jacobi formulation is completely equivalent to the Newtonian equation of motion, Eq. (4.30) can also be written:

$$M_a \frac{d\mathbf{R}_{aj}}{dt} = -\nabla_{\mathbf{R}_a} H_{mm}(\underline{\mathbf{R}})\,. \tag{4.32}$$

This is a set of equations describing *classical trajectories.* For each electronic state m the nucleus a moves on the (adiabatic) *potential energy surface (PES)* defined by $H_{mm}(\underline{\mathbf{R}})$.

For Eq. (4.29) the classical limit $\hbar \to 0$ cannot be performed, because of the presence of $\hbar$ at the denominator of the last term in square brackets. Nevertheless, the equation can be further developed by using the relation in Eq. (4.31) to write

$$\frac{\partial A_m}{\partial t} + \sum_a \nabla_{\mathbf{R}_a} A_m \cdot \mathbf{p}_{am} + \sum_a A_m \nabla \cdot \mathbf{p}_{aj} + \sum_{n \neq m} A_n \left[\sum_a \mathbf{D}^a_{mn} \cdot \mathbf{v}_{an} + \frac{\mathrm{i}}{\hbar} H_{mn} \right] e^{\frac{\mathrm{i}}{\hbar}\int_{t_0}^{t}(H_{nn}-H_{mm})d\tau} = 0\,. \tag{4.33}$$

If we neglect the last non-classical term, the set of Eq. (4.33) is isomorphic to the *Navier–Stokes equations* [69]

$$\frac{\partial \rho_m}{\partial t} + \nabla \rho_m \cdot \mathbf{p}_m + \rho_m \nabla \cdot \mathbf{p}_m = 0\,, \tag{4.34}$$

with the connecting transformation $A_m \leftrightarrow \rho_m$. We can thus interpret the functions A_j in terms of *classical fluxes* and, consequently, $\int |A_m(\underline{\mathbf{R}}, t)|^2 d^{3M}\underline{\mathbf{R}}$ as the population of the (adiabatic) state m at time t. In this context the non-classical terms in Eq. (4.33) are coupling terms governing the exchange of flux between different trajectories or, equivalently, the change of population between different (adiabatic) states.

To gain more insight into the non-classical coupling terms governing the population exchange between different electronic states, we fix a precise form for the electronic wave functions requiring that they are solutions of the *electronic Schrödinger equation*

$$\left[-\frac{\hbar^2}{2m}\sum_{i=1}^{N}\nabla^2_{\mathbf{r}_i} + V_{int}(\underline{\mathbf{r}}, \underline{\mathbf{R}})\right]\Phi_n(\underline{\mathbf{r}}, \underline{\mathbf{R}}) = E_n\Phi_n(\underline{\mathbf{r}}, \underline{\mathbf{R}}) \,. \qquad (4.35)$$

If we consider all wave functions Φ_n having the same spin and no spin-orbit interaction term in V_{int}, then we have $H_{nm} = E_n\delta_{nm}$ and the coupling terms read:

$$\Lambda_{mn} = \sum_a \mathbf{D}^a_{mn} \cdot \mathbf{v}_{an} \,. \qquad (4.36)$$

These are the *nonadiabatic coupling terms* controlling the internal conversion rate [70–72].

If we consider electronic wave functions Φ_n and Φ_m having different spins, all terms defining the coupling term vanish, unless we consider a spin-orbit interaction V_{SO} in the Hamiltonian of Eq. (4.21) (but not in Eq. (4.35)). In this case we have that $H_{nm} = \langle\Phi_n|V_{SO}|\Phi_m\rangle$ while $\mathbf{D}^a_{mn} = 0$. The coupling term governing the intersystem crossing is thus:

$$\Lambda_{ji} = \frac{\mathrm{i}}{\hbar}\langle\Phi_m|V_{SO}|\Phi_n\rangle \,. \qquad (4.37)$$

Note that the inclusion of the spin-orbit term in Eq. (4.35) would lead to $H_{nm} = E_n{}'\delta_{nm}$ (the H_{mn} term would thus not contribute to the coupling), but in this case $\mathbf{D}^a_{mn} \neq 0$, because the electronic wave functions would not be anymore pure spin states.

4.4 Computational Methods for Excited States

The simulation of molecular excited states is, still nowadays, a challenging task in computational chemistry. In fact, an ideal theoretical method, aiming at a broad applicability in the field, should be able to meet a series of requirements:

- It should be able to describe with comparable accuracy a large variety of possible excitations with different character (see Sec. 4.2). In fact, in photophysical applications often several excited states of different character and their mutual interactions are simultaneously involved in the process. A balanced description of all states is therefore fundamental to correctly describe the photoinduced phenomena and the corresponding rates, as well as to avoid artifacts. The correct description of many different states is also very important in spectral simulations, where several tens to hundreds of excited states are simultaneously considered.

- To be a practical tool in the comparison with experiments, the method of choice should have an absolute accuracy for the excitation energies of less than 0.1–0.2 eV (for all kinds of excitations). It shall also be able to provide a reliable estimation of transition moments, with relative errors smaller than 20%.
- It should provide easy access to all standard one-electron molecular properties and to analytical gradients of the excitation energies. The latter in particular are very important to perform geometry optimizations of excited states, vibrational analyses, and to compute nonadiabatic coupling terms.
- Of course, the ideal theoretical method should possess all those properties that are generally desirable for any computational method, such as a weak basis-set dependence, systematic increasing of accuracy, ease of interpretation (e.g., through a molecular-orbital picture), size consistency, respect of formal known constraints.
- In order to be applicable to medium- and large-size molecules, which are actually the systems of interest for real applications, any method must be extremely efficient. In particular, it must have low CPU and memory requirements (comparable with those of ground-state methods) and a favorable scaling with the system size.
- It should provide as much as possible a "black box" applicability, without the need for a large human effort to preliminary tune technical parameters. This requirement is especially important for excited states calculations where the large variety of possible excitations and the complexity of the problem make very difficult the manual set up of the calculations. For the same reasons however, it is very

difficult to find theoretical methods able to automatically include in an efficient way all the effects required.

Of course, actually there exist no method which can fulfill all the preceding requirements at once. Therefore, a wise choice of the computational method must be performed in each situation, carefully assessing the advantages and limitations of different approaches and the peculiar requirements of the problem at hand.

To aid the choice of the most appropriate method, we can briefly classify the presently available approaches for the treatment of molecular excitations into few main groups. In the following the reader is supposed to be familiar with the basics of the Hartree–Fock method [33], density functional theory [73, 74], and perturbation theory [33].

- **Variational approaches**. In these methods the total electronic wave function is expressed using a suitable ansatz and then optimized by minimizing the expectation value of the electronic energy for the ground and excited states. The most popular methods belonging to this group are the CI methods with various levels of truncation [75], where a superposition of Slater determinants is used to build the wave-function ansatz. Some additional details about CI methods are given in Sec. 4.4.1.
- **Linear response methods**. These methods make use of the fact that the linear response of the electron density has poles (diverges) at the exact excitation energies of the system [76]. For a given ansatz of the ground-state wave function, the excitation energies can therefore be obtained by solving algebraic equations that yield the poles of the frequency-dependent electric polarizability. The corresponding transition moments can be instead found by computing the residue associated with each pole [76]. The most important approach belonging to this class is TD-DFT [2, 4, 24, 77]; see Sec. 4.4.2 for more details. Very similar to TD-DFT, but based on the Hartree–Fock ground-state wave-function ansatz, is the *time-dependent Hartree–Fock (TD-HF)* method [78], also known in computational physics as *random phase approximation (RPA)*. Finally, there exist

also linear response methods based on the coupled-cluster reference [7, 8, 27, 79].

- **Multireference approaches**. These methods make use of a multiconfigurational reference ground-state wave function, build from the static superposition of several Slater determinants [80]. They are especially suited for the treatment of molecules whose ground state has a strong multireference character (generally related to the (quasi)degeneracy of low-energy Slater determinants) or for the description of excitations characterized by multiply-excited single-particle transitions. The basic approach in this group of methods is the *multireference CI (MR-CI)* method. This approach allows a very accurate description of different kinds of excited states, provided that the reference multiconfigurational wave function is chosen properly. However, MR-CI calculations easily show an overwhelming computational cost for molecules with more than a few atoms, due to the large dimension of the CI matrix. To reduce the computational cost one possible strategy is to reduce the number of single-particle orbitals which participate in the formation of substituted Slater determinants. This choice of an "active space" can save considerable computer time, but raises the question of whether the selected orbitals can efficiently account for all the correlation effects. To face this problem two different ways are possible. In the *complete active space self-consistent field (CASSCF)* method [80, 81] the active orbitals are optimized in a self-consistent multiconfigurational procedure, to include variationally most of the relevant effects. Alternatively, the starting multireference ground-state wave function is built using single-particle orbitals which already effectively include some correlation effects. Most notable examples of this are the use of orbitals from DFT calculations [13], from semiempirical calculations [82], or even from a previous CASSCF calculation.
- **Perturbation theory methods**. A perturbation approach can be used to include efficiently high-order correlation effect into low-level CI solutions, either from a single-reference or a multireference approach. This gives rise to

a large number of methods differing mainly for the choice of the reference wave function: CIS(D) [79, 83, 84], CIS-MP2 [79], MR-MP2 [85], CASMP2 [86], CASPT2 [12, 87–91], MS-CASPT2 [92].

4.4.1 *Configuration Interaction*

The CI method is based on the expansion of the electronic wave function in a linear combination of *Slater determinants* $\Psi = \Psi(\underline{\mathbf{r}})$, including the ground-state Slater determinant and a number of substituted (or excited) Slater determinants:

$$\Psi_{\mathrm{CI}} = c_0\Psi_0 + \sum_{ia} c_i^a \Psi_i^a + \frac{1}{2!}\sum_{ijab} c_{ij}^{ab}\Psi_{ij}^{ab} + \cdots . \tag{4.38}$$

Here Ψ_0 is the ground-state Hartree–Fock wave function, the indexes $i, j, \ldots$ denote occupied orbitals, the indexes $a, b, \ldots$ denote virtual orbitals, and Ψ_i^a indicates the Slater determinant obtained from Ψ_0 by replacing the occupied orbital i with the virtual orbital a.

The ansatz wave function of Eq. (4.38) must solve the Schrödinger equation

$$\hat{\mathcal{H}}\Psi_{\mathrm{CI}} = E_{\mathrm{CI}}\Psi_{\mathrm{CI}} . \tag{4.39}$$

Multiplication on the left by Ψ_0, Ψ_i^a, Ψ_{ij}^{ab}, ...and integration over all variables yields a system of linear equations that can be written in matrix form:

$$\begin{pmatrix} \langle\Psi_0|\hat{\mathcal{H}}|\Psi_0\rangle & 0 & \langle\Psi_0|\hat{\mathcal{H}}|D\rangle & 0 & \cdots \\ 0 & \langle S|\hat{\mathcal{H}}|S\rangle & \langle S|\hat{\mathcal{H}}|D\rangle & \langle S|\hat{\mathcal{H}}|T\rangle & \cdots \\ \langle D|\hat{\mathcal{H}}|\Psi_0\rangle & \langle D|\hat{\mathcal{H}}|S\rangle & \langle D|\hat{\mathcal{H}}|D\rangle & \langle D|\hat{\mathcal{H}}|T\rangle & \cdots \\ \vdots & \vdots & \vdots & \vdots & \ddots \end{pmatrix} \begin{pmatrix} c_0 \\ C_S \\ C_D \\ C_T \\ \vdots \end{pmatrix} = \mathbb{E} \begin{pmatrix} c_0 \\ C_S \\ C_D \\ C_T \\ \vdots \end{pmatrix}, \tag{4.40}$$

where S, D, and T denote the single-, double-, and triple-substituted determinant subspaces respectively, $\mathbb{E}$ is the diagonal matrix of the energies, we made use of the Brillouin's theorem [33], which states that the is no coupling between the Hartree–Fock ground state and single excitations (i.e., $\langle\Psi_0|\hat{\mathcal{H}}|S\rangle = 0$), and we exploited the fact that the matrix elements between Slater determinants differing by more than two single-particle orbitals are zero.

The excitation energies can be found, within the CI approach, by simply diagonalizing the CI matrix in the left hand side of Eq. (4.40). Moreover, solutions of increasing accuracy can be systematically obtained by including higher excitation levels in the expansion in Eq. (4.38). When all possible substituted Salter determinant are considered (full CI limit), the exact solution, within the given atomic-orbital space, is obtained.

Despite its conceptual simplicity, the CI method is often computationally unaffordable for most practical applications, because the dimension of the CI matrix becomes rapidly huge, if high-order substitutions are considered for the Slater determinants in the expansion in Eq. (4.38). For this reason, calculations are generally performed at the CI single (CIS) or CI single and double (CISD) level of theory. In particular, despite its limited accuracy, the CIS approach is a rather common choice for the calculation of excited states because of its conceptual simplicity and computational efficiency.

Limiting the expansion of Eq. (4.38) to singly-substituted Slater determinants allows to write the CI matrix ($\mathbb{A}_{ia,jb}$) explicitly in the basis of atomic orbitals $\{\phi_l\}$ as:

$$\mathbb{A}_{ia,jb} = (E_0 + \epsilon_a - \epsilon_i)\delta_{ij}\delta_{ab} + (ia||jb)\,, \tag{4.41}$$

with ia and jb super-indexes, E_0 the ground-state energy, ϵ the molecular orbital energies, and

$$(ia||jb) = \iint \mathrm{d}^3\mathbf{r}\mathrm{d}^3\mathbf{r}' \frac{\phi_i(\mathbf{r})\phi_a(\mathbf{r})\phi_j(\mathbf{r}')\phi_b(\mathbf{r}') - \phi_i(\mathbf{r})\phi_j(\mathbf{r})\phi_a(\mathbf{r}')\phi_b(\mathbf{r}')}{|\mathbf{r}-\mathbf{r}'|}\,. \tag{4.42}$$

The CIS problem is thus given by the system of linear equations:

$$\sum_{ia}\left[(E_0 + \epsilon_a - \epsilon_i)\delta_{ij}\delta_{ab} + (ia||jb)\right] c_i^a = E_{\mathrm{CIS}} \sum_{ia} c_i^a \delta_{ij}\delta_{ab}\,. \tag{4.43}$$

Multiplication on the left by c_j^b and sum over jb yields the CIS energies:

$$E_{\mathrm{CIS}} = E_0 + \sum_{ia}(c_i^a)^2(\epsilon_a - \epsilon_i) + \sum_{ia,jb} c_i^a c_j^b (ia||jb)\,, \tag{4.44}$$

where we used the normalization of the CIS wave function.

The CIS method gives in general excitation energies of rather poor quality, with errors ranging typically from 0.5 to 2.0 eV [5, 93, 94]. This depends partly from the fact that the Hartree–Fock orbital energies, whose differences are the leading terms in Eq. (4.44), are very poor approximations of the true excitation energies. Partly, from the inclusion of only singly-substituted Slater determinants in the calculation, which causes the neglect of correlation effects for the excited states. This later problem can be partially corrected by including perturbatively second-order correlation corrections through the CIS(D) approach [79, 83, 84]. Moreover, the CIS method does not obey the *Thomas–Reiche–Kuhn dipole sum rule* , which states that the sum of all transition dipole moments must equal the number of electrons, and thus yields very poor estimates for the transition dipole moments.

On the other hand, the CIS method constitutes an efficient way for calculating the excitation energies of molecules with up to several hundreds of atoms. In addition it provides, through Eq. (4.44), an analytical expression for the excitation energies, therefore analytical gradients and second derivatives are readily obtained. Finally, contrary to other higher order CI treatments, it is size consistent and allows for pure singlet and triplet spin solutions.

4.4.2 *Time-Dependent Density Functional Theory*

TD-DFT is the extension of the conventional ground-state DFT formulation to a time-dependent domain. The TD-DFT formalism is thus suited for the description of an electronic system under the influence of a time-dependent external potential and for the treatment of excitations.

TD-DFT is grounded on the *Runge–Gross theorem* [95], which is the time-dependent analogous of the first Hoehnberg and Kohn theorem [96], and on the variational principle for the action integral [3], which provides the variational principle for the theory, in analogy with the second Hoehnberg and Kohn theorem [96].

The Runge–Gross theorem states that for a many-body system evolving from a fixed initial state there is a one-to-one correspondence between the external time-dependent potential and the (time-dependent) electron density $\rho(\mathbf{r}) = \rho(\mathbf{r}, t)$. Therefore, the behavior of any time-dependent system is completely determined by its time-dependent electron density. In particular, the many-body wave function is a *functional* of the density up to a time-dependent phase factor, i.e.:

$$\Psi(\underline{\mathbf{r}}, t) = \Psi[\rho(t)](t)e^{-\mathrm{i}\alpha(t)}\,, \tag{4.45}$$

where $d\alpha(t)/dt = C(t)$, with $C(t)$ an arbitrary time-dependent function; the expectation value of an arbitrary operator $\hat{O}$ is a unique functional of the density:

$$\begin{aligned}\langle\Psi(\underline{\mathbf{r}}, t)|\hat{O}(t)|\Psi(\underline{\mathbf{r}}, t)\rangle &= \langle\Psi[\rho(t)](t)e^{\mathrm{i}\alpha(t)}|\hat{O}(t)|\Psi[\rho(t)](t)e^{-\mathrm{i}\alpha(t)}\rangle \\ &= \langle\Psi[\rho(t)](t)|\hat{O}(t)|\Psi[\rho(t)](t)\rangle = O[\rho](t)\,.\end{aligned} \tag{4.46}$$

The variational principle for the action integral is derived starting from the observation that, if the time-dependent wave function $\Psi(\underline{\mathbf{r}}, t)$ is a solution of the time-dependent Schrödinger equation, then it corresponds to a stationary point of the quantum-mechanical action integral:

$$A = \int_{t_0}^{t_1} \mathrm{d}t \left\langle \Psi(\underline{\mathbf{r}}, t) \left| \mathrm{i}\frac{\partial}{\partial t} - \hat{\mathcal{H}}(\underline{\mathbf{r}}, t) \right| \Psi(\underline{\mathbf{r}}, t) \right\rangle\,. \tag{4.47}$$

Since, because of the Runge–Gross theorem, the wave function is a functional of the electron density, the action integral of Eq. (4.47) is also a unique functional of the density:

$$A[\rho] = \int_{t_0}^{t_1} \mathrm{d}t \left\langle \Psi[\rho] \left| \mathrm{i}\frac{\partial}{\partial t} - \hat{\mathcal{H}}(\underline{\mathbf{r}}, t) \right| \Psi[\rho] \right\rangle\,, \tag{4.48}$$

and the exact time-dependent electron density corresponds to a stationary point of the quantum-mechanical action integral, i.e., it is a solution of the *Euler equation*:

$$\frac{\delta A[\rho]}{\delta\rho(\mathbf{r}, t)} = 0\,. \tag{4.49}$$

Note that the action integral of Eq. (4.48) can be partitioned into an universal part:

$$B[\rho] = \int_{t_0}^{t_1} \mathrm{d}t \left\langle \Psi[\rho] \left| \mathrm{i}\frac{\partial}{\partial t} - \hat{\mathcal{T}} - \hat{V}_{e-e} \right| \Psi[\rho] \right\rangle\,, \tag{4.50}$$

with $\hat{T}$ the many-body kinetic energy operator and $\hat{V}_{e-e}$ the many-body electron-electron interaction operator, and a reminder that depends only on the external time-dependent potential, so that:

$$A[\rho] = B[\rho] - \int_{t_0}^{t_1} \mathrm{d}t \int \mathrm{d}^3\mathbf{r}\, \rho(\mathbf{r}, t) v_{ext}(\mathbf{r}, t)\,. \qquad (4.51)$$

Equipped with the Runge–Gross theorem and the variational principle for the action integral, we can now develop the analogous of the standard *Kohn–Sham formalism* [97], in time-dependent form. We assume that there exists a *non-interacting reference system*, with external one-particle potential $v_S(\mathbf{r}, t)$, whose electron density is equal to the exact time-dependent electron density $\rho(\mathbf{r}, t)$. Because the reference system consists of non-interacting particles it can be described by a single Slater determinant built from single-particle orbitals $\phi_i(\mathbf{r}, t)$. The electron density is thus given by:

$$\rho(\mathbf{r}, t) = \sum_i^N |\phi_i(\mathbf{r}, t)|^2\,, \qquad (4.52)$$

with N the number of electrons, and the single-particle orbitals will obey the one-particle Schrödinger equation:

$$\mathrm{i}\frac{\partial}{\partial t}\phi_i(\mathbf{r}, t) = \left[-\frac{1}{2}\nabla_i^2 + v_S(\mathbf{r}, t)\right]\phi_i(\mathbf{r}, t)\,. \qquad (4.53)$$

At the same time the electron density is also given by the solution of the Euler Equation in Eq. (4.49), with the action integral given by:

$$A_S[\rho] = B_S[\rho] - \int_{t_0}^{t_1} \mathrm{d}t \int \mathrm{d}^3\mathbf{r}\, \rho(\mathbf{r}, t) v_S(\mathbf{r}, t)\,. \qquad (4.54)$$

Thus, we find:

$$0 = \frac{\delta A_S[\rho]}{\delta\rho(\mathbf{r}, t)} = \frac{\delta B_S[\rho]}{\delta\rho(\mathbf{r}, t)} - v_S(\mathbf{r}, t)\,, \qquad (4.55)$$

and we can recognize that the Kohn–Sham time-dependent potential is given by:

$$v_S(\mathbf{r}, t) = \frac{\delta B_S[\rho]}{\delta\rho(\mathbf{r}, t)} \qquad (4.56)$$

evaluated at the exact density.

To obtain a better description of the functional B_S we consider the action integral of the true interacting system and write:

$$\begin{aligned} A[\rho] &= B[\rho] - \int_{t_0}^{t_1} \mathrm{d}t \int \mathrm{d}^3\mathbf{r}\, \rho(\mathbf{r},t) v_{ext}(\mathbf{r},t) \\ &= B_S[\rho] - \int_{t_0}^{t_1} \mathrm{d}t \int \mathrm{d}^3\mathbf{r}\, \rho(\mathbf{r},t) v_{ext}(\mathbf{r},t) \\ &\quad -\frac{1}{2}\int_{t_0}^{t_1} \mathrm{d}t \int \mathrm{d}^3\mathbf{r}\,\mathrm{d}^3\mathbf{r}'\, \frac{\rho(\mathbf{r},t)\rho(\mathbf{r}',t)}{|\mathbf{r}-\mathbf{r}'|} - A_{XC}[\rho], \end{aligned} \tag{4.57}$$

where we made explicit the Coulomb energy term and we defined the exchange-correlation (XC) part of the action integral as

$$A_{XC}[\rho] = B_S[\rho] - B[\rho] - \frac{1}{2}\int_{t_0}^{t_1} \mathrm{d}t \int \mathrm{d}^3\mathbf{r} \int \mathrm{d}^3\mathbf{r}' \frac{\rho(\mathbf{r},t)\rho(\mathbf{r}',t)}{|\mathbf{r}-\mathbf{r}'|}. \tag{4.58}$$

The exchange-correlation part of the action integral (also termed the XC kernel) includes all the quantum effects of the electron-electron time-dependent interaction and is the analogous of the exchange-correlation functional of the standard time-independent Kohn–Sham formalism. Using the action integral (Eq. (4.57)) into the Euler equation (Eq. (4.49)) gives

$$\frac{\delta B_S[\rho]}{\delta\rho(\mathbf{r},t)} = v_{ext}(\mathbf{r},t) + \int \mathrm{d}^3\mathbf{r}' \frac{\rho(\mathbf{r},t)}{|\mathbf{r}-\mathbf{r}'|} + \frac{\delta A_{XC}[\rho]}{\delta\rho(\mathbf{r},t)}. \tag{4.59}$$

Using Eqs. (4.53), (4.56), and (4.59) we can finally write the *time-dependent Kohn–Sham equations*

$$\mathrm{i}\frac{\partial}{\partial t}\phi_i(\mathbf{r},t) = \left[-\frac{1}{2}\nabla_i^2 + v_{ext}(\mathbf{r},t) + \int \mathrm{d}^3\mathbf{r}' \frac{\rho(\mathbf{r}',t)}{|\mathbf{r}-\mathbf{r}'|} + \frac{\delta A_{XC}[\rho]}{\delta\rho(\mathbf{r},t)}\right]\phi_i(\mathbf{r},t) \tag{4.60}$$

and the electron density is obtained via Eq. (4.52). In analogy with the ground-state case, the time-dependent Kohn–Sham equations provide a formally exact solution of the time-dependent many-body problem. However, the exact form of the A_{XC} kernel is not known and approximations are needed. One common approximation is the so called adiabatic local approximation in which the original XC kernel, which is in principle non-local in time and space, is replaced by a time-independent one. This assumption allows the use in the time-dependent Kohn–Sham equations of the standard ground-state exchange-correlation functionals.

4.4.3 *Linear Response TD-DFT*

To obtain excitation energies and properties within the time-dependent Kohn–Sham framework, it is possible to propagate in time the time-dependent electron density, through the solution of Eq. (4.60), and then extract energies and oscillator strengths from a Fourier analysis of the results [98–102]. Alternatively, the excited-state properties can be determined through the linear response theory. This is an efficient approach which avoids the direct solution of the time-dependent Kohn–Sham equations and is often used in practical applications.

As shown in Sec. 1.2.3, if we consider a small time-dependent external perturbation $v_{ext}(\mathbf{r},\omega)$ we can write the linear variation of the electron density as (in the Fourier domain):

$$\delta\tilde{\rho}(\mathbf{r},\omega) = \int \tilde{\Pi}\left(\mathbf{r},\mathbf{r}',\omega\right)\delta\tilde{v}_{ext}(\mathbf{r}',\omega)\mathrm{d}^3\mathbf{r}', \qquad (4.61)$$

where $\tilde{\Pi}$ is the linear density response function. Analogously, we can calculate the variation of the electron density in the Kohn–Sham non-interacting reference system as:

$$\delta\tilde{\rho}(\mathbf{r},\omega) = \int \tilde{\Pi}_s\left(\mathbf{r},\mathbf{r}',\omega\right)\delta\tilde{v}_S(\mathbf{r}',\omega)\mathrm{d}^3\mathbf{r}', \qquad (4.62)$$

where the variation of the Kohn–Sham potential is:

$$\delta\tilde{v}_S(\mathbf{r},\omega) = \delta\tilde{v}_{ext}(\mathbf{r},\omega) + \int\left[\frac{1}{|\mathbf{r}-\mathbf{r}'|} + \frac{\delta^2\tilde{A}_{XC}[\tilde{\rho}]}{\delta\tilde{\rho}(\mathbf{r},\omega)\delta\tilde{\rho}(\mathbf{r}',\omega)}\right]\delta\tilde{\rho}(\mathbf{r}',\omega)d^3\mathbf{r}' \qquad (4.63)$$

and the linear density response function of the non-interacting system is given by [76]:

$$\tilde{\Pi}_s\left(\mathbf{r},\mathbf{r}',\omega\right) = \sum_{ij}(f_i - f_j)\frac{\phi_i^*(\mathbf{r})\phi_j(\mathbf{r})\phi_j^*(\mathbf{r}')\phi_i(\mathbf{r}')}{\omega - (\epsilon_j - \epsilon_i) + \mathrm{i}\eta}, \qquad (4.64)$$

with f_i the occupation number of orbital i, ϵ_i the orbital energy of orbital i, and $\eta \to 0^+$. Since in the non-interacting linear density response function only particle-hole $(f_i > f_j)$ and hole-particle $(f_i < f_j)$ terms contribute, it is convenient to rewrite the density variation in terms of particle-hole and hole-particle contributions as:

$$\delta\tilde{\rho}(\mathbf{r},\omega) = \sum_{ia}\left[\tilde{P}_{ia}(\omega)\phi_a^*(\mathbf{r})\phi_i(\mathbf{r}) + \tilde{P}_{ai}(\omega)\phi_i^*(\mathbf{r})\phi_a(\mathbf{r})\right], \qquad (4.65)$$

where we adopted the convention of labeling occupied orbitals with indexes i, j and virtual orbitals with indexes a, b. Substitution of Eq. (4.65) into Eq. (4.62) and use of Eqs. (4.63, 4.64) yields a set of coupled equations

$$\left[\delta_{ij}\delta_{ab}(\epsilon_a - \epsilon_i + \omega) + \tilde{K}_{iajb}\right]\tilde{P}_{jb} + K_{iabj}\tilde{P}_{bj} = -\tilde{v}_{ia}\,, \quad (4.66)$$

$$\left[\delta_{ij}\delta_{ab}(\epsilon_a - \epsilon_i + \omega) + \tilde{K}_{aibj}\right]\tilde{P}_{bj} + K_{aijb}\tilde{P}_{jb} = -\tilde{v}_{ai}\,, \quad (4.67)$$

with

$$\tilde{K}_{klmn} = \int\int \mathrm{d}^3\mathbf{r}\,\mathrm{d}^3\mathbf{r}'\phi_k^*(\mathbf{r})\phi_l(\mathbf{r})\left[\frac{1}{|\mathbf{r}-\mathbf{r}'|} + \frac{\delta^2\tilde{A}_{XC}[\tilde{\rho}]}{\delta\tilde{\rho}(\mathbf{r},\omega)\delta\tilde{\rho}(\mathbf{r}',\omega)}\right]\phi_n^*(\mathbf{r}')\phi_m(\mathbf{r}')\,, \quad (4.68)$$

$$\tilde{v}_{kl} = \int \mathrm{d}^3\mathbf{r}\,\phi_k^*(\mathbf{r})\delta\tilde{v}_{ext}(\mathbf{r},\omega)\phi_l(\mathbf{r})\,. \quad (4.69)$$

Equations (4.66) and (4.67) can be finally written in the more compact form:

$$\left[\begin{pmatrix}\tilde{\mathbb{L}} & \tilde{\mathbb{M}}\\ \tilde{\mathbb{M}}^* & \tilde{\mathbb{L}}^*\end{pmatrix} - \begin{pmatrix}-\mathbb{I} & 0\\ 0 & \mathbb{I}\end{pmatrix}\right]\begin{pmatrix}\tilde{\mathbb{X}}\\ \tilde{\mathbb{Y}}\end{pmatrix} = -\begin{pmatrix}\tilde{\mathbb{V}}\\ \tilde{\mathbb{V}}^*\end{pmatrix} \quad (4.70)$$

where $\tilde{\mathbb{X}}_{ia} = \tilde{P}_{ia}$, $\tilde{\mathbb{Y}}_{ia} = \tilde{P}_{ai}$, and $\tilde{\mathbb{V}}_{ia} = v_{ia}$, and we defined:

$$\tilde{\mathbb{L}}_{iajb} = \delta_{ij}\delta_{ab}(\epsilon_a - \epsilon_i) + \tilde{K}_{iajb}\,, \quad (4.71)$$

$$\tilde{\mathbb{M}}_{iajb} = \tilde{K}_{iabj}\,. \quad (4.72)$$

From the linear response theory [76] it is known that the linear response function of Eq. (4.61) must have poles at the excitation energies of the system. However, since the external potential does not have any special pole structure, this implies that also the density variation $\delta\rho$ must have poles at the excitation energies of the system. The same must be obviously true also for the particle-hole and hole-particle components $\tilde{\mathbb{X}}$ and $\tilde{\mathbb{Y}}$. Therefore, at the excitation energies, the left hand side of Eq. (4.70), diverges while the right hand side remains finite. This implies that the matrix on the left hand side must have zero eigenvalues at the excitation energies of the system. That is, the excitation energies Ω of the system are defined by the following equation:

$$\begin{pmatrix}\tilde{\mathbb{L}} & \tilde{\mathbb{M}}\\ \tilde{\mathbb{M}}^* & \tilde{\mathbb{L}}^*\end{pmatrix}\begin{pmatrix}\tilde{\mathbb{X}}\\ \tilde{\mathbb{Y}}\end{pmatrix} = \Omega\begin{pmatrix}-\mathbb{I} & 0\\ 0 & \mathbb{I}\end{pmatrix}\begin{pmatrix}\tilde{\mathbb{X}}\\ \tilde{\mathbb{Y}}\end{pmatrix}\,. \quad (4.73)$$

Equation (4.73) has the same form as that of the random-phase approximation (or TD-HF) problem and its solutions were extensively studied in literature [103].

TD-DFT is nowadays the most used computational approach for the study of the excited states of molecular systems. This is due to its efficiency and relatively good accuracy. TD-DFT has in fact a computational cost comparable with the CIS approach, is size-consistent, respects the *Thomas–Reiche–Kuhn dipole sum rule*, and yields pure spin-state solutions. On the other hand it has in general an accuracy much superior to simple approaches such as CIS, although the exact performance of TD-DFT calculations depends quite significantly on the choice of the XC functional employed to compute the ground state and generate the XC kernel. At present the best results are obtained by using hybrid functionals (e.g., B3LYP [104, 105] or PBE0 [106]) or even orbital-dependent functionals (e.g., LHF [39, 43]). In these cases, for singlet valence excitations TD-DFT yields errors in the range 0.1–0.3 eV, while only slightly larger errors (0.3–0.5 eV) are found for triplet states [43, 108]. These results compare favorably with those from high level correlated methods such as coupled cluster or CASPT2, which are only slightly superior but are obtained with a much higher computational cost.

On the other hand, there are several known drawback of TD-DFT. It shows in fact severe problems in the correct description of valence states of molecules with extended conjugated π-systems [109, 110], doubly excited states [111, 112], long-range charge-transfer excitations [47, 113–116], and excitations of Rydberg character [42, 43]. For such systems quantitative errors of several eVs are possible and even a wrong qualitative description of the states can be found. For extended π-conjugated systems the problems of TD-DFT are related to the wrong asymptotic behavior of the standard exchange-correlation potentials. A similar origin have the limitations in the description of Rydberg states. In fact, both problems are significantly reduced when asymptotically corrected or orbital-dependent exact-exchange approaches are used [43, 66, 116, 117]. For excited states with double excitation character the failure of TD-DFT is related to the adiabatic local approximation, because in this case the linear response formalism contains only single-excitation contributions [25, 111, 112]. For long-range charge-transfer states, finally, too low excitation energies are found in TD-DFT when local (in space) functionals are employed. This effect is simply explained looking at Eq. (4.68) and noting that, when a transition involves orbitals with vanishing overlap, the K terms are zero and so in Eq. (4.70) only the eigenvalue differences determine the excitation energies. A significant improvement in the description of the charge-transfer states is obviously achieved when hybrid functionals, including a fraction of the Hartree–Fock non-local exchange, are used. Even better results can be found by employing range separated hybrid functionals [118, 119], which partition the non-local contribution into a short- and long-range part.

Acknowledgments

This work was funded by the ERC Starting Grant FP7 Project DEDOM, Grant Agreement No. 207441.

References

1. Born, M., and Oppenheimer, R. (1927) Zur Quantentheorie der Molekeln, *Ann. Phys. (NY)*, **84**, 457–484.
2. Marquez, M. A. L., and Gross, E. K. U. (2004) Time-dependent density functional theory, *Annu. Rev. Phys. Chem.*, **55**, 427–455.
3. Dreuw, A., and Head-Gordon, M. (2005) Single-reference *ab initio* methods for the calculation of excited states of large molecules, *Chem. Rev.*, **105**, 4009–4037.
4. Elliott, P., Furche, F., and Burke, K. (2008) Excited states from time-dependent density functional theory, in *Reviews in Computational Chemistry*, vol. 26 (ed. Lipkowitz, K. B., and Cundari, T. R.), John Wiley & Sons, Hoboken, NJ, pp. 91–166.
5. Foresman, J. B., Head-Gordon, M., Pople, J. A., and Frisch, M. J. (1992) Toward a systematic molecular orbital theory for excited states, *J. Phys. Chem.*, **96**, 135–149.
6. Sekino, H., and Bartlett, R. J. (1984) A linear response, coupled-cluster theory for excitation energy, *Int. J. Quantum Chem.*, **26**(S18), 255–265.

7. Koch, H., and Jörgensen, P. (1990) Coupled cluster response functions, *J. Chem. Phys.*, **93**, 3333–3344.
8. Koch, H., Jörgen Aa. Jensen, H., Jörgensen, P., and Helgaker, T. (1990) Excitation energies from the coupled cluster singles and doubles linear response function (CCSDLR). Applications to Be, CH^+, CO, and H_2O, *J. Chem. Phys.*, **93**, 3345–3350.
9. Hättig, C., and Weigend, F. (2000) CC2 excitation energy calculations on large molecules using the resolution of the identity approximation, *J. Chem. Phys.*, **113**, 5154–5161.
10. Stanton, J. F., and Bartlett, R. J. (1993) The equation of motion coupled–cluster method. A systematic biorthogonal approach to molecular excitation energies, transition probabilities, and excited state properties, *J. Chem. Phys.*, **98**, 7029–7039.
11. Krylov, A. I. (2008) Equation-of-motion coupled-cluster methods for open-shell and electronically excited species: The Hitchhiker's guide to fock space, *Annu. Rev. Phys. Chem.*, **59**, 433–462.
12. Roos, B. O. (1999) Theoretical studies of electronically excited states of molecular systems using multiconfigurational perturbation theory, *Accounts Chem. Res.*, **32**, 137–144.
13. Grimme, S., and Waletzke, M. (1999) A combination of Kohn–Sham density functional theory and multi-reference configuration interaction methods, *J. Chem. Phys.*, **111**, 5645–5655.
14. Vaval, N., and Pal, S. (1999) Molecular property calculations for excited states using a multireference coupled-cluster approach, *Chem. Phys. Lett.*, **300**, 125–130.
15. Ivanov, V. V., Lyakh, D. I., and Adamowicz, L. (2011) State-specific multireference coupled-cluster theory of molecular electronic excited states, *Annu. Rep. Prog. Chem. C*, **107**, 169–198.
16. Li, X., and Paldus, J. (2011) Multi-reference state-universal coupled-cluster approaches to electronically excited states, *J. Chem. Phys.*, **134**, 214118–214133.
17. Jaffe, H. H., and Orchin, M. (eds) (1962) *Theory and Applications of Ultraviolet Spectroscopy*, Wiley, New York.
18. Murrel, J. N. (ed.) (1963) *The Theory of the Electronic Spectra of Organic Molecules*, Methuen, London.
19. Mataga, N., and Kubota, A. (eds) (1970) *Molecular Interactions and Electronic Spectra*, Marcel Dekker, New York.
20. Turro, N. J. (ed.) (1978) *Modern Molecular Photochemistry*, Benjamin Cummings, Menlo Park, CA.

21. Michl, J., and Banacic-Koutecky, V. (eds) (1990) *Electronic Aspects of Organic Photochemistry*, Wiley, New York.
22. Casida, M. E., Jamorski, C., Bohr, F., Guan, J., and Salahub, D. R. (1994) Optical properties from density-functional theory, in *Theoretical and Computational Modeling of NLO and Electronic Materials* (ed. Karna, S. P., and Yeates, A. T.), ACS Press, Washington, D.C., p. 145.
23. Klessinger, M., and Michl, J. (eds) (1995) *Excited States and Photochemistry of Organic Molecules*, VHC, New York.
24. Casida, M. E. (1995) Time-dependent density-functional response theory for molecules, in *Recent Advances in Density Functional Methods*, Part I (ed. Chong, D. P.), World Scientific, Singapore, p. 155.
25. Casida, M. E. (1996) Time-dependent density functional response theory of molecular systems: Theory, computational methods, and functionals, in *Recent Developments and Applications of Modern Density Functional Theory* (ed. Seminario, J. M.), Elsevier, Amsterdam, p. 391.
26. Peyerimhoff, S. D. (1998) Spectroscopy: Computational methods, in *Encyclopedia of Computational Chemistry*, Wiley, New York, pp. 2646–2664.
27. Helgaker, T., Jörgensen, P., and Olsen, J. (eds) (2000) *Molecular Electronic-Structure Theory*, Wiley, New York.
28. Grabo, T., Pertersilka, M., and Gross, E. K. U. (2000) Molecular excitation energies from time-dependent density functional theory, *J. Mol. Struct. (Theochem)*, **501**, 353–367.
29. Grimme, S. (2004) Calculation of the electronic spectra of large molecules, in *Reviews in Computational Chemistry*, vol. 20 (ed. Lipkowitz, K. B., Larter, R., and Cundari, T. R.), John Wiley & Sons Inc., Hoboken, New Jersey, pp. 153–218.
30. Casida, M. E., Ipatov, A., and Cordova, F. (eds) (2006) Linear-response time-dependent density-functional theory for open-shell molecules in *Time-Dependent Density-Functional Theory* (ed. Marques, M. A. L., Ullrich, C., Nogueira, F. , Rubio, A., and Gross, E. K. U.) Springer, Berlin, pp. 243–257.
31. Dorogan, I. V. (2007) Theoretical methods of investigation of excited states of organic molecules, *Russ. J. Gen. Chem.*, **78**, 774–783.
32. Casida, M. E. (2009) Time-dependent density-functional theory for molecules and molecular solids, *J. Mol. Struct. (Theochem)*, **914**, 3–18.
33. Szabo, A., and Ostlund, N. S. (eds) (1989) *Modern Quantum Chemistry: Introduction to Advanced Electronic Structure Theory*, Dover Publications Inc., Mineola, New York.

34. Engemann, C., Kohring, G., Pantelouris, A., Hormes, J., Grimme, S., Peyerimhoff, S. D., Clade, J., Frick, F., and Jansen, M. (1997) Experimental and theoretical investigations of the X-ray absorption near edge spectra (XANES) of P_4O_6 and P_4O_6X (X = O, S, Se), *Chem. Phys.*, **221**, 189–198.

35. Minkov, I., Gel'mukhanov, F., Agren, H., Friedlein, R., Suess, C., and Salaneck, W. R. (1995) Core excitations of biphenyl, *J. Phys. Chem. A*, **109**, 1330–1336.

36. Besley, N. A., and Asmuruf, F. A. (2010) Time-dependent density functional theory calculations of the spectroscopy of core electrons, *Phys. Chem. Chem. Phys.*, **12**, 12024–12039.

37. Stener, M., Fronzonia, G., and De Simone, M. (2003) Time dependent density functional theory of core electrons excitations, *Chem. Phys. Lett.*, **373**, 115–123.

38. Eisberg, R., and Resnick, R. (eds) (1985) *Quantum Physics of Atoms, Molecules, Solids, Nuclei, and Particles*, John Wiley & Sons, New York.

39. Della Sala, F., and Görling, A. (2001) Efficient localized Hartree–Fock methods as effective exact-exchange Kohn–Sham methods for molecules, *J. Chem. Phys.*, **115**, 5718–5732 .

40. Wasserman, A., Maitra, N. T., and Burke, K. (2003) Accurate rydberg excitations from the local density approximation, *Phys. Rev. Lett.*, **91**, 263001–263005.

41. Ciofini, I., and Adamo, C. (2007) Accurate evaluation of valence and low-lying Rydberg states with standard time-dependent density functional theory, *J. Phys. Chem. A*, **111**, 5549–5556.

42. Peach, M. J. G., Benfield, P., Helgaker, T., and Tozer, D. J. (2008) Excitation energies in density functional theory: An evaluation and a diagnostic test, *J. Chem. Phys.*, **128**, 044118/1–8.

43. Della Sala, F., and Fabiano, E. (2011) Accurate singlet and triplet excitation energies using the Localized Hartree–Fock Kohn–Sham potential, *Chem. Phys.*, **391**, 19–26.

44. Kochi, J. K. (1991) Charge-transfer excitation of molecular complexes in organic and organometallic chemistry, *Pure Appl. Chem.*, **63**, 255–264.

45. Vogler, A., and Kunkely, H. (2004) Charge transfer excitation of organometallic compounds: Spectroscopy and photochemistry, *Coordin. Chem. Rev.*, **248**, 273–278.

46. Tozer, D. J. (2003) Relationship between long-range charge-transfer excitation energy error and integer discontinuity in Kohn–Sham theory, *J. Chem. Phys.*, **119**, 12697–12699.

47. Dreuw, A., and Head-Gordon, M. (2004) Failure of time-dependent density functional theory for long-range charge-transfer excited states: The Zincbacteriochlorin–Bacteriochlorin and Bacteriochlorophyll–Spheroidene Complexes, *J. Am. Chem. Soc.*, **126**, 4007–4016.

48. Autschbach, J. (2009) Charge-transfer excitations and time-dependent density functional theory: Problems and some proposed solutions, *Chem. Phys. Chem*, **10**, 1757–1760.

49. Dreuw, A., Plötner, J., Wormit, M., Head-Gordon, M., and Dean Dutoi, A. (2010) An additive long-range potential to correct for the charge-transfer failure of time-dependent density functional theory, *Z. Phys. Chem.*, **224**, 311–324.

50. Norman, P., and Ruud, K. (2006) Microscopic theory of nonlinear optics, in *Non-Linear Optical Properties of Matter* (ed. Papadopoulos, M. G., *et al.*) Springer, pp. 1–49.

51. Marian, C. M. (2001) Spin-orbit coupling in molecules, in *Reviews in Computational Chemistry*, vol. 17 (ed. Lipkowitz, K. B., and Boyd, D. B.), John Wiley & Sons, New York, USA.

52. Yarkony, D. R. (1996) Diabolical conical intersections, *Rev. Mod. Phys.*, **68**, 985–1013.

53. Yarkony, D. R. (2001) Conical intersections: The new conventional wisdom, *J. Phys. Chem. A*, **105**, 6277–6293.

54. Kästner, J., Carr, J. M., Keal, T. W., Thiel, W. Wander, A., and Sherwood, P. (2009) DL-FIND: An open-source geometry optimizer for atomistic simulations, *J. Phys. Chem. A*, **113**, 11856–11865.

55. Ozkan, I., and Goodman, L. (1979) Coupling of electronic and vibrational motions in molecules, *Chem. Rev.*, **79**, 275–285.

56. Lucas, N. J. D. (1973) The Franck–Condon principle for polyatomic molecules, *J. Phys. B: At. Mol. Phys.*, **6**, 155–163.

57. Kupka, H., and Cribb, P. H. (1986) Multidimensional Franck–Condon integrals and Duschinsky mixing effects, *J. Chem. Phys.*, **85**, 1303–1326.

58. Dierksen, M., and Grimme, S. (2004) Density functional calculations of the vibronic structure of electronic absorption spectra, *J. Chem. Phys.*, **120**, 3544–3555.

59. Santoro, F., Improta, R., Lami, A., Bloino, J., and Barone, V. (2007) Effective method to compute Franck–Condon integrals for optical spectra of large molecules in solution, *J. Chem. Phys.*, **126**, 084509/1–13.

60. Santoro, F., Lami, A., Improta, R., and Barone, V. (2007) Effective method to compute vibrationally resolved optical spectra of large

molecules at finite temperature in the gas phase and in solution, *J. Chem. Phys.*, **126**, 184102/1–11.

61. Santoro, F., Lami, A., Improta, R., Bloino, J., and Barone, V. (2008) Effective method for the computation of optical spectra of large molecules at finite temperature including the Duschinsky and Herzberg-Teller effect: The Qx band of porphyrin as a case study, *J. Chem. Phys.*, **128**, 224311/1–17.
62. Simons, J. (ed.) (1983) *Energetic Principles of Chemical Reactions*, Jones and Bartlett, Boston.
63. Domcke, W., Yarkony, D., and Köppel, H. (eds) (2004) *Conical Intersections: Electronic Structure, Dynamics and Spectroscopy*, World Scientific, Singapore.
64. Domcke, W., Yarkony, D., and Köppel, H. (eds) (2011) *Conical Intersections: Theory, Computation and Experiment* , World Scientific, Singapore.
65. Etinski, M., Tatchen, J., and Marian, C. M. (2011) Time-dependent approaches for the calculation of intersystem crossing rates, *J. Chem. Phys.*, **134**, 154105–154114.
66. Fabiano, E., Della Sala, F., Cingolani, R., Weimer, M., and Görling, A. (2005) Theoretical study of singlet and triplet excitation energies in oligothiophenes, *J. Phys. Chem. A*, **109**, 3078–3085.
67. Messiah, A. (ed.) (1966) *Quantum Mechanics*, vol. 1, John Wiley & Sons, New York.
68. Landau, L. D., and Lifschitz, E. M. (eds) (1976) *Mechanics*, vol. 1 of Course of Theoretical Physics, Pergamon Press, Oxford.
69. Landau, L. D., and Lifschitz, E. M. (eds) (1959) *Fluid Mechanics*, vol. 6 of Course of Theoretical Physics, Pergamon Press, Oxford.
70. Hammes-Schiffer, S., and Tully, J. C. (1994) Proton transfer in solution: Molecular dynamics with quantum transitions, *J. Chem. Phys.*, **101**, 4657–4668.
71. Fabiano, E., Keal, T. W., and Thiel, W. (2008) Implementation of surface hopping molecular dynamics using semiempirical methods, *Chem. Phys.*, **349**, 334–347.
72. Fabiano, E., Lan, Z., Lu, Y., and Thiel, W. (2011) Nonadiabatic trajectory calculations with ab initio and semiempirical methods, in *Conical intersections: Theory, Computation and Experiment* (ed. Domcke, W., Yarkony, and D., Köppel), World Scientific.
73. Parr, R. G., and Yang, W. (1989) *Density-Functional Theory of Atoms, and Molecules*, Oxford University Press, New York.

74. Koch, W., and Holthausen, M. C. (eds) (2000) *A Chemist's Guide to Density Function al Theory*, Wiley-VCH, Weinheim.

75. Sherrill, C. D., and Schaefer III, H. F. (1999) The configuration interaction method: Advances in highly correlated approaches, *Adv. Quantum Chem.*, **34**, 143–269.

76. Fetter, A. L., and Walecka, J. D. (2003) *Quantum theory of many-particle systems*, Dover Publications, New York.

77. Gross, E. K. U., Dobson, J. F., and Petersilka, M. (1996) Density functional theory of time-dependent phenomena, in *Density Functional Theory, Topics in Current Chemistry*, vol. 181 (ed. Nalewajski, R. F.), Springer-Verlag Berlin Heidelberg, p. 81.

78. Dirac, P. A. M. (1930) Note on exchange phenomena in the Thomas atom, *Proc. Cambridge Philos. Soc.*, **26**, 376–385.

79. Rico, R. J., and Head-Gordon, M. (1993) Single-reference theories of molecular excited states with single and double substitutions, *Chem. Phys. Lett.*, **213**, 224–232.

80. Lischka, H., Müller, T., Szalay, P. G., Shavitt, I., Pitzer, R. M., and Shepard, R. (2011) Columbus—a program system for advanced multireference theory calculations, *Wiley Interdisciplinary Reviews: Computational Molecular Science*, **1**, 191–199.

81. Roos, B. O. (1987) The Complete Active Space Self-Consistent Field Method and its Applications in Electronic Structure Calculations, in *Advances in Chemical Physics* (ed. Lawly, K. P.), John Wiley & Sons, Chichester, England.

82. Silva-Junior, M. R., and Thiel, W. (2010) Benchmark of electronically excited states for semiempirical methods: MNDO, AM1, PM3, OM1, OM2, OM3, INDO/S, and INDO/S2, *J. Chem. Theory Comput.*, **6**, 1546–1564.

83. Head-Gordon, M., Rico, R. J., Oumi, M., and Lee, T. J. (1994) A doubles correction to electronic excited states from configuration interaction in the space of single substitutions, *Chem. Phys. Lett.*, **219**, 21–29.

84. Head-Gordon, M., Grana, A. M., Maurice, D., and White, C. A. (1995) Analysis of electronic transitions as the difference of electron attachment and detachment densities, *J. Phys. Chem.*, **99**, 14261–14270.

85. Grimme, S., and Waletzke, M. (2000) Multi-reference Møller-Plesset theory: computational strategies for large molecules, *Phys. Chem. Chem. Phys.*, **2**, 2075–2081.

86. McDoull, J. J. W., Peasley, K., and Robb, M. A. (1998) A simple MC SCF perturbation theory: Orthogonal valence bond Møller-Plesset 2 (OVB MP2), *Chem. Phys. Lett.*, **148**, 183–189.
87. Roos, B. O., Linse, P., Siegbahn, P. E. M., and Blomberg, M. R. A. (1982) A simple method for the evaluation of the second-order-perturbation energy from external double-excitations with a CASSCF reference wavefunction, *Chem. Phys.*, **66**, 197–207.
88. Anderson, K., Malmqvist, P.-Å., Roos, B. O., Sadlej, A. J., and Wolinski, K. (1990) Second-order perturbation theory with a CASSCF reference function, *J. Phys. Chem.*, **94**, 5483–5488.
89. Anderson, K., Malmqvist, P.-Å., and Roos, B. O. (1992) Second-order perturbation theory with a complete active space self-consistent field reference function, *J. Chem. Phys.*, **96**, 1218–1227.
90. Anderson, K., and Roos, B. O. (1993) Multiconfigurational second-order perturbation theory: A test of geometries and binding energies, *Int. J. Quantum Chem.*, **45**, 591–607.
91. Ghigo, G., Roos, B. O., and Malmqvist, P.-Å. (2004) A modified definition of the zeroth-order Hamiltonian in multiconfigurational perturbation theory (CASPT2), *Chem. Phys. Lett.*, **396**, 142–149.
92. Finley, J., Malmqvist, P.-Å., Roos, B. O., and Serrano-Andrés, L. (1998) The multi-state CASPT2 method, *Chem. Phys. Lett.*, **288**, 299–306.
93. Hirata, S., Head-Gordon, M., and Bartlett, R. J. (1999) Configuration interaction singles, time-dependent Hartree–Fock, and time-dependent density functional theory for the electronic excited states of extended systems, *J. Chem. Phys.*, **111**, 10774–10791.
94. Stanton, J. F., Gauss, J., Ishikawa, N., and Head-Gordon, M. (1995) A comparison of single reference methods for characterizing stationary points of excited state potential energy surfaces, *J. Chem. Phys.*, **103**, 4160–4175.
95. Runge, E., and Gross, E. K. U. (1984) Density-functional theory for time-dependent systems, *Phys. Rev. Lett.*, **52**, 997–1000.
96. Hohenberg, P., and Kohn, W. (1964) Inhomogeneous electron gas, *Phys. Rev.*, **136**, B864–B871.
97. Kohn, W., and Sham, L. J. (1965) Self-consistent equations including exchange and correlation effects, *Phys. Rev.*, **140**, A1133–A1138.
98. Iwata., J. I., Yabana, K., and Bertsch, G. F. (2001) Real-space computation of dynamic hyperpolarizabilities, *J. Chem. Phys.*, **115**, 8773–8784.

99. Marques, M. A. L., Castro, A., Bertsch, G. F., and Rubio, A. (2003) octopus: A first-principles tool for excited electron-ion dynamics, *Comput. Phys. Commun.*, **151**, 60–78.

100. Castro, A., Marques, M. A. L., and Rubio, A. (2004) Propagators for the time-dependent Kohn–Sham equations, *J. Chem. Phys.*, **121**, 3425–3434.

101. Takimoto, Y., Vila, F. D., and Rehr, J. J. (2007) Real-time time-dependent density functional theory approach for frequency-dependent non-linear optical response in photonic molecules, *J. Chem. Phys.*, **127**, 154114/1–10.

102. Akama, T., and Nakai, H. (2010) Short-time Fourier transform analysis of real-time time-dependent Hartree–Fock and time-dependent density functional theory calculations with Gaussian basis functions, *J. Chem. Phys.*, **132**, 054104/1–11.

103. McWeeny, R. (ed.) (1992) *Methods of Molecular Quantum Mechanics*, Accademic Press, London.

104. Becke, A. D. (1993) Density-functional thermochemistry. III. The role of exact exchange, *J. Chem. Phys.*, **98**, 5648–5651.

105. Lee, C., Yang, W., and Parr, R. G. (1988) Development of the Colle-Salvetti correlation-energy formula into a functional of the electron density, *Phys. Rev. B*, **37**, 785–789.

106. Perdew, J. P., Ernzerhof, M., and Burke, K. (1996) Rationale for mixing exact exchange with density functional approximations, *J. Chem. Phys.*, **105**, 9982–9986.

107. Della Sala, F., and Görling, A. (2001) Efficient localized Hartree–Fock methods as effective exact-exchange Kohn–Sham methods for molecules, *J. Chem. Phys.*, **15**, 5718–5732.

108. Silva-Junior, M. R., Schreiber, M., Sauer, S. P. A., and Thiel, W. (2008) Benchmarks for electronically excited states: Time-dependent density functional theory and density functional theory based multireference configuration interaction, *J. Chem. Phys.*, **129**, 104103/1–14.

109. Cai, Z.-L., Sendt, K., and Reimers, J. R. (2002) Failure of density-functional theory and time-dependent density-functional theory for large extended π-systems, *J. Chem. Phys.*, **117**, 5543–5550.

110. Grimme, S., and Parac, M. (2003) Substantial errors from time-dependent density functional theory for the calculation of excited states of large π-systems, *Chem. Phys. Chem.*, **4**, 292–295.

111. Cave, R. J., Zang, F., Maitra, N. T., and Burke, K. (2004) A dressed TDDFT treatment of the 2^1A_g states of butadiene and hexatriene, *Chem. Phys. Lett.*, **389**, 39–42.

112. Maitra, N. T., Zhang, F., Cave, R. J., and Burke, K. (2004) Double excitations within time-dependent density functional theory linear response, *J. Chem. Phys.*, **120**, 5932–5938.

113. Tozer, D. J., Amos, R. D., Handy, N. C., Roos, B. J., and Serrano-Andrés, L. (1999) Does density functional theory contribute to the understanding of excited states of unsaturated organic compounds, *Mol. Phys.*, **97**, 859–868.

114. Dreuw, A., Weisman, J. L., and Head-Gordon, M. (2003) Long-range charge-transfer excited states in time-dependent density functional theory require non-local exchange, *J. Chem. Phys.*, **119**, 2943–2947.

115. Sobolewski, A. L., and Domcke, W. (2003) Ab initio study of the excited-state coupled electron-proton-transfer process in the 2-aminopyridine dimer, *Chem. Phys.*, **294**, 73–83.

116. Fabiano, E., and Della Sala, F. (2007) Localized exchange-correlation potential from second-order self-energy for accurate Kohn–Sham energy gap, *J. Chem. Phys.*, **126**, 214102/1-15.

117. Della Sala, F., and Görling, A. (2003) Excitation energies of molecules by time-dependent density functional theory based on effective exact exchange Kohn–Sham potentials, *Int. J. Quantum Chem.*, **91**, 131–138.

118. Tawada, Y., Tsuneda, T., Yanagisawa, S., Yanai, T., and Hirao, K. (2004) A long-range-corrected time-dependent density functional theory, *J. Chem. Phys.*, **120**, 8425–8434.

119. Yanai, T., Tew, D. P., and Handy, N. C. (2004) A new hybrid exchange-correlation functional using the Coulomb-attenuating method (CAM-B3LYP), *Chem. Phys. Lett.*, **393**, 51–57.

Chapter 5

Metal–Molecule Electrodynamic Coupling

S. Corni
Centro S3, CNR - Istituto Nanoscienze, v. Campi 213/A, Modena, 41125, Italy
stefano.corni@unimore.it

In this chapter, an introduction to the problem of studying the electromagnetically mediated coupling between a molecule and a metal nanoparticle is given. Various models proposed in the literature for this problem are presented, and the most common relations needed to obtain molecular photophysical quantities from the solution of proper electrodynamics problems are discussed.

5.1 Introduction

Molecular plasmonics involves the interaction between molecules and plasmonics systems such as metal nanoparticles. Such interaction is always mediated by electromagnetic (EM) fields, which determine how the optical properties of the molecules are modified by the presence of the plasmonics system, and vice versa. In this chapter, we shall describe the origin of the coupling and how the relevant quantities can be calculated. In particular regimes

Handbook of Molecular Plasmonics
Edited by Fabio Della Sala and Stefania D'Agostino

ISBN 978-981-4303-20-0 (Hardcover), 978-981-4303-21-7 (eBook)
www.panstanford.com

(e.g. short metal–molecule distances), other interactions are also relevant, namely those grouped under the name of *chemical effects*. They include changes of the molecular ground state (and thus of molecular response properties) due to short range interactions such as local polarizations, dispersion and, when present, orbital mixing (i.e. incipient or fully developed chemical bonds) with or without charge transfer in the ground state. Chemical effects also include possible charge transfer between the excited molecule and the plasmonics system. While all these interactions are pivotal for a complete understanding of some molecular plasmonics phenomena, the rest of this chapter is focused on pure electromagnetic effects, and chemical effects will not be discussed further. Unraveling chemical effects and quantifying their importance is a challenging and largely unexplored field of research [1].

The interaction between plasmons and molecular excitations is involved in several different phenomena. The list below is an attempt to enumerate the experimentally observed surface-enhanced properties (as molecular plasmonics phenomena are often termed), and includes bibliographic references to representative works:

(1) **Electronic absorption (energies and band shapes)** [2–6]. In the case of dyes deposited on silver island films [2–4], when the absorption spectrum of the molecule and that of the metal specimen alone overlap, the spectrum of the composite system is not just the sum of the two spectra, but a different complex structure of peaks. The case of metal colloids, in which different molecules seem to show different behaviors [5, 6], is less clear.

(2) **Luminescence (lifetimes and energies)** [2–4, 7–12]. This is one of the most studied phenomena for the metal–molecule systems. The presence of a metal nanoparticle close to the molecule strongly modifies both the total decay rate and the radiative rate (see Sec. 1.7). Effects on excitation energies are also possible [13].

(3) **Vibrational absorption** [14, 15]. Molecules adsorbed on metal nanoparticles show an infrared light absorption (due to vibrational transitions) which is amplified of a factor 10^2–10^3 compared to the molecule without the metal. This phenomenon is called *surface-enhanced infrared absorption* (SEIRA).

(4) **Vibrational Raman scattering** [16–22]. *Surface-enhanced Raman scattering* (SERS) refers to the observation that certain molecules adsorbed on specially prepared metal surfaces or on metal colloids show Raman spectra with some bands enhanced by a factor 10^5–10^6 in comparison with the ones of non-adsorbed molecules. Recently, huge enhancement factors of 10^{10}–10^{12} have been also reported [23–33]. The metals that give SERS more easily are Ag, Au, Cu (coinage metals), although SERS has been reported for a large variety of metals (e.g. Li, Na, K, In, Pt, Pd and Rh) [34].

(5) **Second harmonic generation (SHG) and sum frequency generation (SFG)** [35–38]. As the linear process quoted in the previous items, also non-linear processes such as second harmonic generation (the creation of light at frequency 2ω when a sample is irradiated with light at ω) or sum frequency generation (creation of light at $\omega = \omega_1 + \omega_2$ from light at ω_1 and ω_2) can be strongly amplified by the presence of a metal surface. The terms used to indicate such amplified non-linear processes are *surface-enhanced second harmonic generation* (SESHG) and *surface-enhanced sum frequency generation* (SESFG).

(6) **Two-photon absorption** [39]. This process is strictly related to SHG and, as that one, it is found to be amplified (five orders of magnitude enhancement factors have been reported for dyes adsorbed on colloidal silver aggregate [39]).

(7) **Vibrational hyper-Raman scattering** [40–43]. Another non-linear phenomenon observed for metal/molecule system is the *surface-enhanced hyper-Raman scattering* (SEHRS), consisting in the emission from a sample irradiated with light at a frequency ω of a radiation with frequency $2\omega \pm \omega_m$, where ω_m is a frequency typical of the molecule (such as a vibrational frequency). For hyper-Raman signals enhancement factors of 10^{13} have been measured [41].

(8) **Photochemistry**. While plasmon effects on photochemistry have been proposed earlier [44–46], recently the process of charge separation induced by light, amplified by plasmons, has gained much attention for the potential use in photovoltaics applications [47].

(9) **Chiro-optical properties**. The optical properties specific of chiral molecules (optical rotation, circular dichroism, Raman optical activity (ROA)) may also be enhanced by the interaction with plasmons. This very interesting field is much less developed than others, although theoretical contributions have already been published (e.g. for surface-enhanced Raman optical activity, SEROA) [48, 49].

The investigation of electrodynamic coupling between molecules and metal nanoparticles, and therefore of the phenomena enumerated above, has exploited different levels of approximations so far. We can classify such approximations referring to two criteria. The first is the use of quantum electrodynamics (QED) *versus* classical electrodynamics to describe the coupling. The second is related to the levels of description used for the molecule and the metal nanoparticle. For this, three different options can be found in the literature:

- The molecule is a classical oscillating charge density (usually a point dipole) and the metal nanoparticle is a continuous body characterized by the frequency-dependent dielectric function (see Chapter 1). This is by far the most common description of the metal–molecule electrodynamic coupling problem in the literature. Notably, sometimes even the metal nanoparticle is reduced to a polarizable dipole. Depending on the phenomenon under study, this may be acceptable or results in an oversimplification [50].
- While the continuous body description of the metal is exploited, the molecule is treated atomistically by standard electronic structure techniques, such as time-dependent Hartree–Fock (TD-HF) or time-dependent density functional theory (TD-DFT) (see Sec. 4.4.2), and the electromagnetic interaction is included in the molecular Hamiltonian. This is a promising route not only to bypass inaccuracies related to the classical dipole model for the molecule, but also to go toward an *ab initio molecular plasmonics*. At present, this model has been explored mostly in the *polarizable continuum model* (PCM) group [51, 52, 54–58], but recently other implementations have been proposed [59].

- Finally, the whole system (molecule + metal nanoparticle) can be treated atomistically via TD-DFT or other quantum chemical methods. The interaction between the metal nanoparticle and the molecule are treated on the same foot as the intra-molecule and intra-nanoparticle ones. This method is therefore able to include much more than just the electrodynamics coupling, as it can include mutual polarization, chemical bonding, charge transfers (also in excited states). On the down side, at present this approach is limited to very small metal particles (a few tens of atoms, a few nm in size). Moreover, electrodynamics coupling is limited to the quasi-static limit, as standard molecular Hamiltonian includes only non-retarded Coulombic potential. Nevertheless, this method represents a fully ab initio approach to molecular plasmonics.

As for the quantum *versus* classical electrodynamics, QED description is of course the correct and complete theory to describe all the molecular plasmonics phenomena. Nevertheless, it requires the definition and the manipulation of quantities that are often not as intuitive as their classical counterparts. Moreover, classical electrodynamics is able to explain most of the molecular plasmonics phenomena, and, with a few expedients, even intrinsically quantum-mechanical phenomena such as spontaneous emission. Therefore, in the following we shall stick to a classical electrodynamics description of the system, and we refer the reader to other works for a QED treatment [60].

5.2 The Quasi-Static Limit

Propagation of EM fields is described by the Maxwell's equations [61]:

$$\nabla \cdot \tilde{\mathbf{D}} = 4\pi\tilde{\rho}\,, \tag{5.1}$$

$$\nabla \cdot \tilde{\mathbf{B}} = 0\,, \tag{5.2}$$

$$\nabla \times \tilde{\mathbf{E}} = \mathrm{i}\frac{\omega}{c}\tilde{\mathbf{B}}\,, \tag{5.3}$$

$$\nabla \times \tilde{\mathbf{H}} = \frac{4\pi}{c}\tilde{\mathbf{J}} - \mathrm{i}\frac{\omega}{c}\tilde{\mathbf{D}}\,, \tag{5.4}$$

where $\tilde{\mathbf{D}}, \tilde{\mathbf{E}}, \tilde{\mathbf{B}}, \tilde{\mathbf{H}}, \tilde{\mathbf{J}}$ and c have the usual meanings (see Sec. 1.1), while ω is the frequency of the oscillating fields (we are using a frequency-domain representation of the field time dependence). We used the notation introduced in Chapter 1.

As mentioned above, the molecule is typically treated as a field source (i.e. as a charge density and charge current), while the metal nanoparticle is considered as a region of space with different dielectric properties with respect to the medium that hosts the molecule and the nanoparticle.

Strictly speaking, the plasmonics nanoparticle–molecule coupling should be always described by such a set of equation. While it can be demonstrated that not all the components of $\tilde{\mathbf{E}}, \tilde{\mathbf{B}}, \tilde{\mathbf{H}}, \tilde{\mathbf{D}}$ fields are independent, it is still necessary to solve for at least one vectorial and one scalar field. However, Eqs. (5.1–5.4) greatly simplify when the intrinsic spatial variations of EM fields are smooth on the scale of the studied systems, i.e. when the latter are much smaller that the wavelength of the free propagating light at the relevant frequencies. In this situation we can in fact assume a wavevector $k = \omega/c \to 0$. Under this limit, a few terms in the equations can be disregarded, and they simplify to:

$$\nabla \cdot \tilde{\mathbf{D}} = 4\pi\tilde{\rho}\,, \tag{5.5}$$

$$\nabla \cdot \tilde{\mathbf{B}} = 0\,, \tag{5.6}$$

$$\nabla \times \tilde{\mathbf{E}} = 0\,, \tag{5.7}$$

$$\nabla \times \tilde{\mathbf{H}} = \frac{4\pi}{c}\tilde{\mathbf{J}}\,, \tag{5.8}$$

i.e. $\tilde{\mathbf{E}}$ is purely longitudinal and can thus be fully expressed in terms of a scalar electrostatic potential. This limit is often called *quasi-static approximation*, as the equations are the same for the static case (i.e. the Poisson problem for electrostatic potential derived from Eq. (5.5) and Eq. (5.7) but are written for time-dependent quantities. In the field of molecular plasmonics, practical tests have demonstrated that the quasi-static approximation provides reliable results for nanoparticle diameters around 1/10 of the wavelengths, i.e. up to 40–60 nm for visible light excitation. Of course, the quasi-static approximation does not break-down suddenly at these sizes, but electrodynamics correction terms (such as quadrupolar resonances or radiation reaction corrections) [62] increase gradually upon enlarging the metal particle and quasi-static results becomes less and less reliable.

5.3 The Point-Dipole Model of the Molecule, and the Classical Metal Nanoparticle

Among the three models mentioned in Sec. 5.1 for molecular plasmonics, that using the point dipole model for the molecule is the most apt to introduce molecule-electrodynamics coupling concepts [110, 111]. In this model, the molecule is punctiform, and only its dipolar properties are considered. In particular, the molecule is considered to have the same properties as in the vacuum (i.e. same transition moments, same polarizability).

The metal nanoparticle enters the optical phenomena by modifying the EM field acting in the position occupied by the molecule (for phenomena involving incident light) or by modifying the fields emitted by the molecule (for phenomena involving emitted light). In fact, while the molecular plasmonics phenomena listed in Sec. 5.1 differ in many details, the number of the involved EM fields in all these phenomena is quite limited. In particular, such EM fields may

(i) come from an external source (e.g. a laser probing the system, sunlight for photovoltaics applications) and be incident on the molecule + metal system. These EM fields affect all the phenomena where an external excitation source is needed, such as absorption, scatterings and photochemistry. It is to be remembered that also some phenomena such as enhanced fluorescence or phosphorescence may be affected by external fields, if the latter are used to take the molecule to the excited states;

(ii) be emitted by the molecule + metal system, such as in luminescence or scattering, where the far field is collected;

(iii) mediate the interaction between molecule and metal: these are present in all molecular plasmonics phenomena, but are not necessarily the most important to explain the phenomenon.

This is pictorially represented in Fig. 5.1, where all the involved fields are represented.

In the following we shall describe the various EM fields appearing in Fig. 5.1 and their role in plasmonics phenomena. To organize the discussion, we shall make reference to the absorption, the emission and the scattering of light.

5.3.1 *Light Absorption*

In the absorption of light by the molecule only the EM fields depicted in the panel "Incident" of Fig. 5.1 are involved. In this panel, four EM fields are defined: (i) $\tilde{\mathbf{E}}_{inc}$ is the *incident field*, unchanged with respect to the field in the bulk of the medium that hosts the molecule

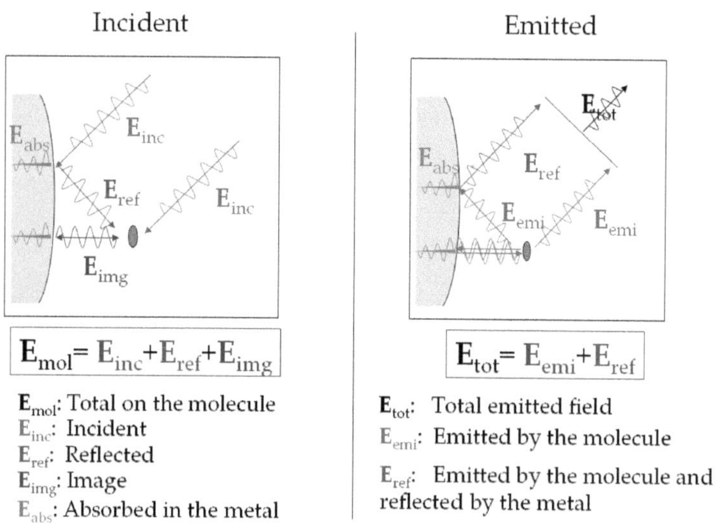

Figure 5.1 Pictorial representation of EM fields coming into play in optical phenomena involving metal–molecule systems. Note that in the right ("emitted") panel, the image field $\tilde{\mathbf{E}}_{img}$ is represented in terms of $\tilde{\mathbf{E}}_{emi}$ and $\tilde{\mathbf{E}}_{ref}$.

and the nanoparticle; (ii) $\tilde{\mathbf{E}}_{ref}$ is the *reflected field*, i.e. the field originating from the polarization induced in the metal nanoparticle by $\tilde{\mathbf{E}}_{inc}$; (iii) $\tilde{\mathbf{E}}_{img}$ is the field that is originated by the interaction with the metal of the field emitted by the molecule and it is called the *image field*; (iv) $\tilde{\mathbf{E}}_{abs}$ is the *field absorbed* in the metal, due to penetration in the metal of either the incident or the image field.

In terms of the Maxwell's equations, we can think to (i) $\tilde{\mathbf{E}}_{inc}$ as the solution of the homogeneous (i.e. no $\tilde{\rho}$ and $\tilde{\mathbf{J}}$ sources within the considered region) Maxwell's equations for the bulk matrix that hosts the molecule and the metal particle; (ii) $\tilde{\mathbf{E}}_{ref}$ as the difference between $\tilde{\mathbf{E}}_{inc}$ and the solution of the homogeneous Maxwell's equation in the presence of the metal nanoparticle described as a dielectric object, neglecting the molecule (boundary conditions far from the system must be set so to have an incoming field equal to $\tilde{\mathbf{E}}_{inc}$) and finally (iii) $\tilde{\mathbf{E}}_{img}$ as the solution of the inhomogeneous Maxwell's equation for the system, having as source terms the charge density and the current associated with the molecule. In particular, for a point dipole molecule, it turn out that the density is represented by the oscillating dipole term only, $\tilde{\mu}$, placed in the position $\mathbf{r}_{mol}$ and $\tilde{\mathbf{J}} = -i\omega\tilde{\mu}\delta(\mathbf{r} - \mathbf{r}_{mol})$. Which $\tilde{\mu}$ to be used (e.g, the transition dipole) will be discussed later.

Of course, a basic, yet unexpressed, assumption is the linearity of the electromagnetic response of the medium that contains the nanoparticle and the molecule (a solvent or a dielectric matrix), as we are assuming additivity of the EM fields.

$\tilde{\mathbf{E}}_{ref}$ is called "reflected" to recall the macroscopic optics (objects large with respect to wavelength), but it should be clear that for objects as nanoparticles, whose size is smaller or comparable to EM field wavelength, "reflected" field should be meant in a broad sense. In fact, the interaction of EM fields with objects smaller than the wavelength gives different results than with larger objects. $\tilde{\mathbf{E}}_{ref}$ is the EM field that includes the effect of the resonance of the nanoparticle surface plasmon, and that can be highly enhanced with respect to $\tilde{\mathbf{E}}_{inc}$ (electromagnetic enhancement). An example of the distribution of electric field ($\tilde{\mathbf{E}}_{inc} + \tilde{\mathbf{E}}_{ref}$) around a metal nanoparticle of complex shape is depicted in Fig. 5.2.

The image field $\tilde{\mathbf{E}}_{img}$ is called in this way after the method of images [61], i.e. the classical electrostatic technique of solving a

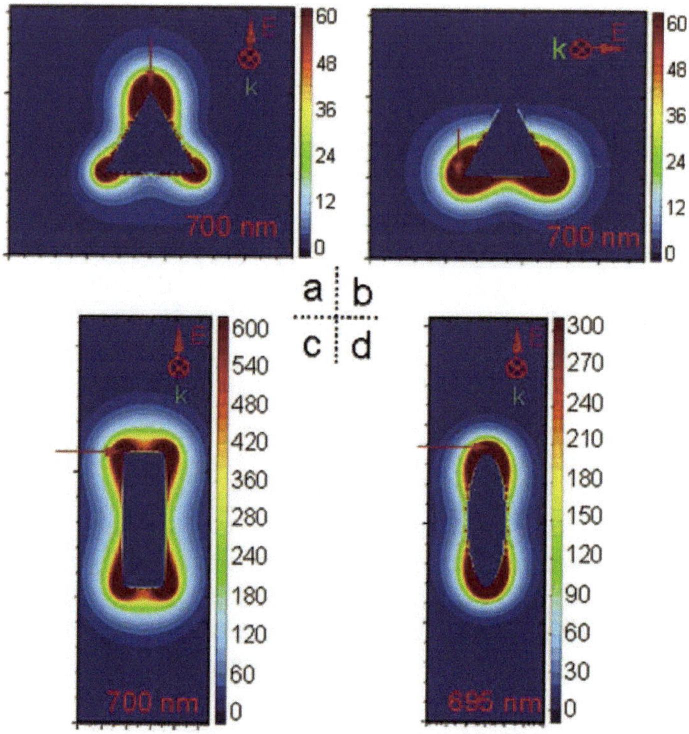

Figure 5.2 E-field enhancement contours external to monomers with different shapes. (a) and (b) are the E-field enhancement contours external to a triangular prism polarized along the two different primary symmetry axes, (c) and (d) are the E-fields enhancement contours for a rod and spheroid polarized along their long axes. The arrows show where is the maximum of E-filed. Results are obtained numerically via the discrete dipole approximation (DDA) method. Reprinted with permission from Ref. [64]. Copyright [2004], American Institute of Physics.

Poisson problem in the presence of conductors or dielectrics by adding point charges of proper position and magnitude (mirror, opposite charges for a semi-infinite planar conductor).

The total field acting on the molecule ($\tilde{\mathbf{E}}_{mol}$) is obtained as the sum of $\tilde{\mathbf{E}}_{inc}$, $\tilde{\mathbf{E}}_{ref}$ and $\tilde{\mathbf{E}}_{img}$. As mentioned above, $\tilde{\mathbf{E}}_{img}$ is often neglected as it results to be small with respect to $\tilde{\mathbf{E}}_{ref}$. When $\tilde{\mathbf{E}}_{img}$ is neglected, the molecule does not intervene directly in the definition of $\tilde{\mathbf{E}}_{mol}$, beside defining the position where $\tilde{\mathbf{E}}_{mol}$ should be calculated. Therefore, the problem is effectively transformed to the electromagnetic problem of determining the total EM field acting around a dielectric body (the metal nanoparticle). Various computational techniques have been developed to this aim (e.g. T-matrix method, DDA, *boundary elements method* (BEM), generalized Mie theory, finite-difference time domain (FDTD) method, finite elements method (FEM)) and used in the field of molecular plasmonics. They are treated in Chapter 2, and recent reviews are available [65, 66], so we shall not go into details here.

As mentioned above, $\tilde{\mathbf{E}}_{ref}$ (and thus $\tilde{\mathbf{E}}_{mol}$) can be highly amplified when the incident field has a frequency in resonance with a plasmon excitation. For metal nanoparticle small with respect to the wavelength, only the dipolar plasmon can be excited. It is educative to recall (see Sec. 1.4.1) the very simple case (a small spherical metal nanoparticle described by the Drude dielectric constant in the vacuum) and compare its absorption spectrum (dominated by the plasmon band) and the intensity of $\tilde{\mathbf{E}}_{ref}$ at a fixed point along the direction of oscillation of the dipolar plasmon. The absorption cross-section C_{abs} is given by (see Eq. (1.299)):

$$C_{abs} = \frac{4\pi\omega}{c}\mathrm{Im}\left[\tilde{\alpha}_{sph}(\omega)\right], \tag{5.9}$$

where $\tilde{\alpha}_{sph}(\omega)$ is the frequency-dependent isotropic polarizability of the nanoparticle, given by (see Eq. (1.193)):

$$\tilde{\alpha}_{sph}(\omega) = R^3\frac{\tilde{\varepsilon}_p(\omega) - 1}{\tilde{\varepsilon}_p(\omega) + 2}, \tag{5.10}$$

with the Drude dielectric constant given by (see Eq. (1.117)):

$$\tilde{\varepsilon}_p(\omega) = 1 - \frac{\omega_p^2}{\omega(\omega + \mathrm{i}\Gamma_0)}, \tag{5.11}$$

where ω_p is the plasma frequency and $\Gamma_0 = 1/\tau_0$ is the inverse of the mean relaxation time for the bulk metal.

The ratio $\left|\left|\tilde{\mathbf{E}}_{ref}\right|\right|^2 / \left|\left|\tilde{\mathbf{E}}_{inc}\right|\right|^2$ at a distance d from the metal surface along the incident field direction is instead given by:

$$\frac{\left|\left|\tilde{\mathbf{E}}_{ref}\right|\right|^2}{\left|\left|\tilde{\mathbf{E}}_{inc}\right|\right|^2} = \frac{4|\tilde{\alpha}_{sph}(\omega)|^2}{(R+d)^6}. \tag{5.12}$$

Substituting Eq. (5.11) in Eq. (5.10) and Eq. (5.12) we get:

$$C_{abs} = \frac{4\pi}{c} R^3 \omega_p^2 \frac{3\Gamma}{(\omega_p^2 - 3\omega^2)^2 + 9\Gamma^2\omega^2}, \tag{5.13}$$

$$\frac{\left|\left|\tilde{\mathbf{E}}_{ref}\right|\right|^2}{\left|\left|\tilde{\mathbf{E}}_{inc}\right|\right|^2} = 4\omega_p^4 \frac{R^6}{(R+d)^6} \frac{1}{(\omega_p^2 - 3\omega^2)^2 + 9\Gamma^2\omega^2}. \tag{5.14}$$

As the frequency dependence in the denominators are the same, for Drude spherical particles the absorption due to the plasmon excitation and the enhancement of the reflected field follow the same frequency dependence (in particular, they both peak at $\approx \omega_p/\sqrt{3}$).

However, it is important to remark that for more realistic cases, this overlap between absorption and enhancement is lost, and for complex shaped particles for particles arrays and even for spherical particles outside the quasi-static limit, the maximum absorption frequency of the plasmon and the maximum available total field may appear at quite different frequencies. This is shown for example in Fig. 5.3 for a silver spherical particles.

Once the total electric field acting on the molecule is known ($\tilde{\mathbf{E}}_{mol} = \tilde{\mathbf{E}}_{ref} + \tilde{\mathbf{E}}_{inc}$ as $\tilde{\mathbf{E}}_{img}$ has been neglected), the absorption

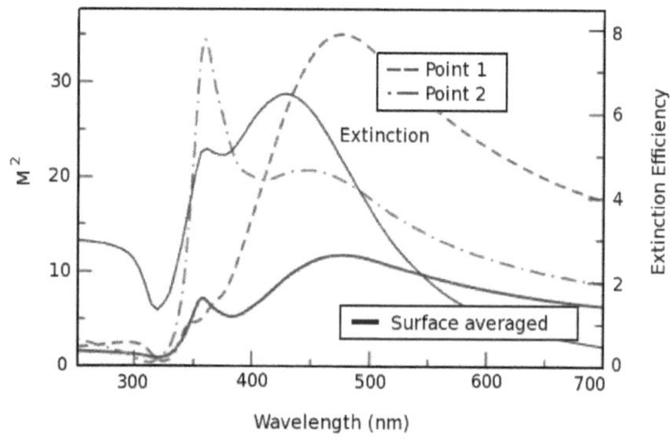

Figure 5.3 Near-field intensity enhancement $M^2 = \left|\left|\tilde{\mathbf{E}}_{mol}\right|\right|^2 / \left|\left|\tilde{\mathbf{E}}_{inc}\right|\right|^2$ (averaged on the surface and on two different single point at the surface) for a silver sphere (radius 60 nm at the bottom) as a function of the wavelength. Calculation performed with the Mie theory (see Ref. [67] for details). Also shown is the extinction cross section. Reprinted with permission from Ref. [67]. Copyright (2003) American Chemical Society.

cross-section for the molecule close to the plasmonics particle σ_{abs}^{env} with respect to the absorption cross-section for the free-standing molecule σ_{abs}^{0} (see Eq. (4.8) is (see also Sec. 1.7.3):

$$\sigma_{abs}^{env} = \sigma_{abs}^{0} \frac{\left|\left|\tilde{\mathbf{E}}_{mol}\right|\right|^2}{\left|\left|\tilde{\mathbf{E}}_{inc}\right|\right|^2}. \qquad (5.15)$$

An important caveat to this expression is that it is assuming that the light absorption by the molecule is far from saturation (i.e. the population in the excited state remains much smaller than that of the ground state). While this is generally the case, care must be taken as absorption enhancement can reach 4–5 orders of magnitude.

We end this section by remarking that neglecting $\mathbf{E}_{img}$ in the absorption does not allow to predict an interesting effects of the electrodynamics coupling between molecule and plasmon excitations: the shifts of the absorption frequencies, of both the molecule and the plasmonics nanoparticle.

5.3.2 *Light Emission*

The panel "Emitted" in Fig. 5.1 represents the EM fields that are involved in any process where the molecule is acting as a source of light. $\tilde{\mathbf{E}}_{emi}$ is in fact the electric field emitted by the molecule as in the absence of the metal nanoparticle, while $\tilde{\mathbf{E}}_{ref}$ is the field originated by the interaction of $\tilde{\mathbf{E}}_{emi}$ with the metal nanoparticle. We have also highlighted the fact that part of the emitted field is actually penetrating the metal particle and is dissipated there. The field that results from the superposition of $\tilde{\mathbf{E}}_{emi}$ and $\tilde{\mathbf{E}}_{ref}$ is the total field $\tilde{\mathbf{E}}_{tot}$ that can be measured by a far- or a near- field probe.

As mentioned before, our discussion is following a classical electrodynamics treatment of the EM fields. Spontaneous emission requires QED for a proper description. However, with some ad hoc assumptions, spontaneous emission can be described in a classical framework. Within this framework, spontaneous emission is understood as the emission of light due to an electrical dipole oscillating at a frequency ω corresponding to the de-excitation from the molecular state n to the ground state 0. To calculate the radiative decay rate γ_r of an excited molecule within this theory, one has to preliminary calculate the total power emitted by the dipole $\boldsymbol{\mu}_{cl}$.

Then, to heuristically introduce quantization of EM fields, one can assume that this power is due to the emission by the molecule of photons each carrying an energy $\hbar\omega$. By applying this procedure to a free-standing molecule (i.e. no metal nanoparticle), one would find that (see Sec. 1.5):

$$P_{rad} =< P^0_{em} >_T = \frac{\omega^4}{3c^3} ||\boldsymbol{\mu}_{cl}||^2 , \tag{5.16}$$

and

$$\gamma_r^{cl} = \frac{P_{rad}}{\hbar\omega} = \frac{\omega^3}{3\hbar c^3} ||\boldsymbol{\mu}_{cl}||^2 . \tag{5.17}$$

We now compare this expression with the proper one coming from QED:

$$\gamma_r^{\mathrm{QED}} = \frac{4\omega^3}{3\hbar c^3} ||\boldsymbol{\mu}_{0n}||^2 , \tag{5.18}$$

where $\boldsymbol{\mu}_{0n}$ is the transition dipole for the excitation from ground state 0 to state n. It is clear that the classical dipole $\boldsymbol{\mu}_{cl}$ should be chosen to be *twice* the transition dipole $\boldsymbol{\mu}_{0n}$ to obtain the correct radiative decay rate by Eq. (5.17) [61]:

$$\boldsymbol{\mu}_{cl} = 2\boldsymbol{\mu}_{0n} . \tag{5.19}$$

Many treatments of molecular plasmonics phenomena based on classical electrodynamics bypasses this possible source of mistakes by focusing on the ratio between the rates for the molecule close to the nanoparticle and the free-standing molecules, in such a way that the molecular dipole cancels out (for linear media).

Having clarified which value of the dipole should be used within a classical model of the emission, the problem of calculating the decay rate via photon emission has been recast in the classical electrodynamics problem of finding the power emitted by an oscillating dipole in a complex dielectric environment. This is in fact the energy flux method presented by Chance, Prock and Silbey [7], that was applied to the study of molecular fluorescence near metal interfaces (see Sec. 1.7.4). The technique is based on calculating (analytically or numerically, depending on the specific problem) the flux of the complex Poynting vector **S**, i.e. (in the vacuum):

$$\tilde{\mathbf{S}} = \frac{c}{8\pi} \tilde{\mathbf{E}}_{tot} \times \tilde{\mathbf{H}}^*_{tot} \tag{5.20}$$

$$P_{rad} = \mathrm{Re}\left[\oint_A \tilde{\mathbf{S}} \cdot \widehat{\mathbf{n}} \, dA\right] , \tag{5.21}$$

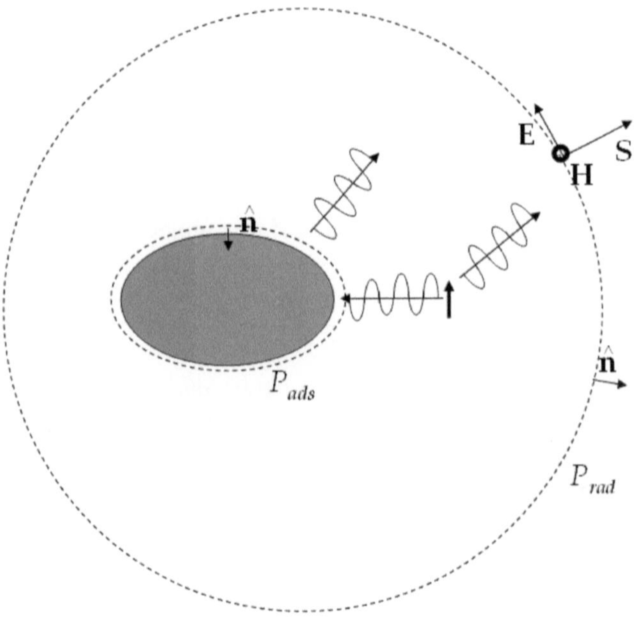

Figure 5.4 Pictorial representation of the surfaces that may be used to implement the energy flux method for a molecule + metal nanoparticle system.

on a suitable surface that enclose the molecule + metal system (see Fig. 5.4). $\hat{\mathbf{n}}$ is the normal pointing outside the region where the dipole is located.

Enhancement of the decay rate by photon emission is achieved as $\tilde{\mathbf{E}}_{emi}$ is exciting the oscillation of the nanoparticle plasmon, which in turn emits $\tilde{\mathbf{E}}_{ref}$ that sums up to $\tilde{\mathbf{E}}_{emi}$ to give and enhanced $\tilde{\mathbf{E}}_{tot}$ and, in turn, P_{rad}.

Part of the energy transferred from the molecule to the excitation of plasmon is actually dissipated within the metal nanoparticle. This is an additional channel of de-excitation of the molecule that is not leading to photon emission, and it is not included in P_{rad}. The energy flux method calculates the decay rate associated to this quenching by the net flux of $\tilde{\mathbf{S}}$ (equal to the absorbed power P_{abs}) through a closed surface that contains the metal nanoparticle but not the molecule (see Fig. 5.4). The non-radiative decay γ_{nr} is then calculated as for

γ_r, i.e.

$$\gamma_{nr} = \frac{P_{abs}}{\hbar\omega} \,. \tag{5.22}$$

Notably, the sum of $P_{abs} + P_{rad}$ corresponds to the total power emitted by the molecule, and therefore from this quantity the total decay rate induced by the metal ($\gamma_r^{tot} + \gamma_{nr}^{env}$) is obtained. As such, the luminescence quantum yield q^{tot} of a molecule close to a metal nanoparticle can be written as (see Sec. 1.7.3):

$$q^{tot} = \frac{\gamma_r^{tot}}{\gamma_r^{tot} + \gamma_{nr}^{env} + \gamma_{nr}^0} \,, \tag{5.23}$$

where γ_{nr}^0 is the intrinsic non radiative decay time for the free-standing molecule. Depending on the intrinsic quantum yield of the molecule ($q^0 = \gamma_r^0/(\gamma_r^0 + \gamma_{nr}^0)$) and the details of the studied system, the presence of the metal nanoparticle may result in a quenching or an enhancement of the quantum yield. Of course, for good fluorophores (i.e. those with a high q^0) enhancement of q^{tot} is more difficult than for poor fluorophores, as the quantum yield is already close to the maximum value (1). Quantities related to q that take into account also the enhanced absorption, and therefore the enhanced population in the excited state, can be defined, such as the *relative brightness* Φ^{RB} [56]:

$$\Phi^{\mathrm{RB}} = q^{tot} \frac{\sigma_{abs}^{env}}{\sigma_{abs}^0} \,. \tag{5.24}$$

While the energy flux method requires the calculation of a flux through a surface, other equations can be derived that make reference only to quantities defined for the molecule and the nanoparticle. In fact, P_{rad} can be considered as the power emitted by the oscillating molecular dipole and by the oscillating charge density induced in the nanoparticle by the field emitted by the molecule. If the latter is approximated by the dipole term only ($\boldsymbol{\mu}_{met,0n}$), P_{rad} is just the power of the light emitted by a total dipole $\boldsymbol{\mu}_{0n} + \boldsymbol{\mu}_{met,0n}$, i.e.:

$$\gamma_r^{tot} = \frac{4\omega^3}{3\hbar c^3} \left|\left|\boldsymbol{\mu}_{0n} + \boldsymbol{\mu}_{met,0n}\right|\right|^2 \,. \tag{5.25}$$

Note that Eq. (5.25) contains a *coherent sum* of two dipoles. In the case $\boldsymbol{\mu}_{met,0n}$ is *antiparallel* to $\boldsymbol{\mu}_{0n}$, the radiative rate may actually be depressed, instead of enhanced, by the presence of the metal

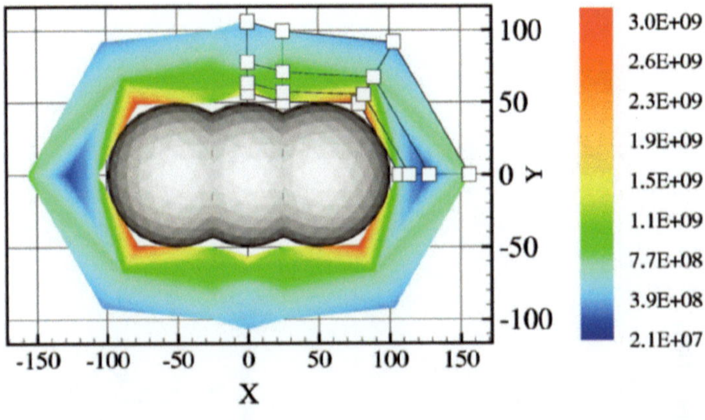

Figure 5.5 Radiative decay rate (in s^{-1}) calculated for a coumarin-type molecule near a silver aggregate composed of three identical interlocking spheres placed in a linear configuration. Results are depicted by using a color scale. The white squares indicate the molecule positions for which calculations have been performed (Cartesian axes are in Å). The molecular transition dipole moment is directed perpendicular to the metal surface. Details on model can be found in the original work. Reprinted with permission from Ref. [54]. Copyright [2004], American Institute of Physics.

nanoparticle. An example of this phenomenon (predicted by a more advanced model for the molecule, see Sec. 5.4) is depicted in Fig. 5.5. In fact, the blue spot (a "coldspot") along the direction of the particle main axis corresponds to the situation where the molecular dipole is almost compensated by the induced dipole, and the radiative emission is depressed.

The induced dipole being antiparallel to the original dipole may depend on (i) geometrical factors or on (ii) the frequency dependence of the metal response (see Fig. 5.6). The former refers to the fact that an oscillating dipole tangential to a surface of a non-absorbing dielectric (or a conductor) induces an antiparallel dipole in the dielectric, as the method of images reveals. The latter is a more subtle effect. For frequencies somewhat larger than the plasmon absorption frequency, the real part of the polarizability of the metal nanoparticle may become negative, while the imaginary part may already be small. This means that the induced dipole is in counter-phase with respect to its normal behavior, and therefore becomes

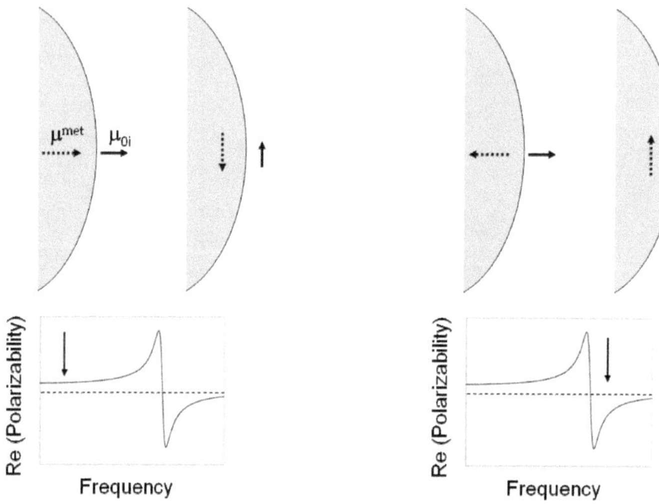

Figure 5.6 Pictorial representation of geometrical and frequency-related effects on the relative orientation of the molecular dipole and the induced dipole in the plasmonics particle. For frequencies far from the absorption edge (left panel), molecular transition dipoles tangential to the surface induces antiparallel dipole in the nanoparticle. For frequencies just above the surface plasmon excitation, the opposite may be true. Note that the magnitude of the induced dipole and the exact position where the inversion takes place depend on the dielectric function and the shape of the metal nanoparticle.

antiparallel to the molecular dipole when the latter is perpendicular to the particle surface.

Interestingly, the radiative rate suppression due to the geometrically-induced counter-phase behavior has been experimentally evidenced (see Fig. 5.7) [12].

Not only γ_r^{tot}, but also γ_{nr}^{env} can be obtained by skipping the calculation of the Poynting vector flux. To this aim it is central the standard electrodynamics equation:

$$\frac{1}{2}\int_V \tilde{\mathbf{J}}^* \cdot \tilde{\mathbf{E}} dV + 2i\omega \int_V (\tilde{w}_e - \tilde{w}_m) dV + \oint_A \tilde{\mathbf{S}} \cdot \hat{\mathbf{n}}\, dA = 0, \quad (5.26)$$

where $\tilde{w}_e$ and $\tilde{w}_m$ are the density of the electric and magnetic field energies, defined as:

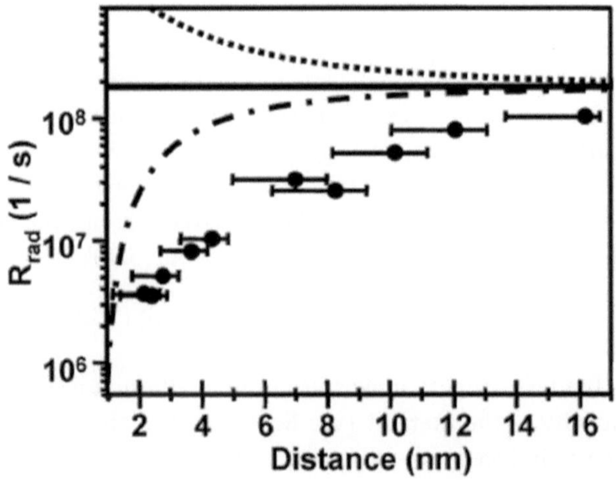

Figure 5.7 Measured radiative rate for Cy5 molecules as a function of their distance to the surface of a gold nanoparticle 12 nm in diameter. The molecule is bound to the nanoparticle via a DNA spacer, and the distance is fixed by changing the load of the nanoparticle and the length of the spacer (see the original work for details). Dotted and dashed-dotted line refers to computational results obtained by the Gersten–Nitzan model. Reprinted with permission from Ref. [70]. Copyright [1981], American Institute of Physics.

$$\tilde{w}_e = \frac{1}{16\pi}\tilde{\mathbf{E}} \cdot \tilde{\mathbf{D}}^*, \tag{5.27}$$

$$\tilde{w}_m = \frac{1}{16\pi}\tilde{\mathbf{B}} \cdot \tilde{\mathbf{H}}^*. \tag{5.28}$$

Equation (5.26) can be obtained by Maxwell's equations as shown, e.g. in Jackson [61]. V is a portion of space bounded by the surface A; $\hat{\mathbf{n}}$ is the outward normal to such surface. The real part of this equation reads:

$$\frac{1}{2}\mathrm{Re}\left[\int_V \tilde{\mathbf{J}}^* \cdot \tilde{\mathbf{E}} dV\right] - 2\omega\,\mathrm{Im}\left[\int_V (\tilde{w}_e - \tilde{w}_m) dV\right] + \mathrm{Re}\left[\oint_A \tilde{S} \cdot \hat{\mathbf{n}} dA\right] = 0. \tag{5.29}$$

If we choose for the surface A that depicted for P_{abs} in Fig. 5.4, the volume V coincides with the nanoparticle, and, considering that the normal $\hat{\mathbf{n}}$ used to defined P_{abs} in Fig. 5.4 is pointing inward, we obtain:

$$P_{abs} = \frac{1}{2}\mathrm{Re}\left[\int_V \tilde{\mathbf{J}}^* \cdot \tilde{\mathbf{E}}\, dV\right] - 2\omega\, \mathrm{Im}\left[\int_V (\tilde{w}_e - \tilde{w}_m) dV\right]. \quad (5.30)$$

Depending on how we interpret the dielectric response of the nanoparticle, Eq. (5.30) may be simplified in two different ways. If we assume a real dielectric function for the response of the metal electrons, and consider explicitly as free currents the electrical current $\tilde{\mathbf{J}}$ induced by the electric field, then Eq. (5.30) becomes:

$$P_{abs} = \frac{1}{2}\mathrm{Re}\left[\int_V \tilde{\mathbf{J}}^* \cdot \tilde{\mathbf{E}}\, dV\right] = \frac{1}{2}\int_V [\sigma(\omega)] \left|\left|\tilde{\mathbf{E}}\right|\right|^2 dV, \quad (5.31)$$

where the last equality makes use of the definition of the complex conductivity of the metal $\tilde{\sigma}$ (see Eq. 5.16). The absorption of light (and the so-induced non-radiative decay of the molecule) is thus due to Ohmic losses in the metal nanoparticle.

On the contrary, if we embed the conductive response of the nanoparticle in the dielectric function (i.e. without distinguishing between absorption due to free or bound charges), then $\tilde{\mathbf{J}} = 0$ and

$$P_{abs} = -2\omega\mathrm{Im}\left[\int_V (\tilde{w}_e - \tilde{w}_m) dV\right] = \frac{\omega}{8\pi}\int_V \mathrm{Im}[\tilde{\epsilon}(\omega)] \left|\left|\tilde{\mathbf{E}}\right|\right|^2 dV. \quad (5.32)$$

Both the expressions in Eq. (5.31) and Eq. (5.32) have been proposed and used to calculate $\gamma_{nr} = P_{abs}/\hbar\omega$.

We can do a step further and, in the quasi-static limit, express γ_{nr} in terms of the molecular transition dipole. To this aim, let us apply Eq. (5.26) to the volume W comprised between the two dashed surfaces in Fig. 5.4. We get:

$$P_{abs} + P_{rad} + \frac{1}{2}\mathrm{Re}\left[\int_V \tilde{\mathbf{J}}_{mol} \cdot \tilde{\mathbf{E}}^* dV\right] = 0, \quad (5.33)$$

where $\tilde{\mathbf{J}}_{mol}$ is the current density associated with the molecular transition dipole (see Sec. 1.5):

$$\tilde{\mathbf{J}}_{mol} = -\mathrm{i}\omega\, \boldsymbol{\mu}_{cl}\, \delta(\mathbf{r} - \mathbf{r}_{mol}). \quad (5.34)$$

Equation (5.33) and Eq. (5.34) lead to:

$$P_{abs} + P_{rad} = \frac{1}{2}\omega\mathrm{Im}\left[\boldsymbol{\mu}_{cl} \cdot \tilde{\mathbf{E}}(\mathbf{r}_{mol})\right], \quad (5.35)$$

which is therefore a further equation to calculate the total decay time of the molecule.

If we now apply the quasi-static (qs) approximation (i.e. considering $\omega/c \to 0$), it is easy to show that $P_{rad}^{qs} = 0$. In fact, taking the surface on which the flux of $\tilde{\mathbf{S}}$ is calculated large enough, only the dipole terms from the molecule and the induced density in the nanoparticle survive, and we can therefore write (from Eq. (1.216) and Eq. (1.217)):

$$\tilde{\mathbf{S}}^{qs} = \frac{i\omega}{8\pi} \frac{3\hat{\mathbf{n}}(\hat{\mathbf{n}} \cdot \boldsymbol{\mu}_{tot}) - \boldsymbol{\mu}_{tot}}{r^3} \times \frac{\hat{\mathbf{n}} \times \boldsymbol{\mu}_{tot}}{r^2} \propto \frac{1}{r^5} \tag{5.36}$$

where r is the distance from the molecular dipole. S^{qs} is decaying at least as $1/r^5$, and as we can choose a spherical surface centered on the dipole of arbitrarily large size:

$$P_{rad}^{qs} = \mathrm{Re}\left[\oint_A \tilde{\mathbf{S}}^{qs} \cdot \hat{\mathbf{n}}\, dA\right] \propto 1/r^3 \to 0\,. \tag{5.37}$$

Therefore:

$$\gamma_{nr}^{env,qs} = \frac{P_{abs}^{qs}}{\hbar\omega} = \frac{1}{2\hbar}\mathrm{Im}\left[\boldsymbol{\mu}_{cl} \cdot \tilde{\mathbf{E}}\right], \tag{5.38}$$

which can be further simplified by noting that only the image field $\tilde{\mathbf{E}}_{img}$ can contribute to Eq. (5.38):

$$\gamma_{nr}^{env,qs} = \frac{1}{2\hbar}\mathrm{Im}\left[\boldsymbol{\mu}_{cl} \cdot \tilde{\mathbf{E}}_{img}\right]. \tag{5.39}$$

Finally, we recall that the classical dipole moment $\boldsymbol{\mu}_{cl}$ should be twice the transition dipole $\mathbf{m}_{0n}$. As $\mathbf{E}_{img}$ is obtained in the linear response framework from the molecular dipole moment, it is also twice the reflected field calculated from the proper transition dipole, $\mathbf{E}_{img,0n}$. Therefore, the proper γ_{nr}^{qs} expression in terms of $\mathbf{m}_{0n}$ reads:

$$\gamma_{nr}^{env,qs} = \frac{2}{\hbar}\mathrm{Im}\left[\boldsymbol{\mu}_{0n} \cdot \tilde{\mathbf{E}}_{img,0n}\right]. \tag{5.40}$$

It is worth to be remarked that other classical treatments are possible, such as those based on the local density of state of the EM fields and dyadyc Green's functions. We refer to Refs. [7, 71, 72] for such an option.

5.3.3 *Light Scattering*

Once again, for the purpose of studying the effects of electrodynamics coupling between a molecule and a metal nanoparticle, the classical dipole model for the molecule sets an intuitive framework [73]. Scattering of light by a polarizable punctiform dipole can be seen as the emission of a dipole $\tilde{\boldsymbol{\mu}}_{sc}$ induced by the incident EM field at the frequency ω:

$$\tilde{\boldsymbol{\mu}}_{sc} = \tilde{\alpha}(-\omega, \omega)\tilde{\mathbf{E}}_{mol} \,. \tag{5.41}$$

If the polarizability of the dipole is modulated by a molecular vibration having frequency ω_v associated to a normal mode Q, then:

$$\tilde{\boldsymbol{\mu}}_{sc} = \left(\tilde{\alpha}_0(-\omega, \omega) + \frac{\mathrm{d}\tilde{\alpha}}{\mathrm{d}Q}\cos(\omega_v t) + \ldots\right)\tilde{\mathbf{E}}_{mol} \,, \tag{5.42}$$

where the term $\tilde{\alpha}_0(-\omega, \omega)$ is responsible for Rayleigh scattering (i.e. scattering at the same frequency of the incident light), while the second term is responsible for light scattered at frequencies $\omega - \omega_v$ (Stokes) and $\omega + \omega_v$ (anti-Stokes). This classical picture cannot explain, among other things, why anti-Stokes is much less intense that Stokes. However, if we stick to the relative enhancement of the signals due to the plasmon–molecule electrodynamics coupling, the classical view is acceptable.

The scattering of light by a molecule close to a metal nanoparticle involves all the fields depicted in Fig. 5.1 Therefore, from the point of view of the electrodynamic coupling between the molecule and the metal nanoparticle, we can see the light scattering as a problem involving:

(i) a molecule whose charge density is put into oscillatory motion by a local field $\tilde{\mathbf{E}}_{mol}$, enhanced with respect to the probing field due to the effects of the nearby plasmonics nanoparticle (mediated by $\tilde{\mathbf{E}}_{ref}$ in Fig. 5.1, left panel);
(ii) the EM field emitted by the oscillating molecular charge density, $\tilde{\mathbf{E}}_{emi}$, which is in turn amplified by interaction with the nanoparticle to yield a total far field $\tilde{\mathbf{E}}_{tot}$.

As such, we can use the results already discussed in the previous paragraphs, and we find that the enhancement G^{tot} of the total

Raman Stokes-scattered intensity for the mode Q reads:

$$G^{tot} = \frac{||\boldsymbol{\mu}_{met} + \boldsymbol{\mu}_{sc}||^2}{||\boldsymbol{\mu}_{sc}^0||^2}, \tag{5.43}$$

where $\boldsymbol{\mu}_{met}$ is the oscillating dipole induced in the nanoparticle at frequency $\omega - \omega_v$ by the oscillating charge distribution of the molecule. The latter has a dipole term $\boldsymbol{\mu}_{sc}$. $\boldsymbol{\mu}_{sc}^0$ is the dipole induced in the free-standing molecule by the incident field $\tilde{\mathbf{E}}_{inc}$.

Considering that the emitted intensity is proportional to the squared electric field modulus in the far-field region (see Eq. (1.222)), we can also write the enhancement factor G along a direction θ, ϕ as [74]:

$$G(\theta, \phi) = \frac{n_s c \, ||\tilde{\mathbf{E}}_{tot}(r, \theta, \phi)||^2}{8\pi < I_{emi}^0(r) >}, \tag{5.44}$$

and therefore

$$G(\theta, \phi) = \frac{n_s c \, ||\tilde{\mathbf{E}}_{tot}(r, \theta, \phi)||^2}{8\pi < I_{emi}(r) >} \frac{< I_{emi}(r) >}{< I_{emi}^0(r) >}, \tag{5.45}$$

where r is large enough to indicate a point far enough from the molecule to be in the radiation region; $< I_{emi}^0(r) >$ is the intensity emitted by the free-standing molecule averaged over all the directions (0 means that the molecule is polarized by $\tilde{\mathbf{E}}_{inc}$ only) and $< I_{emi}(r) >$ is the average intensity emitted by the molecule polarized by $\tilde{\mathbf{E}}_{mol}$, i.e. in the presence of the nanoparticle. Note that $< I(r) >= P/(4\pi r^2)$ where P is the emitted power (see Sec. 1.5). n_s is the refractive index of the matrix.

For the sake of clarity, from now on we shall consider an isotropic Raman tensor (i.e. $d\bar{\alpha}/dQ$ is behaving as a scalar quantity), and omit r from the notation. As the molecule and the metal are responding linearly to the field, we get that:

$$\frac{I_{emi}}{I_{emi}^0} = \frac{||\tilde{\mathbf{E}}_{mol}||^2}{||\tilde{\mathbf{E}}_{inc}||^2}. \tag{5.46}$$

The estimate of the enhancement factor due to the first term in Eq. (5.45) is often performed by invoking the *optical reciprocity theorem* (ORT) [75]. It states that: *the interaction of a dipole* $\boldsymbol{\mu}$ *placed in position O with an EM field* $\tilde{\mathbf{E}}_2$ *created by a dipole* $\boldsymbol{\mu}_2$ *in position M is equal to the interaction of the dipole* $\boldsymbol{\mu}_2$ *with the EM field* $\tilde{\mathbf{E}}$ *created by* $\boldsymbol{\mu}$, i.e.:

$$\boldsymbol{\mu} \cdot \tilde{\mathbf{E}}_2 = \boldsymbol{\mu}_2 \cdot \tilde{\mathbf{E}} \,. \tag{5.47}$$

We now consider $\boldsymbol{\mu}_2$ as a probe of the far field $\tilde{\mathbf{E}}_{tot}$ emitted by the total dipole $\boldsymbol{\mu} = \boldsymbol{\mu}_{met} + \boldsymbol{\mu}_{sc}$. The square modulus of this field (proportional to the light intensity) can be in general rewritten as

$$\left|\left|\tilde{\mathbf{E}}_{tot}(r,\theta,\phi)\right|\right|^2 = \frac{\left|\left|(\mu_2\widehat{\mathbf{e}}^{\theta})\cdot\tilde{\mathbf{E}}_{tot}\right|\right|^2 + \left|\left|(\mu_2\widehat{\mathbf{e}}^{\phi})\cdot\tilde{\mathbf{E}}_{tot}\right|\right|^2}{||\boldsymbol{\mu}_2||^2}, \tag{5.48}$$

where $\widehat{\mathbf{e}}^{\theta}$ and $\widehat{\mathbf{e}}^{\phi}$ are the unit vectors tangential to the propagation direction of the emitted fields in the far field region. If we now choose μ_2 to produce a field with modulus $E_{inc}(\omega - \omega_v)$ at the position of the molecule in the situation with no metal, i.e.

$$\mu_2 = \frac{rE_{inc}}{k^2\exp(ikr)}, \tag{5.49}$$

we get:

$$\left|\left|\tilde{\mathbf{E}}_{tot}(r,\theta,\phi)\right|\right|^2 = \frac{\left|\left(\frac{k^2\exp(ikr)}{r}\mu_2\widehat{\mathbf{e}}^{\theta}\right)\cdot\tilde{\mathbf{E}}_{tot}\right|^2 + \left|\left(\frac{k^2\exp(ikr)}{r}\mu_2\widehat{\mathbf{e}}^{\phi}\right)\cdot\tilde{\mathbf{E}}_{tot}\right|^2}{\left|\left|\tilde{\mathbf{E}}_{inc}(\omega-\omega_v)\right|\right|^2}. \tag{5.50}$$

Using the ORT to rewrite the numerator of this equation, we get:

$$\left|\left|\tilde{\mathbf{E}}_{tot}(r,\theta,\phi)\right|\right|^2 = \frac{\left|\frac{k^2\exp(ikr)}{r}\boldsymbol{\mu}_{sc}\cdot\tilde{\mathbf{E}}^{PW-\theta}\right|^2 + \left|\frac{k^2\exp(ikr)}{r}\boldsymbol{\mu}_{sc}\cdot\tilde{\mathbf{E}}^{PW-\phi}\right|^2}{\left|\left|\tilde{\mathbf{E}}_{inc}(\omega-\omega_v)\right|\right|^2}, \tag{5.51}$$

where $\tilde{\mathbf{E}}^{PW-\theta}$ is the local electric field $\tilde{\mathbf{E}}^{PW-\theta}_{mol} = \tilde{\mathbf{E}}^{PW-\theta}_{inc} + \tilde{\mathbf{E}}^{PW-\theta}_{ref}$ in the position of the molecule for an incident field with intensity E_{inc}, frequency $\omega - \omega_v$ and polarization along $\widehat{\mathbf{e}}^{\theta}$. Similar definitions hold for $\tilde{\mathbf{E}}^{PW-\phi}$. These fields may be obtained by solving the electrodynamics problems of a plane-wave EM field incident on the metal nanoparticle. If we now consider that:

$$< I_{emi} >= \left(\frac{2}{3}\right)\frac{n_s c}{8\pi}\left|\left|\frac{k^2\exp(ikr)}{r}\boldsymbol{\mu}_{sc}\right|\right|^2, \tag{5.52}$$

we get:

$$\frac{n_s c \left|\left|\tilde{\mathbf{E}}_{tot}(r,\theta,\phi)\right|\right|^2}{8\pi < I_{emi} >} = \frac{3}{2}\frac{(|\widehat{\mathbf{e}}_{sc}\cdot\tilde{\mathbf{E}}^{PW-\theta}|^2 + |\widehat{\mathbf{e}}_{sc}\cdot\tilde{\mathbf{E}}^{PW-\phi}|^2)}{||\mathbf{E}_{inc}(\omega-\omega_v)||^2}, \quad (5.53)$$

where $\widehat{\mathbf{e}}_{sc}$ is the unit vector associated with $\boldsymbol{\mu}_{sc}$. Therefore,

$$G(\theta,\phi) = \frac{3}{2}\frac{|\hat{\mathbf{e}}_{sc}\cdot\mathbf{E}^{PW-\theta}|^2 + |\hat{\mathbf{e}}_{sc}\cdot\mathbf{E}^{PW-\phi}|^2}{||\mathbf{E}_{inc}(\omega-\omega_v)||^2}\frac{||\mathbf{E}_{mol}(\omega)||^2}{||\mathbf{E}_{inc}(\omega)||^2}. \quad (5.54)$$

As discussed in the paper where this derivation is reported [74], this expression is not exactly equivalent to the one often used for SERS enhancement:

$$G_{approx} \approx \frac{\left|\left|\tilde{\mathbf{E}}_{mol}(\omega)\right|\right|^4}{\left|\left|\tilde{\mathbf{E}}_{inc}(\omega)\right|\right|^4}. \quad (5.55)$$

There are two reasons for this discrepancy. The first, which is widely recognized, is that computing $\tilde{\mathbf{E}}_{mol}$ at the frequency ω, instead of $\omega-\omega_v$ may not be a good approximation as the metal response (and thus $\tilde{\mathbf{E}}_{ref}$) can be quite different at the two frequencies. The second reason is that the terms $|\widehat{\mathbf{e}}_{sc}\cdot\tilde{\mathbf{E}}^{PW-\theta}|^2 + |\widehat{\mathbf{e}}_{sc}\cdot\tilde{\mathbf{E}}^{PW-\phi}|^2$ is not equal to $|\left|\left|\tilde{\mathbf{E}}_{mol}\right|\right|^2$, even if the modulus of the incident field $\tilde{\mathbf{E}}_{inc}^{PW}$ is E_{inc} by construction [74]. Differences among G and G_{approx} can be relevant, as shown in Fig. 5.8

5.3.4 *The Dielectric Function of the Classical Metal Nanoparticle*

For time-dependent electrical perturbation, the typical assumption is that the metal nanoparticle behaves as a dielectric, characterized by a frequency-dependent permittivity $\tilde{\varepsilon}(\omega)$. Permittivities experimentally determined on bulk sample are almost invariably used. They need to be corrected with terms depending on the particle size. In fact, when the size of the metal particle has the same order of magnitude of the mean free path of conduction electrons in the bulk of the solid (tens of nanometers), it is necessary to take into account the scattering of the electrons at the metal particle surface. This is one aspect of a more general class of phenomena, known as *quantum size effects*. They are tightly related to the confinement of electrons in the metal particle and hence to the loss of the band structures typical of a bulk metal. Since this phenomenon regards mainly the valence

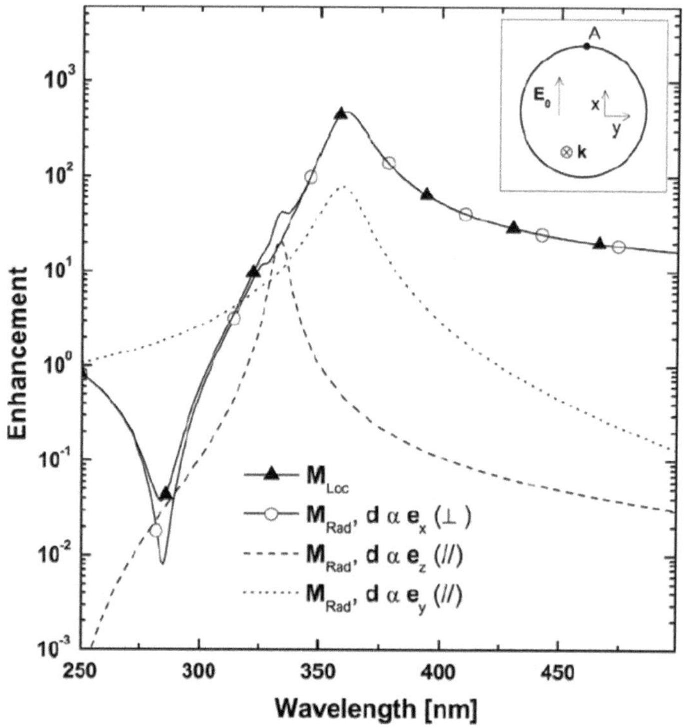

Figure 5.8 Enhancement of the local field $M_{Loc} = ||\tilde{\mathbf{E}}_{mol}||^2/||\tilde{\mathbf{E}}_{inc}||^2$ and of the scattered field $M_{Rad'} = (|\hat{\mathbf{e}}_{sc} \cdot \mathbf{E}^{PW-\theta}|^2 + |\hat{\mathbf{e}}_{sc} \cdot \mathbf{E}^{PW-\phi}|^2)/||\mathbf{E}_{inc}(\omega - \omega_v)||^2$ for a dipole $\mathbf{d}$ in a position A (see the inset) close to a spherical metal nanoparticle. The various M_{Rad} curves refer to different orientations of the dipole. Reprinted from Ref. [74], Copyright (2006), with permission from Elsevier.

electrons, it is necessary to decompose the dielectric constant into two contributions, one arising from the core (ions and bounded electrons) which will not be modified and the other arising from the valence (free) electrons:

$$\tilde{\varepsilon}(\omega) = \tilde{\varepsilon}_b(\omega) + \tilde{\varepsilon}_f(\omega) - 1\,. \qquad (5.56)$$

This partition has been used for example by Zeman and Schatz [76]. In the classical Drude free-electron model (see Sec. 1.2) $\tilde{\varepsilon}_f$ can be expressed as:

$$\tilde{\varepsilon}_f(\omega) = 1 - \frac{\omega_p^2}{\omega(\omega + \mathrm{i}/\tau_0)}\,, \qquad (5.57)$$

where ω_p is the plasma frequency and τ_0 the relaxation time suitable for the bulk of the metal.

The effects related to the reduced mean free path can be treated in the case of spherical particles with a relatively simple classical model [77], which gives essentially the same results of more rigorous quantum theories [78, 79] (see Sec. 2.3).

Other quantum size effects [79] (e.g. the variation in the resonance frequency of the surface plasmon or in the polarizability of the particle) are often neglected as they require to go beyond the classical, local continuum dielectric model.

The permittivity discussed so far depends only on frequencies and, through the relaxation times, on the particle size. It is well-known that the dielectric response of materials, in particular metal, is *non-local*, i.e. the polarization vector induced at a certain point depends on the values of the electric field in all other points. In the reciprocal space language, we can say that $\tilde{\varepsilon}(\omega)$ depends also on the wavevector of the various plane-waves in which the probing electric field can be decomposed. The permittivity of metals such as Ag, Au and Cu at optical frequencies mainly depends on the behavior of both the valence electrons, which is close to that of a free-electron gas, and the core of the metal. As we did before, the total dielectric constant of the metal $\tilde{\varepsilon}(\omega, q)$, depending on both the frequency ω and the wavevector modulus q, can be written as:

$$\tilde{\varepsilon}(\omega, q) = \tilde{\varepsilon}_b(\omega, q) + \tilde{\varepsilon}_f(\omega, q) - 1\,, \tag{5.58}$$

where $\tilde{\varepsilon}_f$ is the free-electron contribution and $\tilde{\varepsilon}_b$ is the remaining part, collecting among others the contributions from the "bound" core electrons. It is reasonable to assume that the dependency on q is more pronounced for $\tilde{\varepsilon}_f$ than for $\tilde{\varepsilon}_b$. In fact, by definition, $\tilde{\varepsilon}_f$ is due to the most delocalized ("free") electrons in the metal, which sample an applied electric field in a quite large region and are thus the most sensitive to the spatial variation of the field. Hence, one can assume that $\tilde{\varepsilon}_b(\omega, q) \approx \tilde{\varepsilon}_b(\omega, 0)$. However, note that the interband transitions contribution to $\tilde{\varepsilon}_{met}$ is always accounted for by $\tilde{\varepsilon}_b$, despite of the kind of electrons involved in the transitions. Thus, by approximating $\tilde{\varepsilon}_b(\omega, q)$ with $\tilde{\varepsilon}_b(\omega, 0)$ we are implicitly neglecting non-local effects on interband transitions, whatever electrons they involve. The quantity $\tilde{\varepsilon}_b(\omega, 0)$ can be found from the experimental dielectric constant of the bulk metal: since the probing light is characterized by small q values, the experiments give $\tilde{\varepsilon}_{met}(\omega, q \approx 0)$; then, by assuming a Drude form for $\tilde{\varepsilon}_f(\omega, 0)$, the $\tilde{\varepsilon}_b$ contribution reads:

$$\tilde{\varepsilon}_b(\omega, q) \approx \tilde{\varepsilon}_b(\omega, 0) = \tilde{\varepsilon}(\omega, 0) + \frac{\omega_p^2}{\omega(\omega + \mathrm{i}/\tau_0)} . \tag{5.59}$$

In order to introduce the q-dependence in the free-electron part of the metal permittivity, the so-called *hydrodynamic dielectric constant* [81] has been used:

$$\tilde{\varepsilon}_f(\omega, q) = 1 - \frac{\omega_p^2}{\omega(\omega + \mathrm{i}/\tau_0) - \beta^2 q^2} , \tag{5.60}$$

β is related to the Fermi velocity v_F of the electrons in the metal. For $\omega \rightarrow 0$, $\tilde{\varepsilon}_f$ should reduce to the usual Thomas–Fermi dielectric constant, i.e. β^2 should be $v_F^2/3$, but in order to reproduce the plasmon dispersion relation obtained at the *random phase approximation* (RPA) level it is necessary that at optical frequency $\beta^2 = 3v_F^2/5$ [81]. Eventually, the non local hydrodynamic permittivity that can be used to describe non-local effects, is:

$$\tilde{\varepsilon}(\omega, q) = \tilde{\varepsilon}(\omega, 0) + \frac{\omega_p^2}{\omega(\omega + \mathrm{i}/\tau_0)} - \frac{\omega_p^2}{\omega(\omega + \mathrm{i}/\tau_0) - \beta^2 q^2} . \tag{5.61}$$

The modified hydrodynamic dielectric constant of Eq. (5.61) is a useful first approximation and allows estimating the importance of non-locality. However, it takes into account at the lowest order only the dependence on q. In particular, it lacks an effect which is, very likely, important for the determination of the molecule decay rate: the excitation of electron–hole pairs. Such excitations are a way to transfer energy from the molecule to the metal, and then to dissipate such energy since the excited electron–hole pairs recombine mainly non-radiatively. Both energy and momentum must be conserved in the excitation of electron–hole pairs. Two sources can supply the momentum needed for the electron–hole pair excitation: one is the spatial variation of the oscillating electric field acting in the metal and the other is the metal surface potential, which is able to provide momentum in the direction perpendicular to the surface. When the electric field is originated by a molecule which is far from the metal, the excitation of electron–hole pairs due to the field spatial variation is inefficient, since the field is almost homogeneous in space (the carried momentum is almost zero) while its frequency is in the optical range. Thus it is not possible to satisfy conservation of both momentum and energy. For dipole transitions of the molecule, the quenching rate due to the electron–hole pair excitations made possible by the surface potential decays asymptotically (in the quasi-static framework) as $1/d^4$ [82, 83, 95], while the quenching rate due to the bulk local response of the metal decays slower, as $1/d^3$. Hence, for large metal–molecule distances, the quenching rate due to electron–hole pair excitations becomes negligible in comparison with that originating from the bulk local response.

When the molecule is close to the metal, the spatially inhomogeneous field can carry the needed momentum. To take into account these electron–hole pair excitations and, more generally, to improve the description of the q-dependence of the dielectric constant, one can use a *modified Lindhard–Mermin electric permittivity* [84, 85]. The Lindhard–Mermin dielectric function is the *Lindhard function* [86] corrected for a finite-electron relaxation time. This constant accounts only for the free electron behavior of the metal. We can include the (local) bound-electron term phenomenologically as done with the hydrodynamic constant. In this way, we obtain precisely the dielectric constant used by Ford and Weber in their pioneering work on SERS [87]. In particular, to be coherent with the quasi-static framework for the metal–molecule interaction, the longitudinal dielectric constant should be used. Following the proposal of Ref. [87]:

$$\tilde{\varepsilon}_{LM}(\omega, q) = \tilde{\varepsilon}(\omega, 0) + \frac{\omega_p^2}{\omega(\omega + \mathrm{i}/\tau_0)} + \frac{3\omega_p^2 u^2 f_l(z,u)}{(\omega + \mathrm{i}/\tau)(\omega + \mathrm{i}(f_l(z,u)/f_l(z,0))/\tau_0)}, \tag{5.62}$$

where z and u are defined as:

$$z = \frac{q}{2k_F} \quad , \quad u = \frac{(\omega + i/\tau_0)}{qv_F}, \tag{5.63}$$

and

$$f_l(z,u) = \frac{1}{2} + \frac{1}{8z}\left((1-(z-u)^2)\log\frac{z-u+1}{z-u-1} + (1-(z+u)^2)\log\frac{z+u+1}{z+u-1}\right). \tag{5.64}$$

In Eq. (5.63), k_F is the Fermi wavevector. This dielectric constant can be used in the framework of the specular scattering or *semiclassical infinite-barrier* (SCIB) model for planar metal surface [88]. Within the framework of such approach, the different descriptions of the metal response discussed so far (local dielectrics, hydrodynamic dielectric function and Lindhard–Mermin dielectric function) can be compared. This has been done in Ref. [89] for the metal-induced non radiative rate of a molecule close to an Ag surface. The results are summarized in Fig. 5.9.

For metal nanoparticle, only hydrodynamic corrections to the dielectric constant have been used so far in the calculations of plasmons effects on molecular luminescence [90–92]. As for planar surfaces the hydrodynamic constant and the more accurate Lindharm–Mermin one gave qualitative different results, it would be interesting to see if the same happens for nanoparticles as well. Quite generic non-local dielectric constant can be used with BEM [92, 93].

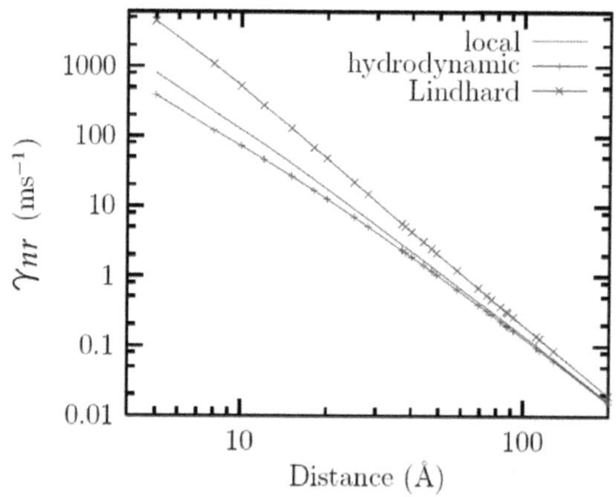

Figure 5.9 Non radiative decay rate γ_{nr} of biacetyl in an ammonia matrix as a function of the metal–molecule distance from a silver surface described through a local, a modified hydrodynamic [Eq. (5.61)] and a modified Lindhard–Mermin [Eq. (5.62)] dielectric constant. The molecular plane is parallel to the metal surface. Reprinted with permission from Ref. [89]. Copyright [2003], American Institute of Physics.

5.4 Toward an *ab initio* Molecular Plasmonics

In the classification of models for metal–molecule electrodynamic coupling that we have done in Sec. 5.1, the model that we have described so far (a classical punctiform dipole close to a metal nanoparticle described as a continuous medium) is the simplest. While it has proven to be extremely useful, not only as a mean to grasp the basic physics of molecular plasmonics phenomena, but also to provide semi-quantitative and, sometimes, even quantitative results, it still remains a model empirical in nature. In this section we shall briefly describe models that goes beyond such an approach.

5.4.1 *Coupling the* ab-initio *Description of the Molecule with a Continuous Metal*

A model where an ab initio description of the molecule is coupled to a continuous description of the metal nanoparticle and the hosting matrix has been proposed in Ref. [51] and further developed in the course of the years [52, 54–58]. Here, we briefly present the basics of such model, referring to the original papers for details.

The core idea is that the problem of a proper description of molecule-metal nanoparticle systems strongly resembles the one for solvated molecules. In fact, a molecule in solution weakly interacts with a great number of other electrons and nuclei, feeling all the fields of external as well as internal origin, as a molecule physisorbed on a metal surface does. A successful model for solutions, used also in the study of solute properties [94], considers the solvent as a *continuum dielectric medium* in which a cavity housing the solute is built [95]. The latter is described at a high quantum-mechanical level. This model is based on the idea that it is not necessary to describe the whole system (the solution) at the same degree of accuracy, but it is convenient to focus on the portion of the system we want to study (e.g. the solute molecule) and to consider only the features of the remaining part of the system which are involved in the interaction with the studied portion. The experience in describing optical effects of chromophores dispersed into a homogeneous dielectric phase [96, 97] through the use of the PCM [98] made such an approach natural also for the investigation of the properties of chromophores located near a metal surface. Thus the metal has been described as a continuous body characterized by electric response properties only. It behaves as a perfect conductor for static fields and as a dielectric for time-dependent fields. Infinite planar metal surfaces, complex shaped nanoparticles and aggregates of nanoparticles have been dealt with.

The chromophore has been treated at the Hartree–Fock or density functional theory level (see Chapter 4), in the determination of both its ground state and its properties. This model for the molecule represents a remarkable progress in the accuracy of the description of the molecular chromophore compared to polarizable point dipole model. The solvation effects have been described with the PCM. More in details, the solvent is described as a continuum dielectric which occupies all the space free from the metal specimen and the molecule is hosted in a molecular shaped cavity inside such dielectric.

Let us now pass to the description of the interactions between the various portions of the system. In the model of Ref. [51], the only considered interactions are electrostatics/electrodynamics in origin. The metal-molecule and solvent-molecule electrostatic and interactions have been treated through the *integral equation formalism* (IEF) [99], numerically solved by a BEM. The IEF is based on the possibility of writing the electrostatic potential inside a given region of space in terms of a fictitious surface density of charge placed on the region boundary. This density is obtained by solving an integral equation defined on such boundary. The BEM consists in translating the integral equation into a matrix equation by discretizing the boundary.

The electromagnetic interaction with a probing field and among the various oscillating densities of charge in the system is treated in the quasi-static approximation, and the resulting Poisson problems are once again solved with a IEF/BEM methodology.

From the point of view of the molecule, the electrostatic and/or electrodynamics interactions with the metal nanoparticle enter directly an effective Hamiltonian. This is formally identical to the problem of a molecule in solution, when the solvent is described by a continuum dielectric [98]. Thus, the ground state wavefunction should minimize the molecular electronic density functional, $\mathcal{G}_{tot}[\rho_M]$:

$$\mathcal{G}_{tot}[\rho_M] = \mathcal{E}_{vac}[\rho_M] + \frac{1}{2}\int d^3\mathbf{r}\, V(\rho_M;\mathbf{r})\rho_p(\mathbf{r})\,, \qquad (5.65)$$

where $V(\rho_M;\mathbf{r})$ is the potential generated in $\mathbf{r}$ by the polarized molecule, $\rho_p(\mathbf{r})$ is the metal polarization charge density and $\mathcal{E}_{vac}[\rho_M]$ is the energy of the molecule calculated as in the vacuum but for the molecular electronic density proper for the molecule close to the metal ρ_M. The metal polarization $\rho_p(\mathbf{r})$ is due to the "reaction" to the presence of the molecule and it is the origin of the image field $\tilde{\mathbf{E}}_{img}$ in Fig. 5.1. As it is a linear function of ρ_M, the second term in Eq. (5.65) is a quadratic functional of the density.

Minimization of $\mathcal{G}_{tot}[\rho_M]$ leads to an effective Schrödinger equation where the Hamiltonian is:

$$\hat{H}_{eff} = \hat{H}_0 + \hat{V}_{pol}(\rho_M)\,. \qquad (5.66)$$

$\hat{H}_0$ is the Hamiltonian in the vacuum and $\hat{V}_{pol}(\rho_M)$ is the metal–molecule interaction operator which depends on the charge distribution of the molecule ρ_M.

Applying a variational procedure, defining a basis set for molecular orbitals and using standard approximations, one is able to obtain a matrix equation with the standard Hartree–Fock or Kohn–Sham form [100], which however also contains terms dependent on the metal–molecule interaction

$$\mathbb{F}'\mathbb{C} = \mathbb{S}\mathbb{C}\boldsymbol{\varepsilon}\,, \qquad (5.67)$$

with

$$\mathbb{F}' = \mathbb{H} + \frac{1}{2}(\mathbb{J} + \mathbb{Y}) + \mathbb{G}(\mathbb{P}) + \mathbb{X}(\mathbb{P})\,. \qquad (5.68)$$

The matrix indexes run on the basis set orbitals. In Eq. (5.68) $\mathbb{P}$ is the density matrix and $\mathbb{H} + \mathbb{G}(\mathbb{P})$ is the vacuum Fock operator $\mathbb{F}$. The other terms represent the interaction of the metal polarization with the charge distribution of the molecule. The interaction of the electrons of the molecule with the nuclei-induced metal polarization is accounted for $\mathbb{J}$, the nuclei–electrons-induced polarization interaction for $\mathbb{Y}$ and the electrons–electrons-induced interaction for $\mathbb{X}(\mathbb{P})$. They can be recasted in terms on the numerical solution of the electrostatic problems for the electronic and nuclear charge density, separately, in the presence of the metal nanoparticle and the solvent. No simplification of the molecule to a point dipole is needed.

Once the ground state is obtained, the molecular quantities relevant for the various molecular plasmonics phenomena studied so far with this model (SERS, SEIRA, enhanced luminescence) can be calculated by using the response theory. Again, the response equations are modified to include the effect of the metal polarization, induced either by the incident electric field or by the molecule itself. Moreover, the polarization induced in the metal by the oscillating charge density of the molecule is needed to calculate the total emitted EM field. We refer to the original articles for a description of this modified response theory [51, 89].

The increased power of modern computers and the extension to semi-empirical methods for the molecular response [55] allows to routinely perform with this approach systematic studies of molecular plasmonics phenomena involving many different nanoparticle specimens, sizes and relative molecular positions/orientations [56–58]. An example, for surface-enhanced fluorescence, is given in Fig. 5.10; see also Chapter 8.

Recently, other proposals that implement this mixed quantum mechanical/continuum electrodynamics approach appeared in the literature [59].

5.4.2 *Fully* ab initio *Description of the Molecule–Metal Nanoparticle Systems*

The system composed by the molecule and the metal nanoparticle can be seen as a giant supermolecule, in which both the molecule and the metal can be treated at the same, *ab initio*, level. As such, molecular plasmonics phenomena are interpreted as the results of the excitations of this supermolecular system. One obvious limitation of this approach is the size of the systems that can be treated: even with nowadays computers, noble metal particles larger than tens of atoms represent a challenge. Further limitations include those inherent the chosen *ab initio* method and the fact that this supermolecular approach intrinsically assumes the quasi-static approximation. In fact, the electrons of the molecule and the metal

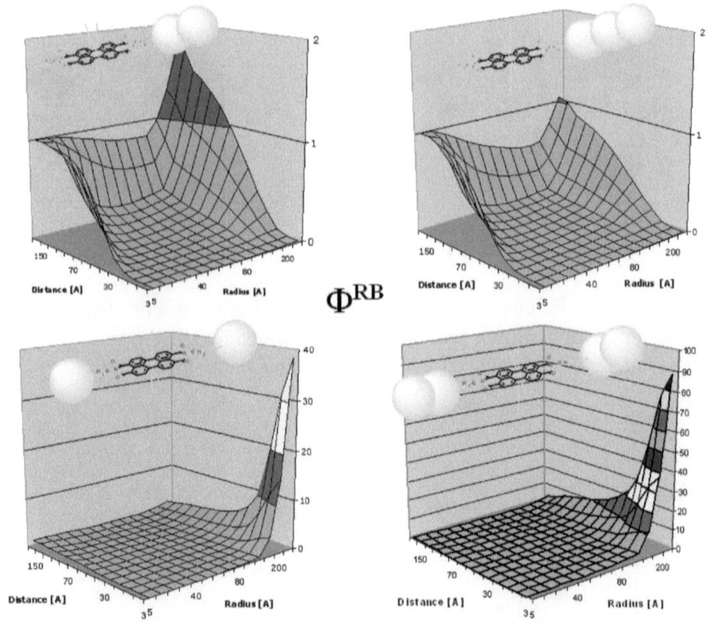

Figure 5.10 Relative brightness Φ^{RB} (a measure of fluorescence enhancement) for an organic dye, Perylene DiImide, in various configurations with respect to different kinds of gold nanoparticles. Reprinted with permission from Ref. [56]. Copyright (2003) American Chemical Society.

nanoparticle interact only via the non-retarded Coulombic repulsion contained in the supermolecule Hamiltonian.

Early attempts of applying this approach to molecular plasmonics (namely to SERS) were done already in the '80s and '90s [101–107]. They were however limited to metal clusters composed of a few atoms. More recently [108, 109], the power of this approach has been fully disclosed by treating metal nanoparticle (Ag_{20}) that has excitation with a qualitative character of plasmons. In fact, a study of chemical effects was possible with this approach, that is precluded to the simpler models discussed in the previous sections. For example, the importance of charge transfer could be qualitatively investigated (despite some limitations of the underlying ab initio methods), and involved orbitals could be identified (see Fig. 5.11).

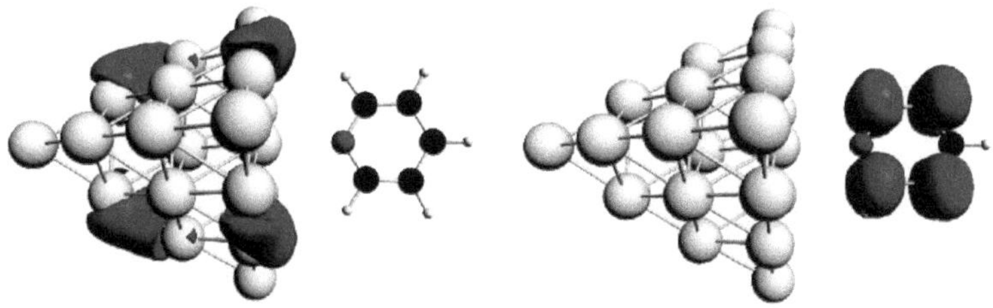

(a) Ag_{20} HOMO (b) Pyridine LUMO+1

Figure 5.11 Relevant orbitals involved in the major charge-transfer excitation of the shown pyridine-Ag_{20} complex. Reprinted with permission from Ref. [108]. Copyright (2006) American Chemical Society.

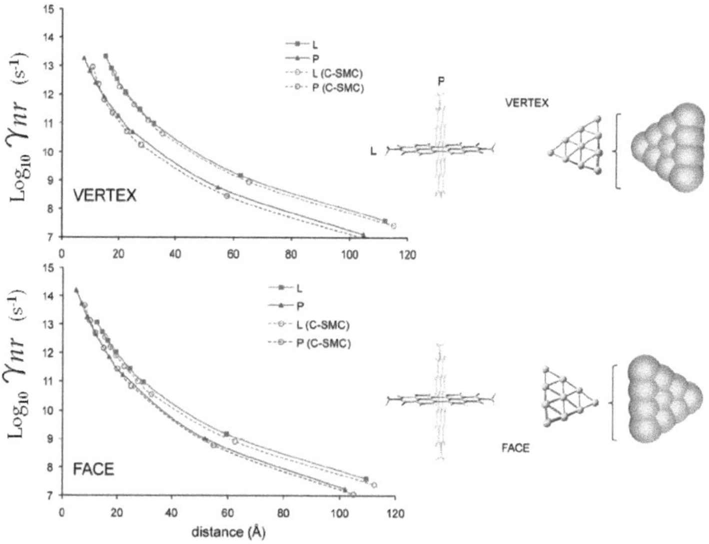

Figure 5.12 Decimal logarithm of the non-radiative decay rate for the PDI molecule as a function of the distances from a Au_{20} metal cluster. The various relative orientations considered are depicted on the right panel. Dotted lines refer to the continuous small metal cluster (C-SMC) representation of the metal nanoparticle, while solid lines to full ab initio results. Reprinted with permission from Ref. [57]. Copyright (2009) American Chemical Society.

Interesting informations can be gained by comparing the results of the full ab initio model with those of the more approximate methods. In particular, it is very convenient to keep the description level of the molecule exactly the same, while choosing different levels for the metal nanoparticle: continuous dielectric or *ab initio*. By this approach, in Ref. [57] it was found that even for a very small nanoparticle, where excitations are of molecular character, quenching efficiency is as high as when plasmon resonances are present. We report in Fig. 5.12 the non-radiative decay rate of a dye molecule (PDI) as a function of its distance from a metal cluster composed of 20 Au atoms. For such a small cluster, ab initio results predict a molecular-like excitation spectrum, while applying a continuum model for the molecule is the same as assuming that the Au surface plasmon excitations of larger metal nanoparticle are conserved also for this size.

Finally, we also remark an interesting intermediate level of description for the metal nanoparticle that bridges continuous dielectric with atomistic *ab initio* description. This is the use of jellium nanoparticle, i.e. exploiting a continuous background of positive charge to simulate nuclei, keeping a quantum mechanical descriptions for electrons. This approach has a venerable tradition [110, 111], and has also been recently employed to study optical properties of nanoshells [112].

References

1. Jensen, L., Aikens, C. M., and Schatz, G. C. (2008) Electronic structure methods for studying surface-enhanced Raman scattering, *Chem. Soc. Rev.*, **37**, 1061–1073.
2. Glass, A. M., Liao, P. F., Bergman, J. G., and Olson, D. H. (1980) Interaction of metal particles with adsorbed dye molecules: Absorption and luminescence, *Opt. Lett.*, **5**, 368–370.
3. Craighead, H. G., and Glass, A. M. (1981) Optical absorption of small metal particles with adsorbed dye coats, *Opt. Lett.*, **6**, 248–250.
4. Glass, A. M., Wokaun, A., Heritage, J. P., Bergman, J. G., Liao, P. F., and Olson, D. H. (1981) Enhanced two-photon fluorescence of molecules adsorbed on silver particle films, *Phys. Rev. B*, **24**, 4906–4909.

5. Makarova, O. V., Ostafin, A. E., Miyoshi, H., Norris, J. R., and Meisel, D. (1999) Adsorption and encapsulation of fluorescent probes in nanoparticles, *J. Phys. Chem. B*, **103**, 9080–9084.
6. Franzen, S., Folmer, J. C. W., Glomm, W. R., and O'Neal, R. (2002) Optical properties of dye molecules adsorbed on single gold and silver nanoparticles, *J. Phys. Chem. A*, **106**, 6533–6540.
7. Chance, R. R., Prock, A., and Silbey, R. (1978) Molecular fluorescence and energy transfer near interfaces, *Adv. Chem. Phys.*, **37**, 1–65.
8. Alivisatos, A. P., Waldeck, D. H., and Harris, C. B. (1985) Nonclassical behavior of energy transfer from molecules to metal surfaces: Biacetyl($^3n\pi^*$)/Ag(111), *J. Chem. Phys.*, **82**, 541–547.
9. Pal, T., Jana, N. R., and Sau, T. (1997) Nanoparticle induced fluorescence quenching, *Radiat. Phys. Chem.*, **49**, 127–130.
10. Thomas, K. G., and Kamat, P. V. (2000) Making gold nanoparticles glow: enhanced emission from a surface-bound fluoroprobe, *J. Am. Chem. Soc.*, **122**, 2655–2656.
11. Trabesinger, W., Kramer, A., Kreiter, M., Hecht, B., and Wild, U. P. (2002) Single-molecule near-field optical energy transfer microscopy, *Appl. Phys. Lett.*, **81**, 2118–2120.
12. Dulkeith, E., Morteani, A. C., Niedereichholz, T., Klar, T. A., Feldmann, J., Levi, S. A., van Veggel, F. C. J. M., Reinhoudt, D. N., Möller, M., and Gittins, D. I. (2002) Fluorescence quenching of dye molecules near gold nanoparticles: radiative and nonradiative effects, *Phys. Rev. Lett.*, **89**, 203002/1-4.
13. Caruge, J.-M., and Orrit, M. (2001) Probing local currents in semiconductors with single molecules, *Phys. Rev. B*, **64**, 205202/1–13.
14. Hartstein, A., Kirtley, J. R., and Tsang, J. C. (1980) Enhancement of the infrared absorption from molecular monolayers with thin metal overlayers, *Phys. Rev. Lett.*, **45**, 201–204.
15. Osawa, M., and Ikeda, M. (1991) Surface-enhanced infrared absorption of p-nitrobenzoic acid deposited on silver island films: contributions of electromagnetic and chemical mechanisms, *J. Phys. Chem.*, **95**, 9914–9919.
16. Fleischmann, M., Hendra, P. J., and McQuillan, A. J. (1974) Raman spectra of pyridine adsorbed at a silver electrode, *Chem. Phys. Lett.*, **26**, 163–166.
17. Albrecht, M. G., and Creighton, J. A. (1977) Anomalously intense Raman spectra of pyridine at a silver electrode, *J. Am. Chem. Soc.*, **99**, 5215–5217.

18. Jeanmaire, D. L., and Van Duyne, R. P. (1977) Surface Raman spectroelectrochemistry: Part I. Heterocyclic, aromatic, and aliphatic amines adsorbed on the anodized silver electrode, *J. Electroanal. Chem.*, **84**, 1–20.
19. Moskovits, M. (1985) Surface-enhanced spectroscopy, *Rev. Mod. Phys.*, **57**, 783–826.
20. Otto, A., Mrozek, I., Grabhorn, H., and Akemann, W. (1992) Surface-enhanced Raman scattering, *J. Phys.: Condens. Matter*, **4**, 1143–1212.
21. Campion, A., and Kambhampati, P. (1998) Surface-enhanced Raman scattering, *Chem. Soc. Rev.*, **27**, 241–250.
22. Schatz, G. C., and Van Duyne, R. P. (2002) Electromagnetic mechanism of surface-enhanced spectroscopy, in *Handbook of Vibrational Spectroscopy*, Vol. 1 (ed. Chalmers, J., and Griffiths, P. R.) Wiley, New York, pp. 759–774.
23. Emory, S. R., and Nie, S. (1997) Near-field surface-enhanced Raman spectroscopy on single silver nanoparticles, *Anal. Chem.*, **69**, 2631–2635.
24. Nie, S., and Emory, S. R. (1997) Probing single molecules and single nanoparticles by surface-enhanced Raman scattering, *Science*, **275**, 1102–1106.
25. Kneipp, K., Wang, Y., Kneipp, H., Perelman, L. T., Itzkan, I., Dasari, R. R., and Feld, M. S. (1997) Single molecule detection using surface-enhanced Raman scattering (SERS), *Phys. Rev. Lett.*, **78**, 1667–1670.
26. Kneipp, K., Kneipp, H., Deinum, G., Itzkan, I., Dasari, R. R., and Feld, M. S. (1998) Single-molecule detection of a cyanine dye in silver colloidal solution using near-infrared surface-enhanced Raman scattering, *Appl. Spectrosc.*, **52**, 175–178.
27. Kneipp, K., Kneipp, H., Manoharan, R., Hanlon, E. B., Itzkan, I., Dasari, R. R., and Feld, M. S. (1998) Extremely large enhancement factors in surface-enhanced Raman scattering for molecules on colloidal gold clusters, *Appl. Spectrosc.*, **52**, 1493–1497.
28. Kneipp, K., Kneipp, H., Itzkan, I., Dasari, R. R., and Feld, M. S. (1999) Ultrasensitive chemical analysis by Raman spectroscopy, *Chem. Rev.*, **99**, 2957–2976.
29. Xu, H., Bjerneld, E. J., Kall, M., and Borjesson, L. (1999) Spectroscopy of single hemoglobin molecules by surface enhanced Raman scattering, *Phys. Rev. Lett.*, **83**, 4357–4360.

30. Michaels, A. M., Nirmal, M., and Brus, L. E. (1999) Surface enhanced Raman spectroscopy of individual rhodamine 6G molecules on large Ag nanocrystals, *J. Am. Chem. Soc.*, **121**, 9932–9939.
31. Michaels, A. M., Jiang, J., and Brus, L. (2000) Ag nanocrystal junctions as the site for surface-enhanced Raman scattering of single rhodamine 6G molecules, *J. Phys. Chem. B*, **104**, 11965–11971.
32. Bosnick, K. A., Jiang, J., and Brus, L. E. (2002) Fluctuations and local symmetry in single-molecule rhodamine 6G Raman scattering on silver nanocrystal aggregates, *J. Phys. Chem. B*, **106**, 8096–8099.
33. Maruyama, Y., Ishikawa, M., and Futamata, M. (2001) Surface-enhanced Raman scattering of single adenine molecules on silver colloidal particles, *Chem. Lett.*, **30**, 834–835.
34. Tian, Z.-Q., Ren, B., and Wu, D.-Y. (2002) Surface-enhanced Raman scattering: from noble to transition metals and from rough surfaces to ordered nanostructures, *J. Phys. Chem. B*, **106**, 9463–9483.
35. Chen, C. K., de Castro, A. R. B., and Shen, Y. R. (1981) Surface-enhanced second-harmonic generation, *Phys. Rev. Lett.*, **46**, 145–148.
36. Chen, C. K., Heinz, T. F., Ricard, D., and Shen, Y. R. (1981) Detection of molecular monolayers by optical second-harmonic generation, *Phys. Rev. Lett.*, **46**, 1010–1012.
37. Haller, K. L., Bumm, L. A., Altkorn, R. I., Zeman, E. J., Schatz, G. C., and Van Duyne, R. P. (1989) Spatially resolved surface enhanced second harmonic generation: Theoretical and experimental evidence for electromagnetic enhancement in the near infrared on a laser microfabricated Pt surface, *J. Chem. Phys.*, **90**, 1237–1252.
38. Baldelli, S., Eppler, A. S., Anderson, E., Shen, Y.-R., and Somorjai, G. A. (2000) Surface enhanced sum frequency generation of carbon monoxide adsorbed on platinum nanoparticle arrays, *J. Chem. Phys.*, **113**, 5432–5438.
39. Wenseleers, W., Stellacci, F., Meyer-Friedrichsen, T., Mangel, T., Bauer, C. A., Pond, S. J. K., Marder, S. R., and Perry, J. W. (2002) Five orders-of-magnitude enhancement of two-photon absorption for dyes on silver nanoparticle fractal clusters, *J. Phys. Chem. B*, **106**, 6853–6863.
40. Murphy, D. V., Von Raben, K. U., Chang, R. K., and Dorain, P. B. (1982) Surface-enhanced hyper-Raman scattering from SO_{32}-adsorbed on Ag powder, *Chem. Phys. Lett.*, **85**, 43–47.
41. Golab, J. T., Sprague, J. R., Carron, K. T., Schatz, G. C., and Van Duyne, R. P. (1988) A surface enhanced hyper-Raman scattering study of pyridine adsorbed onto silver: Experiment and theory, *Chem. Phys.*, **88**, 7942–7951.

42. Yang, W.-H., and Schatz, G. C. (1992) Ab initio and semiempirical molecular orbital studies of surface enhanced and bulk hyper-Raman scattering from pyridine, *J. Chem. Phys.*, **97**, 3831–3845.

43. Yang, W.-H., Hulteen, J., Schatz, G. C., and Van Duyne, R. P. (1996) A surface-enhanced hyper-Raman and surface-enhanced Raman scattering study of trans-1,2-bis(4-pyridyl)ethylene adsorbed onto silver film over nanosphere electrodes, *J. Chem. Phys.*, **104**, 4313–4323.

44. Nitzan, A., and Brus, L. E. (1981) Can photochemistry be enhanced on rough surfaces?, *J. Chem. Phys.*, **74**, 5321–5322.

45. Nitzan, A., and Brus, L. E. (1981) Theoretical model for enhanced photochemistry on rough surfaces, *J. Chem. Phys.*, **75**, 2205–2214.

46. Yang, X. M., Tryk, D. A., Hashimoto, K., and Fujishima, A. (1998) Examination of the photoreaction of p-nitrobenzoic acid on electrochemically roughened silver using surface-enhanced Raman imaging (SERI), *J. Phys. Chem. B*, **102**, 4933–4943.

47. Atwater, H. A., and Polman, A. (2010) Plasmonics for improved photovoltaic devices, *Nat. Mater.*, **9**, 205–213.

48. Efrima, S. (1983) The effect of large electric field gradients on the Raman optical activity of molecules adsorbed on metal surfaces, *Chem. Phys. Lett.*, **102**, 79–82.

49. Janesko, B. G., and Scuseria, G. E. (2006) Surface enhanced Raman optical activity of molecules on orientationally averaged substrates: Theory of electromagnetic effects, *J. Chem. Phys.*, **125**, 124704/1–12.

50. Pascal, A., Bharadwaj, P., and Novotny, L. (2006) Enhancement and quenching of single-molecule fluorescence, *Phys. Rev. Lett.*, **96**, 113002/1–4.

51. Corni, S., and Tomasi, J. (2001) Enhanced response properties of a chromophore physisorbed on a metal particle, *J. Chem. Phys.*, **114**, 3739–3751.

52. Corni, S., and Tomasi, J. (2001) Theoretical evaluation of Raman spectra and enhancement factors for a molecule adsorbed on a complex-shaped metal particle, *Chem. Phys. Lett.*, **342**, 135–140.

53. Corni, S., and Tomasi, J. (2002) Surface enhanced Raman scattering from a single molecule adsorbed on a metal particle aggregate: A theoretical study, *J. Chem. Phys.*, **116**, 1156–1164.

54. Andreussi, O., Corni, S., Mennucci, B., and Tomasi, J. (2004) Radiative and nonradiative decay rates of a molecule close to a metal particle of complex shape, *J. Chem, Phys.*, **121**, 10190–10202.

55. Caricato, M., Andreussi, O., and Corni, S. (2006) Semiempirical (ZINDO-PCM) approach to predict the radiative and nonradiative decay rates of a molecule close to metal particles, *J. Phys. Chem. B*, **110**, 16652–16659.
56. Vukovic, S., Corni, S., and Mennucci, B. (2009) Fluorescence enhancement of chromophores close to metal nanoparticles. Optimal setup revealed by the polarizable continuum model, *J. Phys. Chem. C*, **113**, 121–133.
57. Muñoz-Losa, A., Vukovic, S., Corni, S., and Mennucci, B. (2009) Nonplasmonic metal particles as excitation energy transfer acceptors: an unexpected efficiency revealed by quantum mechanics, *J. Phys. Chem. C*, **113**, 16364–16370.
58. Sánchez-González, A., Muñoz-Loza, A., Vukovic, S., Corni, S., and Mennucci, B. (2010) Quantum mechanical approach to solvent effects on the optical properties of metal nanoparticles and their efficiency as excitation energy transfer acceptors, *J. Phys. Chem. C*, **114**, 1553–1561.
59. Chen, H., McMahon, J. M., Ratner, M. A., and Schatz, G. C. (2010) Classical electrodynamics coupled to quantum mechanics for calculation of molecular optical properties: a RT-TDDFT/FDTD approach, *J. Phys. Chem. C*, **114**, 14384–14392.
60. Hohenester, U., and Trügler, A. (2008) Interaction of single molecules with metallic nanoparticles, *IEEE J. Sel. Top. Quant. Elec.*, **14**, 1430–1440.
61. Jackson, J. D. (ed.) (1975) *Classical Electrodynamics*, Wiley, New York.
62. Draine, B. T. (1988) The discrete-dipole approximation and its application to interstellar graphite grains, *Astrophys. J.*, **333**, 848–872.
63. Metiu, H. (1984) Surface enhanced spectroscopy, *Prog. Surf. Sci.*, **17**, 153–320.
64. Hao, E., and Schatz, G. C. (2004) Electromagnetic fields around silver nanoparticles and dimers, *J. Chem. Phys.*, **120**, 357–366.
65. Zhao, J., Pinchuk, A. O., McMahon, J. M., Li, S., Ausman, L. K., Atkinson, A. L., and Schatz, G. C. (2008) Methods for describing the electromagnetic properties of silver and gold nanoparticles, *Accounts Chem. Res.*, **41**, 1710–1720.
66. Myroshnychenko, V., Rodriguez-Fernandez, J., Pastoriza-Santos, I., Funston, A. M. , Novo, C., Mulvaney, P., Liz-Marzan, L. M., and Garcia de Abajo, F. J. (2008) Modelling the optical response of gold nanoparticles, *Chem. Soc. Rev.*, **37**, 1792–1805.

67. Kelly, K. L., Coronado, E., Zhao, L. L., and Schatz, G. C. (2003) The optical properties of metal nanoparticles: The influence of size shape, and dielectric environment, *The Journal of Physical Chemistry B*, **107**, 668–677.
68. Xu, H., and Kall, M. (2006) Estimating SERS properties of silver-particle aggregates through generalized Mie theory, *Top. Appl. Phys.*, **103**, 87–104.
69. Dulkeith, E., Ringler, M., Klar, T. A., Feldmann, J., Munoz Javier, A., and Parak, W. J. (2005) Gold nanoparticles quench fluorescence by phase induced radiative rate suppression, *Nano Lett.*, **5**, 585–589.
70. Gersten, J., and Nitzan, A. (1981) Spectroscopic properties of molecules interacting with small dielectric particles, *J. Chem. Phys.*, **75**, 1139–1152.
71. Joulain, K., Carminati, R., Mulet, J.-P., and Greffet, J.-J. (2003) Definition and measurement of the local density of electromagnetic states close to an interface, *Phys. Rev. B*, **68**, 245405/1-10.
72. Novotny, L., and Hecht, B. (eds) (2006) *Principles of Nano-Optics*, Cambridge University Press, Cambridge, UK.
73. Long, D. A. (ed.) (2002) *The Raman Effect*, John Wiley & Sons, Chichester.
74. Le Ru, E. C., and Etchegoin, P. G. (2006) Rigorous justification of the E4 enhancement factor in surface-enhanced Raman spectroscopy, *Chem. Phys. Lett.*, **423**, 63–66.
75. Landau, L. D., Lifchitz, E. M., and Pitaevskii, L. P. (eds) (1984) *Electromagnetics of Continuous Media*, Pergamon, Oxford SS.
76. Zeman, E. J., and Schatz, G. C. (1987) An accurate electromagnetic theory study of surface enhancement factors for silver, gold, copper, lithium, sodium, aluminum, gallium, indium, zinc, and cadmium, *J. Phys. Chem.*, **91**, 634–643.
77. Euler, J. (1954) Infrared properties of metals and the mean free paths of conduction electrons, *Z. Phys.*, **137**, 318–332.
78. Yannouleas, C., and Broglia, R. A. (1992) Landau damping and wall dissipation in large metal clusters, *Ann. Phys. (NY)*, **217**, 105–141.
79. Kreibig, U., and Vollmer, M. (eds) (1995) *Optical Properties of Metal Cluster*, Springer-Verlag, Berlin.
80. Coronado, E. A., and Schatz, G. C. (2003) Surface plasmon broadening for arbitrary shape nanoparticles: A geometrical probability approach, *J. Chem. Phys.*, **119**, 3926–3934.

81. Lundqvist, S., and March, N. H. (eds) (1983) *Theory of the inhomogeneus electron gas*, Plenum Press, New York.

82. Persson, B. N. J., and Lang, N. D. (1982) Electron-hole-pair quenching of excited states near a metal, *Phys. Rev. B*, **26**, 5409–5415.

83. Persson, B. N. J., and Andersson, S. (1984) Dynamical processes at surfaces: Excitation of electron-hole pairs, *Phys. Rev. B*, **29**, 4382–4394.

84. Kliewer, K. L., and Fuchs, R. (1969) Lindhard dielectric functions with a finite electron lifetime, *Phys. Rev.*, **181**, 552–558.

85. Mermin, N. D. (1970) Lindhard dielectric function in the relaxation-time approximation, *Phys. Rev. B*, **1**, 2362–2363.

86. Lindhard, J. (1954) On the properties of a gas of charged particles, *Kgl. Danske Videnskab. Selskab Mat.-Fys. Medd.*, **28**, 1–57.

87. Ford, G. W., and Weber, W. H. (1981) Electromagnetic effects on a molecule at a metal surface: I. Effects of nonlocality and finite molecular size, *Surf. Sci.*, **109**, 451–481.

88. Fuchs, R., and Barrera, R. G. (1981) Dynamical response of a dipole near the surface of a nonlocal metal, *Phys. Rev. B*, **24**, 2940–2950.

89. Corni, S., and Tomasi, J. (2003) Lifetimes of electronic excited states of a molecule close to a metal surface, *J. Chem. Phys.*, **118**, 6481–6494.

90. Leung, P. T. (1990) Decay of molecules at spherical surfaces: Nonlocal effects, *Phys. Rev. B*, **42**, 7622–7625.

91. Leung, P. T., and Hider, M. H. (1993) Nonlocal electrodynamic modeling of frequency shifts for molecules at rough surfaces, *J. Chem. Phys.*, **98**, 5019–5022.

92. García de Abajo, F. J. (2008) Nonlocal effects in the plasmons of strongly interacting nanoparticles, dimers, and waveguides, *J. Phys. Chem. C*, **112**, 17983–17987.

93. Corni, S. (2005) The reorganization energy of azurin in bulk solution and in the electrochemical scanning tunneling microscopy setup, *J. Phys. Chem. B*, **109**, 3423–3430.

94. Tomasi, J., Cammi, R., and Mennucci, B. (1999) Medium effects on the properties of chemical systems: An overview of recent formulations in the polarizable continuum model (PCM), *Int. J. Quantum Chem.*, **75**, 783–803.

95. Tomasi, J., and Persico, M. (1994) Molecular interactions in solution: an overview of methods based on continuous distributions of the solvent, *Chem. Rev.*, **94**, 2027–2094.

96. Cammi, R., Mennucci, B., and Tomasi, J. (1998) Solvent effects on linear and nonlinear optical properties of donor-acceptor polyenes: investigation of electronic and vibrational components in terms of structure and charge distribution changes, *J. Am. Chem. Soc.*, **120**, 8834–8847.

97. Cammi, R., Cossi, M., Mennucci, B., and Tomasi, J. (1996) Analytical Hartree-Fock calculation of the dynamical polarizabilities alpha, beta, and gamma of molecules in solution, *J. Chem. Phys.*, **105**, 10556–10564.

98. Miertuš, S., Scrocco, E., and Tomasi, J. (1981) Electrostatic interaction of a solute with a continuum. A direct utilizaion of ab initio molecular potentials for the prevision of solvent effects, *Chem. Phys.*, **55**, 117–129.

99. Hackbusch, W. (ed.) (1995) *Integral Equations - Theory and Numerical Treatment*, Birkhauser, Basel.

100. Jensen, F. (ed.) (2006) *Introduction to Computational Chemistry*, John Wiley & Sons, New York.

101. Pandey, P. K. K., and Schatz, G. C. (1982) Time-dependent Hartree-Fock calculations of surface-enhanced Raman intensities. H_2 adsorbed on a model Li cluster, *Chem. Phys. Lett.*, **88**, 193–197.

102. Pandey, P. K. K., and Schatz, G. C. (1984) A detailed analysis of the Raman enhancement mechanisms associated with the interaction of a Raman scatterer with a resonant metal cluster: Results for Li_n-H_2, *J. Chem. Phys.*, **80**, 2959–2972.

103. Yang, W.-H., and Schatz, G. C. (1992) Ab initio and semiempirical molecular orbital studies of surface enhanced and bulk hyper-Raman scattering from pyridine, *J. Chem. Phys.*, **97**, 3831–3845.

104. Nakai, H., and Nakatsuji, H. (1995) Electronic mechanism of the surface enhanced Raman scattering, *J. Chem. Phys.*, **103**, 2286–2294.

105. Kitao, O., and Nakatsuji, H. (1988) Cluster expansion of the wave function. Valence excitations and ionizations of pyridine, *J. Chem. Phys.*, **88**, 4913–4925.

106. Arenas, J. F., Tocon, I. L., Otero, J. C., and Marcos, J. I. (1995) A priori scaled quantum mechanical vibrational spectra of trans- and cis-stilbene, *J. Phys. Chem.*, **99**, 11392–11398.

107. Arenas, J. F., Soto, J., López Tocón, I., Fernndez, D. J., Otero, J. C., and Marcos, J. I. (2002) The role of charge-transfer states of the metal-adsorbate complex in surface-enhanced Raman scattering, *J. Chem. Phys.*, **116**, 7207–7216.

108. Zhao, L., Jensen, L., and Schatz, G. C. (2006) Pyridine-Ag_{20} cluster: a model system for studying surface-enhanced Raman scattering, *J. Am. Chem. Soc.*, **128**, 2911–2919.

109. Zhao, L. L., Jensen, L., and Schatz, G. C. (2006) Surface-enhanced Raman scattering of pyrazine at the junction between two Ag_{20} nanoclusters, *Nano Lett.*, **6**, 1229–1234.

110. Korzeniewski, G., Maniv, T., and Metiu, H. (1980) The interaction between an oscillating dipole and a metal surface described by a jellium model and the random phase approximation, *Chem. Phys. Lett.*, **73**, 212–217.

111. Maniv, T., and Metiu, H. (1980) Electron gas effects in the spectroscopy of molecules chemisorbed at a metal surface. I. Theory, *J. Chem. Phys.*, **72**, 1996–2006.

112. Prodan, E., Nordlander, P., and Halas, N. J. (2003) Electronic structure and optical properties of gold nanoshells, *Nano Lett.*, **3**, 1411–1415.

Part II

Applications and Experimental Aspects

Chapter 6

Near-Fields in Assembled Plasmonic Nanostructures

P. K. Jain[a] and C. Deeb[b]

[a]*Departments of Chemistry and Physics, Beckman Institute for Advanced Science and Technology, University of Illinois Urbana-Champaign, 600 S Mathews Avenue, Urbana IL, 61801, USA*

[b]*Department of Chemistry, Northwestern University, 2145 Sheridan Road, Evanston, IL 60208, USA*

jain@illinois.edu

Metal nanoparticles exhibit, on account of their localized surface plasmon oscillations, strong tunable absorption and scattering resonances and surface-enhanced electric fields. These resonances and near-fields can be enhanced and engineered further by coupling metal nanoparticles into defined groupings or assemblies. In this chapter, we discuss a) the conceptual foundation of plasmonic coupling between metal nanoparticles, b) how this coupling influences the near- and far-field optical properties of assembled metal nanostructures, and c) implications for field-enhanced light–matter interactions, spectroscopy, nanophotonics, and sensing.

Handbook of Molecular Plasmonics

Edited by Fabio Della Sala and Stefania D'Agostino

ISBN 978-981-4303-20-0 (Hardcover), 978-981-4303-21-7 (eBook)

www.panstanford.com

6.1 Introduction

The interaction of light with molecules is primarily entailed by excitation of electronic and vibration modes of the molecule by the electromagnetic (EM) field of light. The characteristic length scale of electronic and vibrational excitations is typically on the molecular size scale (ca. 1 Å), whereas the characteristic length scale of the EM field is defined for a plane-wave by its wavelength λ (which is on the order of few 100 nm for visible light). This disparity in length scales between a molecule and the EM field puts a limit on the possible realm of light–matter interactions beyond common linear dipole processes. For instance, processes involving higher-order multipolar excitations or "helical" electronic excitations (e.g. circular dichroism, magnetic optical activity) scale in their strength as a function of a/λ, where a represents the size of the molecule. The typical efficiency of such light–molecule interactions is thus 10^{-3} or smaller. Such processes rely on having a strongly "squeezed" EM field that supports large photon wave-vectors or field gradients. At the same time large electric field intensities are desirable for non-linear optical processes. Light-matter interactions can therefore be significantly enhanced in EM fields that are strongly confined to a size that is comparable to that of a molecule. Unfortunately, the diffraction limit does not allow light to be focused or squeezed below a size scale smaller than $\lambda/2$. Research over the past decade has shown that metal nanostructures, due to their strong interaction with light, allow us to break the diffraction limit and confine EM fields to the nanoscale. We therefore attempt to characterize the optical response of metal nanostructures, so as to be able to design nanostructures for squeezing the EM field of light for enhancement of light–matter interactions and applications in photonics and sensing.

6.2 Optical Properties of Metal Nanoparticles

Since historical times, noble metal nanoparticles have been known to display brilliant colors due to intense visible light absorption and scattering [1]. The strong interaction of metal nanoparticles with light can be understood from the work of Gustav Mie, who solved Maxwell's electrodynamic equations for a homogeneous sphere (see Sec. 1.6.5). The solution consisted of electromagnetic waves on the surface of the sphere of different orders ranging from the lowest dipole to higher order multipoles, depending on the size of the nanoparticle relative to the wavelength of light. In the case of particles of size much smaller than the wavelength of light, (i.e. radius $R \ll \lambda$), only the dipole mode contributes considerably to the light–nanoparticle interaction. For a nanoparticle in the size range of few tens of nm, it is sufficient to consider only the dipolar mode of interaction (dipole approximation) and to assume that the electric field of light is constant across the nanostructure (quasi-static approximation). As shown in Sec. 1.4.1, in the dipolar quasi-static limit, the polarizability of a particle of volume V is given by the *Clausius–Mossotti polarizability* [2]:

$$\tilde{\alpha}(\omega) = \frac{3}{4\pi} V \frac{\tilde{\varepsilon}_p(\omega) - \varepsilon_{out}}{\tilde{\varepsilon}_p(\omega) + 2\,\varepsilon_{out}} . \tag{6.1}$$

ε_{out} is the dielectric constant of the non-dissipative surrounding medium and $\tilde{\varepsilon}_p(\omega)$ is the dielectric function of the metal which is complex and frequency-dependent, expressed as $\tilde{\varepsilon}_p(\omega) = \varepsilon_1(\omega) + i\varepsilon_2(\omega)$ where ε_1 is the real part of the dielectric function and ε_2 is the imaginary part. From Eq. (6.1), we see that the polarizability has a strong maximum when the following relation is satisfied:

$$\varepsilon_1(\omega) = -2\varepsilon_{out} , \tag{6.2}$$

where it is assumed here that ε_2 is small and/or weakly dependent on frequency. For Eq. (6.2) to be satisfied, it is required that the real part $\varepsilon_1(\omega)$ is negative, which is possible for some metals (gold, silver, and copper) at optical frequencies. At the frequency ω_{sp} where the condition in Eq. (6.2) is satisfied, the metal nanoparticle interacts very strongly with the light, resulting in the collective coherent oscillation of the conduction electrons in resonance with the oscillating EM field of the light. This oscillation is bound to the surface of the nanoparticle and hence is known as the *localized surface plasmon resonance* (LSPR). The LSPR process can be visualized as light confined to a nanoscale size defined by the diameter of the nanoparticle.

The oscillation induced by the light results in a strong electric field at the surface of the nanoparticle (see Sec. 1.4.1 and Sec. 5.3.1). At resonance ($\varepsilon_1 = -2\varepsilon_{out}$), a strong enhancement in the field intensity is expected. For instance, the resonant field enhancement on the surface of a 20-nm silver nanoparticle has been calculated by electrodynamic simulations to be as high as 200 [3]. The strong electric field decays rapidly (as $1/d^3$ in the dipolar limit) away from the particle surface, and is termed as the *near-field*. The potential for achieving interesting optical effects [4] using this strongly confined field is tremendous, the most important being the large enhancement of light scattering and absorption cross-sections of the metal nanoparticle itself (see Fig. 6.1).

Plasmon oscillations can decay radiatively giving rise to scattering of light [5]. On the other hand, the oscillating electrons suffer collisions with other electrons resulting in the excitation of inter-band and intra-band excitations in the metal [6]. Additional damping and dephasing processes include collisions with the lattice phonons, the nanoparticle surface, defects, and surface ligands [7, 8]. These non-radiative pathways result in the generation of heat, and constitute light absorption by the particle. Absorption and scattering together constitute the light extinction of the particle. The extinction and scattering cross-sections, C_{ext} and C_{sca} are given in the terms of

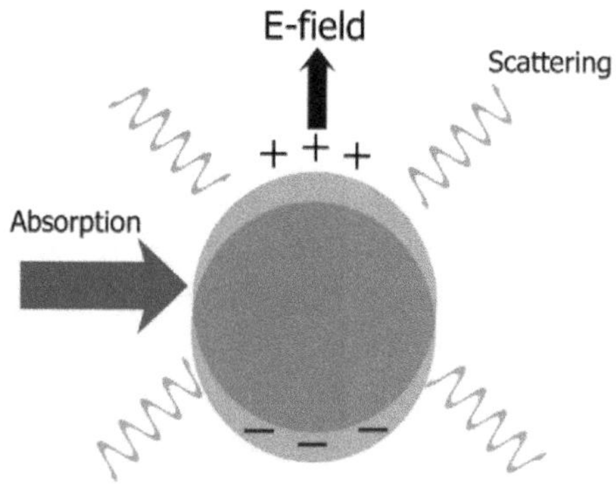

Figure 6.1 LSPR oscillation of electrons of a metal nanoparticle, resulting in the strong enhancement of the surface electric field, and light scattering and absorption cross-sections.

the dipolar polarizability as [9]:

$$C_{ext} = 4\pi k \mathrm{Im}[\tilde{\alpha}] , \tag{6.3}$$

$$C_{sca} = \frac{8\pi}{3} k^4 |\tilde{\alpha}|^2 , \tag{6.4}$$

where k is the wavevector of light in the medium:

$$k = \frac{2\pi\sqrt{\varepsilon_{out}}}{\lambda} . \tag{6.5}$$

Since the cross-sections of scattering and absorption depend directly on the optical polarizability of the metal nanoparticle, these processes are strongly enhanced at the LSPR frequency. For instance, colloidal gold nanoparticles in the 40-nm size range exhibit strong absorption and scattering resonances centered around 530 nm with cross-sections that are 5–6 orders larger than those of molecules [10]. The strong light absorption has been utilized for localized laser photothermal heating of the nano-environment around a nanoparticle, with applications in selective nano-surgery and tumor therapy [11–15]. Scattering on the other hand, has been exploited in the use of metal nanoparticles as sensitive molecularly-targeted labels for biological imaging [12, 16–18]. While absorption and scattering are far-field properties, they serve as useful reporters of the LSPR and spectral attributes of the near-field.

6.2.1 *Nanoplasmonic Field Enhancement*

Plasmonic near-fields have the ability to enhance spectroscopic signals from molecules in the vicinity of the nanostructure [19, 20]. The most common example is the plasmonic enhancement of *Raman scattering*. Raman scattering spectroscopy can provide detailed information about molecular structure, thus offering molecular-specific signatures for chemical imaging. However, molecules typically have small Raman scattering cross-sections, resulting in weak signals. Raman cross-sections of molecules at the metal nanostructure surface can be enhanced 5 orders or more, as first shown by Van Duyne and coworkers [21]. This phenomenon is known as *surface-enhanced Raman scattering* (SERS).

While a part of this enhancement is attributed to changes in the chemical polarizability of the molecule due to interactions with the metal surface, a major part comes from the near-field enhancement (see Sec. 5.3.3 and Chapter 8). The electromagnetic part of the SERS enhancement is proportional to the square of the field intensity ($|E|^4$) at the nanostructure, and it shows an intense maximum at the LSPR. Non-linear optical effects [22] including second harmonic generation (SHG), the cross-section of which is proportional to $|E|^4$, have also been enhanced by utilizing the intense near-field of noble metal nanoparticles [23]. Plasmonic field-enhanced emission from fluorophores [24] resulting from a Purcell enhancement of the radiative decay of the excited state has been demonstrated in several systems. In such a case, the radiative rate enhancement is proportional to the field intensity $|E|^2$.

While relatively unexplored, the near-field is also promising for enhancing processes such as natural and *magnetic circular dichroism*. These processes involve a circularly polarized beam of light inducing a helical electronic motion constituted by a simultaneous magnetic dipole and electric dipole excitation. A molecule placed in the near-field of a nanostructure can effectively sample the helicity of the EM field due to the large wavevector of the confined mode, resulting in the enhancement of the circular dichroism cross-section. Lieberman *et al.* have observed a 100-fold enhancement in the circular dichroism of L-glutathione molecules adsorbed to silver nanoparticles [25]. As another example, Jain *et al.* have shown a resonant enhancement of magnetic optical rotation in iron-oxide nanoparticles coated with a gold shell [26]. The enhancement results in the emergence of an otherwise unresolved inter-band transition in the iron oxide in the spectral vicinity of the LSPR frequency of the gold shell.

The enhanced near-field of plasmon resonant metal nanoparticles has also been employed to induce local *photolithography* [27]. Due to the strongly confined nature of the near-field, sub-diffraction resolution is achievable. Plasmon enhanced nanolithography will be discussed in a later section. There have also been proposals on plasmon field-enhanced photochemistry and photocatalysis.

6.2.2 *Tunability of the LSPR and Near-Field*

For realizing phenomena that employ the near-field of metal nanoparticles, it is often necessary to engineer the near-field in terms of its spectrum, enhancement, and spatial profile. In the case of a metal particle, the frequency and quality of the LSPR mode (as expressed by the optical polarizability in Eq. (6.1)), and the resulting far- and near-field properties depend sensitively on the nanoparticle size, geometry, metal composition, and refractive index (RI) of the local environment [9, 10, 28–30]. For instance, by changing the shape of the nanoparticle (see Chapter 3) from a sphere to a rod, the LSPR response can be tuned [31]. Rod-shaped nanoparticles have two resonances: one due to plasmon oscillation along the nanorod short axis and another due to plasmon oscillation along the long axis [32]. The resonant frequency of the long-axis oscillation depends strongly on the nanorod aspect ratio, that is, the length-to-width ratio. Depending on the aspect ratio, the Coulombic restoring force on the electrons polarized along the long axis is much lower than that in the case of a sphere. Therefore, when the nanorod aspect ratio is increased from 1 to higher values, there is a red-shift in the long axis LSPR and an increase in its polarizability. In the case of nanorods of silver of aspect ratio 2.8, the resonant field enhancement can be as high as 3500 at the rod tips, as compared to a value of 200 for silver spheres [3]. For a detailed analysis of the shape dependence of LSPRs, see Chapter 3.

The nature of the metal also influences LSPR quality (defined as the energy position of the LSPR band divided by its linewidth). Inherent Ohmic losses in a metal result in damping or broadening of the LSPR. Silver, which is the most conductive metal, has a much higher plasmon quality as compared to gold [33]. Because of this much stronger plasmon fields are produced in silver nanostructures. The plasmonic fields are relatively damped in gold nanostructures due to the proximity of the LSPR to the inter-band absorption edge.

The LSPR frequency is also tunable via changes in the dielectric properties of the medium surrounding the nanoparticle. As seen from Eq. (6.2), an increase in the medium dielectric constant ε_{out} (or RI $n_{out} = \sqrt{\varepsilon_{out}}$) results in an increase in the negative value of ε_1 required to satisfy the resonance condition, which causes a red-shift of the LSPR. In a physical sense, the increase in the medium dielectric constant results in a lowering of the Coulombic restoring force on the polarized electron cloud, thus lowering the LSPR frequency. The LSPR is thus sensitive to any RI change in the local medium around the nanoparticles, including the presence of solvent and adsorbate molecules [8, 30].

6.3 Optical Properties of Coupled Nanoparticles

6.3.1 *Nanoparticle Assemblies*

While nanostructure size, shape, and medium can be employed as handles to tune the LSPR, the most effective approach to customize the near- and far-field field response is via the arrangement of nanoparticles into complex groupings.

6.3.1.1 LSPR response of assemblies

As discussed earlier, the LSPR of a plasmonic nanoparticle is highly sensitive to the dielectric nature of the medium around the nanoparticle. While the latter typically represents a response to a non-resonant medium, the effect can be much more dramatic when the nanoparticle is subject to the presence of a resonant absorber or emitter, especially another plasmon resonant nanoparticle. Experimental and simulation studies by Mirkin and Schatz groups have shown that the assembly of spherical Au nanoparticles into a close-packed structure results in a strong red shift of the LSPR (relative to the isolated nanosphere LSPR band centered around 520 nm) [34]. The LSPR red-shift is attributed to the favorable coupling of the plasmon oscillations on vicinal nanoparticles. As expected, the extent of the coupling-induced red shift increases with decreasing inter-particle distance and increasing assembly size. Thus, assembling nanoparticles together allows strong modulation of the LSPR response.

6.3.1.2 Field enhancement

At the same time, coupling nanoparticles together presents an opportunity to enhance and engineer near-fields. As an example, the assembly or aggregation of nanoparticles has been shown to result in huge SERS enhancements [35]. The inter-particle junctions formed by the assembly of nanoparticles are sites for electric field enhancements that are several orders larger than those in isolated nanoparticles. This is due to the favorable or constructive coupling of the plasmon resonances of adjacent nanoparticles. These junction electric fields can increase very sharply with decreasing inter-particle separation. In fact, single molecule experiments by Nie and coworkers have shown up to 14 orders of magnitude enhancement in SERS of R6G molecules on silver nanoparticles, which has been attributed by Brus and coworkers to be from the molecules at hot spots formed at the junction of two or more nearly-touching nanoparticles [36].

6.3.1.3 Plasmonic wave-guiding

By means of the strong near-field coupling between metal nanoparticles, the plasmon oscillation of a nanoparticle can induce an oscillation in a neighboring particle in close proximity. This has great potential for designing plasmonic waveguides consisting of a linear chain or array of closely spaced nanoparticles. It has been theoretically shown that the near-field coupling between the nanoparticles results in coupled plasmon polariton modes leading to the coherent transport of electromagnetic energy along the nanoparticle chain [37]. This makes possible wave-guiding of photons at a lateral confinement much below the diffraction limit, which is unattainable using conventional photonics. Near-field coupling-based photonic transport has been experimentally demonstrated in a lithographically fabricated plasmonic waveguide consisting of a linear array of silver spheroids interacting along their short axes parallel to the chain axis [38]. In this demonstration by the Atwater group, localized excitation of the waveguide was achieved by the tip of a near-field scanning optical microscope, while photonic transport was evidenced by observation of distant (up to 500 nm) excitation of fluorophores by the transported plasmons.

6.3.2 *Polarization Dependence of Coupling in a Dimer*

As discussed in the previous section, in assemblies or groupings of metal nanoparticles, the LSPR and near-field response is strongly affected by the coupling of the resonances of the individual particles, henceforth referred to as plasmon coupling. A conceptual quantitative understanding of plasmonic coupling between nanoparticles is instrumental for developing strategies [39] for systematic engineering of the spectrum, strength, as well as the spatial distribution and polarization of local electric fields in and around a nanostructure. It is most instructive to consider the simplest case of a pair or dimer of coupled nanoparticles.

The most informative experiments have been those on arrays of metal nanodisc dimers fabricated by electron beam lithography [33, 40–42], a technique which allows control over the inter-particle distance and orientation of the interacting particles. LSPR absorption spectra of such arrays have shown that the plasmon coupling is polarization-dependent (see Fig. 6.2). When the light excitation is polarized parallel to the inter-particle axis, the LSPR of the nanodisc pair is red-shifted relative to the LSPR of an individual nanodisc. The smaller the gap between the discs, the larger is the red-shift. In contrast, when the incident light polarization is perpendicular to the inter-particle axis, there is a small blue-shift of the LSPR. Thus, the coupling between the particles can be favorable or unfavorable depending on the excitation polarization direction relative to the inter-particle axis.

The colloidal assembly of gold nanorods presents an elegant demonstration of the polarization and orientation dependence of

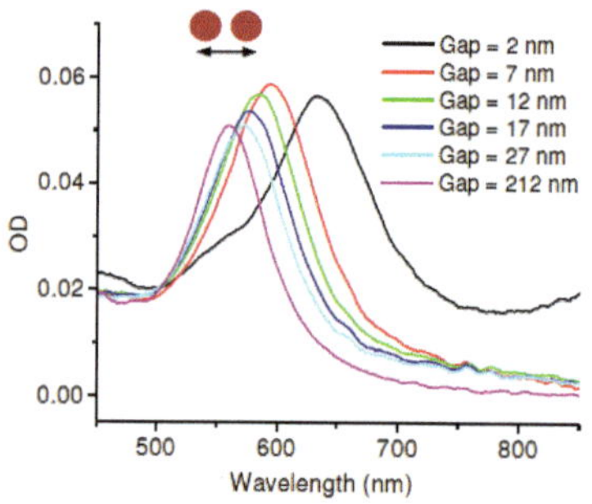

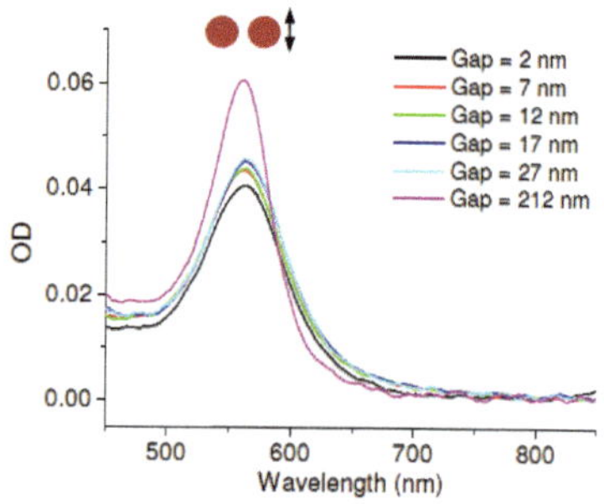

Figure 6.2 Experimental extinction spectra of gold nanodisc pair arrays show that the LSPR mode red-shifts with decreasing inter-particle gap for incident light polarization parallel to the inter-particle axis (left) and slightly blue-shifts for perpendicular polarization (right). Reprinted with permission from Ref. [33]. Copyright (2007) American Chemical Society.

plasmon coupling [43]. On account of their shape anisotropy, there are two orientations along which nanorods can be assembled: end-to-end or side-by-side. Gold nanorods have been linked in an end-to-end manner in colloidal solution using bifunctional thioalkyl carboxylic acid–based molecular linkers. The end-to-end linkage has been seen to result in a strong red-shift of the long-axis LSPR band. Alternatively, gold nanorods can be assembled by using negatively charged citrate linkers which bind the cationic surfactant bilayer on the sides of the rods, linking the nanorods together in a side-by-side orientation. In direct contrast to the end-to-end linkage case, the side-by-side linkage results in a blue-shift of the long-axis LSPR. This observation is consistent with the polarization dependence of plasmon coupling seen in lithographically fabricated dimers. The long-axis plasmon oscillations are polarized, in the end-to-end case, along the inter-particle axis, giving rise to the red-shift upon coupling. In the side-by-side case, the long-axis plasmons are polarized perpendicular to the inter-particle axis, giving rise to the blue-shift upon coupling. Exactly opposite shifts are observed for the short-axis mode, which is polarized 90° to the long-axis mode. However, the LSPR shifts are much smaller for the short-axis mode. This is due to the lower polarizability of the short-axis mode, resulting in relatively weaker coupling. The above two experiments lead us to a general rule: plasmon oscillations polarized along the inter-particle axis couple "favorably" causing a reduction in the LSPR frequency, whereas those polarized perpendicular to the inter-particle axis, interact "unfavorably" causing an increase in the LSPR frequency.

6.3.3 *Dipolar-Coupling Model*

The observed trends in plasmon coupling in a dimer can be explained by a simple model where each of the interacting particles is considered to be a point dipole [44]. In a dimer, the electric field E_1 (or E_2) felt by particle 1 (or 2) is the sum of the incident light field E_0 and the near-field of the electric dipole μ_2 (or μ_1) on the neighboring particle, which for a dipole decays as the inverse cube of the distance d from the particle. While the assumption of a dipolar near-field is not strictly valid, as we shall see later, the dipolar-coupling treatment suffices for a qualitative understanding.

$$E_1 = E_0 + \frac{\kappa \mu_2}{\varepsilon_{out} d^3} \,. \tag{6.6}$$

$$E_2 = E_0 + \frac{\kappa \mu_1}{\varepsilon_{out} d^3} \,. \tag{6.7}$$

κ is an orientation factor, which depends on the alignment of the two single-particle dipoles:

$$\kappa = 3\cos(\theta_1)\cos(\theta_2) - \cos(\theta_{12}) \,, \tag{6.8}$$

where θ_{12} is the angle between the directions of the two dipoles, θ_1 is the angle between the direction of dipole 1 and the inter-dipole axis, and θ_2 is the angle between the direction of dipole 2 and the inter-dipole axis. Since $\mu = \varepsilon_{out}\alpha E$, we get from Eq. (6.6) and Eq. (6.7):

$$E_1 = E_0 + \frac{\kappa \alpha_2 E_2}{d^3} \,, \tag{6.9}$$

$$E_2 = E_0 + \frac{\kappa \alpha_1 E_1}{d^3} \,. \tag{6.10}$$

The net polarizability α' of the two-particle system is given as:

$$\alpha' = \frac{\mu_1 + \mu_2}{2\varepsilon_{out} E_0} = \frac{\alpha_1 E_1 + \alpha_2 E_2}{2E_0} \,, \tag{6.11}$$

From Eq. (6.9) and Eq. (6.10), we obtain:

$$\alpha' = \frac{\alpha_1 + \alpha_2}{2} + \frac{\kappa \alpha_1 \alpha_2}{d^3} \frac{(E_1 + E_2)}{2E_0} \tag{6.12}$$

$$= \frac{\alpha_1 + \alpha_2}{2} + \frac{\kappa \alpha_1 \alpha_2}{d^3} \left[1 + \frac{\kappa}{d^3} \left(\frac{\mu_1 + \mu_2}{2\varepsilon_{out} E_0} \right) \right] . \tag{6.13}$$

Using Eq. (6.11) into Eq. (6.13) we obtain:

$$\alpha' = \frac{\alpha_1 + \alpha_2}{2} + \frac{\kappa \alpha_1 \alpha_2}{d^3} \left[1 + \frac{\kappa \alpha'}{d^3} \right] . \tag{6.14}$$

From Eq. 6.14 we finally get:

$$\alpha' = \frac{\alpha_1 \left(1 + \frac{\kappa \alpha_2}{d^3}\right) + \alpha_2 \left(1 + \frac{\kappa \alpha_1}{d^3}\right)}{2\left[1 - \left(\frac{\kappa}{d^3}\right)^2 \alpha_1 \alpha_2\right]} \,. \tag{6.15}$$

For the special case of dimer of identical particles, i.e. a homodimer, $\alpha_1 = \alpha_2 = \alpha$:

$$\alpha' = \frac{\alpha}{1 - \frac{\kappa \alpha}{d^3}} \,. \tag{6.16}$$

The coupled-particle polarizability α' is maximum when the denominator vanishes. Combining with Eq. (6.1) we get the resonance condition:

$$\varepsilon_1 = -\varepsilon_{out}\left(\frac{2+\kappa\frac{R^3}{d^3}}{1-\kappa\frac{R^3}{d^3}}\right). \tag{6.17}$$

It must be noted that for an isolated nanoparticle ($d \to \infty$), the resonance condition becomes the expected $\varepsilon_1 = -2\varepsilon_{out}$. The resonance condition for a coupled-particle system depends on the orientation factor κ and the distance d between the particles (assumed to be point dipoles): $\kappa = 2$ for dipoles polarized along the interaction axis (end-to-end), whereas $\kappa = -1$ for dipoles polarized perpendicular to the interaction axis (side-by-side). Thus, from Eq. (6.17), it can be seen that for parallel polarization ($\kappa > 0$), a higher negative value ε_1 of would be required to achieve resonance at a shorter distance d. This gives an LSPR red-shift, the magnitude of which increases with decreasing d. For perpendicular polarization ($\kappa < 0$), an LSPR blue-shift is expected as the particles come closer. Since the absolute value of κ is smaller for the latter case, the magnitude of LSPR blue-shift under perpendicular polarization is expected to be weaker than that of the red-shift under parallel polarization. Thus, a simple dipolar-coupling model qualitatively explains the general polarization dependence observed, however it must be noted that the model is valid only for $d \gg R$ and as we shall see later, predicted shift magnitudes are expected to deviate even at moderate inter-particle spacing. Besides, since the particles have a finite size, electromagnetic retardation has to be accounted for, which for particle sizes approaching the wavelength of light can give rise to severe deviations [45].

6.3.4 *Analogy to Exciton Coupling in Molecular Aggregates*

The polarization and orientation dependence of plasmon coupling described above is very similar to absorption spectra shifts observed in organic molecules upon their dimerization or aggregation. Organic chromophores dimerized head-to-tail with respect to their transition dipole moment direction (known as *J-dimers*) show an absorption band that is red-shifted compared to that of the isolated

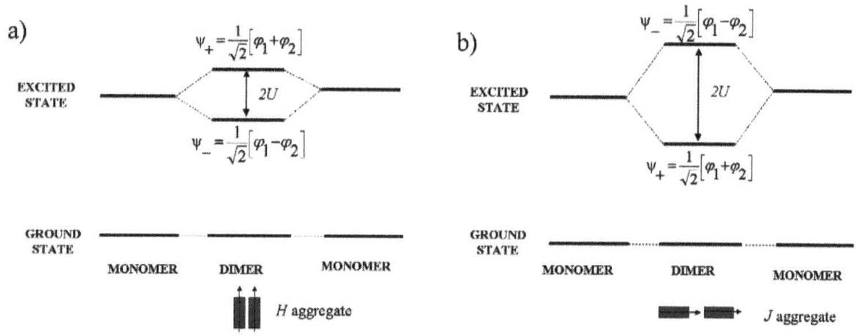

Figure 6.3 The energy level splitting resulting from the dipolar coupling in a dimer, showing symmetric (ϕ^+) and anti-symmetric coupling (ψ^-) of excitons for (a) H-aggregate geometry or side-by-side dimer and (b) J-aggregate geometry or end-to-end dimer. Reprinted with permission from Ref. [43]. Copyright (2006) American Chemical Society.

chromophore, whereas parallel dimers of organic chromophores (known as *H-dimers*) show a blue-shifted absorption spectrum. The *exciton-coupling model*, which was developed by Michael Kasha to explain shifts in the spectra of dimerized organic molecules [46], therefore, applies very well to the case of coupling between two plasmon resonant nanoparticles [43]. As per the exciton theory (see Fig. 6.3), the excited-state levels of the monomer split in two levels upon dimerization, a lower energy level and a higher energy level relative to the monomer excited state. These correspond to two possible arrangements of the transition dipoles of the chromophores in the dimer: in-phase or symmetric and out-of-phase or anti-symmetric. The energy splitting $2U$ between the two modes is given by the interaction energy between the chromophores which is approximated by the Coulombic interaction between the transition dipole moments of the monomers. While this treatment strictly applies to molecular excitons, in the limit where the particle size is small compared to the inter-particle distance d, each of the interacting particles can be considered to be a dipolar exciton.

For the head-to-tail case, the interaction is attractive for dipoles aligned symmetrically. As a result, we see a red-shift in the resonance of the dimer. The configuration corresponding to the dipoles aligned anti-symmetrically is higher in energy, but is optically forbidden, since the dipoles, being equal and opposite, cancel out. This represents therefore a dark mode in a dimer consisting of two identical particles (homodimer). For the case where the incident field polarization (and hence the direction of both dipoles) is orthogonal to the inter-particle axis, the interaction is repulsive for a symmetric configuration of the dipoles. As a result, a blue shift of the resonance frequency is observed in this case. The anti-symmetric configuration is lower in energy, but dark in the optical spectrum, due to the dipoles canceling out in a homodimer.

In a heterodimer, where the two interacting particles are nonidentical in terms of their transition dipole moments (for instance, two nanorods of dissimilar aspect ratio), it is expected that both the symmetric and anti-symmetric modes would be present in the optical spectrum. Of course, the anti-symmetric mode, otherwise dark in the homodimer, would be significantly weaker in intensity compared to the symmetric mode.

6.3.5 *Bonding and Anti-Bonding of Plasmons*

Plasmon coupling in nanoparticles assemblies can also be visualized using the *plasmon hybridization theory* (see also Sec. 3.3.5.1), analogous to the hybridization of atomic orbitals in molecules. When two metal nanoparticles approach each other, the plasmon modes supported by the two surfaces hybridize, resulting in a lower energy bonding plasmon mode and a higher energy anti-bonding plasmon mode. It is instructive to apply the molecular hybridization analogy to the case of plasmon coupling between gold nanorods [43]. The coupled long-axis mode for the end-to-end dimer has a bonding nature, analogous to the formation of a σ bond from two π_z orbitals. Consistent with the molecular analog of this bonding mode, the resulting electric field intensity for this mode peaks in the junction between the interacting nanorods. On the other hand, the coupled long-axis mode for the side-by-side dimer has an anti-bonding nature analogous to the formation of a π^* bond from $\pi_{x/y}$ orbitals, with the electric field concentrated on either side of the inter-particle junction. It must be noted that the σ^* mode in the end-to-end dimer and the π mode in the side-by-side dimer are dark in the nanorod homodimers. The knowledge of the bonding or anti-bonding nature of the coupled plasmon modes is potentially useful

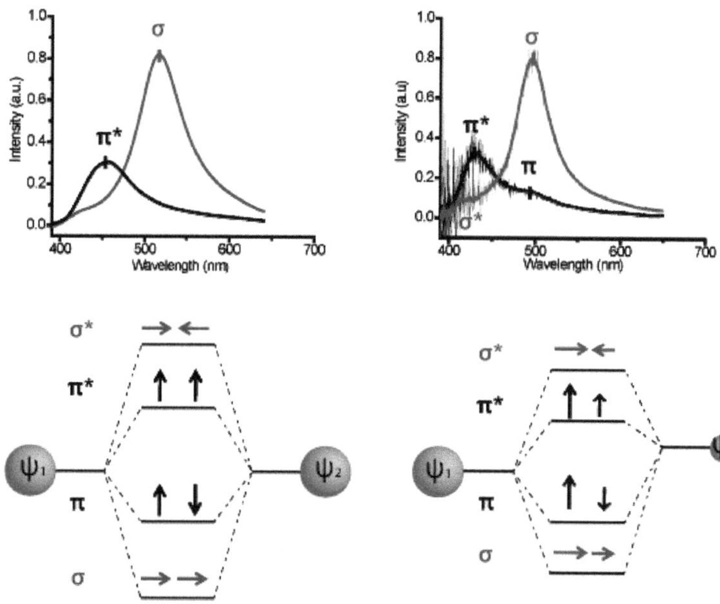

Figure 6.4 Coupled LSPR modes (top) and bonding/anti-bonding model (bottom) for homodimers (left) and heterodimers (right). Scattering spectra of a silver nanoparticle homodimer show only the symmetric mode: σ for parallel polarization (gray) and σ^* for perpendicular polarization (black). For a heterodimer of a 20-nm silver nanoparticle and a 40-nm silver nanoparticle, both symmetric and anti-symmetric modes are seen: σ and σ^* for parallel polarization (gray) and π^* and π for perpendicular polarization (black). Reprinted with permission from Ref. [47]. Copyright (2010) American Chemical Society.

for designing assembled plasmonic nanostructures, where the near-field can be squeezed in a confined region of the nanostructure, for field-enhanced spectroscopy of absorbers or emitters placed in this region.

The applicability of the bonding/anti-bonding model of plasmon resonances is verified fully in the case of a heterodimer (see Fig. 6.4 from Ref. [47]), where due to symmetry breaking it is expected that all four modes red-shifted σ and blue-shifted σ^* for parallel polarization and red-shifted π and blue-shifted π^* for perpendicular polarization would be bright and observable. This has been found to be indeed the case in single-particle spectra of silver heterodimers composed of a 20-nm and a 40-nm particle. As expected, for polarization along the dimer axis, an anti-symmetric σ^* mode is observed that is weaker and higher in energy than the symmetric σ mode. Likewise, for polarization perpendicular to the dimer axis, an anti-symmetric π mode is observed, which is weaker and lower in energy than the π^* mode. A similar result can, in principle, be expected for a heterodimer composed of a silver and a gold nanoparticle of the same size. However, in practice, this case is complicated by the coupling of the inter-band transitions with the LSPR mode of the silver particle giving rise to anomalous shifts.

It must be mentioned that a plasmon hybridization model was first developed by Nordlander and Halas in order to explain the LSPR resonances properties of metal nanoshell structures [48]. In a later section, we shall show the analogy between the hybridization in the two-particle dimer with that of a metal nanoshell.

6.4 Spatial Profile of the Near-Field

While we have been assuming that the near-field of a particle decays with distance as $1/d^3$, it is crucial to characterize the true spatial extent of the plasmonic near-field. Such characterization would allow reliable estimation of the field enhancement acting on molecule/emitter placed in the vicinity of a resonant metal nanoparticle. Knowing the actual distance dependence of the near-field is also important for reliably estimating the field gradient experienced by such a molecule.

The most definitive characterization of the distance-dependence of the near-field has been obtained from the investigation of plasmon coupling in metal nanodisc pairs lithographically fabricated with systematically varying inter-particle separations [33]. In these studies, only the bonding mode (parallel polarization) is of interest. As the inter-particle separation between the nanodiscs is reduced, the near-field coupling becomes stronger resulting in an increasing LSPR red-shift. The LSPR red-shift is thus a direct measure of the near-field coupling strength.

6.4.1 *Distance-Dependence of the Near-Field*

The inter-particle distance dependence of the near-field coupling would therefore reflect the distance decay of the near-field itself. In other words, each particle senses the near-field due to the other particle. By varying the distance of the other particle and monitoring the LSPR response, the spatial profile of the near-field can be deduced. The plot of the LSPR red-shift as a function of inter-particle gap (surface-to-surface separation) shows a much more rapid decay of the near-field than predicted by the dipolar model. This is because the dipolar model does not take into account the multipolar interactions between the particles, which become increasingly important at smaller and smaller inter-particle gaps. Plasmon coupling is therefore a multipolar interaction and its true distance-dependence can be quantitatively reproduced only by a complete treatment that includes all modes of interaction (dipolar, quadrupolar, octupolar). Computational electrodynamics methods such as discrete dipole approximation (DDA) (see Chapter 2) and finite-difference-time-domain (FDTD), which include a full multipolar treatment in addition to finite-size retardation effects, fit experimental trends well.

6.4.2 *Size-Scaling of Near-Field Decay*

The near-field decay shows a *size-scaling* behavior (see Fig. 6.5): when the fractional LSPR shift is plotted against inter-particle gap d normalized by particle size, plots for different diameter nanodiscs collapse to the same trend [33, 40, 41]. This implies that the near-field decays away from a particle surface over a distance proportional to the particle size. In other words, a larger size particle has a larger distance range of its near-field, in proportion to its size. Such a size-scaling is expected from the dipolar model: in Eq. (4.24), the resonance frequency of the dimer can be reduced to a function of the size-normalized distance d/R.

The origin of the size-scaling derived from the dipolar model is still valid in the case of a full multipolar interaction. It is known that a quadrupolar field decays as $1/d^5$ while the quadrupole polarizability is proportional to R^5, an octupolar field decays as $1/d^7$

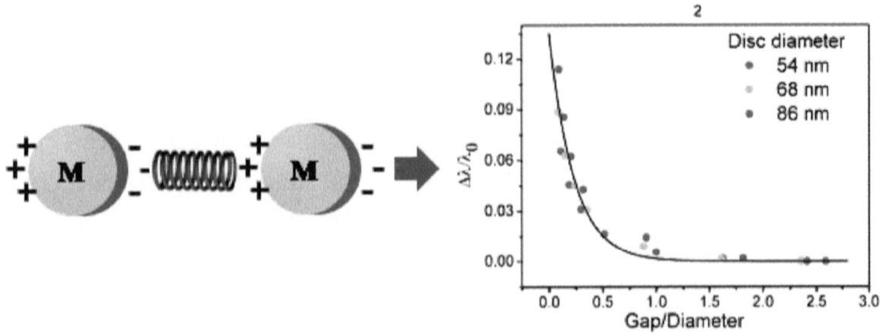

Figure 6.5 The near-field coupling between two plasmonic nanoparticles (as indicated by the fractional LSPR red-shift) follows the same trend with inter-particle gap regardless of particle size, when the inter-particle gap is scaled by the particle size. Reprinted with permission from Ref. [33]. Copyright (2007) American Chemical Society.

while the octupolar polarizability is proportional to R^7, and so on for higher order modes. Thus, when higher-order coupling terms, e.g. dipole–quadrupole $(1/d^4)$, dipole–octupole $(1/d^5), \ldots$ enter into the interaction, the resulting distance-decay is significantly steeper than that predicted by the purely dipolar model, and can be closely modeled by a single-exponential decay, rather than any inverse power law. It must however be acknowledged that the exponential trend used to describe the distance-decay of the near-field coupling is not an analytic dependence but only an empirical fit best suited for comparing the near-field coupling for different systems. Besides, such a treatment is limited to gaps larger than $0.05R$. When particles almost touch (separation < 1 nm), anomalously large shifts and new bands are seen that need a treatment that incorporates non-local effects and quantum mechanical corrections [49].

The distance-dependence of the near-field is quite universal [33]. When approximated to an exponential decay, the near-field coupling in the dimer system is found to have a decay length of $\approx$0.2 in units of the particle diameter D, regardless of the nanostructure size, shape, metal, or medium. That is, as a function of the distance s away from the nanoparticle surface, the near-field strength E can be expressed as:

$$E(s) \approx K \exp\left(-\frac{s}{0.2D}\right). \tag{6.18}$$

While K may vary with a change in any of the above-mentioned parameters, the decay length is unaffected. For instance, colloidal gold nanosphere dimers show a similar decay length of $\approx$0.2D as that shown by lithographically patterned gold nanodisc dimers. In the case of nanorods, the near-field corresponding to the long-axis mode decays over a length given by 0.2D, provided D is the long-axis dimension of the nanorod [50]. While the magnitude of the near-field increases with increasing end curvature or aspect ratio of the nanorod, the size-scaled decay length is always the same. Furthermore, silver nanodisc pairs show a similar near-field decay length of $\approx$0.2D as gold nanodisc pairs. The absolute LSPR shifts are, however, three times larger in the silver case, which can be attributed to the earlier discussed fact that fields are stronger for silver. Regardless of the difference in the absolute strength of the near-field, the distance-decay of the field itself appears to be independent of the nature of the metal.

6.4.3 *Direct Mapping of the Near-Field*

Although we concluded from the above discussion that the distance decay of the near-field can be indirectly deduced from the far-field LSPR response of a pair of coupled metal nanoparticles, a direct measurement of the near-field profile is desirable. Not only would this enable verification of the near-field distance-dependence derived above, but also allow us to image spatial profiles of near-fields in more complex nanostructure geometries. A primary pre-requisite for such a near-field profiling approach is nanometer-scale resolution.

A common approach for imaging near-fields relies on *scanning near-field microscopy* (SNOM) [51, 52], which offers a lateral resolution, down to tens of nanometers (20–100 nm). However, the measurement involves the insertion of a probe in the immediate vicinity of the object under study, to locally either detect or excite an evanescent field. The presence of the probe generally perturbs the physics of the sample to be characterized and the effective object becomes a complex probe-sample nanosystem, whose physics strongly depends on probe features such as geometry, material, etc. Additionally, producing high quality SNOM probes in a reproducible way remains a critical issue. Photoemission electron microscopy (PEEM) [53] and electron energy-loss spectroscopy (EELS) techniques [54], while powerful, are indirect in nature. For instance, a PEEM measurement involves the conversion of photons into electrons, which are then detected. Even though high-resolution and full spectrum spatial maps can be obtained using this technique, absolute quantitation of the near-field is not possible.

A promising method that is non-perturbative, quantitative, and offers high-resolution relies on *photochemistry enhanced* in the near-field of a nanostructure [55, 56]. It is possible to employ the resonant field enhancement around the nanostructure to structure a photopolymer resulting in a molecular mold or map of the near-field intensity around the particle. In a typical procedure (see Fig. 6.6), the plasmon resonant nanostructure is immobilized or fabricated on a glass substrate and photoexcited at its resonance in the presence of a photolithographic formulation [56]. The photopolymerizable system is designed to meet three requirements: a) the photoinitiator is required to have sensitivity (as evidenced by its absorption spectrum) in the visible range overlapping the LSPR spectrum of the nanostructure, b) the photopolymerizable system should exhibit a non-linear threshold response. The threshold dose, below which no photopolymerization can occur, needs to be well characterized as a function of excitation wavelength *a priori*, and c) the diffusion of the molecules in the formulation should be ideally low. During photoexcitation, the laser dose is set at a fraction f of the threshold dose. Thus, photopolymerization can only be initiated at regions (within few nm around the nanoparticle) where the near-field enhancement is $1/f$ or higher. Following a washing step, an AFM image of the composite nanoparticle + polymerized structure is acquired, from which a background AFM image of the nanoparticle preceding the photoexcitation is subtracted. The differential imaging provides a high signal-to-noise, drift-free and tip convolution-free image of the nanoscale polymer structures fabricated around the nanoparticle. For instance, in the case of a silver nanosphere excited with linearly polarized light, polymer wings are seen along the polarization direction, clearly reflecting the dipolar lobes of the plasmon field around the excited nanoparticle. Low diffusion and a sharp threshold response of the formulation are crucial for achieving a high resolution of this near-field enhanced photolithography. Systems that offer a sub-10-nm resolution have been specially designed.

When the excitation dose (as a fraction f of the threshold) is increased, the thickness w of the polymer wings around the nanoparticle increases, since the enhancement at regions further away from the nanoparticle surface is able to overcome the threshold dose. The field enhancement at the edge of the polymer wing can thus be estimated to be $1/f$. By plotting $1/f$ as a function of w, we essentially obtain a plot of the near-field intensity enhancement as a function of the distance away from the nanoparticle. Deeb *et al.* have characterized the near-field intensity around a silver nanosphere of D = 60 nm using this approach [56]. In this case, the field intensity enhancement is found to be described as a function of the distance s away from the nanoparticle surface as:

$$E(s) \approx K \exp\left(-\frac{s}{l}\right), \tag{6.19}$$

where $K = 39$ represents the maximum enhancement, i.e. at the surface of the nanoparticle and $l = 11$ nm represents the decay

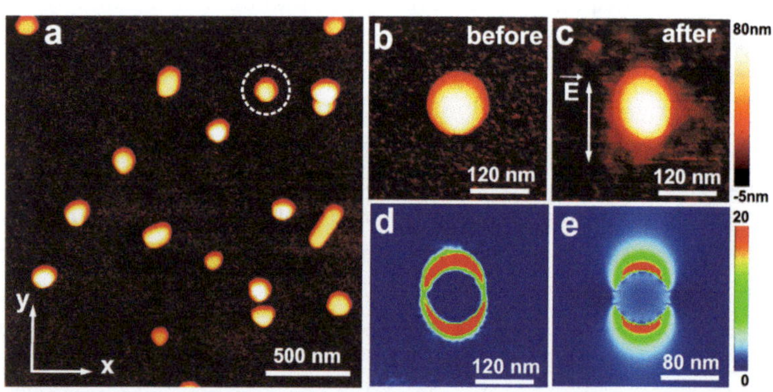

Figure 6.6 Near-field photopolymerization based on the resonant excitation of the dipolar LSPR mode of silver nanoparticles. (a) Topographic AFM image of silver nanoparticles before photoexcitation. (b) Close-up image of nanoparticle selected in panel a. (c) Close-up topographic image of the same silver nanoparticle after photopolymerization. (d) Differential image of panels c and d. (e) Spatial map of near-field intensity calculated around a 60-nm silver nanoparticle using the FDTD method. Reprinted with permission from Ref. [56]. Copyright (2010) American Chemical Society.

length of the near-field. Thus, a quantitative spatial profile (with an unprecedented sub-10 nm resolution) of a plasmonic near-field has been obtained for the first time. This direct measurement confirms the rapid near-exponential decay of the near-field deduced from far-field measurements of coupled dimer LSPRs [33]. Furthermore, it is found that the decay length is very close to 0.2 times the diameter of the nanoparticle, in agreement with the indirect far-field measurements.

The knowledge of the threshold dose as a function of wavelength also allows the measurement of the full spectrum of the near-field enhancement. Whereas this near-field molecular mapping technique has so far been demonstrated for the well known case of a spherical silver nanoparticle, it can be extended to more complex particle geometries that exhibit interesting resonance modes and spatial field profiles. It is worth noting that this method can also be potentially used to probe a larger fraction of the full three-dimensional (3-D) near-field intensity distribution, whereas scanning probe microscopies are generally sensitive only to the near-field distribution at the tip apex.

6.5 Applications of Near-Field Coupling Concepts

6.5.1 *Plasmon Ruler*

Since the strength of plasmon coupling and the LSPR shift induced by it is dependent on inter-particle distance, the LSPR spectrum of a dimer of coupled (silver or gold) nanoparticles can be used to report on the inter-particle distance, as first shown by Alivisatos and coworkers [57]. In a scheme in which a biomolecular structure is used to link the two particles together, dynamic distance changes in the biomolecular structure (triggered by external biochemical stimuli/binding/signaling events) could be probed by following the LSPR shift of the dimer. Since plasmonic nanoparticles scatter light very strongly, scattering spectra can be collected from single dimers using dark-field spectroscopy, allowing sensitive optical measurement of biomolecular distances and probing of distance changes at the nanoscale. Plasmonic nanoparticles do not suffer from the drawback of dye photobleaching, a major problem in the earlier method of biomolecular distance measurement fluorescence resonance energy transfer (FRET). For the plasmon ruler application, the near-field distance-dependence discussed earlier provides a calibration of the LSPR shift as a function of inter-particle distance. LSPR shifts of gold nanoparticle plasmon rulers assembled by DNA linkers of different lengths can be fit reasonably well with Eq. (6.18) [33, 58]. Thus, the distance decay of the near-field offers a guideline for the design of plasmon rulers with a desired dynamic distance range and sensitivity for the given biomolecular application. Whereas the FRET ruler has a maximum range of 10 nm due to a steep $1/d^6$-dependence between two molecular species, the plasmon coupling distance range can be increased directly by increasing the particle dimension. For a dimer of 40 nm gold nanoparticles, the range is estimated to be around 70 nm.

6.5.2 *Metal Nanoshells*

Metal nanoshells (see also Sec. 3.3.5.1) are nanostructures that consist of a dielectric or hollow core surrounded by a thin (1020 nm) shell of metal, in most cases gold. The Halas group developed this tunable nanostructure and showed that the LSPR of the metal shell nanostructure depends strongly on the dimensions of the metal shell and the dielectric core [59]. As the gold shell thickness is decreased, there is red-shift in the nanoshell LSPR from the visible region ($\lambda_{sp} = 520$ nm for a solid gold nanosphere) towards the near-infrared region. LSPRs of the metal nanoshells always show the same trend when they are plotted as a function of shell thickness scaled by the core size [60]. The scaling trend is universal regardless of inner core material, the shell metal (gold or silver), or the surrounding medium. The size scaling shown by the nanoshell structure is similar in principle to that in the two-particle system.

The metal shell has two surfaces, an inner one and an outer one, each with its own plasmon mode. These two plasmon modes couple across the thickness of the shell. The strength of the coupling is therefore determined by the field decay across the thickness (as $1/t^3$ assuming dipolar behavior) and the core polarizability, which is proportional to R^3, resulting in a similar size-scaling as in the two-particle system. There is, however, one major difference between the nanoshell structure and the two-particle system. In the case of the two-particle system, as discussed earlier, the dipole–dipole model under-estimates the steepness of the distance-decay of the near-field coupling. However, in the nanoshell case, a dipolar model gives the same trend as the full electrodynamic treatment. This is due to the spherical symmetry in the nanoshell system, which allows only modes of the same angular momentum to interact. This is the reason why the dipolar nanoshell resonance resulting from the coupling is describable by a purely dipole–dipole interaction and does not involve interactions of dipole–multipole type.

6.5.3 *Coupling in Larger Arrays*

The distance-dependence and size-scaling of near-field coupling is also found to be valid in a system of three interacting nanospheres, a first step towards extending this model to chains, arrays, or assemblies of metal nanoparticles [50]. In 3-D assemblies, the volume fraction of the nanoparticles could be the scaling variable in lieu of the inter-particle separation scaled by size used for the psuedo-1-D coupling cases.

6.5.4 *Molecular Sensing*

The dependence of the LSPR on the medium dielectric constant discussed earlier becomes very useful in the optical sensing of chemical and biomolecular analytes [61]. When the RI of the medium is increased, there is a linear red shift in the LSPR wavelength maximum position. By following this shift, it is possible to optically probe changes in the local environment of the nanoparticle. For instance, the adsorption or binding of molecules on or near the nanoparticle surface can cause a RI change and a shift in the LSPR wavelength maximum. In order to have sensitivity down to few molecules per nanoparticle, it is desired that the magnitude of shift resulting from a small RI change in the local medium be as high as possible. It is advantageous that the geometry of plasmonic nanostructures can be tuned to achieve enhanced plasmon sensitivity [62]. LSPRs of polarizable geometries such as nanorods, nanoshells, and nanotriangles are much more sensitive to the medium as compared to nanospherical particles. There is a strong correlation between geometries that support strong near-field confinement and those that offer high sensitivity.

In accordance, the nanoparticle dimer geometry offers tunable sensitivity [63]. Similar to an isolated nanoparticle, the LSPR wavelength maximum of a dimer increases near-linearly with increasing medium RI. As the inter-particle gap in the dimer is decreased, a near-exponential increase in the sensitivity of the LSPR mode (polarized along the inter-particle axis) is predicted. Thus, assemblies of strongly coupled particles are much better candidates for plasmonic nanosensing as compared to colloids of non-interacting particles. This was recently verified experimentally in lithographically fabricated dimers of gold nanodiscs [64]. Nanoparticle dimers with small gaps allow highly enhanced detection of proteins, with sensitivity tending towards single-molecule detection. A similar effect has been theoretically and experimentally observed in gold nanoshells, where a reduction in the shell thickness-to-core size ratio results in a near-exponential enhancement of the LSPR sensitivity. In a physical sense, the near-field interaction in the coupled-particle system can be seen to result in a strong confinement of the electric field within a nanovolume/junction of the nanostructure, leading to an enhanced sensitivity of the optical resonance to the dielectric environment of this volume.

6.6 Future Outlook

A number of phenomena and applications in spectroscopy, photonics, imaging, and sensing, which rely on strong EM field and field gradients, can benefit from our knowledge of the enhanced tunable near- and far-field properties of metal nanoparticle assemblies. Basic conceptual principles and simulation tools have advanced enough to allow the design of customized near-fields by assembling desired groupings of metal nanoparticles. While lithographic fabrication is still the method of choice due to the exquisite control achievable, it is limited by low throughput, larger size, and poor crystalline quality of fabricated nanostructures. Bottom-up methods of programmed assembly, employing molecular recognition (e.g. DNA, peptides) [65] may overcome such limitations. Another challenge lies in the controlled placement of molecules and emitters at specific positions and within the nanostructure. Advances in the latter will allow us to utilize the full potential of tightly confined fields in assemblies and revisit near-field enhancement investigations that have met very limited success with isolated metal nanoparticles.

Acknowledgments

The contribution of collaborators and colleagues, especially Mostafa El-Sayed, Paul Alivisatos, Adam Cohen, Wenyu Huang, Sassan Sheikholeslami, Young-Wook Jun, Renaud Bachelot, Jerome Plain, Alexandre Bouhelier, Libai Huang, and Olivier Soppera is acknowledged. P.J. dedicates this chapter to the inspiration of late Shreepal B. Jain.

References

1. Faraday, M. (1857) The Bakerian lecture: Experimental relations of gold (and other metals) to light, *Phil. Trans. R. Soc.*, **147**, 145–181.
2. Kreibig, U., and Vollmer, M. (eds) (1995) *Optical Properties of Metal Clusters*, Springer.
3. Hao, E., and Schatz, G. C. (2004) Electromagnetic fields around silver nanoparticles and dimers, *J. Chem. Phys.*, **120**, 357–366.
4. Jain, P. K., Huang, X., El-Sayed, I. H., and El-Sayed, M. A. (2008) Noble metals on the nanoscale: Optical and photothermal properties and some applications in imaging, sensing, biology, and medicine, *Accounts Chem. Res.*, **41**, 1578–1586.
5. Sonnichsen, C., Franzl, T., Wilk, T., von Plessen, G., Feldmann, J., Wilson, O., and Mulvaney, P. (2002) Drastic reduction of plasmon damping in gold nanorods, *Phys. Rev. Lett.*, **88**, 077402/1–4.

6. Link, S., and El-Sayed, M. A. (2003) Optical properties and ultrafast dynamics of metallic nanocrystals, *Annu. Rev. Phys. Chem.*, **54**, 331–366.
7. Jain, P. K., Qian, W., and El-Sayed, M. A. (2006) Ultrafast electron relaxation dynamics in coupled metal nanoparticles in aggregates, *J. Phys. Chem. B*, **110**, 136–142.
8. Jain, P. K., Qian, W., and El-Sayed, M. A. (2006) Ultrafast cooling of photoexcited electrons in gold nanoparticle-thiolated DNA conjugates involves the dissociation of the gold-thiol bond, *J. Am. Chem. Soc.*, **128**, 2426–2433.
9. Kelly, K. L., Coronado, E., Zhao, L. L., and Schatz, G. C. (2003) The optical properties of metal nanoparticles: The influence of size, shape, and dielectric environment, *J. Phys. Chem. B*, **107**, 668–677.
10. Jain, P. K., Lee, K. S., El-Sayed, I. H., and El-Sayed, M. A. (2006) Calculated absorption and scattering properties of gold nanoparticles of different size, shape, and composition: Applications in biological imaging and biomedicine, *J. Phys. Chem. B*, **110**, 7238–7248.
11. Hirsch, L. R., Stafford, R. J., Bankson, J. A., Sershen, S. R., Rivera, B., Price, R. E., Hazle, J. D., Halas, N. J., and West, J. L. (2003) Nanoshell-assisted tumor ablation using near infrared light under magnetic resonance guidance, *Proc. Natl. Acad. Sci. USA*, **100**, 13549–13554.
12. Loo, C. A., Lowery, A., Halas, N., West, J., and Drezek, R. (2005) Immunotargeted nanoshells for integrated cancer imaging and therapy, *Nano Lett.*, **5**, 709–711.
13. El-Sayed, I. H., Huang, X., and El-Sayed, M. A. (2006) Selective laser photo-thermal therapy of epithelial carcinoma using anti-EGFR antibody conjugated gold nanoparticles, *Cancer Lett.*, **239**, 129–135.
14. Zharov, V. P., Galitovskaya, E. N., Johnson, C., and Kelly, T. (2005) Synergistic enhancement of selective nanophotothermolysis with gold nanoclusters: Potential for cancer therapy, *Lasers Surg. Med.*, **37**, 219–226.
15. Zharov, V. P., Mercer, K. E., Galitovskaya, E. N., and Smeltzer, M. S. (2006) Photothermal nanotherapeutics and nanodiagnostics for selective killing of bacteria targeted with gold nanoparticles, *Biophys. J.*, **90**, 619–627.
16. El-Sayed, I. H., Huang, X., and El-Sayed, M. A. (2005) Surface plasmon resonance scattering and absorption of anti-EGFR antibody conjugated gold nanoparticles in cancer diagnostics: Applications in oral cancer, *Nano Lett.*, **5**, 829–834.

17. Sokolov, K., Follen, M., Aaron, J., Pavlova, I., Malpica, A., Lotan, R., and Richards-Kortum, R. (2003) Real-time vital optical imaging of precancer using anti-epidermal growth factor receptor antibodies conjugated to gold nanoparticles, *Cancer Res.*, **63**, 1999–2004.
18. Yguerabide, J., and Yguerabide, E. E. (1998) Light-scattering submicroscopic particles as highly fluorescent analogs and their use as tracer labels in clinical and biological applications I. and II., *Anal. Biochem.*, **262**, 137–176.
19. Boyd, G. T., Rasing, T., Leite, J. R. R., and Shen, Y. R. (1984) Local-field enhancement on rough surfaces of metals, semimetals, and semiconductors with the use of optical 2nd-harmonic generation, *Phys. Rev. B*, **30**, 519–526.
20. Chen, C. K., Heinz, T. F., Ricard, D., and Shen, Y. R. (1983) Surface-enhanced 2nd-harmonic generation and Raman-scattering, *Phys. Rev. B*, **27**, 1965–1979.
21. Dieringer, J. A., McFarland, A. D., Shah, N. C., Stuart, D. A., Whitney, A. V., Yonzon, C. R., Young, M. A., Zhang, X., and Van Duyne, R. P. (2006) Surface enhanced Raman spectroscopy: New materials, concepts, characterization tools, and applications (introductory lecture), *Faraday Discuss.*, **132**, 9–26.
22. Shalaev, V. M., Poliakov, E. Y., and Markel, V. A. (1996) Small-particle composites. II. Nonlinear optical properties, *Phys. Rev. B*, **53**, 2437–2449.
23. Schatz, G. C. (1984) Theoretical studies of surface enhanced Raman scattering, *Acc. Chem. Res.*, **17**, 370–376.
24. Lakowicz, J. R. *et al.* (2006) Plasmon-controlled fluorescence: A new detection technology, *Proc. SPIE*, **6099**, 609909/1–15.
25. Lieberman, I., Shemer, G., Fried, T., Kosower, E., and Markovich, G. (2008) Plasmon-resonance-enhanced absorption and circular dichroism, *Angew. Chem. Int. Edit.*, **47**, 4855–4857.
26. Jain, P. K., Xiao, Y., Walsworth, R., and Cohen, A. E. (2009) Surface plasmon resonance enhanced magneto-optics (SuPREMO): Faraday rotation enhancement in gold-coated iron oxide nanocrystals, *Nano Lett.*, **9**, 1644–1650.
27. Sundaramurthy, A., Schuck, P. J., Conley, N. R., Fromm, D. P., Kino, G. S., and Moerner, W. E. (2006) Toward nanometer-scale optical photolithography: Utilizing the near-field of bowtie optical nanoantennas, *Nano Lett.*, **6**, 355–360.

28. El-Sayed, M. A. (2001) Some interesting properties of metals confined in time and nanometer space of different shapes, *Accounts Chem. Res.*, **34**, 257–264.
29. Mulvaney, P. (1996) Surface plasmon spectroscopy of nanosized metal particles, *Langmuir*, **12**, 788–800.
30. Underwood, S., and Mulvaney, P. (1994) Effect of the solution refractive index on the color of gold colloids, *Langmuir*, **10**, 3427–3430.
31. Gans, R. (1912) The form of ultramicroscopic gold particles, *Ann. Phys. (NY)*, **37**, 881–900.
32. Link, S., Mohamed, M. B., and El-Sayed, M. A. (1999) Simulation of the optical absorption spectra of gold nanorods as a function of their aspect ratio and the effect of the medium dielectric constant, *J. Phys. Chem. B*, **103**, 3073–3077.
33. Jain, P. K., Huang, W., and El-Sayed, M. A. (2007) On the universal scaling behavior of the distance decay of plasmon coupling in metal nanoparticle pairs, *Nano Lett.*, **7**, 2080–2088.
34. Storhoff, J. J., Lazarides, A. A., Mucic, R. C., Mirkin, C. A., Letsinger, R. L., and Schatz, G. C. (2000) What controls the optical properties of DNA-linked gold nanoparticle assemblies?, *J. Am. Chem. Soc.*, **122**, 4640–4650.
35. Nie, S., and Emory, S. R. (1997) Probing single molecules and single nanoparticles by surface-enhanced Raman scattering, *Science*, **275**, 1102–1106.
36. Michaels, A. M., Jiang, J., and Brus, L. (2000) Ag nanocrystal junctions as the site for surface-enhanced Raman scattering of single rhodamine 6G molecules, *J. Phys. Chem. B*, **104**, 11965–11971.
37. Brongersma, M. L., Hartman, J. W., and Atwater, H. H. (2001) Plasmonics. Electromagnetic energy transfer and switching in nanoparticle chain-arrays below the diffraction limit, *MRS Symp. Proc.*, **582**, H105/1-7.
38. Maier, S. A., Kik, P. G., Atwater, H. A., Meltzer, S., Harel, E., Koel, B. E., and Requicha, A. A. G. (2003) Local detection of electromagnetic energy transport below the diffraction limit in metal nanoparticle plasmon waveguides, *Nat. Mater.*, **2**, 229–232.
39. D'Agostino, S., and Della Sala, F. (2010) Electromagnetic modelling of the optical behaviour of silver nanospheres on dielectric substrates: The role of a silver buffer layer, *Superlattices Microst.*, **47**, 55–59.
40. Gunnarsson, L., Rindzevicius, T., Prikulis, J., Kasemo, B., Käll, M., Zou, S., and Schatz, G. C. (2005) Confined plasmons in nanofabricated single

silver particle pairs: Experimental observations of strong interparticle interactions, *J. Phys. Chem. B*, **109**, 1079–1087.

41. Su, K. H., Wei, Q.-H., Zhang, X., Mock, J. J., Smith, D. R., and Schultz, S. (2003) Interparticle coupling effects on plasmon resonances of nanogold particles, *Nano Lett.*, **3**, 1087–1090.
42. Rechberger, W., Hohenau, A., Leitner, A., Krenn, J. R., Lamprecht, B., and Aussenegg, F. R. (2003) Optical properties of two interacting gold nanoparticles, *Opt. Commun.*, **220**, 137–141.
43. Jain, P. K., Eustis, S., and El-Sayed, M. A. (2006) Plasmon coupling in nanorod assemblies: Optical absorption, discrete dipole approximation simulation, and exciton-coupling model, *J. Phys. Chem. B*, **110**, 18243–18253.
44. Jain, P. K., and El-Sayed, M. A. (2010) Plasmonic coupling in noble metal nanostructures, *Chem. Phys. Lett.*, **487**, 153–164.
45. Kottmann, J. P., and Martin, O. J. F. (2001) Retardation-induced plasmon resonances in coupled nanoparticles, *Opt. Lett.*, **26**, 1096–1098.
46. Kasha, M., Rawls, H. R., and El-Bayoumi, M. A. (1965) Exciton model in molecular spectroscopy, *Pure Appl. Chem.*, **11**, 371–392.
47. Sheikholeslami, S., Jun, Y., Jain, P. K., and Alivisatos, A. P. (2010) Coupling of optical resonances in a compositionally asymmetric plasmonic nanoparticle dimer, *Nano Lett.*, **10**, 2655–2660.
48. Prodan, E., Radloff, C., Halas, N. J., and Nordlander, P. (2003) A hybridization model for the plasmon response of complex nanostructures, *Science*, **302**, 419–422.
49. Garciá de Abajo, F. J. (2008) Nonlocal effects in the plasmons of strongly interacting nanoparticles, dimers, and waveguides, *J. Phys. Chem. C*, **112**, 17983–17987.
50. Jain, P. K., and El-Sayed, M. A. (2008) Surface plasmon coupling and its universal size scaling in metal nanostructures of complex geometry: Elongated particle pairs and nanosphere trimers, *J. Phys. Chem. C*, **112**, 4954–4960.
51. Bachelot, R., Gleyzes, P., and Boccara, A. C. (1997) Reflection-mode scanning near-field optical microscopy using an apertureless metallic tip, *Appl. Opt.*, **36**, 2160–2170.
52. Hecht, B., Bielefeldt, H., Novotny, L., Inouye, Y., and Pohl, D. W. (1996) Local excitation, scattering, and interference of surface plasmons, *Phys. Rev. Lett.*, **77**, 1889–1892.
53. Douillard, L., Charra, F., Fiorini, C., Adam, P. M., Bachelot, R., Kostcheev, S., Lerondel, G., Lamy de la Chapelle, L., and Royer, P. (2007) Optical

properties of metal nanoparticles as probed by photoemission electron microscopy, *J. Appl. Phys.*, **101**, 083518/1-5.

54. Hohenester, U., Ditlbacher, H. and Krenn, J. R. (2009) Electron-energy-loss spectra of plasmonic nanoparticles, *Phys. Rev. Lett.*, **103**, 106801/1-4.
55. Hubert, C., Rumyantseva, A., Lerondel, G., Grand, J., Kostcheev, S., Billot, L., Vial, A., Bachelot, R., Royer, P., Chang, S., Gray, S. K., Wiederrecht, G. P., and Schatz, G. C. (2005) Near-field photochemical imaging of noble metal nanostructures, *Nano Lett.*, **5**, 615–619.
56. Deeb, C., Bachelot, R., Plain, J., Baudrion, A., Jradi, S., Bouhelier, A., Soppera, O., Jain, P. K., Huang, L., Ecoffet, C., Balan, L., and Royer, P. (2010) Quantitative analysis of localized surface plasmons based on molecular probing, *ACS Nano*, **4**, 4579–4586.
57. Sonnichsen, C., Reinhard, B. M., Liphardt, J., and Alivisatos, A. P. (2005) A molecular ruler based on plasmon coupling of single gold and silver nanoparticles, *Nat. Biotechnol.*, **23**, 741–745.
58. Reinhard, B. M., Siu, M., Agarwal, H., Alivisatos, A. P., and Liphardt, J. (2005) Calibration of dynamic molecular rulers based on plasmon coupling between gold nanoparticles, *Nano Lett.*, **5**, 2246–2252.
59. Oldenburg, S. J., Averitt, R. D., Westcott, S. L., and Halas, N. J. (1998) Nanoengineering of optical resonances, *Chem. Phys. Lett.*, **28**, 243–247.
60. Jain, P. K., and El-Sayed, M. A. (2007) Universal scaling of plasmon coupling in metal nanostructures: Extension from particle pairs to nanoshells, *Nano Lett.*, 2854–2858.
61. Haes, A. J., and Van Duyne, R. P. (2002) A nanoscale optical biosensor: Sensitivity and selectivity of an approach based on the localized surface plasmon resonance spectroscopy of triangular silver nanoparticles, *J. Am. Chem. Soc.*, **124**, 10596–10604.
62. Jain, P. K., and El-Sayed, M. A. (2007) Surface plasmon resonance sensitivity of metal nanostructures: Physical basis and universal scaling in metal nanoshells, *J. Phys. Chem. C*, **111**, 17451–17454.
63. Jain, P. K., and El-Sayed, M. A. (2008) Noble metal nanoparticle pairs: effect of medium for enhanced nanosensing, *Nano Lett.*, **8**, 4347–4352.
64. Acimovic, S. S., Kreuzer, M. P., Gonzalez, M. U., and Quidant, R. (2009) Plasmon near-field coupling in metal dimers as a step toward single-molecule sensing, *ACS Nano*, **3**, 1231–1237.

65. Mirkin, C. A. (2000) Programming the assembly of two- and three-dimensional architectures with DNA and nanoscale inorganic building blocks, *Inorg. Chem.*, **39**, 2258–2272.

Chapter 7

Noble Metal Nanostructure Enhancement of Fluorescence

R. J. Phaneuf
Department of Materials Science and Engineering and University of Maryland, College Park, MD 20742, USA
phaneuf@lps.umd.edu

In this chapter we will see that there is a multiplicity of effects which are important in determining the enhancement of fluorescence available from proximity of molecules or nanocrystals to metal nanostructures. The possibility of enhancing light absorption via the intense field generated by plasmon excitation at a fluorescent species leads to a dependence on the size, shape and dielectric surroundings of a metallic nanostructure, while the possibility of scattering of a plasmon into far field radiation leads to a dependence on additional factors, including the periodicity of the array. Substrate effects are strong and lead to an additional level of tunability which promises to be useful in a number of applications.

7.1 Introduction

Strong resonant coupling between light and collective oscillations of the conduction electrons, or "plasmons" in noble metallic nanostruc-

Handbook of Molecular Plasmonics
Edited by Fabio Della Sala and Stefania D'Agostino

ISBN 978-981-4303-20-0 (Hardcover), 978-981-4303-21-7 (eBook)
www.panstanford.com

tures [1–3] leads to a number of striking and technologically important optical effects, among them surface-enhanced Raman scattering (SERS) [6–8] and the *metal-enhanced fluorescence* (MEF) from nearby molecules [4, 5]. While each of these show great promise for the development of highly sensitive biomolecule detectors [9–11], fluorescence remains the technique of choice for many biological assays. Significant fluorescence enhancement would greatly enhance the sensitivity of these assays to a host of target biomolecules. As yet the maximum enhancement available in fluorescence has not been definitively established. This is largely due both to difficulties in controlling the size and shape of the particles, and to the multiplicity of contributing factors including increased radiative decay rate and enhanced electric fields at resonance, as well as the possible role of *hot spots*, or regions of high field between closely spaced particles.

As discussed in Chapter 4, fluorescence is the phenomenon in which a material is excited by the absorption of a photon, followed by the emission of a photon at a different wavelength [12]. In the case of molecular fluorescence the difference between the peak in the absorption spectrum and that in the emission spectrum, termed the *Stokes' shift*, is due to the *Franck–Condon effect* [13–15]. Physically it occurs due to the relatively slow motion of the nuclei compared to the rate of electronic transitions, as well as the higher probability of transitions between states in which there is substantial overlap of the wavefunctions. In fluorescence the initial excited state usually has not only a higher electronic quantum number, but also a higher vibrational number. This state quickly decays into the ground vibrational level of the excited electronic state, releasing heat, and possibly decays to a vibrationally excited, but ground electronic state via the emission of a photon. A Stokes' shift also occurs in fluorescence from *semiconductor quantum dots*, with a red shift whose magnitude decreases with dot radius, and which disappears beyond a certain value [16–19]. It results from a different mechanism, believed to be associated with a splitting of the exciton states by the electron–hole exchange interaction [19].

Because fluorescence involves light-induced excitation followed by radiative decay, it is possible to enhance either, or both, depending on the overlap of the absorption spectrum, emission spectrum and plasmon resonance [20], the separation between the fluorescent species and the metal nanostructures [21–23] and whether the nanostructures are arranged periodically [21, 22, 24–26]. Fluorescence competes with other mechanisms by which an excited molecule or nanocrystal can relax to the ground state. As discussed in 1.7.3, this competition can be taken into account in defining the *fluorescence quantum yield*, equivalent to the ratio of the number of photons emitted to the number absorbed:

$$q = \frac{\gamma_r}{\gamma_r + \gamma_{nr}}, \tag{7.1}$$

where γ_r is the radiative rate of the molecule or nanocrystal (or equivalently the inverse of the natural lifetime of the excited state) and γ_{nr} is the nonradiative rate. Effects which increase the nonradiative rate result in a decrease in fluorescence, or *quenching*. Proximity of a molecule to a flat metal surface has long been known to produce quenching, but recent work has shown that proximity of a flat metal surface to a metallic nanostructure can have the opposite effect [25, 27]; examples of this will be described in detail below.

As fluorescence involves a radiative transition from an excited state, the associated radiative rate is predicted to vary as the square of the local field at the position of a molecule or nanocrystal. A useful approach in understanding how changes in geometrical or dielectric parameters effect fluorescence is thus to calculate the field generated by the superposition of the incident light and the response of the system, including nanostructures, substrate, and surrounding medium, and integrate its square across the volume of the fluorescent molecules or nanocrystals [24, 27, 28]. A good deal of work has been devoted to enhanced fluorescence in colloidal metal nanoparticle systems and in metal nanostructures which form due to deposition onto substrates in which nonwetting occurs, due to the relative simplicity of the synthesis and/or fabrication processes involved [29–33]. Although much of what we discuss below is relevant to such systems, we will confine most of our discussion to illustrative examples in which a "top down" approach allows for independent control of individual parameters such as nanostructure size, shape and spacing, and thus a clearer understanding of their individual effects.

7.2 Nanostructure Size, Shape and Spacing Dependence

Arguably the highest level of control over nanostructure enhancement of fluorescence is achieved via a *lithographic approach* for the fabrication of arrays of noble metal nanostructures in which the size, shape and separation can be independently varied. A number of recent studies have adopted such an approach to allow for a systematic determination of how the enhancement depends upon these geometrical parameters [24, 25, 34, 35]. As an example we consider the work of Corrigan *et al.* [34, 36] in which electron beam lithography was used to create Ag nanopillars on an aluminum oxide (25 nm)-coated Si(001) substrate. In this study they fabricated multiple cells of Ag nanopillars on a single substrate, varying the lateral dimensions from cell to cell, with the lateral particle size ranging from 50 nm to 130 nm and the center to center spacing from 150 nm to 390 nm, and also varying the shape of the pillars; the height of the pillars was fixed at 75 nm. Also fixed were the thickness and composition of the fluor-pillar spacer layer, required to avoid fluorescence quenching [37–39]. This was created by initial deposition of BSA-biotin protein layer, approximately 3–4 nm thick onto the nanopillar arrays, followed by steptavidin which had been tagged with Cy3 and Cy5 fluorophores. The overall fluor-nanopillar separation was approximately 8 nm. Fluorescence images were collected using a laser scanning microscope employing either an Ar ion laser ($\lambda = 514$ nm) or a HeNe laser ($\lambda = 633$ nm) for excitation, along with suitable filters for measurement of fluorescence from Cy3 or Cy5, respectively.

Figures 7.1(a) and 7.1(b) show fluorescence microscopy images excitation at 514 nm and at 633 nm, respectively. From these images it is possible to identify the combination of pillar width and spacing producing greatest enhancement by visual inspection. Plots of the enhancement ratio R *versus* the width of the nanopillars are shown beneath for comparison (see Fig. 7.1(c), Fig. 7.1(d)), where the enhancement ratio is defined as:

$$R \equiv \frac{I_{nf} - I_B}{I_f - I_B}, \tag{7.2}$$

and where I_{nf} is the measured fluorescent intensity from within a fluor-coated Ag nanopillar array, I_f is the intensity measured

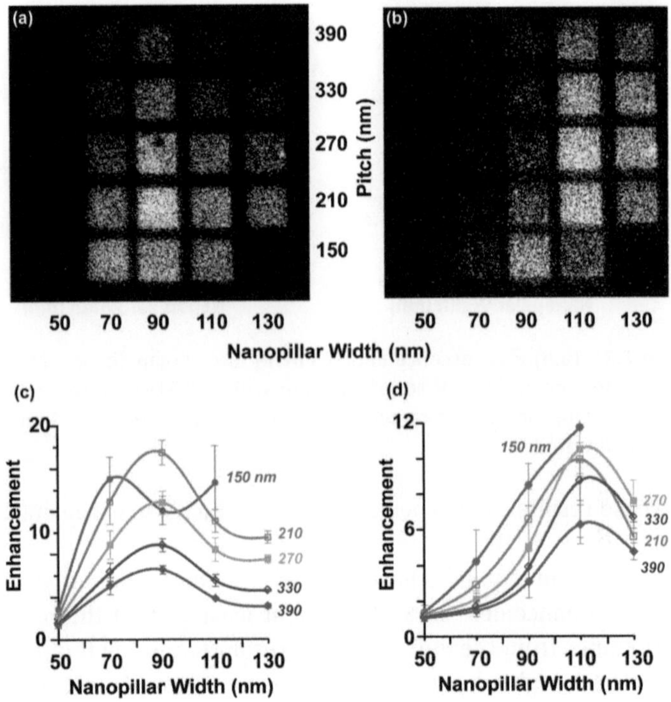

Figure 7.1 (a,b) Fluorescence images from combinatorial pattern with excitation wavelength of 514 nm and 633 nm, respectively, (c,d) plots of fluorescence enhancement *versus* width of square section nanopillars for varying spacing for the same patterns as in (a) and (b), respectively. Center-to-center spacings in nm indicated; Nanopillar heights 75 nm [36].

from a fluor-coated region containing no Ag nanopillars, and I_B is the background intensity measured from a region which contains neither fluor, nor Ag nanoparticles. For excitation at 514 nm, the maximum in R occurs for a lateral size of 90 nm, and a height/width aspect ratio c/a = 0.83 while for excitation at 633 nm the maximum is shifted to approximately 110 nm, or c/a = 0.68. The wavelength dependence of the observed optimum size is qualitatively consistent with a red shift in the lateral plasmon–polariton oscillation modes for oblate spheroid particles of decreasing height/width ratio [2], although the shapes in Corrigan's work were more complex, resulting in additional modes.

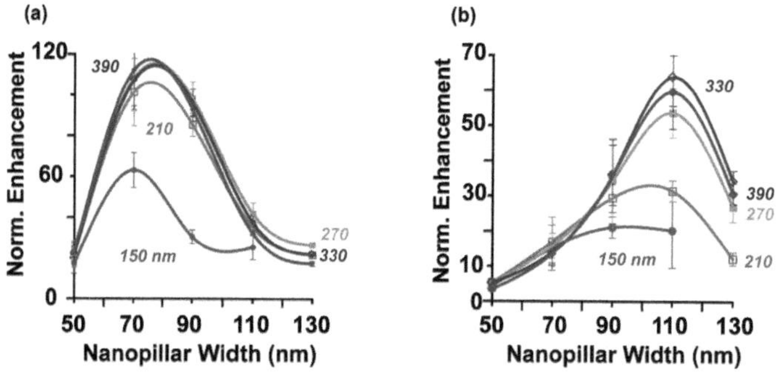

Figure 7.2 (a,b) Fluorescence enhancement after normalizing to the area of silver for nanopillars of square-section with (a) 514 nm excitation, (b) 633 nm excitation. Center-to-center spacings in nm indicated; Nanopillar heights 75 nm [36].

These results show a nearly monotonic increase in the enhancement ratio R as the spacing decreases. Although regions of small separation and high field have been identified as a potential source of huge enhancement in SERS, here at least part of the observed effect comes from the limited spatial resolution in the fluorescence images. A weighted average of intensities from Ag-covered and substrate regions is measured within each cell, and larger Ag area fractions in cells with smaller spacings gives a higher effective enhancement. A simple way to remove this contribution from the observed ratio is to replace R by the area-corrected ratio N, defined as:

$$N \equiv \frac{I_c - I_B}{I_f - I_B}, \tag{7.3}$$

with $I_c \equiv \frac{I_{nf}-(1-A_n)I_f}{A_n}$, where I_c is the average fluorescent intensity corrected for the area fraction, An, of the surface covered by the silver nanostructures [36], I_{nf} is the local raw fluorescence intensity measured from the area covered with fluorophore/spacer coated silver nanostructures, I_f is the fluorescent intensity measured from an area covered with a fluorophore/spacer layer, but no silver nanostructures, I_B is the "background" intensity, measured from an area where neither fluorophore nor silver nanostructures are present. As seen in Fig. 7.2, this area-normalized ratio is independent of the spacing beyond approximately 270 nm for excitation at 514 nm, I_f and beyond approximately 330 nm for excitation at 633 nm. Corrigan *et al.* found that within the range explored, the effect of decreasing the spacing was to decrease the normalized enhancement ratio, which is the opposite of what might be expected if "hot spots" played a significant role. However, it is consistent with a shift in the plasmon–polariton frequency away from that of the fluorophore at relatively small particle spacings, i.e. less than 1.5 times the lateral size of the nanoparticles. Indeed such an effect had been reported in optical absorption measurements from Ag dot arrays at comparable spacings by Gotschy *et al.* [40] Corrigan's results showed that for center-to-center spacings of more than ~2–3 times the lateral size of the nanoparticles, the optimum size for fluorescence no longer depends on spacing, indicating weak interpillar coupling beyond this. It should be noted that the peaks shown in Fig. 7.2 are local maxima in the fluorescence enhancement: Corrigan *et al.* found a monotonic increase in the peak in the enhancement ratio with height over the range they studied, from 25 nm to 75 nm. Although they did not observe evidence for hot spot-enhancements between particles for edge-to-edge spacings down to approximately 20 nm, Corrigan *et al.* [36] did report larger area-corrected enhancements for particle shapes producing higher local electric fields. They also found that in comparing results for pillars of equilateral triangular cross-section to those of square-cross section, the lateral size of the particle that gave rise to the optimal area-normalized enhancement for excitation at 514 nm increased from ~75 nm to ~90 nm, consistent with reports by Jensen *et al.* of a red shift in the optical extinction wavelength of triangular-section pillars compared to elliptical particles [41]. Pompa *et al.* reported similar results for fluorescence of CdSe/ZnS core-shell nanocrystals in proximity to nanopillars of triangular *versus* cylindrical shapes [35]. These results are also consistent with theoretical modeling by Kottman *et al.* [42] who found that the plasmon–polariton resonance for triangular-section nanopillars was red-shifted from other shapes, including square cross-section pillars. Their calculations also predicted that the peak values of the enhancement of the electromagnetic fields associated with triangular cross-section pillars is larger than for square and other cross-sectional shapes studied [42] and that the field was strongest at specific positions around and close to the nanopillar.

7.3 Role of Substrate

The substrate has long been known to play a role in the excitation of localized particle plasmons, shifting the resonant frequencies due to a change in the permittivity of the medium on which the metal nanostructures are located [28]. In addition, recent work has shown that certain substrates play an active role in light-plasmon coupling [24, 27, 43].

As an illustrative example we consider the results of Guo *et al.* [24], who examined the effect of the substrate on the size dependence of the enhancement of fluorescence from Cy3 and Cy5-tagged streptavidin/biotin-coated monodisperse silver nanospheres. In this case a comparison between scanning electron microscopy images and fluorescence microscopy images from size-selected silver nanoparticles [44, 46, 47] on a silicon substrate showed individual nanospheres were well separated from their neighbors, so that the measured fluorescent enhancement was apparently predominantly from effects associated with individual silver particles. No intensely bright spots were evident in scanned fluorescence images, contrary to what would be expected from so-called *hot spots* between particles in near contact [48] or in agglomeration. Summaries of the fluorescence enhancement *versus* average nanosphere diameter for excitation at 514 nm and 633 nm from that work are shown in Figs. 7.3(a) and 7.3(b), respectively. The quantity plotted in this case is the coverage corrected enhancement, defined above in Eq. (7.3). For excitation at 514 nm the enhancement initially increases with particle diameter, and reaches a maximum of 240-fold at d $\sim$85 nm; there is a window of particle diameters between $\sim$85 nm and $\sim$150 nm for which the enhancement exceeds 100 fold. For excitation at a wavelength of 633 nm, a maximum enhancement of 350 was measured at a slightly larger diameter, $\sim$150 nm, and the enhancement exceeded 100 for diameters from $\sim$95 nm to $\sim$160 nm.

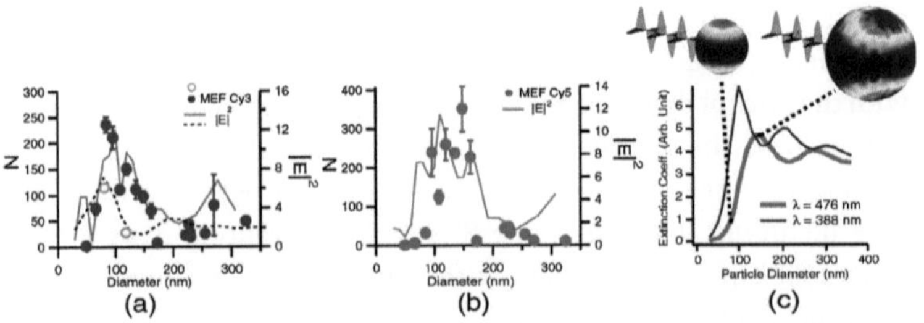

Figure 7.3 Measured normalized enhancement of the fluorescence, N on silicon substrate (filled circles) or glass substrate (open circles), and the calculated average E-field intensity (solid curve) near the Ag particles as a function of Ag particle diameter: (a) 514 nm excitation (Cy3 fluorophore), (b) 633 nm excitation (Cy5 fluorophore). The dashed curve in (a) shows the same averaged E-field intensity calculated for the case of glass substrate. (c) Calculated extinction coefficient and electrical field intensity, $|E|^2$, (with $|E|$ normalized to incident amplitude) for silver nanoparticles. Extinction coefficients are plotted as a function of particle diameter for incident light wave length fixed at 476 nm (633 nm in air) in red curve and 388 nm (514 nm in air) in blue curve. Insets show the $|E|^2$ distribution over a spherical surface 8 nm from the surface of silver particles with diameter of 80 nm and 150 nm, respectively. The incident light represented by the sinusoidal modulations is traveling toward the particle, and the E-field is polarized in the vertical direction. The color scale bar indicates the field intensity normalized to that of the incident light [43].

A signature of the enhancement of the absorption of incident light by fluorescent species by proximity to metallic nanostructures is the proportionality of the fluorescence to the local field intensity [24, 27, 28, 49]. Guo *et al.* [24] modeled the response of silver nanospheres to incident light (approximated as a linearly polarized plane-wave), using the discrete dipole approximation [50]. To understand the effect of the substrate, for contrast they first calculated the response of a freestanding silver nanosphere in water [51] to account for the fact that the samples were maintained in an aqueous buffer solution. Figure 7.3(c) shows the calculated extinction coefficients, normalized to the geometric cross-section of the particle as a function of particle diameter. The thin blue curve is for an excitation wavelength of 388 nm in water, corresponding to 514 nm in air. The extinction coefficient first increases with particle diameter, reaching a maximum, at ~100 nm for the former case and ~140 nm for the latter case, but then oscillates around a slow decaying tail. The latter behavior is qualitatively inconsistent with the abrupt drop in enhancement beyond the high-enhancement windows in the fluorescence. They next calculated the square of the electric field at the location of the fluorescent tags, in this case including the underlying Si substrate, treating it as a finite-square slab [52] placed behind the particle in the direction of the wavevector of the incident light.

Figure 7.4 shows how the calculated E-field intensity distribution in the azimuth of incident E-field polarization changes with the size of the silver nanoparticles; here a 150-nm thick Si substrate and an incident light wavelength of 388 nm in water (514 nm in air) are used. Figures 7.4(a) and 7.4(e) are three-dimensional (3-D) and cross-sectional renderings, respectively, of the calculated intensity for a relatively small diameter of 60 nm. As can be seen most clearly in the cross-sectional view, the Si substrate, because of its large dielectric function, has a profound effect on the field; for this size the strongest field enhancement occurs beneath the particle. Increasing the particle size to 100 nm, close to the observed optimum in fluorescence enhancement causes regions of strong E-field enhancement to sweep upward, to the regions at and above the equator of the particle. A further slight increase in particle diameter, to 120 nm, results both in a slight downward sweep of the high field regions and a decrease in the measured enhancement. Increasing the particle size further both sweeps the regions of high E-field back into the region beneath the particle, and causes a dramatic drop in the measured fluorescence.

This redistribution of regions of high field, and its correlation to the observed fluorescence enhancement demonstrate the profound effect of coupling to a Si substrate. Guo *et al.* [24] found good quantitative agreement between the measured enhancement and the calculated field-squared, averaged over the region of the fluorescent molecular-tags for which no shadowing occurs, shown as solid curves in Figs. 7.3(a) and 7.3(b). This result showed that the distribution of the enhanced E-field intensity is crucial to the measured fluorescence enhancement, and leads to an important insight: as the particle size varies, regions of electrical field enhancement shift position; maximum fluorescence enhancement

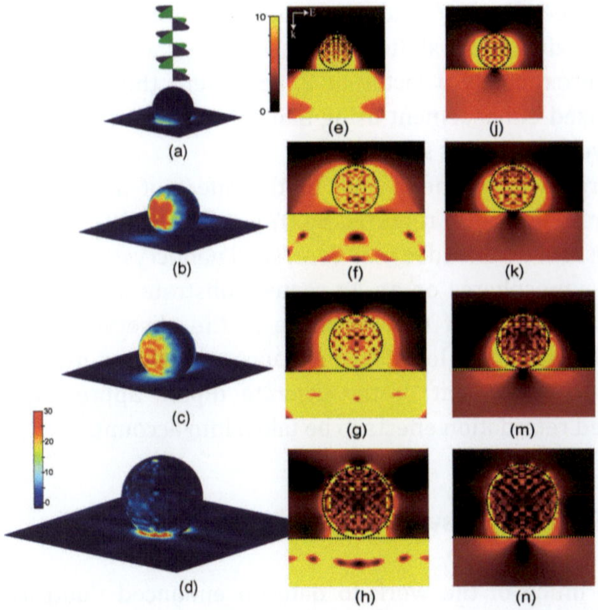

Figure 7.4 Calculated electric field intensity ($|E|^2$) distribution at the distance 80nm from the surface of the silver nanoparticle and the Si substrate. The particle diameters are 60 nm, 100 nm, 120 nm, and 210 nm for panels (a), (b), (c), and (d), respectively. As shown in panel (a), the incident light propagates downward and the polarization of the E-field is parallel to the substrate (colored in green). The wave length of the incident light is 388 nm wavelength (514 nm in air). The color table indicates the field intensity normalized to the incident light intensity. Panels (e), (f), (g), (h) are cross-sectional renderings of electric field intensity ($|E|^2$) for particle diameters are 60 nm, 100 nm, 120 nm, and 210 nm, respectively; panels (j), (k), (m), (n) are cross-sectional renderings of electric field intensity ($|E|^2$) for the same particle diameters, but on a SiO_2 substrate [43].

occurs when a large fraction of those fluorophores which are not shadowed by either the particles or substrate are immersed in regions of high field; low or no enhancement otherwise.

This effect is even more striking when compared to the field intensity which results from a more conventional, passive substrate, SiO_2. Simulations by Guo *et al.* showed that in that case the substrate perturbs the field much less strongly, resulting in a conventional dipole-like distribution for smaller particles, and mixed dipolar/quadrupolar distribution at larger diameters. The maximum is shifted to a smaller diameter, and reduced by approximately a factor of 2, consistent the observed relative optimized enhancement using a Si substrate compared to that for a conventional glass substrate.

Large enhancements due to the effect of a high-permittivity substrate were also reported by D'Agostino *et al.* [27]; they studied fluorescence from CdSe/ZnS core/shell nanocrystals in proximity to silver nanospheres on an Ag-coated substrate. D'Agostino demonstrated the role of this substrate in the observed enhancement by calculating the local field-enhancement factor using a Green's function refinement of the discrete dipole approximation; this allowed retardation effects to be taken into account.

7.4 Standing Wave Surface Plasmons

While much of the work to date on enhanced fluorescence has involved resonant coupling to particle plasmons, the use of lithography to produce periodic arrays of noble metal nanostructures produces what might be termed *grating modes*, which can also couple to light. Guo *et al.* found that these modes can also produce enhanced fluorescence in proximal molecules [24]. In this work they again used electron beam lithography to fabricate both square arrays of square-cross-section Ag nanopillars, and parallel arrays of long Ag nanowires on aluminum oxide-coated Si (001) substrates. For simplicity they maintained the period at twice the structure width in each case and varied the column/wire widths over an order of magnitude, from 64 to 633 nm.

As in studies described above [24, 34, 36] they maintained the height of the nanostructures at 75 nm, used an 8 nm thick biotin/streptavidin spacer layer tagged with both Cy3 and Cy5 fluors, and excited fluorescence using both 514 nm and 633 nm wavelength normally incident light. They also probed the effect of the polarization of the incident light by fabricating nanowires with two perpendicular orientations on the same sample; an example is shown in the SEM image of Fig. 7.5(a).

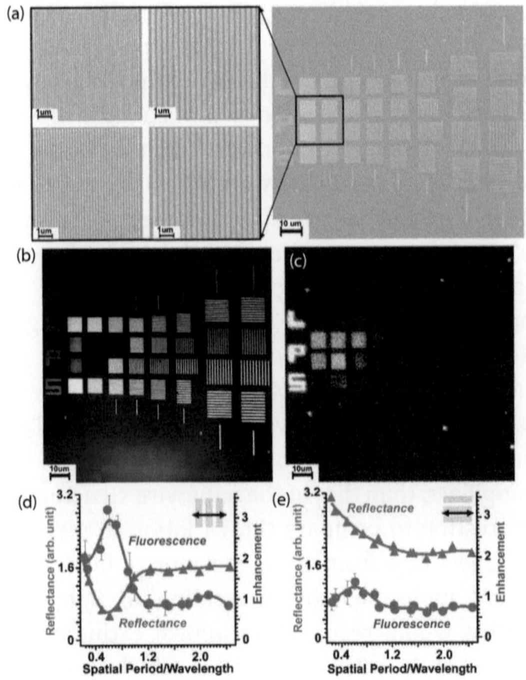

Figure 7.5 (a) Scanning electron microscope image with enlargements of select areas shown to the left (a1–a4). (b) Reflectance image with incident light of 514 nm. (c) Corresponding fluorescence map. (d, e) Summary of fluorescence enhancement (line with circles) and reflectance intensity (line with triangles) along two different orientations of excitation. The period is defined as the line width plus edge-to-edge spacing, and in this case line width equals edge-to-edge spacing. Chosen line widths vary by 1/8 the wavelength of the excitation light, for both wavelengths used (514 nm for the top half, and 633 nm for the bottom half). Line widths in the top two rows are 64 nm, 128 nm, 192 nm, etc. up to 514 nm, and in the bottom rows are 79 nm, 158 nm, 237 nm, etc. up to 633 nm for the largest square. Each half consists of longitudinal and transverse orientations plus a single vertical line. For transverse excitation polarization, there is an inverse correlation between reflectance measurements and fluorescence enhancement; when reflectance is lowest, fluorescence intensity is highest. Error bars represent the standard deviation for several measurements of different samples [24].

For a relatively thin (25 nm) Al_2O_3 Si substrate–Ag nanowire spacer layer they found that the optimum polarization for fluorescence enhancement was perpendicular to the long axis of the wires. They also found that there were two maxima in the fluorescence intensity *versus* nanowire period for this polarization: the largest enhancement is at approximately a 0.65 fraction of the wavelength of the incident light in air, or approximately a 0.8 fraction of the wavelength in solution, and the second, less intense maximum at approximately three times that period.

The optimum period for arrays of square Ag nanopillars was smaller than for arrays of nanowires of the same height, approximately a 0.3 fraction of the incident light wavelength in air or a 0.4 fraction of the wavelength in solution. The maximum fluorescence enhancement ratio was approximately 4 times greater for the nanopillars, than that of the nanowire structures.

It is interesting to compare the variation in measured fluorescence and reflectance from these structures. Even in the absence of fluorescent molecules the interaction of light with metallic nanostructures would result in an optical extinction spectrum in which resonances occur at frequencies which depend on the size and period of the particles [45, 53], and indeed it is common to use extinction spectra to quantify particle–light interactions. Since Guo *et al.* [24] fabricated nanowire and nanopillar arrays on non-transparent silicon wafers they instead compared fluorescence and reflectance images. They found that conditions producing minimum reflectance from the nanowire arrays (see Fig. 7.5(b)) matched those for maximum fluorescence enhancement: both occurred at a lateral spatial period of approximately half of the incident light wavelength, and for transverse polarization (long wire axes perpendicular to incident light E-field). By contrast, all longitudinal patterns (long wire axes parallel to incident light E-field) produce reflectance greater than one; no significant fluorescence enhancements occur for this polarization. For the square nanopillar arrays, patterns with periods exhibiting a local minimum in reflectance show a local maximum in enhancement. However, in this case the optimum in fluorescence enhancement did not coincide with the overall lowest reflectance.

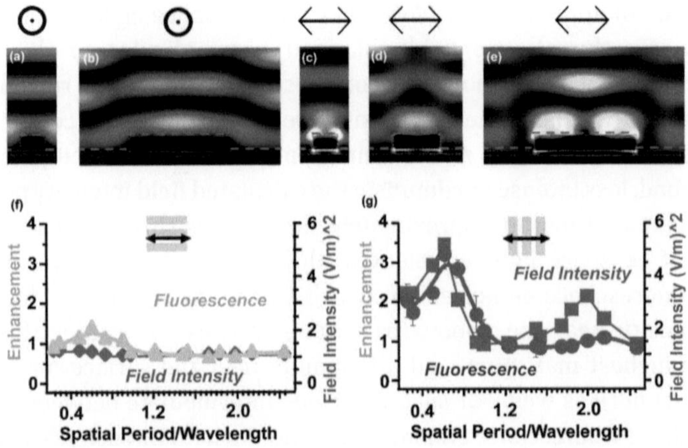

Figure 7.6 Field intensity calculation by 2-D FDTD method and examples are as shown in (a–e). The long axis of the nanobars is normal to the image plane; this is an end-on view of the long structures positioned on a Si substrate, with a layer of aluminum oxide between the substrate and the silver bars, and water filling the top portion of the frame. Bright areas show the positions of high local electric field. (f, g) Summary plots show the average electric field intensity 8 nm above the surface (at the location of the dashed lines in (a–e)), compared with fluorescence enhancement along two different polarizations. Excitation polarization compared to bar orientation is indicated by the illustrations within the graphs. Intensity plots (a–e) represent the visualization of calculations for points noted by the labels within graphs (f–g), at periods of 316, 1028, 316, 512, and 1028 nm, respectively. Error bars represent the standard deviation for several measurements of different samples [24].

Insight into the mechanism for the fluorescence enhancement from these arrays comes from comparison of the measured dependence on spatial period with that of the square of the local field. Guo *et al.* [24] carried out simulations of the latter using the finite-difference time domain (FDTD) method, illustrated in Fig. 7.6.

A comparison of the measured fluorescence and calculated field intensity at the position of the fluorescent tags (8 nm above the surface) showed qualitative consistency, that both are larger for the polarization perpendicular to the nanowires. The FDTD calculations showed that the largest calculated field intensity coincided with the largest fluorescence enhancement: this occurred for a spatial period of approximately 0.65 times the free space light wavelength, and for the electric field polarized perpendicular to the long direction of the nanowires. For these conditions, the regions of highest calculated field intensity are near the upper corners of the nanowires, with a local minimum at the surface midplane. A second, less intense maximum in the calculated field intensity occurs at a spatial period approximately three times this, i.e. near twice the free space light wavelength; there is also a weak maximum in fluorescence enhancement near this spatial period. The highest intensity near the nanowires surface is closer to center than at the highest maximum, and a minimum near the surface midplane exists here as well. For spatial periods intermediate between these both the fluorescence enhancement and the field intensity reaches a minimum; the calculated field intensities near this minimum shows highest intensity near the midplane of the nanowire.

These observations indicate the excitation of plasmon modes consisting of standing waves at the surface of the nanowires. Such a standing wave consists of two *surface plasmons* traveling in opposite directions, each of which has a dispersion ω *versus* k_x given by (see 1.3):

$$k_x = \pm\frac{\omega}{c}\sqrt{\frac{\tilde{\varepsilon}_{Ag}(\omega)\tilde{\varepsilon}_d}{\tilde{\varepsilon}_{Ag}(\omega)+\tilde{\varepsilon}_d}}, \tag{7.4}$$

where ω is the frequency, c is the speed of light, $\tilde{\varepsilon}_{Ag}(\omega)$ is the frequency-dependent complex dielectric function of the Ag nanowire, and $\tilde{\varepsilon}_d$ is the dielectric function of the surrounding dielectric medium. In these experiments the thickness of the nanowire is 75 nm, which exceeds the skin depth of Ag ($\delta \approx 23$ nm at optical frequencies) by approximately a factor of 3; thus to a good approximation ϵ_w is that of the surrounding solution, which is mainly water. Using values of the dielectric functions of Ag and water at a light excitation frequency of 583 THz yields a magnitude of k equal to approximately 1.44 times the value for light propagating in air, or equivalently, a spatial period of 0.69 times the light wavelength in air. This is in very good agreement with the spatial period at which the largest fluorescence enhancement is observed.

The condition for a standing surface plasmon across the width of a nanowire is:

$$k_x W \approx m\pi \, , \tag{7.5}$$

where W is the nanowire width and m is an integer. The approximately equal sign in this equation comes from the fact that the edge is not a hard boundary. The actual boundary condition may differ for various reasons: the plasmon might "wrap" around the edge of the wire, in which case $W_{eff} > W$. Alternatively the effective width may be smaller because of surface defects near the edges. The lowest order standing wave surface plasmon mode, corresponding to $m = 1$, produces a left–right polarization of charge at the wire edges; this would couple readily to the oscillating E-field of an incident light wave polarized along this direction. In Guo *et al.*'s [24] investigations the center-to-center separation of adjacent nanowires was fixed at twice the individual wire width, so that this first mode also satisfied the condition for the Wood's anomaly; this latter effect corresponds to diffraction of surface waves along the grating [54–57], i.e. $n\lambda = a \sin(\theta)$, where a is the *grating period*, and in this case $\theta = \pi/2$. For sufficiently deep gratings a *Wood's anomaly* is also expected for the opposite polarization [56]; this might explain the (considerably weaker) peak seen at this same period for longitudinal polarization. The second standing wave mode corresponding to $m = 2$ would give a charge distribution which is symmetric about the midplane, and thus would not couple to the incident light wave; this in agreement with the absence of a peak in the fluorescence enhancement at this condition. The third mode, i.e. for $m = 3$, also produces a net polarization, but greatly reduced compared to that for $m = 1$; this is also in agreement with the relatively small peak observed in the fluorescence enhancement.

A departure of the observed fluorescence enhancement from a simple proportionality to the local field intensity at the fluorescent species as the parameters of a grating of nanostructures are varied might be an indication of the contribution of the scattering of plasmons into far field radiation by the grating [1, 21, 22, 26, 58]. In the work by Guo [24], the particle plasmon modes interact strongly with the grating as well as with the fluorescent molecule itself. In related work by Steele *et al.* [26] on enhanced fluorescence from Atto 680 molecules on Au nanowire gratings the scattering of plasmons into far field radiation was dominant; the resulting angular dependence of the measured output indicated the conservation of wavevector occurred via the contribution of grating reciprocal lattice vectors.

7.5 Spacer Layer Effect

Earlier in this chapter we pointed out that the substrate plays an important role in metallic nanostructure-enhancement of fluorescence. In this section we consider the role of coupling between arrays of fluorescent molecule-coated Ag nanostructures and an underlying metal film.

To probe the effect of this coupling Guo *et al.* [25] again used electron beam lithography to create arrays of Ag nanowires, in this case using a 75 nm Ag film on Si as a substrate, and spacing arrays of 75 nm high parallel Ag nanowires above this substrate using an aluminum oxide layer, whose thickness they varied systematically. As in their previous studies they employed a biotin/streptavidin spacer layer to position Cy3 and Cy5 tags 8 nm above the nanowire-patterned surfaces and excited fluorescence using a scanning laser microscope with excitation at 514 nm and 633 nm. Consistent with the results discussed in the previous section, for a Si substrate Guo *et al.* [25] found that deposition of a very thin aluminum oxide spacer layer (between 0 and 25 nm thick) between the nanowire arrays and a silver film produced higher fluorescence enhancement for light polarized perpendicular to the nanowires. Increasing the oxide spacer thickness to 80 nm however resulted in the opposite behavior, i.e. a larger enhancement occurred when the incident light was polarized parallel to the nanowires. They were thus able to switch the orientation of the optimum polarization for fluorescence. Their results in fact showed that with increasing spacer layer thickness the optimum polarization direction switches twice between parallel and perpendicular to the silver nanowires in the thickness range explored. The optimum polarization switches again to perpendicular for spacer layers thicker than 140 nm.

To interpret the origin of this switching they again carried out numerical calculations of the local electric field intensity based on the FDTD method for range of geometries investigated in our measurements. Not surprisingly the field intensity distributions were qualitatively different for incident light polarized parallel and perpendicular to the nanowire axes. For the perpendicular polarization there is more intense field above the nanowires than between them, while the opposite is true for the parallel polarization. For polarization of the incident light parallel to the wires, the position of the fluorophores, i.e. 8 nm above the surface placed them well beneath the regions of maximum field intensity, consistent with the observation of relatively little fluorescence enhancement for this case. For perpendicular incident polarization on the other hand, the position of the fluors above the wire places them very close to the regions of maximum field intensity near the corners. Significantly, the simulations for a spacer layer 25 nm thick showed evidence for strong coupling between the nanowires and underlying silver layer for the perpendicular polarization in the form of regions of intense local field, while for spacer layer thickness of 80 nm the coupling was much weaker, and the intensity at the position of the fluors was relatively weak. For the parallel polarization on the other hand the simulations indicated that the increased spacer thickness lifts the fluorophores into the regions of intense field in the space between the nanowires, explaining the switching of the incident polarization producing higher enhancement for these two different oxide thicknesses.

The measurements done by Guo *et al.* [25] showed decay in the enhancement beyond approximately 80 nm. Their FDTD calculations instead showed a continued oscillation in the average field-squared at the position of the fluors out to at least 600 nm; these oscillations correlate with Fabry–Perot resonances within the oxide, an observation which has been made previously by Andrewartha [59] for a similar type of structure (a "bottle grating"); interestingly calculations of the reflectivity from this related structure showed similar sharp oscillations [59]. The calculated oscillations originate from the interference between diffraction from the nanowire array and multiple reflections from the aluminum oxide interfaces. The large amplitude and shape of the oscillations is seemingly connected with the geometry of the nanowire array: we find that replacing the array by a continuous silver film of thickness less than the skin depth result in much smaller amplitude, broader resonances.

Earlier observations by Cesario *et al.* [60] of a decay in fluorescence for arrays of Au nanoparticles spaced above a Ag film by a SiO_2 layer of increasing thickness, were interpreted as due to the finite vertical extent of the evanescent fields associated with a surface plasmon. In this model the coupling results in an enhanced interaction between individual localized plasmons at individual nanostructures [61] and thus an enhancement in the radiative efficiency; increasing the spacer layer thickness moves the nanowires out of the evanescent field of the surface plasmon. A possible physical mechanism for the experimentally observed decay involves nonradiative decay of the excited states. The aluminum oxide deposited in these experiments was likely to be nonstoichiometric, and defects in the oxide could act as recombination centers. Thicker oxides would result in higher areal densities of defects, and decay in fluorescence. A definitive assignment of the mechanism for the observed fall off of fluorescence would require determination of the complex dielectric function of our oxides (as deposited onto an Ag film), and inclusion in the field-square calculations. Finally it should be noted that at very small thicknesses quenching of the fluorescence is expected [38, 62] consistent with observations of an optimum nanowire-substrate spacer thickness.

Acknowledgments

The author is pleased to acknowledge contributions to the examples presented in this chapter from D.G. Britti, R. Chiuri, T.D. Corrigan, S. D'Agostino, F. Della Sala, S.-H. Guo, J.J. Heetderks, H.-C. Kan, P.P. Pompa, H. Szmacinski, S.-J. Tsai, R. Rinaldi, and R. Cingolani, and support from The Laboratory for Physical Sciences, the CNR-INFM, a NSF International Grant (OISE0242579) and a NSF-MRSEC (DMR-0080008).

References

1. Raether, H. (ed.) (1988) *Surface Plasmons on Smooth and Rough Surfaces and on Gratings*, Springer Tracts on in Modern Physics, vol. 111, Springer, Berlin.
2. Kreibig, U., and Vollmer, M. (eds) (1995) *Optical Properties of Metal Clusters*, Springer, Berlin.
3. Maier, S. A. (ed.) (2007) *Plasmonics: Fundamentals and Applications*, Springer, Berlin.
4. Lakowicz, J. R. (2001) Radiative decay engineering: Biophysical and biomedical applications, *Anal. Biochem.*, **298**, 1–24.
5. Fort, E., and Grosillon, S. (2008) Surface enhanced fluorescence, *J. Phys. D: Appl. Phys.*, **41**, 013001/1-31.
6. Jeanmaire, D. L., and Van Duyne, R. P. (1977) Surface Raman spectro-electrochemistry: Part I. Heterocyclic, aromatic, and aliphatic amines adsorbed on the anodized silver electrode, *J. Electroanal. Chem.*, **84**, 1–20.
7. Moskovits, M. (1985) Surface-enhanced spectroscopy, *Rev. Mod. Phys.*, **57**, 783–826.
8. Kneipp, K., Moskovits, M., and Kneipp, H. (eds) (2006) *Surface Enhanced Raman Scattering: Physics and Applications*, Springer, Berlin.
9. Zhao, J., Zhang, X., Yonzon, C. R., Haes, A. J., and Van Duyne, R. P. (2006) Localized surface plasmon resonance biosensors, *Nanomedicine-UK*, **1**, 219–228.
10. Hirsch, L. R., Jackson, J. B., Lee, A., Halas, N. J., and West, J. L. (2003) A whole blood immunoassay using gold nanoshells, *Anal. Chem.*, **75**, 2377–2381.
11. Bishnoi, S. W., Rozell, C. J., Levin, C. S., Gheith, M. K., Johnson, B. R., Johnson, Don H., and Halas, N. J. (2006) All-optical nanoscale pH meter, *Nano Lett.*, **6**, 1687–1692.
12. Lakowicz, J. R. (ed.) (2006) *Principles of Fluorescence Spectroscopy*, 3rd edn, Springer, New York.
13. Franck, J., and Dymond, E. G. (1926) Elementary processes of photo-chemical reactions, *Trans. Faraday Soc.*, **21**, 536–542.
14. Condon, E. U. (1926) A theory of intensity distribution in band systems, *Phys. Rev.*, **28**, 1182–1201.
15. Condon, E. U. (1928) Nuclear motions associated with electron transitions in diatomic molecules, *Phys. Rev.*, **32**, 858–872.

16. Nirmal, M., Norris, D. J., Kuno, M., Bawendi, M. G., Efros, A. L., and Rosen, M. (1995) Observation of the dark exciton in CdSe quantum dots, *Phys. Rev. Lett.*, **75**, 3728–3731.

17. Efros, A. L., Rosen, M., Kuno, M., Nirmal, M., Norris, D. J., and Bawendi, M. (1996) Band-edge exciton in quantum dots of semiconductors with a degenerate valence band: Dark and bright exciton states, *Phys. Rev. B*, **54**, 4843–4856.

18. Banin, U., Lee, J. C., Guzelian, A. A., Kadavanich, A. V., and Alivisatos, A. P. (1997) Exchange interaction in InAs nanocrystal quantum dots, *Superlattice Microstr.*, **22**, 559–568.

19. Bagga, A., Chattopadhyay, P. K., and Ghosh, S. (2006) Origin of Stokes shift in InAs and CdSe quantum dots: Exchange splitting of excitonic states, *Phys. Rev. B*, **74**, 035341/1-7.

20. Chen, Y., Munechika, K., and Ginger, D. S. (2007) Dependence of fluorescence intensity on the spectral overlap between fluorophores and plasmon resonant single silver nanoparticles, *Nano Lett.*, **7**, 690–696.

21. Kitson, S. C., Barnes, W. L., and Sambles, J. R. (1996) Photoluminescence from dye molecules on silver gratings, *Opt. Commun.*, **122**, 147–154.

22. Kitson, S. C., Barnes, W. L., Sambles, J. R., and Cotter, N. P. K. (1996) Excitation of molecular fluorescence via surface plasmon polaritons, *J. Mod. Optics*, **43**, 573–582.

23. Ray, K., Szmacinski, H., Enderlein, J., and Lakowicz, J. R. (2007) Distance dependence of surface plasmon-coupled emission observed using Langmuir-Blodgett films, *Appl. Phys. Lett.*, **90**, 251116/1-3.

24. Guo, S.-H., Heetderks, J. J., Kan, H-C., and Phaneuf, R. J. (2008) Enhanced fluorescence and near-field intensity for Ag nanowire/nanocolumn arrays: Evidence for the role of surface plasmon standing waves, *Opt. Express*, **16**, 18417–18425.

25. Guo, S.-H., Britti, D. G., Heetderks, J. J., Kan, H-C., and Phaneuf, R. J. (2009) Spacer layer effect in fluorescence enhancement from silver nanowires over a silver film: Switching of optimum polarization, *Nano Lett.*, **9**, 2666–2670.

26. Steele, J. M., Gagnidze, I., and Wiele, S. M. (2010) Efficient extraction of fluorescence emission utilizing multiple surface plasmon modes from gold wire gratings, *Plasmonics*, **5**, 319–324.

27. D'Agostino, S., Pompa, P. P., Chiuri, R., Phaneuf, R. J., Britti, D. G., Rinaldi, R., Cingolani, R., and Della Sala, F. (2009) Enhanced fluorescence by metal nanospheres on metal substrates, *Opt. Lett.*, **34**, 2381–2383.

28. Kelly, K. L., Coronado, E., Zhao, L. L., and Schatz, G. C. (2003) The optical properties of metal nanoparticles: The influence of size, shape, and dielectric environment, *J. Phys. Chem. B*, **107**, 668–677.

29. Weitz, D. A., Garoff, S., Gersten, J. I., and Nitzan, A. (1983) The enhancement of Raman scattering, resonance Raman scattering, and fluorescence from molecules adsorbed on a rough silver surface, *J. Chem. Phys.*, **78**, 5324–5338.

30. Kummerlen, J., Leitner, A., Brunner, H., Aussenegg, F. R., and Wokaun, A. (1993) Enhanced dye fluorescence over silver island films: Analysis of the distance dependence, *Mol. Phys.*, **80**, 1031–1046.

31. Chumanov, G., Sokolov, K., Gregory, B. W., and Cotton, T. M. (1995) Colloidal metal films as a substrate for surface-enhanced spectroscopy, *J. Phys. Chem.*, **99**, 9466–9471.

32. Sokolov, K., Chumanov, G., and Cotton, T. M. (1998) Enhancement of molecular fluorescence near the surface of colloidal metal films, *Anal. Chem.*, **70**, 3898–3905.

33. Geddes, C. D., Cao, H., Gryczynski, I., Gryczynski, Z., Fang, J., and Lakowicz, J. R. (2003) Metal-enhanced fluorescence (MEF) due to silver colloids on a planar surface: Potential applications of indocyanine green to in vivo imaging, *J. Phys. Chem. A*, **107**, 3443–3449.

34. Corrigan, T., Guo, S., Phaneuf, R., and Szmacinski, H. (2005) Enhanced fluorescence from periodic arrays of silver nanoparticles, *J. Fluoresc.*, **15**, 777–784.

35. Pompa, P. P., Matiradonna, L., Della Torre, A., Della Sala, F., Manna, L., De Vittorio, M., Calabi, F., Cingolani, R., and Rinaldi, R. (2006) Metal-enhanced fluorescence of colloidal nanocrystals with nanoscale control, *Nat. Nanotechnol.*, **1**, 126–130.

36. Corrigan, T. D., Guo, S. H., Szmacinski, H., and Phaneuf, R. J. (2006) Systematic study of the size and spacing dependence of Ag nanoparticle enhanced fluorescence using electron-beam lithography, *Appl. Phys. Lett.*, **88**, 101112/1-3.

37. Drexhage, K. (1974) Interaction of light with monomolecular dye layers, *Progress in Optics*, **12**, 163–232.

38. Chance, R. R., Prock, A., and Silbey, R. (1975) Comments on the classical theory of energy transfer, *J. Chem. Phys.*, **62**, 2245–2253.

39. Enderlein, J. (1999) Single-molecule fluorescence near a metal layer, *Chem. Phys.*, **247**, 1–9.

40. Gotschy, W., Vonmetz, K., Leitner, A., and Aussenegg, F. R. (1996) Optical dichroism of lithographically designed silver nanoparticle films, *Opt. Lett.*, **21**, 1099–1101.

41. Jensen, T. R., Malinsky, M. D., Haynes, C. L., and Van Duyne, R. P. (2000) Nanosphere lithography: Tunable localized surface plasmon resonance spectra of silver nanoparticles, *J. Phys. Chem. B*, **104**, 10549–10556.
42. Kottmann, J. P., Martin, O. J. F., Smith, D. R., and Schultz, S. (2001) Plasmon resonances of silver nanowires with a nonregular cross-section, *Phys. Rev. B*, **64**, 235402/1-10.
43. Guo, S.-H., Tsai, S.-J., Kan, H.-C., Tsai, D.-H., Zachariah, M. R., and Phaneuf, R. J. (2008) The effect of an active substrate on nanoparticle-enhanced fluorescence, *Adv. Mater.*, **20**, 1424–1428.
44. Tsai, D-H., Kim, S. H., Corrigan, T. D., Phaneuf, R. J., and Zachariah, M. R. (2005) Electrostatic-directed deposition of nanoparticles on a field generating substrate, *Nanotechnology*, **16**, 1856–1862.
45. Flidj, N., Aubard, J., Lévi, G., Krenn, J. R., Hohenau, A., Schider, G., Leitner, A., and Aussenegg, F. R. (2003) Optimized surface-enhanced Raman scattering on gold nanoparticle arrays, *Appl. Phys. Lett.*, **82**, 3095–3097.
46. Fissan, H. J., Helsper, C., and Thielen, H. J. (1983) Determination of particle size distributions by means of an electrostatic classifier, *J. Aerosol Sci.*, **14**, 354–357.
47. Kim, A. H., and Zachariah, M. R. (2005) In-flight size classification of carbon nanotubes by gas phase electrophoresis, *Nanotechology*, **16**, 2149–2152.
48. Hao, E., and Schatz, G. C. (2004) Electromagnetic fields around silver nanoparticles and dimers, *J. Chem. Phys.*, **120**, 357–366.
49. Vecchi, G., Giannini, V., and Gomez Rivas, J. (2009) Shaping the fluorescent emission by lattice resonances in plasmonic crystals of nanoantennas, *Phys. Rev. Lett.*, **102**, 146807/1-4.
50. Draine, B. T., and Flatau, P. J. (1994) Discrete-dipole approximation for scattering calculations, *J. Opt. Soc. Am. A*, **11**, 1491–1499.
51. Hale, G. M., and Querry, M. R. (1973) Optical constants of water in the 200-nm to 200-μm wavelength region, *Appl. Opt.*, **12**, 555–563.
52. Jellison Jr., G. E. (1992) Optical functions of silicon determined by two-channel polarization modulation ellipsometry, *Opt. Mater.*, **1**, 41–47.
53. Jin, R., Cao, Y., Kelly, K. L., Schatz, G. C., and Zheng., J. G. (2001) Photoinduced conversion of silver nanospheres to nanoprisms, *Science*, **294**, 1901–1903.
54. Wood, R. W. (1902) On a remarkable case of uneven distribution of light in a diffraction grating spectrum, *Phil. Mag. Series 6*, **4**, 396–402.
55. Wood, R. W. (1935) Anomalous diffraction gratings, *Phys. Rev.*, **48**, 928–936.

56. Hessel, A., and Oliner, A. A. (1965) A new theory of Wood's anomalies on optical gratings, *Appl. Opt.*, **4**, 1275–1297.
57. Kobyakov, A., Mafi, A., Zakharian, A. R., Darmanyan, S. A., and Sparks, K. B. (2008) Fundamental and higher-order Bloch surface plasmons in planar bimetallic gratings on silicon and glass substrates, *J. Opt. Soc. Am. B*, **25**, 1414–1421.
58. Zhang, J., Fu, Y., Chowdhury, M. H., and Lakowicz, J. R. (2008) Single-molecule studies on fluorescently labeled silver particles: Effects of particle size, *J. Phys. Chem. C*, **112**, 18–26.
59. Andrewartha, J. R. (1981) A reflection grating interferometer, *J. Opt.*, **12**, 233–240.
60. Cesario, J., Gonzalez, M. U., Cheylan, S., Barnes, W. L., Enoch, S., and Quidant, R. (2007) Coupling localized and extended plasmons to improve the light extraction through metal films, *Opt. Express*, **15**, 10533–10539.
61. Stuart, H. R., and Hall, D. G. (1998) Enhanced dipole-dipole interaction between elementary radiators near a surface, *Phys. Rev. Lett.*, **80**, 5663–5666.
62. Krug, J. T., Sańchez, E. J., and Xie, X. S. (2005) Fluorescence quenching in tip-enhanced nonlinear optical microscopy, *Appl. Phys. Lett.*, **86**, 233102/1–3.

Chapter 8

Surface-Enhanced Raman Scattering

M. Sun
Beijing National Laboratory for Condensed Matter Physics,
Institute of Physics, Chinese Academy of Sciences,
P. O. Box 603-146, Beijing, 100190, P.R. China
mtsun@aphy.iphy.ac.cn

In this chapter, we will review recent studies of the *electromagnetic* (EM) and chemical mechanisms of surface-enhanced Raman scattering (SERS), including experimental and theoretical studies. We will begin with the theory of EM field and the computational methods. Secondly, we will introduce the *charge-transfer* (chemical) mechanism of SERS, where charge transfer between molecule and metal, intramolecular charge redistribution and intramolecular charge transfer will be visualized with the recently developed three-dimensional (3-D) cube representation. Thirdly, we will discuss our recently experimental results on SERS. Lastly, we will discuss recent experimental result on remote SERS excited by propagating dielectric waveguide along *quasi-one-dimensional* (Q1D) transition metal oxides, and by propagating plasmonic waveguide along single Au nanowire with quasi-spherical termini.

Handbook of Molecular Plasmonics
Edited by Fabio Della Sala and Stefania D'Agostino

ISBN 978-981-4303-20-0 (Hardcover), 978-981-4303-21-7 (eBook)
www.panstanford.com

8.1 Introduction

Since the discovery of the SERS [1–3], it has been extensively studied experimentally and theoretically [4–10] because of its extremely high surface sensitivity and powerful application on fingerprint vibrational spectroscopy in qualitative and quantitative analysis, even at the level of single molecules. There is general agreement that SERS occurs by EM *enhancement* and *chemical enhancement* mechanisms [11–18]. The former is caused by local surface plasmon polaritons (SPP) [4–6, 9, 11, 12], which usually enhance the Raman spectrum over a large frequency range, and the latter is due to changes in the electronic structure of molecule adsorbed on metal surfaces, whereby some Raman peaks undergo enormous selective enhancement [7, 8, 13–18]. Chemical enhancement is typically explained by the CT mechanism.

Before the mid-1990s, most investigation focused on the three *coinage metals* (Au, Ag, and Cu), since they can provide the large enhancement via surface plasmon resonance (SPR). Since the mid-1990s, SERS for different transition metals has been investigated gradually, because of the importance of transition metal (VIII B group elements) in electrochemistry and catalyst [19–21]. The SERS signals on the transition metals are rather weak or even below the detection limit from some adsorbates with a very small Raman cross-section or with a low surface coverage because of the lack of SPR features in the visible spectral region. The best way is to further increase the SERS activity of transition metal utilizing the long-range effects of EM enhancements created by SERS-active Au or Ag substrates underneath an ultrathin layer of transition metals [19, 22–27]. To extend the applicability of the SERS effect into ultraviolet (UV), other SERS-active metals besides Ag or Au are required that show a plasmonic resonance in the UV. The investigation of the near- and deep-ultraviolet resonance Raman scattering is rapidly evolving, because of advantages and application in bioscience and material sciences [28–39]. In the search for UV-SERS material, aluminum (Al) is one of best candidates, because it has low absorption down to a wavelength of 200 nm due to its free-electron-like character and plasmon resonance in UV range [28].

One of the most severe restrictions in the application of SERS to a wide variety of problems in ultrahigh vacuum single-crystal surface science, electrochemistry, heterogeneous catalysis, micro-electronics, and tribology is the requirement that the surface be roughened or nanostructured Ag, Au, or Cu. [40] Many approaches have extended the technique to non-traditional substrates, [10, 20, 41] most notably *tip-enhanced Raman spectroscopy* (TERS) [42–44] where the probed substance (molecule or material surface) can be on a generic substrate and where a nanoscale gold tip above the substrate acts as the Raman signal amplifier. The drawback is that the total Raman scattering signal from the tip area is rather weak, thus limiting TERS studies to molecules with large Raman cross-sections. Recently, shell-isolated nanoparticle-enhanced Raman spectroscopy has been investigated on different single-crystal surfaces, which overcome the critical requirement of roughened or nanostructured substrates [45].

All above SERS experiments, the incident light is focus onto the SERS active spots on the nanostructure and the emitted Raman light is directed from the same spots. Recently, works on both propagating SPP and SERS have been reported [46–48]. Compared with local SERS that excites the target directly, this novel high sensitive technique avoids sample damage from the laser, which may be very useful in special samples such as the intercellular molecule of a living cell.

8.2 Electromagnetic Mechanism and Numerical Methods

Raman intensities scale as the product of the incident field intensity and polarizability derivative, so there are two mechanisms for SERS, one is the EM field enhancement as a result of plasmon resonance excitation, and the second one is enhancement in polarizability because of chemical effects such as CT excited states [49]. The *Raman intensity* can be written as:

$$I = |\alpha|^2 \frac{|E_{loc}|^4}{|E_{inc}|^4}, \tag{8.1}$$

where α is polarizability, and where E_{loc} and E_{inc} stand for the norm of the local and incoming field, respectively. There are several computational approaches to estimate the SERS enhancement by EM field mechanism. In the analytical theory, there are Mie theory for spheres and quasi-static approximations for spheroids (see Chapter 1). For the numerical computational method, there are the discrete dipole approximation (DDA) [50, 51], and the *finite-difference time-domain* (FDTD) method [52] to solve Maxwell's equations to determine the local field, which can deal with any shape, and size up to a few hundred nm within standard computational capabilities. For more details, see Chapters 1 and 2.

Here we firstly introduce our theoretical analytical calculations on EM enhancement with Mie theory in near- and deep-ultraviolet region [33]. The near- and deep-ultraviolet resonance research field is rapidly evolving, because of advantages and application in bioscience and material sciences. Three are the reasons: (i) the Raman scattering efficiency scales with the fourth power of the frequency; (ii) scattering efficiency will be dramatic enhanced when molecules are excited at the molecular electronic absorption bands (resonance Raman scattering) in the UV range; for the pre-resonance (the wavelength of the laser is close to this resonance) Raman scattering, the relatively high enhancement factors can also be achieved, since the resonance wavelengths is not sharply defined; (iii) the UV Raman spectroscopy is not interfered by fluorescence; since there is no material that fluoresces below the wavelength of 280 nm, then the fluorescence and Raman spectra are completely separated at an excitation wavelength below 260 nm [53].

Using the Mie theory, far-field extinction spectrum of single aluminum nanosphere and sphere dimer were investigated in near and deep ultraviolet region, because aluminum (Al) is one of best candidates, and it has low absorption down to a wavelength of 200 nm due to its free-electron-like character and plasmon resonance in UV range [53]. The SPR of metal nanoparticles can be effortlessly tuned by varying their size and separation; we can shift their resonance toward to 244 and 325 nm for single sphere and sphere dimer. Far-field extinction spectrum of single aluminum nanosphere and sphere dimer can be seen in Fig. 8.1, where dipolar resonance peaks were tuned toward 244 nm (Fig. 8.1a) and 325 nm (Fig. 8.1b). When the radius of a single sphere is 39 nm and those of the dimer are 20 nm with a gap of 1.5 nm, the SPR peak can be tuned to near

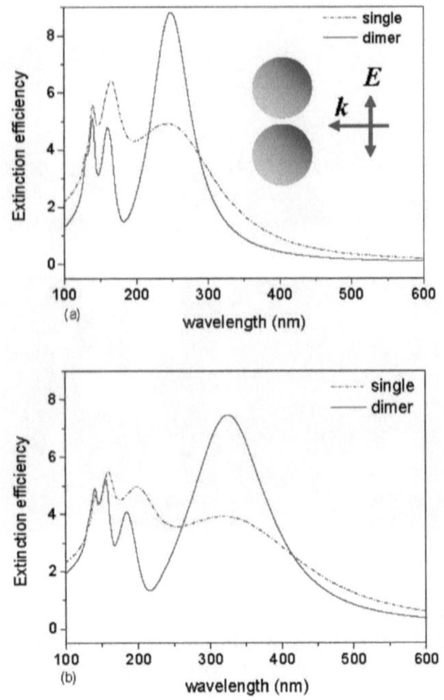

Figure 8.1 Far-field extinction spectrum of single aluminum nanosphere (dashed) and sphere dimer (solid). Dipolar resonance peaks were tuned toward 244 nm (a) and 325 nm (b). The radius of a single sphere is R1 = 39 nm, while those of the dimer are R2 = 20 nm, with gap G = 1.5 nm in (a), in contrast to R1 = 55 nm, R2 = 30 nm, G = 1.2 nm in panel b. The incident plane-wave for particle dimer is shown in the inset. Reprinted with permission from Ref. [33]. Copyright (2009) American Chemical Society.

244 nm. When the radius of a single sphere is 55 nm and the radius of sphere dimer is 30 nm with a gap of 1.2 nm, the SPR peak can be tuned to near 325 nm.

Figures 8.2a,b show the near-field distribution at 244 nm for single sphere and sphere dimer with the same size and gap distance as in Fig. 8.1a.

The strongest EM enhancements are $|M|^2 = 2.9310^1$ ($|M|^2 = |E_{loc}/E_{inc}|^2$) and $|M|^2 = 1.5210^4$ for single sphere and sphere dimer, respectively. The enhancements of SERS are $|M|^4 = 8.5810^2$ and $|M|^4 = 2.3110^8$, respectively. Figures 8.2c,d demonstrates the

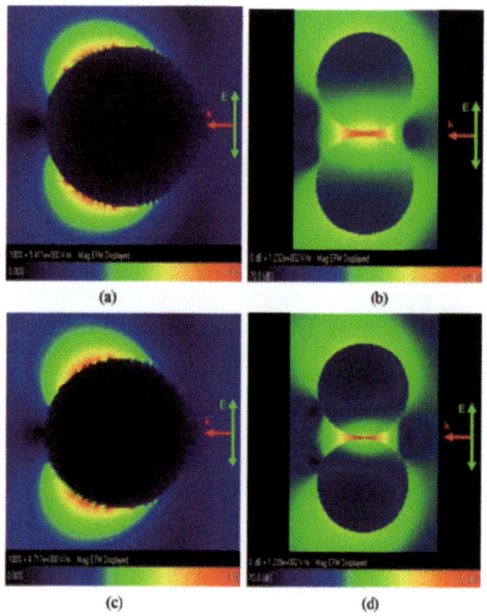

Figure 8.2 Near electric field distribution of aluminum nanosphere: (a) single sphere and (b) sphere dimer at $\lambda = 244$ nm, with the same geometric parameter as in Fig. 8.1a; (c) single sphere and (d) sphere dimer at $\lambda = 325$ nm, with the parameter given in Fig. 8.1b. The incident plane-wave for particle dimer is show in the inset. Reprinted with permission from Ref. [33]. Copyright (2009) American Chemical Society.

near-field distribution at 325 nm for single sphere and sphere dimer with the same size and gap distance as in Fig. 8.1b. The strongest EM enhancements are $|M|^2 = 2.2310^1$ and $|M|^2 = 1.5410^4$ for single sphere and sphere dimer, respectively. The enhancements of SERS are $|M|^4 = 4.9710^2$ and $|M|^4 = 2.3710^8$, respectively.

Secondly, we introduce our recent theoretical numerical calculations, using FDTD method, on EM enhancement for the Au@Pd (core/shell) nano dimer in visual region. Au@Pd nanocubes with well-controllable ultrathin shell has been shown as excellent SERS substrate [54]. Experimental results show that the SERS intensity decays exponentially with the increase in Pd shell thickness, and reaches a constant value at about 2 nm. To quantitatively account for the SERS activity of the Au@Pd coreshell nanocubes system, we use the 3-D-FDTD method to calculate EM field distribution around the illuminated nanocubes [27]. It is well known EM model in

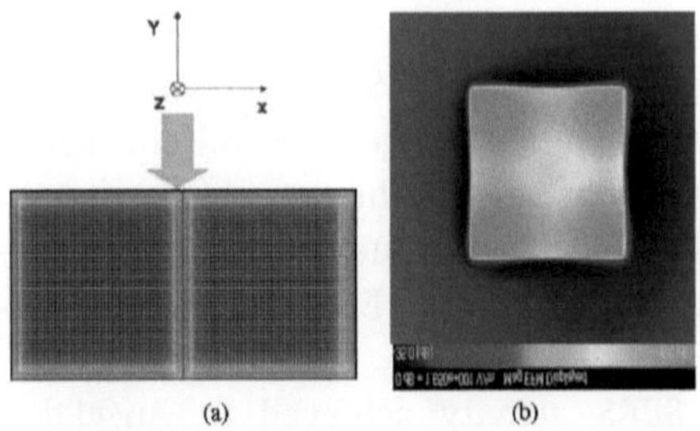

Figure 8.3 FDTD simulations of the near-field distribution in the contact plane. The excitation polarization is along the *x*-axis. Reprinted with permission from Ref. [27]. Copyright [2009], American Institute of Physics.

terms of single particle is a poor model for SERS-active nanoparticle aggregates system, any simulations for the EM enhancement for a real SERS-active system should take the near-field coupling effect into consideration [6]. So, a model consisting of two Au@Pd nanocubes, rather than a single nanocube, was used in our FDTD simulation (see Fig. 8.3a).

In the calculation, the Au cube core is set to be 83 nm, interparticle shell-shell distance 1 nm. The 632.8 nm monochromatic plane-wave is incident from the top (y-axis) with the polarization parallel to the axis along the particle pair (x-axis). The calculation shows the maximum field enhancement factor, defined as the ratio between the maximum local field E_{loc} and the incoming field E_{inc} amplitude, is about 1025 depending on the thickness of Pd shell (05 nm), and the hot site (most enhanced region) of SERS always exists in the gap region. Since the enhanced Raman scattering intensity from a probe molecule at any given position is approximately proportional to the fourth power of the electric field enhancement at the position of the molecule, so the maximum Raman electromagnetic enhancement factor for a Au@Pd particle dimmer can reach from 1.010^4 to 3.910^5 if the thickness value of Pd shell in the range of 05 nm. Figure 8.3b gives a typical field distribution at the junction (yz-plane at $x = 0$) of two nanocubes with a 1 nm Pd shell. It can be seen the maximal field enhancement is about 16.5, corresponding to 7.410^4 of the Raman signal enhancement. This value is higher than that of two pure 83 nm Pd nanocubes. It can be seen clearly that the magnitude of the electric field reaches the maximum in the gap region of the dimer. Since the SERS signal obtained in the experiment should average over all the surfaces, therefore, the enhancement obtained in the experiment may be substantially lower than the calculated value. So, experimental Raman signal enhancement contributed from EM enhancement is estimated in the order of 10^3.

The high SERS activity achieved in Au@Pd ultrathin shell systems is attributed to the long-range effect of the enhanced EM field generated by the Au core. Based on 3D-FDTD method, the dependence of the SERS activity on the Pd shell thickness was quantitatively investigated. The simulated FDTD results (see Fig. 8.4) reveal that the SERS enhancement decreases exponentially with increasing thickness of Pd.

This result shows essentially the same trend as the experimental result reported in Ref. [54].

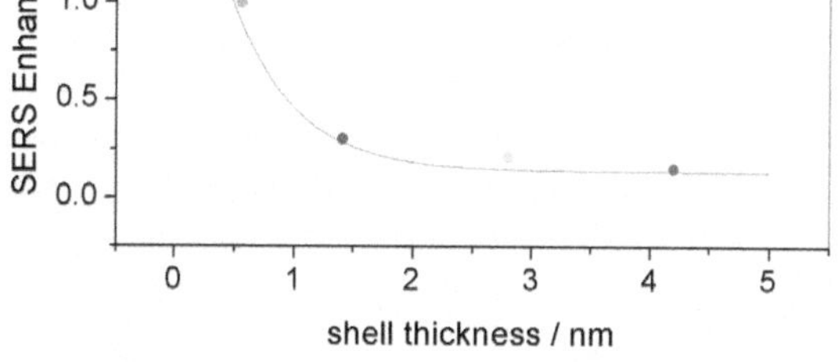

Figure 8.4 Dependence of the SERS intensity of CO adsorbed on 83 nm Au–Pd nanocubes on the thickness of Pd shell (circles) and the corresponding FDTD calculation result (solid gray line). The experimental result is taken from Ref. [54]. Reprinted with permission from Ref. [27]. Copyright [2009], American Institute of Physics.

8.3 Chemical Mechanism and Visualization Method of Charge Transfer

Chemical enhancement is typically explained by the CT mechanism. When a molecule is adsorbed on a metal surface, new electronic states are formed due to chemisorption. The new electronic states may serve as resonant intermediate states in Raman scattering. If the Fermi level of the metal is located between the Highest Occupied Molecular Orbital (HOMO) Lowest Unoccupied Molecular Orbital (LUMO) in energy, CT excitations may likely occur at lower energy than intrinsic intramolecular excitations of the adsorbate [55–58]. According to *Albrechts notation* [55], in the CT mechanism via Albrechts A term (*Franck–Condon term*) only the totally symmetric modes are resonantly enhanced when the laser excitation is close to an allowed electronic transition, and only one excited state is involved. The resonance Raman effects for vibrational modes that are non-totally symmetric, are usually observed when these modes couple two excited states of the chromophore. The product of the symmetry of both excited states should be equal or contain the non-totally symmetry. This mechanism is known as the *Herzberg–Teller mechanism* or B mechanism in Albrechts notation.

It is a great challenge to distinguish chemical enhancement from the EM mechanism through a visual theoretical method. The *charge-density difference* in the 3-D cubic representation can be the best candidate for visualization [17, 18, 59–61], which can reveal visually photo-induced CT between the metal and the molecule or intracluster excitation at resonant electronic transition. The charge transfer between metals and molecules is the evidence of CT mechanism. The intramolecular charge transfer or redistribution is the evidence of molecular resonance. And the intracluster charge redistribution is the evidence of EM mechanism.

We have investigated chemical mechanism of surface-enhanced resonance Raman scattering via charge transfer in *pyridine*-Ag_2 complex. [17] Depending on the nature of the interaction between pyridine and the metal, the new metalpyridine CT excited states could occur. Transition energies and oscillator strengths of pyridine and pyridine-Ag_2 for the lowest six singlet excited states were listed in Table 8.1, as computed from time-dependent density-

Table 8.1 Calculated transition energies (TE) and oscillator strengths (f) of pyridine and pyridine-Ag_2 with from TD-DFT/B3LYP calculations

	Pyridine		Pyridine-Ag_2	
	TE (nm)	f	TE (nm)	f
S_1	265.51	0.0062	638.99	0.0003
S_2	252.42	0.0000	505.82	0.0000
S_3	218.56	0.0211	378.29	0.5784
S_4	187.44	0.0114	316.55	0.2651
S_5	163.32	0.4493	307.79	0.1719
S_6	161.96	0.4575	302.27	0.1537

Source: Reproduced with permission from Ref. [17].

functional theory (TD-DFT), see Chapter 4. It should be noted that the absorption properties of the pyridine-metal cluster depend strongly on the cluster size [62].

The excited state properties of pyridine-Ag_2 are investigated with the charge difference density. From the charge difference densities of pyridine-Ag_2 (listed in Fig. 8.5), S_1, S_2, S_5, and S_6 are the CT excited states in absorption (electrons transfer from Ag_2 to pyridine) since as a result of CT, the holes and electrons are localized on Ag and pyridine, respectively. The transferred electron resides for some femtoseconds in pyridine and after that returns to the metal cluster.

For the pyridine-Ag_2 complex, the resonance Raman spectra (see Fig. 8.5) were calculated with the laser wavelength 780, 630 and 514, 370, 325, and 295 nm, respectively. The energy of the incident light of 780 nm is lower than the S_1 electronic transition energy of pyridine-Ag_2 (638.99 nm), so the calculated result with 785 nm laser excitation is the normal Raman spectrum, and the experimental spectra should be from normal Raman scattering, and the large enhancement in the experiment should mostly result from EM enhancement since the calculated relative intensities of the Raman spectrum of pyridine-Ag_2 with the incident light at 780 nm are almost the same as those of pyridine-Ag_2 without the incident light (normal Raman of pyridine-Ag_2).

When the incident light is at 630 nm, it is almost resonant with the transition energies at the first excited state, which is the CT-excited state. The calculated intensities of them resonance Raman

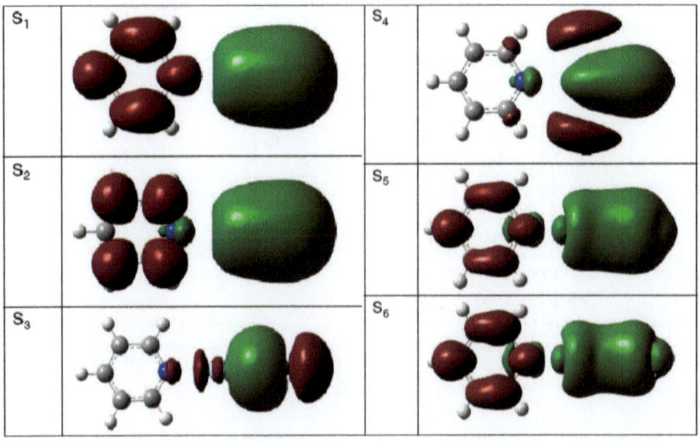

Figure 8.5 The charge density differences of pyridine-Ag_2 for the first six singlet excited states, where the green and red stand for the hole and the electron, respectively. Reprinted with permission from Ref. [17]. Copyright (2008) John Wiley & Sons, Inc.

spectrum of pyridine-Ag_2 is much larger than that of the normal Raman spectrum of pyridine (see Fig. 8.6). It can be seen that the CT enhancement can reach the order of 10^3 (see data in Table 8.2), and the ring stretch mode at 1630 cm^{-1} shows the largest enhancement, which is larger than that of the two ring breathing modes around 1000 cm^{-1}, which is consistent with the order calculated by Zhao *et al.* [8]. Arenas *et al.* [63] also analyzed the results theoretically by considering the displacements between the ground state of pyridine and its anion. Our calculations are in good agreement with their findings, particularly the very strong enhancement of the mode at 1573 cm^{-1}. When the laser wavelength is 514 nm, though this laser energy almost resonates with the transition energy of S_2, the intensity of resonance Raman bands is not enhanced strongly (less than 10 times compared to the intensity of the normal Raman of pyridine (see data in Table 8.2). The reason is that the oscillator strength $f = 0.0000$ for this excited state. The calculated resonance Raman spectrum with the incident light of 295 nm can be seen in Fig. 8.6, and the enhanced rates are listed in Table 8.2. It can be seen that the largest enhancement is of the order of 10^4.

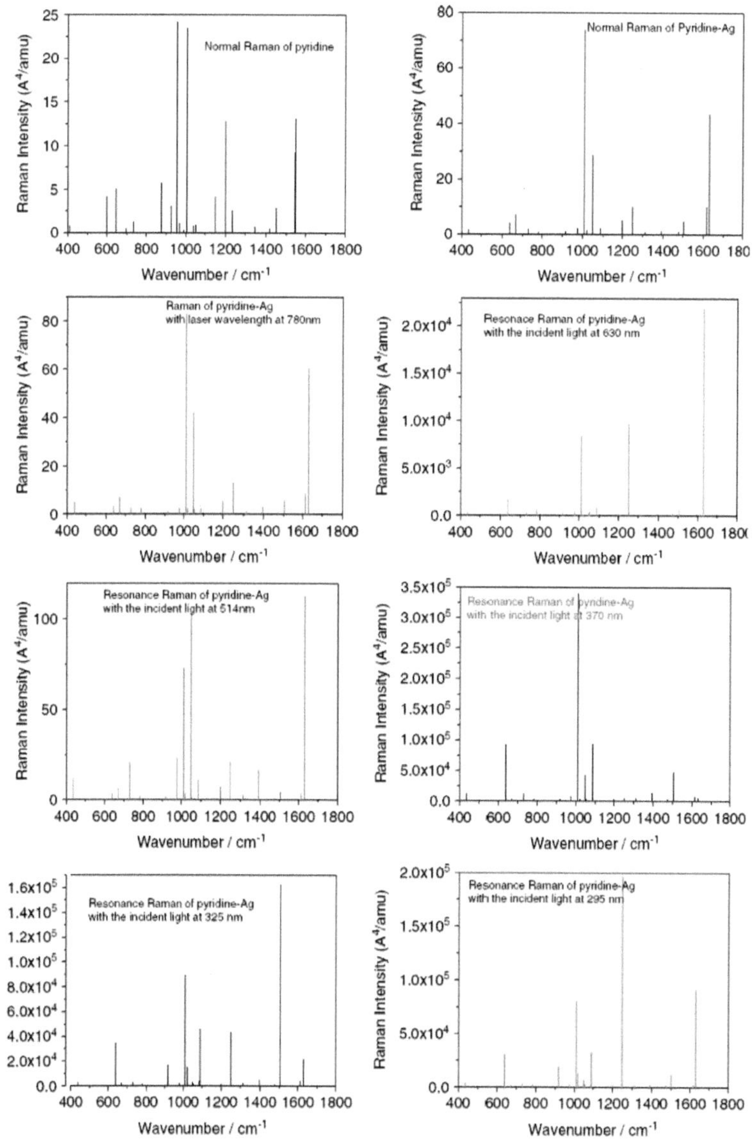

Figure 8.6 The normal Raman spectra of pyridine and pyridine-Ag_2 and the resonance Raman spectra of pyridine-Ag_2 with different incident light wavelengths. Reprinted with permission from Ref. [17]. Copyright (2008) John Wiley & Sons, Inc.

Table 8.2 The enhanced rates of intensities of Raman spectra by static chemical (G_{SC}), CT (G_{CT}), and intracluster excitation (G_{IE}) enhancements. The vibrational modes a–e can be seen from Fig. 4 in Ref. [17]

		G_{CT}				G_{IE}	
Vibrational modes	G_{SC}	785 nm	630 nm	515 nm	295 nm	370 nm	325 nm
a	3.05	3.43	3.4×10^2	3.00	3.3×10^3	1.4×10^4	3.7×10^3
b	1.20	1.78	3.70	4.41	1.9×10^2	1.7×10^3	6.0×10^1
c	0.72	0.95	7.5×10^2	1.57	1.5×10^4	1.1×10^2	3.4×10^3
d	1.56	1.81	1.5×10^2	1.34	3.7×10^3	1.7×10^4	6.3×10^4
e	3.33	4.63	1.7×10^3	8.71	6.8×10^3	1.3×10^2	1.6×10^3

Source: Reproduced with permission from Ref. [17].

8.4 Synthesis and Experiment on SERS

For the SERS activity, the roughness of substrate is one of most important role. We experimentally reported microwave-assisted synthesis of sensitive silver substrate for SERS spectroscopy [64]. We investigated the effect of different heating methods on the morphologies of the samples with SEM. It can be seen clearly from Fig. 8.7a that the shapes of the silver nanoparticles synthesized using formaldehyde as a reductive agent with 5 min of microwave irradiation at a power of 100 W are irregular with particle size ranging from 100 to 1000 nm. Especially, the silver nanoparticles synthesized using formaldehyde as a reductive agent with 5 min of microwave irradiation at a power of 100 W have more well-defined edges, corners, and sharper surface features than those synthesized with 1 h of conventional heating at 40°C, which are spheroidal in shape with particle size ranging from 100 to 500 nm, as shown in Fig. 8.7(b).

We attempted to estimate the *SERS enhancement factor* (EF) using the following equation:

$$EF = \frac{I_{SERS}/N_{SERS}}{I_{NR}/N_{NR}}, \tag{8.2}$$

where I_{SERS} denotes the SERS intensity of the ν_{8a} band of *4-Mercaptobenzoic acid* (4-MBA) adsorbed on the silver nanoparticles synthesized using formaldehyde as a reductive agent with 5 min of microwave irradiation at a power of 100 W and with 1 h

Figure 8.7 FE-SEM images of the silver nanoparticles synthesized using formaldehyde as a reductant: (a) With 5 min of microwave irradiation at a power of 100 W and (b) with 1 h of conventional heating at 40°C. Reprinted with permission from Ref. [64]. Copyright [2008], American Institute of Physics.

of conventional heating at 40°C under the same experimental conditions, I_{NR} denotes the normal Raman (NR) scattering intensity of the same band of dry 4-MBA powder, and N_{SERS} and N_{NR} are the numbers of 4-MBA molecules effectively excited by the laser beam to obtain the corresponding SERS and NR spectra, respectively. According to Fig. 8.8, the EFs of the silver nanoparticles synthesized with microwave radiation and conventional heating estimated using 4-MBA as a probing adsorbate reach $\approx 310^5$ and $\approx 10^4$, respectively.

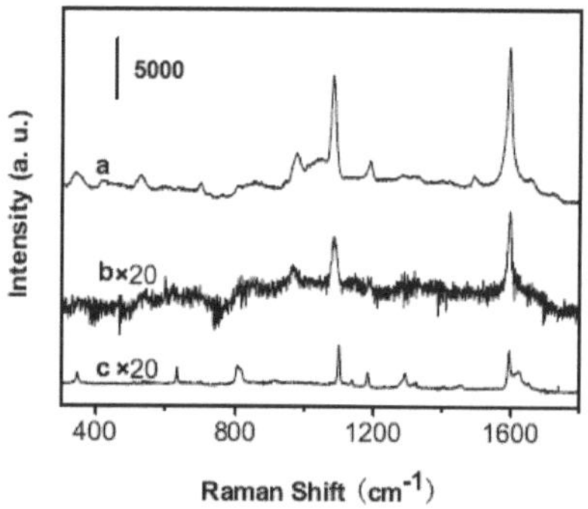

Figure 8.8 SERS spectra of 4-MBA adsorbed on the silver nanoparticles synthesized using formaldehyde as a reductant: (a) With 5 min of microwave irradiation at a power of 100 W, (b) with 1 h of conventional heating at 40°C, and (c) dry 4-MBA powder. Reprinted with permission from Ref. [64]. Copyright [2008], American Institute of Physics.

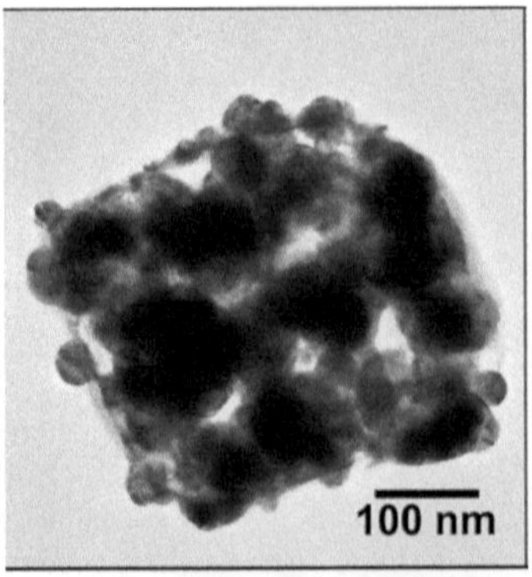

Figure 8.9 TEM image of a silver nanocube synthesized with the directed calcium chloride coalescence process. Reprinted with permission from Ref. [65]. Copyright (2010) Optical Society of America.

Controlling the size, shape, and structure of metal nanoparticles is very important because of the strong correlation between these parameters and their resulting optical, electrical, and catalytic properties [65]. We reported a directed *calcium chloride coalescence method*, based upon incubating a silver colloid with an aqueous solution of $CaCl_2$, for preparation of silver nanocubes with a particle size of 270 to 950 nm.

The method avoids the use of nonvolatile surfactants and polymers, which may be adsorbed onto the silver nanocubes and interfere with their possible applications in catalysis and analytical devices based on SERS spectroscopy. Figure 8.9 shows the transmission electron microscopy (TEM) image of the nanocube obtained using 310^{-2} M $CaCl_2$. It is evident that the nanocube is composed of smaller silver nanoparticles. The nanocubes show good SERS activity in the presence of adsorbed 4-MBA with excitation at 632.8 nm, and the enhancement factor reaches $\approx 7.610^5$. The nanocubes are produced in a simple and cost-effective way, and they are expected to play an important role in the development of SERS-based analytical devices. The method may represent a novel route for preparation of metal nanocubes, which is a subject of intense interest.

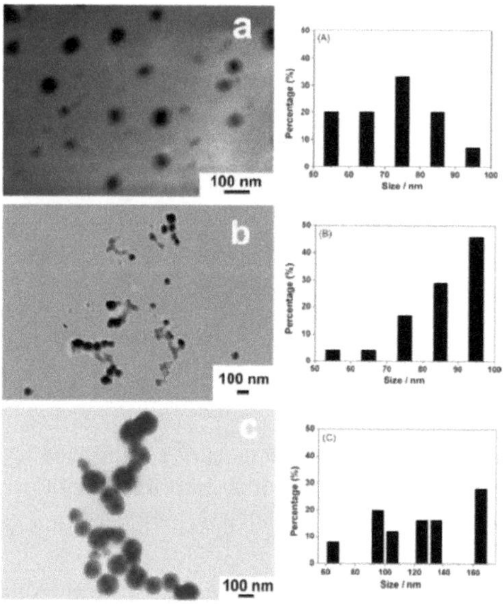

Figure 8.10 TEM micrograph and size distribution of the coreshell nanoparticles. The dosage of Ni^{2+} and Ag^{+} is (a) 0.1 mmol and 0.05 mmol, (b) 0.06 mmol and 0.03 mmol, and (c) 0.24 mmol and 0.12 mmol, respectively. The corresponding histograms of the particle size distribution are indicated as A, B, and C. Reprinted from Ref. [66], Copyright (2010), with permission from Elsevier.

Nickel has been known to be one of the important catalytic, magnetic, and conductive materials. A one-step facile synthesis was devised for preparation of well-dispersed AgNi coreshell nanoparticles with uniform and intact shells [66]. The process is performed by a reduction of silver nitrate and nickel nitrate with sodium borohydride in water-in-oil (W/O) microemulsions of water/polyoxyethylene (4) nonylphenol (OP-4) and polyoxyethylene (7) nonylphenol (OP-7)/n-heptane. TEM (see Fig. 8.10), X-ray diffraction (XRD), X-ray photoelectron spectroscopy (XPS), and UVvis absorption are utilized to characterize the AgNi coreshell nanoparticles. The thickness of Ni layers on the surface of Ag nanoparticles could be controlled by the dosage of Ni^{2+} and Ag^{+}. The AgNi coreshell nanoparticles showed a high catalytic activity for degradation reaction of eosin Y (see Fig. 8.11). The product may also

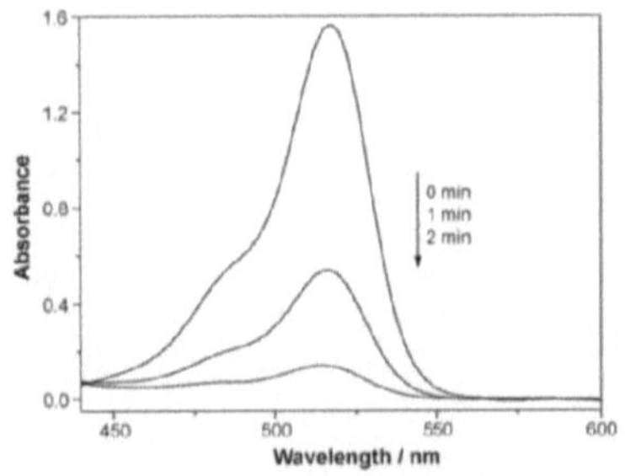

Figure 8.11 UV_{vis} spectra of eosin Y during the reduction catalyzed with the AgNi coreshell nanoparticles. $[NaBH_4]=210^2$ mol L^{-1} and [Eosin Y]$=2\times10^5$ mol L^{-1}. The arrows mark the increase of the reaction time. Reprinted from Ref. [66], Copyright (2010), with permission from Elsevier.

have many potential applications in optical, magnetic, biochemical, and biomedical fields. The synthetic method reported here suggests a very promising route for the preparation of bimetallic coreshell structures, which is a subject of intense interest.

We reported a pelletization method to prepare activity-tunable substrates [67], which is very simple. The results show that SERS activity of the metal pellet varies with the pressure used to make it. Since no chemical process is involved in the preparation of the

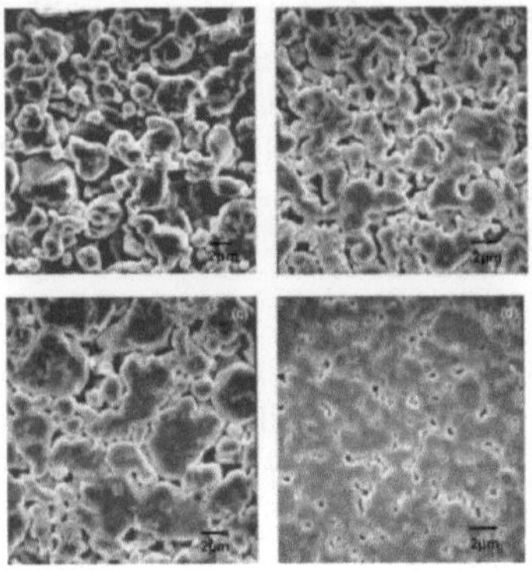

Figure 8.12 SEM images of the metal pellets composed of Ag powder made with (a) 50, (b) 100, (c) 300, and (d) 500 kgf cm^{-2} pressure, respectively. Reprinted with permission from Ref. [67]. Copyright (2010) John Wiley & Sons, Inc.

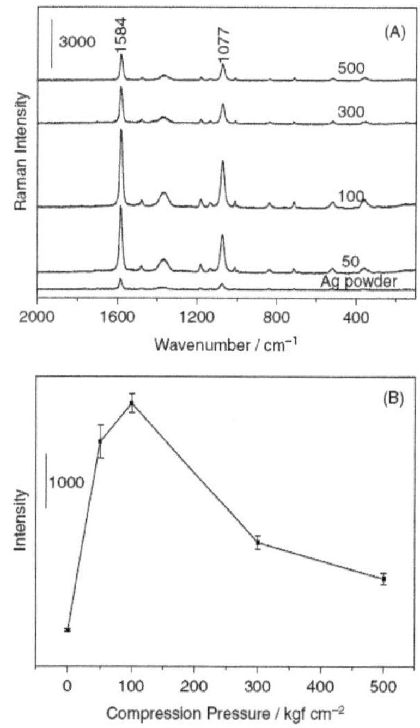

Figure 8.13 (a) SERS spectra of 4-MBA adsorbed on the Ag powder and on metal pellets composed of Ag powder made with 50, 100, 300, and 500 kgf cm^{-2} pressure, respectively using 632.8 nm radiation as an excitation source. (b) The normalized SERS peak intensity of the 1584 cm^{-1} band of 4-MBA adsorbed on the metal pellets composed of Ag powder and Ag powder using 632.8 nm radiation as an excitation source. Reprinted with permission from Ref. [67]. Copyright (2010) John Wiley & Sons, Inc.

metal pellets using a hydraulic press, this must be advantageous in preparing chemically pure SERS-active substrates. Moreover, by controlling the required pressure (see Fig. 8.12), the as-obtained metal pellets show enhanced Raman scattering in the presence of adsorbed 4-MBA with excitation at 632.8 nm. The SERS peak intensity of the 4-MBA band at 1584 cm^{1} adsorbed on the metal pellets varies as a function of the applied pressure, and is about 1.232 times greater than when it is adsorbed on Ag particles (see Fig. 8.13). As a result, the SERS-active metal pellets can be produced

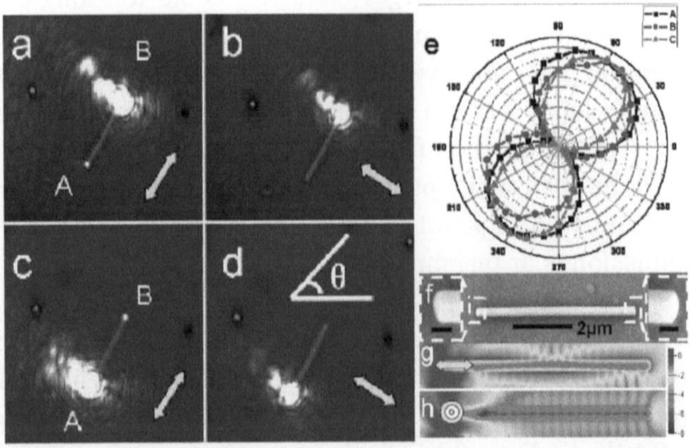

Figure 8.14 (a) Excitation with parallel incident polarization at terminus B. (b) Excitation with perpendicular incident polarization at terminus B. (c) Excitation with parallel incident polarization at terminus A. (d) Excitation with perpendicular incident polarization at terminus A. (e) Incident polarization dependence of emission. Data A (black squares) represented emission at terminus A. Data B (gray circles) represented emission at terminus B. Data C (gray triangles) represented simulation results by FDTD. (f) SEM image of the Au nanowire. The enlarged area indicated the shape of termini, whose scale bar was 100 nm. Distribution of electric field intensity was simulated and plotted in logarithmic scale with parallel incident polarization (g) and with perpendicular incident polarization (h). Reprinted with permission from Ref. [48]. Copyright (2011) American Chemical Society.

by a simple and cost-effective method, and is expected to play an important role in the development of SERS-based analytical devices.

8.5 Remote-Excitation SERS

The *SERS remote excitation* is based on propagating SPP on single well-defined crystal Au nanowire with quasi-spherical termini has been reported [48].

First, we investigate the propagation property of SPP in a single Au nanowire with quasi-spherical termini. The SEM picture in

Fig. 8.14f shows that the target nanowire is 5.5 μm long and has a diameter of about 200 nm. The spherical termini are indicated by a magnified SEM image with high resolution. It is clear that this Au nanowire has a smooth surface without any obvious defect. In Fig. 8.14, remote emission from one terminus of the Au nanowires can clearly be seen while polarized 785 nm laser is focused at the other terminus. In Figs. 8.14a,b, the laser is focused on terminus B and emission is at terminus A. The polarization of incident laser in Fig. 8.14a is parallel to the nanowire (the arrows in Fig. 8.14 mean polarization of the incident laser). It is very clear that terminus A emitted light intensively. However, there is little light while the polarization of the incident laser is perpendicular to the nanowire in Fig. 8.14b. In Figs. 8.14c,d, the laser is focused on terminus A and emission is at terminus B. The situation is almost the same. Figure 8.14e shows the incident polarization dependence of emission in detail. Data A (black squares) are collected when the emission is at terminus A. And data B (red spots) mean the emission is at the other terminus. We use θ (definition of θ is described in Fig. 8.14d, $\theta = 60^\circ$ represents polarization of the incident laser parallel to the nanowire) and normalized intensity of emission to plot polar diagram. We find the experimental data A and data B coincide with the optical image in Fig. 8.14.

Then we measure remote excitation SERS of *malachite green isothiocyanate* (MGITC) on single Au nanowire with quasi-spherical termini. The polarization of the incident laser is parallel to nanowire to receive a maximum emission at the remote terminus according to our result in Fig. 8.15. The sketch of the remote excitation SERS technique is shown in Fig. 8.15a. The red arrows indicate the incident laser. The blue arrow represents propagating SPP on the Au nanowire. Orange arrows mean the Raman signal of MGITC on the terminus of the Au nanowire. The SEM image of this Au nanowire with quasi-spherical termini in Fig. 8.15b shows that its length is 6.5 m and diameter is 210 nm. The optical image of propagation of SPP on a Au nanowire is presented in Fig. 8.15c. The top spectrum (red line) in Fig. 8.15d is remote excitation SERS. The detection point is at terminus A, but the excitation point is at terminus B. The bottom spectrum (black line) in Fig. 8.15d is local SERS, whose

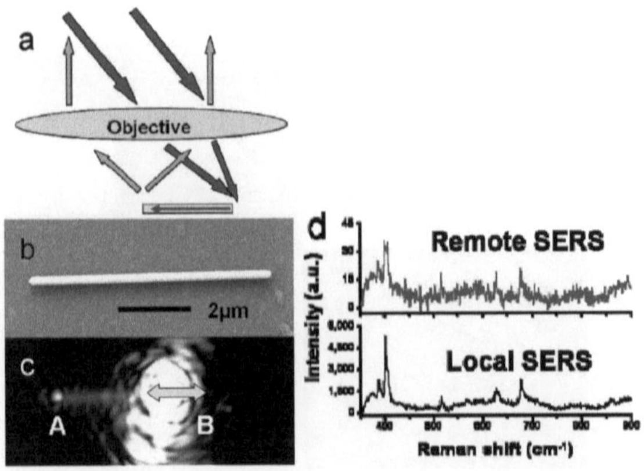

Figure 8.15 (a) Sketch of remote excitation SERS technique. (b) SEM image of Au nanowire. (c) Optical image of propagation of SPP on Au nanowire. (d) Remote excitation SERS and local SERS spectra. Reprinted with permission from Ref. [48]. Copyright (2011) American Chemical Society.

excitation and detection points are at terminus A. We can see that the intensity of local SERS is about 150 times greater than that of remote excitation SERS. We also have tried to measure remote excitation SERS with perpendicular incident polarization. It is very hard to acquire remote excitation SERS spectra due to the very low electric intensity. This result indicates that we have successfully applied this novel technique on Au nanowires. The antioxide property of Au material under ambient conditions ensured the Au nanowire could work for a much longer time compare to a Ag nanostructure. The simple structure of a single Au nanowire also makes it more convenient for application.

The continuous evolution of nanotechnology in these years led to the production of Q1D structures in a variety of morphologies such as nanowires, nanotubes, nanobelts, nanorods, and etc. [68–70]. In particular, Q1D metal oxides are crystalline structures with well-defined chemical composition, surface terminations, free from dislocation and other extended defects, which are attracting an increasing interest for both fundamental and applied science [71–74]. The Q1D transition-metal oxide MoO_3 belt is synthesized

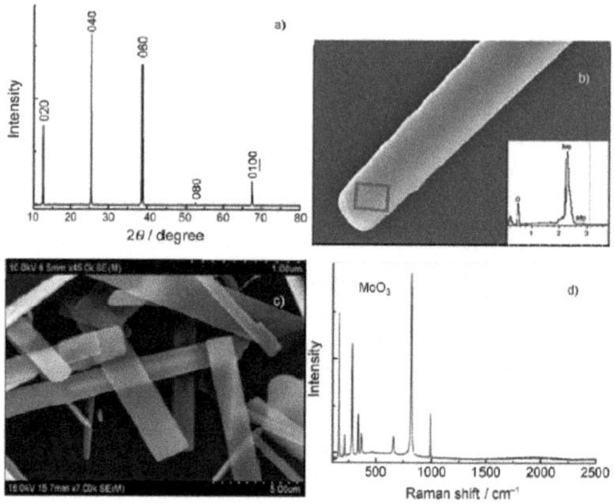

Figure 8.16 (a) The XRD of the MoO_3 belts, (b) and (c) the SEM images of the MoO_3 belts, and (d) Raman spectrum of the MoO_3 belts. The inset in panel (b) is the EDS. Reprinted with permission from Ref. [73]. Copyright (2010) John Wiley & Sons, Inc.

and characterized by us with X-ray diffraction, scanning electron microscopy, and Raman spectroscopy (see Fig. 8.16) [73].

We realized propagating waveguide along Q1D MoO_3 belts [74]. From Fig. 8.17, the light can efficiently propagate along the 1D MoO_3 belt from point A to point C with strong light (the length of this belt is about 52 μm).

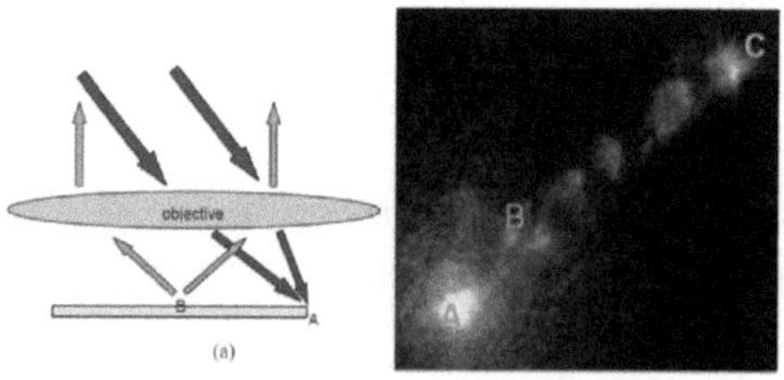

Figure 8.17 (a) The diagram of remote SERS, (b) the propagation of light along MoO_3 belt, the distance between point A and B is about 12.6 m, and the length of this belt is about 52 m. The integral time is 30 seconds and laser power is 5 mW in SERS. Reprinted with permission from Ref. [73]. Copyright (2010) John Wiley & Sons, Inc.

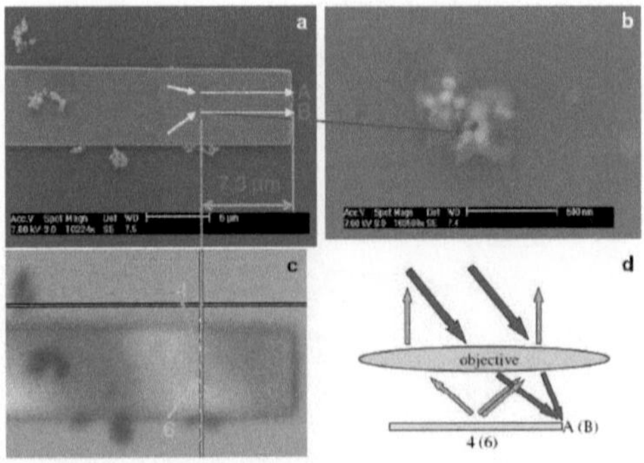

Figure 8.18 (a) SEM image of nanoparticles on the MoO_3 ribbon, where A and B are the points that laser radiates on, 4 and 6 are the collection points of SERS, and distance between A (or B) and 4 (or 6) is 7.3 m, (b) magnified SEM image of Fig. 18a at point 6, (c) optical imaging of nanoparticles on the MoO_3 ribbon, and (d) the scheme of remote SERS measurement, where A and B stand for the point that laser radiates on the edge of ribbon belt and collection point of SERS, and the black and gray arrows stands for the incident and scattering and collection SERS signal, respectively. Reprinted from Ref. [74] with kind permission from Springer Science+Business Media.

We reported experimentally the remote excitation surface plasmon and consequent enhancement of SERS using evanescent wave propagating in Q1D MoO_3 ribbon dielectric waveguide. [74] The remote excitation SERS spectra enhanced by chemical and EM field mechanisms are measured, respectively, where silver (Ag) nanoparticles are excited by EM field after propagating 7.3 μm in the ribbon. Figure 8.18a and b show the morphology investigations, which were characterized with SEM (XL30S-FEG). Figures 8.18c is the optical imaging of Fig. 8.18a. As shown in the Fig. 8.18d, we collected the remote SERS signal (the brown arrows) on the points 4(6) while the laser (the red arrows) was focused on the points A(B). We firstly measured local SERS (light radiation and collection at the same point) spectra at points 4 and 6 (see Fig. 8.19). Secondly, utilizing the properties of evanescent wave propagating in Q1D MoO_3 ribbon dielectric waveguide, the remote SERS spectra were measured at point 4 and in Fig. 8.18c, where the laser radiated at

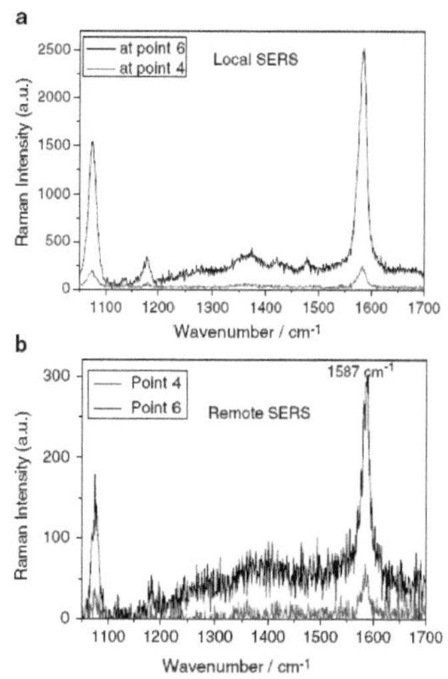

Figure 8.19 (a) Local SERS spectra at points 4 and 6, and (b) the remote SERS at points 4 and 6 in Fig. 8.18a. The 632.8-nm radiation from the HeNe laser was used as an excitation source. The size of the laser spot was small than 1 um. We use the high confocallity mode to acquire the Raman signal and the acquisition range was less than 1 m^2. In the (a), the 10% of the laser intensity was used and the integration time was 10 s; while in the (b) the 50% of the laser intensity was used and the integration time was 20 s. Reprinted from Ref. [74] with kind permission from Springer Science+Business Media.

one side of the MoO_3 belt (point A in Fig. 8.18c). The scheme of the remote SERS measurements can be seen from Fig. 8.18d.

To reveal the nature of remote excitation SERS enhancement by propagating dielectric waveguide along the Q1D MoO_3 ribbon, the FDTD calculations are performed. Figure 8.20 shows the simulated electric field distribution in a MoO_3 belt excited by diffraction limited 632.8 nm spot at left edge of the waveguide. It is obvious that propagating wave is launched. The angle of the propagating wave leaking into the air from the waveguide is larger than the total reflection angle. Hence, the loss in the belt is mainly caused by the evanescent wave. After propagating 7.3 μm in the belt, the intensity decreases to $\approx$1/10 of the incident light. Then, we put a silver nanoparticle on the belt that is 5 m away from the excitation as shown in the

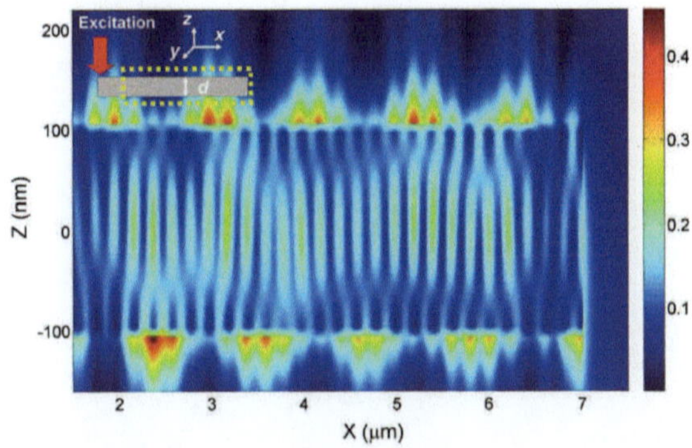

Figure 8.20 Propagating EM field in a MoO_3 belt with a thickness $d = 200$ nm. The xz plane is the cross-section of $y = 0$. Inset: the belt is excited by a 632.8 nm laser focused on the edge. The incident polarization is along x-axis. The yellow square shows the range of the simulation. The refractive index of MoO_3 is 2.06. Reprinted from Ref. [74] with kind permission from Springer Science+Business Media.

inset of Fig. 8.21. It is clear that electric field at the gap (10 nm) between the particle and the surface is enhanced. What should be noted is the surface plasmon of the silver nanoparticle is excited by the evanescent wave leaked from the MoO_3 belt. The highest SERS enhancement at the gap is larger than 10^2, but the averaged SERS enhancement factor for the single nanoparticle is about 10 which is consistent with the value in the experiment. Although the measured SERS is from nanoparticle aggregates, the theoretical results did confirm the physical enhancement originated from the remote excitation by the evanescent wave at the MoO_3 belt surface.

The chemical mechanism for the remote excitation SERS is contributed from the charge transfer between the analyte molecule and MoO_3 ribbon. The EM field mechanism for the remote excitation SERS arises from the energy conversion from the propagating dielectric waveguide to the surface plasmon of Ag nanoparticles on the Q1D MoO_3 ribbon. It is important to reveal the mechanism of energy conversion from the propagating dielectric waveguide to the surface plasmon for potential applications in micro- and nanoscale devices.

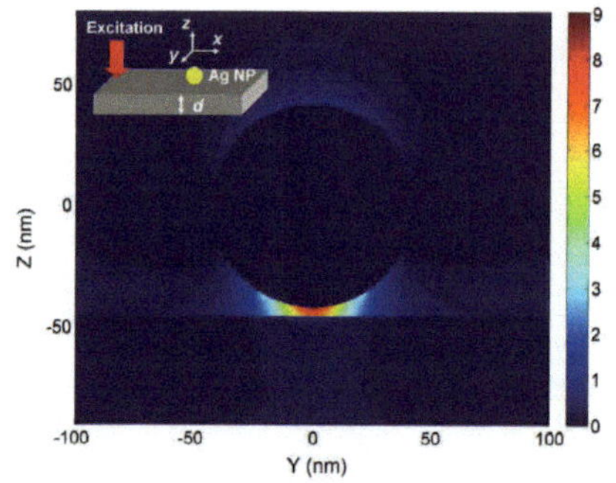

Figure 8.21 Simulated near field distribution of a nanoparticle (the radius 40 nm) on MoO_3 belt. The *yz* plane is the cross-section through the particle center. The particle is separated 10 nm from the surface of the waveguide. Inset: the belt is excited by a 632.8 nm laser focused to the edge. The incident polarization is along *x*-axis. The thickness of the belt is $d = 200$ nm. The particle is 5 m from the excitation. Reprinted from Ref. [74] with kind permission from Springer Science+Business Media.

8.6 Conclusions

Theoretically, the SPR peak can be tuned to 325 and 244 nm by controlling the size and gap distance of nanospheres, and SERS enhancement can reach on the order of 10^2 ($|M|^4$) for single and 10^8 ($|M|^4$) for dimer nanoparticles at 325 and 244 nm. So, Al is a suitable material for near- and deep-UV surface-enhanced resonance Raman scattering. The theoretical study show Au@Pd core/shell nanoparticles are a good substrate for SERS, and the high SERS activity achieved in ultrathin shell systems is contributed mainly from the long range effect of the enhanced EM field generated by the Au core. With the visualization method of charge transfer, the CT mechanism can clearly distinguished from EM mechanism.

Experimentally, the microwave-assisted synthesis of sensitive silver substrate has been shown very good SERS activity. The synthesized nanocube with the directed calcium chloride coalescence method demonstrated a high catalytic activity for degradation reaction of eosin Y, using SERS spectrum. SERS-active substrates can be controlled and tuned to the optimal SERS intensity with pressure. Remote excitation SERS by propagating plasmonic dielectric waveguide realized experimentally, and interpreted theoretically using FDTD method.

Acknowledgments

Mengtao Sun thanks all of my friendly coauthors in my published papers cited in this chapter. This work was supported by the National Natural Science Foundation of China (Grant Nos. 90923003, 10874234, 20703064).

References

1. Fleischmann, M., Hendra, P. J., and McQuillan, A. J. (1974) Raman spectra of pyridine adsorbed at a silver electrode, *Chem. Phys. Lett.*, **26**, 163–166.
2. Jeanmaire, D. L., and Van Duyne, R. P. (1977) Surface Raman spectro-electrochemistry: Part I. Heterocyclic, aromatic, and aliphatic amines adsorbed on the anodized silver electrode, *J. Electroanal. Chem.*, **84**, 1–20.
3. Grant, A. M., and Alan, C. J. (1977) Anomalously intense Raman spectra of pyridine at a silver electrode, *J. Am. Chem. Soc.*, **99**, 5215–5217.
4. Moskovits, M. (1985) Surface-enhanced spectroscopy, *Rev. Mod. Phys.*, **57**, 783–826.
5. Metiu, H., and Das, P. (1984) The electromagnetic theory of surface enhanced spectroscopy, *Annu. Rev. Phys. Chem.*, **35**, 507–536.
6. Xu, H., Bjerneld, E. J., Kall, M., and Borjesson, L. (1999) Spectroscopy of single hemoglobin molecules by surface enhanced Raman scattering, *Phys. Rev. Lett.*, **83**, 4357–4360.
7. Lombardi, J. R., Birke, R. L., Tianhong, L., and Xu, J. (1986) Charge-transfer theory of surface enhanced Raman spectroscopy: Herzberg–Teller contributions, *J. Chem. Phys.*, **84**, 4174–4180.
8. Zhao, L., Jensen, L., and Schatz, G. C. (2006) Pyridine-Ag_{20} cluster: A model system for studying surface-enhanced Raman scattering, *J. Am. Chem. Soc.*, **128**, 2911–2919.
9. Kneipp, K., Kneipp, H., Itzkan, I., Dasari, R. R., and Feld, M. S. (1999) Ultrasensitive chemical analysis by Raman spectroscopy, *Chem. Rev.*, **99**, 2957–2976.

10. Tian, Z.-Q., Ren, B., Li, J.-F., and Yang, Z.-L. (2007) Expanding generality of surface-enhanced Raman spectroscopy with borrowing SERS activity strategy, *Chem. Commun.*, **34**, 3514–3534.
11. Creighton, J. A. (1983), Surface Raman electromagnetic enhancement factors for molecules at the surface of small isolated metal spheres: The determination of adsorbate orientation from SERS relative intensities, *Surf. Sci.*, **124**, 209–219.
12. Kottmann, J. P., Martin, O. J. F., Smith, D. R., and Schultz, S. (2001) Dramatic localized electromagnetic enhancement in plasmon resonant nanowires, *Chem. Phys. Lett.*, **341**, 1–6.
13. Otto, A. (2005) The "chemical" (electronic) contribution to surface-enhanced Raman scattering, *J. Raman Spectrosc.*, **36**, 497–509.
14. Lombardi, J. R., and Birke, R. L. (2008) A unified approach to surface-enhanced Raman spectroscopy, *J. Phys. Chem. C*, **112**, 5605–5617.
15. Wu, D.-Y., Liu, X.-M., Duan, S., Xu, X., Ren, B., Lin, S.-H., and Tian, Z.-Q. (2008) Chemical enhancement effects in SERS spectra: A quantum chemical study of pyridine interacting with copper, silver, gold and platinum metals, *J. Phys. Chem. C*, **112**, 4195–4204.
16. Dong, B., Liu, L., Xu, H., and Sun, M. (2010) Experimental and theoretical evidence for the chemical mechanism in SERRS of rhodamine 6G adsorbed on colloidal silver excited at 1064 nm, *J. Raman Spectrosc.*, **41**, 719–720.
17. Sun, M., Wan, S., Liu, Y., Jia, Y., and Xu, H. (2008) Chemical mechanism of surface-enhanced resonance Raman scattering via charge transfer in pyridine-Ag_2 complex, *J. Raman Spectrosc.*, **39**, 402–408.
18. Sun, M., Liu, S., Chen, M., and Xu, H. (2009) Direct visual evidence for the chemical mechanism of surface-enhanced resonance Raman scattering via charge transfer, *J. Raman Spectrosc.*, **40**, 137–143.
19. Li, C. (2007) UV Raman spectroscopic studies on transition metal-containing microporous and mesoporous materials: Active sites and synthesis mechanism, *Stud. Surf. Sci. Catal.*, **170**, 561–576.
20. Tian, Z.-Q., and Ren, B. (2004) Adsorption and reaction at electrochemical interfaces as probed by surface-enhanced Raman spectroscopy, *Annu. Rev. Phys. Chem.*, **55**, 197–229.
21. Tian, Z.-Q., Yang, Z.-L., Ren, B., Li, J.-F., Zhang, Y., Lin, X.-F., Hu, J.-W., and Wu, D.-Y. (2006) Surface-enhanced Raman scattering from transition metals with special surface morphology and nanoparticle shape, *Faraday Discuss.*, **132**, 159–170.

22. Li, J.-F., Yang, Z.-L., Ren, B., Liu, G.-K., Fang, P.-P., Jiang, Y.-X., Wu, D.-Y., and Tian, Z.-Q. (2006) Surface-enhanced Raman spectroscopy using gold-core platinum-shell nanoparticle film electrodes: Toward a versatile vibrational strategy for electrochemical interfaces, *Langmuir*, **22**, 10372–10379.

23. Zhang, K., Xiang, Y., Wu, X., Feng, L., He, W., Liu, J., Zhou, W., and Xie, S. (2009) Enhanced optical responses of Au@Pd core/shell nanobars, *Langmuir*, **25**, 1162–1168.

24. Chen, J., Wiley, B., McLellan, J., Xiong, Y., Li, Z.-Y., and Xia, Y. (2005) Optical properties of Pd-Ag and Pt-Ag nanoboxes synthesized via galvanic replacement reactions, *Nano Lett.*, **5**, 2058–2062.

25. Cobley, C. M., Campbell, D. J., and Xia, Y. (2008) Tailoring the optical and catalytic properties of gold-silver nanoboxes and nanocages by introducing palladium, *Adv. Mater.*, **20**, 748–752.

26. Zou, S., Williams, C. T., Chen, E. K.-Y., and Weaver, M. J. (1998) Probing molecular vibrations at catalytically significant interfaces: A new ubiquity of surface-enhanced Raman scattering, *J. Am. Chem. Soc.*, **120**, 3811–3812.

27. Yang, Z., Li, Y., Li, Z., Wu, D., Kang, J., Xu, H., and Sun, M. (2009) Surface enhanced Raman scattering of pyridine adsorbed on Au@Pd core/shell nanoparticles, *J. Chem. Phys.*, **130**, 234705/1-7.

28. Asher, S. A. (1993) UV resonance Raman spectroscopy for analytical, physical, and biophysical chemistry. Part 1, *Anal. Chem.*, **65**, 59A–66A.

29. Fodor, S. P. A., and Spiro, T. G. (1986) Ultraviolet resonance Raman-spectroscopy of DNA with 200–266-nm laser excitation, *J. Am. Chem. Soc.*, **108**, 3198–3205.

30. Lin, X.-F., Ren, B., Yang, Z.-L., Liu, G.-K., and Tian, Z.-Q. (2005) Surface-enhanced Raman spectroscopy with ultraviolet excitation, *J. Raman Spectrosc.*, **36**, 606–612.

31. Ren, B., Lin, X.-F., Yang, Z.-L., Liu, G.-K., Aroca, R. F., Mao, B.-W., and Tian, Z.-Q. (2003) Surface-enhanced Raman scattering in the ultraviolet spectral region: UV-SERS on rhodium and ruthenium electrodes *J. Am. Chem. Soc.*, **125**, 9598–9599.

32. Hecht, L., Clarkson, J., Smith, B. J. E., and Springett, R. (2006) A new single grating spectrograph for ultraviolet Raman scattering studies, *J. Raman Spectrosc.*, **37**, 562–573.

33. Sun, M., Zhang, S., Fang, Y., Yang, Z., Wu, D., Dong, B., and Xu, H. (2009) Near- and deep-ultraviolet resonance Raman spectroscopy of

pyrazine-Al4 complex and Al3-pyrazine-Al3 junction, *J. Phys. Chem. C*, **113**, 19328–19334.

34. Taguchi, A., Hayazawa, N., Furusawa, K., Ishitobi, H., and Kawata, S. (2009) Deep-UV tip-enhanced Raman scattering, *J. Raman Spectrosc.*, **40**, 1324–1330.
35. Konorov, S. O., Schulze, H. G., Addison, C. J., Haynes, C. A., Blades, M. W., and Turner, R. F. B. (2009) Ultraviolet resonance Raman spectroscopy of locked single-stranded oligo(dA) reveals conformational implications of the locked ribose in LNA, *J. Raman Spectrosc.*, **40**, 1162–1171.
36. Shafaat, H. S., Sanchez, K. M., Neary, T. J., and Kim, J. E. (2009) Ultraviolet resonance Raman spectroscopy of a β-sheet peptide: A model for membrane protein folding, *J. Raman Spectrosc.*, **40**, 1060–1064.
37. Fujiwara, A., and Mizutani, Y. (2008) Photoinduced electron transfer in glucose oxidase: A picosecond time-resolved ultraviolet resonance Raman study, *J. Raman Spectrosc.*, **39**, 1600–1605.
38. Huang, C.-Y., Balakrishnan, G., and Spiro, T. G. (2006) Protein secondary structure from deep-UV resonance Raman spectroscopy, *J. Raman Spectrosc.*, **37**, 277–282.
39. Shashilov, V. A., and Lednev, I. K. (2008) 2D correlation deep UV resonance Raman spectroscopy of early events of lysozyme fibrillation: Kinetic mechanism and potential interpretation pitfalls, *J. Am. Chem. Soc.*, **130**, 309–317.
40. Yong, M. A., Dieringer, J. A., and Van Duyne, R. P. (2007) Plasmonic materials for surface-enhanced and tip-enhanced Raman spectroscopy, in *Tip Enhancement*, (ed. Kawata, S., and Shalaev, V. M.), Elsevier, Amsterdam.
41. Park, S., Yang, P., Corredor, P., and Weaver, M. J. (2002) Transition metal-coated nanoparticle films: Vibrational characterization with surface-enhanced Raman scattering, *J. Am. Chem. Soc.*, **124**, 2428–2429.
42. Pettinger, B., Ren, B., Picardi, G., Schuster, R., and Ertl, G. (2004) Nanoscale probing of adsorbed species by tip-enhanced Raman spectroscopy, *Phys. Rev. Lett.*, **92**, 096101/1-4.
43. Wu, D.-Y., Li, J.-F., Ren, B., and Tian, Z.-Q. (2008) Electrochemical surface-enhanced Raman spectroscopy of nanostructures, *Chem. Soc. Rev.*, **37**, 1025–1041.
44. Sun, M., Fang, Y., Yang, Z., and Xu, H. (2009) Chemical and electromagnetic mechanisms of tip-enhanced Raman scattering, *Phys. Chem. Chem. Phys.*, **11**, 9412–9419.

45. Li, J. F., Huang, Y. F., Ding, Y., Yang, Z. L., Li, S. B., Zhou, X. S., Fan, F. R., Zhang, W., Zhou, Z. Y., Wu, D. Y., Ren, B., Wang, Z. L., and Tian, Z. Q. (2010) Shell-isolated nanoparticle-enhanced Raman spectroscopy, *Nature*, **464**, 392–395.
46. Hutchison, J. A., Centeno, S. P., Odaka, H., Fukumura, H., Hofkens, J., and Uji-i, H. (2009) Subdiffraction limited, remote excitation of surface enhanced Raman scattering, *Nano Lett.*, **9**, 995–1001.
47. Fang, Y., Wei, H., Hao, F., Nordlander, P., and Xu, H. (2009) Remote-excitation surface-enhanced Raman scattering using propagating Ag nanowire plasmons, *Nano Lett.*, **9**, 2049–2053.
48. Huang, Y., Fang, Y., and Sun, M. (2011) Remote excitation of surface-enhanced Raman scattering on single Au nanowire with quasi-spherical termini, *J. Phys. Chem. C*, **115**, 3558–3561.
49. Schatz, G. C., Yong, M. A., and Van Duyne, R. P. (2006) Electromagnetic Mechanism of SERS, in *Surface-Enhanced Raman Scattering Physics and Applications*, (ed. Kneipp, K., Moskovits, M., and Kneipp, H.), Springer-Verlag, Berlin-Heidelberg.
50. Singham, S. B., and Salzman, G. C. (1986) Evaluation of the scattering matrix of an arbitrary particle using the coupled dipole approximation, *J. Chem. Phys.*, **84**, 2658–2667.
51. Singham, S. B., and Bohren, C. F. (1987) Light scattering by an arbitrary particle: A physical reformulation of the coupled dipole method, *Opt. Lett.*, **12**, 10–12.
52. Kunz, K. S., and Luebber, R. J. (eds) (1993) *The Finite Difference Time Domain Method for Electromagnetics* CRC, Cleveland.
53. Palik, E. D. (ed.) (1985) *Handbook of Optical Constants of Solids*, Academic Press, Orlando.
54. Sheng, J. J., Li, J. F., Yin, B. S., Ren, B., and Tian, Z. Q. (2007) A preliminary study on surface-enhanced Raman scattering from Au and Au@Pd nanocubes for electrochemical applications, *Can. J. Anal. Sci. Spectrosc.*, **52**, 178–185.
55. Frank, J. A. (1982) Charge transfer effects in surface-enhanced Raman scattering, *J. Chem. Phys.*, **77**, 5302–5314.
56. Campion, A., and Kambhampati, P. (1998) Surface-enhanced Raman scattering, *Chem. Soc. Rev.*, **27**, 241–250.
57. Arenas, J. F., Lopez-Tocon, I., Castro, J. L., Centeno, S. P., Lopez-Ramirez, M. R., and Otero, J. C. (2005) Resonant charge transfer on the nanoscale: Studying doublet states of adsorbates by surface-enhanced Raman scattering, *J. Raman Spectrosc.*, **36**, 515–521.

58. Schatz, G. C. (1984) Theoretical studies of surface enhanced Raman scattering, *Accounts Chem. Res.*, **17**, 370–376.
59. Sun, M., and Xu, H. (2009) Direct visualization of the chemical mechanism in SERRS of 4-aminothiophenol/metal complexes and metal/4-aminothiophenol/metal junctions, *Chem. Phys. Chem.*, **10**, 392–399.
60. Sun, M., Liu, S., Li, Z., Duan, J., Chen, M., and Xu, H. (2009) Direct visual evidence for the chemical mechanism of surface-enhanced resonance Raman scattering via charge transfer: (II) Binding-site and quantum-size effects, *J. Raman Spectrosc.*, **40**, 1172–1177.
61. Sun, M., Li, Z., Liu, Y., and Xu, H. (2009) Direct visual evidence for chemical mechanisms of SERRS via charge transfer in Au_{20}-pyrazine-Au_{20} junction, *J. Raman Spectrosc.*, **40**, 1942–1948.
62. Jensen, L., Zhao, L. L., and Schatz, G. C. (2007) Size-dependence of the enhanced Raman scattering of pyridine adsorbed on Ag_n $(n = 2 - 8, 20)$ clusters, *J. Phys. Chem. C*, **111**, 4756–4764.
63. Arenas, J. F., Lopez, T. I., Otero, J. C., and Marcos, J. I. (1996) Charge transfer processes in surface-enhanced Raman scattering. Franck–Condon active vibrations of pyridine, *J. Phys. Chem.*, **100**, 9254–9261.
64. Xia, L., Wang, H., Wang, J., Gong, K., Jia, Y., Zhang, H., and Sun, M. (2008) Microwave-assisted synthesis of sensitive silver substrate for surface-enhanced Raman scattering spectroscopy, *J. Chem. Phys.*, **129**, 134703/1-7.
65. Xia, L., Jia, Y., Liu, G., Li, J., and Sun, M. (2010) Directed calcium chloride coalescence method for preparation of silver nanocubes, *Appl. Spectrosc.*, **64**, 867–870.
66. Xia, L., Hu, X., Kang, X., Zhao, H., Sun, M., and Cihen, X. (2010) A one-step facile synthesis of Ag-Ni core-shell nanoparticles in water-in-oil microemulsions, *Colloids Surf., A,*, **367**, 96–101.
67. Xia, L., Jia, Y., Liu, G., Yang, Z., Chen, X., Zhang, H., and Sun, M. (2010) Adjustment and control of SERS activity of metal substrates by pressure, *J. Raman Spectrosc.*, **41**, 398–405.
68. Comini, E., Baratto, C., Faglia, G., Ferroni, M., Vomiero, A., and Sberveglieri, G. (2009) Quasi-one dimensional metal oxide semiconductors: Preparation, characterization and application as chemical sensors, *Prog. Mater. Sci.*, **54**, 1–67.
69. Kuchibhatla, S. V. N. T., Karakoti, A. S., Bera, D., and Seal, S. (2007) One dimensional nanostructured materials, *Prog. Mater Sci.*, **52**, 699–913.

70. Dong, B., Li, Z. P., Cao, B., Yu, N. S., Sun, M. T. (2011) Quasi-one dimensional $Er^{3+}Yb^{3+}$ codoped single-crystal MoO_3 ribbons: Synthesis, characterization and up-conversion luminescence, *Opt. Commun.*, **284**, 2528–2531.

71. Zhou, J., Xu, N.-S., Deng, S.-Z., Chen, J., She, J.-C., and Wang, Z.-L. (2003) Large-area nanowire arrays of molybdenum and molybdenum oxides: Synthesis and field emission properties, *Adv. Mater.*, **15**, 1835–1840.

72. Gurlo, A., Barsan, N., Ivanovskaya, M., Weimar, U., and Gopel, W. (1998), In_2O_3 and MoO_3-In_2O_3 thin film semiconductor sensors: Interaction with NO_2 and O_3, *Sensor Actuat. B-Chem.*, **47**, 92–99.

73. Dong, B., Huang, Y., Yu, N., Fang, Y., Cao, B., Li, Y., Xu, H., and Sun, M. (2010) Local and remote charge-transfer-enhanced Raman scattering on one-dimensional transition-metal oxides, *Chem.-Asian J.*, **5**, 1824–1829.

74. Dong, B., Zhang, W., Li, Z., and Sun, M. (2011) Remote excitation surface plasmon and consequent enhancement of surface-enhanced Raman scattering using evanescent wave propagating in quasi-one-dimensional MoO_3 ribbon dielectric waveguide, *Plasmonics*, **6**, 189–193.

Chapter 9

Parabolic Mirror–Assisted Gap-Mode Optical Ultramicroscopy

D. Zhang and A. J. Meixner

Institute of Physical and Theoretical Chemistry, University of Tübingen, 72076 Tübingen, Germany

dai.zhang@uni-tuebingen.de

In this chapter we give an introduction to scanning near-field gap-mode microscopy. Our goal is to develop an intuitive picture of this young and fascinating microscopy technique; as well as to discuss the simultaneously obtained topography and optical signal that allow us to extract spectroscopic information with nanometer spatial resolution and sensitivity down to the single-molecule level.

9.1 Introduction

Nowadays various types of ultra-high resolution microscopes are available for every day laboratory use, which offer a resolution on the order of a few nanometers. The most prominent ones probably are the scanning electron microscope (SEM) and the transmission

Handbook of Molecular Plasmonics

Edited by Fabio Della Sala and Stefania D'Agostino

ISBN 978-981-4303-20-0 (Hardcover), 978-981-4303-21-7 (eBook)

www.panstanford.com

electron microscope (TEM). In these microscopes a beam of focused high voltage electrons is raster-scanned over the sample surface and the scattered or transmitted electrons are detected as a function of the beam position. With the invention of the scanning tunneling microscope (STM) and the atomic force microscope (AFM), researchers are able to look into the nanoscale phenomena either via the tunneling current or by the weak force interaction. Each of these techniques has its areas and types of samples for which it is best-suited considering the material property, the conductivity, surface roughness, etc. With the increasing need of characterizing extreme small structures with fine chemical details in nanosciences, microscopy techniques that provide chemical information with nanometer spatial resolution become increasingly eye-catching. Near-field optical microscopy, which is realized by illuminating the tip-sample gap with a tightly focused laser beam and collecting the light emitted from the optical near-field in the nanometer-sized gap, starts to attract more and more attention.

In this chapter, we present a recently developed method for near-field microscopy, using a parabolic mirror (PM) for optical excitation and detection, based on our earlier design of a PM confocal microscope [1]. This microscope allows the use of higher-order laser modes for an efficient excitation of the tip, while enabling studies of non-transparent samples such as metals and semiconductor surfaces, thus holding high potential for large-scale applications in near-field spectroscopic investigations of semiconducting devices. Without the near-field module, the microscope can also be reliably used for confocal studies, delivering high-quality images from opaque samples with high resolution due to its high *numerical aperture* (NA).

9.2 Principles

When the nanometer-sized gap between a metal nanoparticle and a smooth metal surface is resonantly illuminated with light polarized along the junction, a dramatic field enhancement can be achieved. Localized, coupled electron-density oscillations between the metallic nanoparticle and its mirror-image in the metallic substrate are then efficiently excited, and the effect of field-enhancement is similar to the one found in the junction of *nanoantennas* [2, 3] or in the nanometer-sized gap between two colloidal nanoparticles [4]. The system has a large scattering cross section and works as an antenna that transfers the localized mode in the junction into a traveling electromagnetic wave similar to a dipole wave. It can easily be detected in the far-field. This concept is similar to the radio broadcasting antenna systems discussed by Sommerfeld a century ago. Such a configuration serves to collect energy from an electromagnetic wave, to deliver it to an oscillator circuit, and to emit an electromagnetic wave with a signal form that an electronic circuit modulated onto it for broadcasting. Transferring this classic idea to optical microscopy, the central part of the microscope is a laser-illuminated nanometer-wide gap formed between the apex of a sharp metallic tip and the sample surface. The metal tip serves to focus the electric field of the incident radiation into the sub-wavelength-sized gap volume via the coherent light induced oscillation of the conducting electrons.

In Fig. 9.1 we sketch the fundamental configuration of the optical antenna used in near-field optical microscopy, i.e. a sharp noble-metal tip approaching the sample surface down to a few

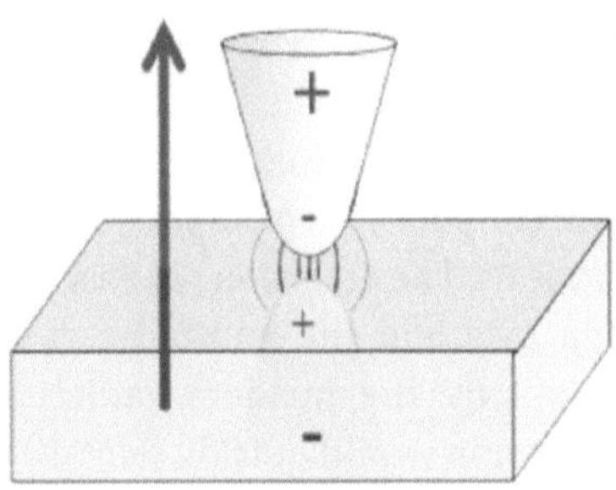

Figure 9.1 Sketch of an optical antenna formed of a metal tip in close proximity to a metallic sample surface. The gap width is on the order of some nanometers, i.e. much smaller than the optical wavelength. The tip is polarized by the electric field of the incident electromagnetic radiation, leading to a surface charge at the tip apex and its mirror on the sample surface. The surface charges establish a strongly enhanced electric field in the gap.

nanometers illuminated by incident laser radiation with an electric field polarized along the tip-axis. Towards the end of the conical tip, its lateral diameter becomes smaller than the penetration depth of the field into the metal and the electric field component of incident radiation induces a coherent oscillation of the conduction electrons, the so-called plasmon oscillation. The energy quanta of the charge oscillation are called plasmons in analogy to the photon as energy quanta of an oscillating electromagnetic field or wave. Since the plasmon oscillation is always coupled with an optical field oscillation it is termed plasmon–polariton. If the incident electric field is oriented along the tip axis the freely moving charge in the metal is periodically driven up and down at the tip shaft bouncing against the tip apex. It leads there to a maximum oscillating surface charge, in turn a greatly enhanced local electric field. The response of the sample to the tip-enhanced field can be easily explained with the concept of mirror charge that induces local polarization acting back to the tip. Hence the tip and the sample form a coupled oscillator system influencing each other via the electric field in the gap as sketched in Fig. 9.1. In the following we will refer to the field distribution formed by the gap as the *gap-mode* in analogy to the field enclosed in a resonator; however the gap width has a dimension much smaller than $\lambda/2$.

The oscillating surface charges of the tip-sample system must again emit radiation with most of it at the incident laser frequency, i.e. elastic- or Rayleigh-scattering. A small fraction is frequency shifted by electron-photon coupling leading to Raman scattering and photoluminescence (PL) emitted from the gap by the radiative decay of electronic excitations. The emission pattern has a radial symmetry about the antenna axis and the emitted power is constantly fed back to the antenna by the incident radiation, else the charge oscillation of the antenna would quickly faint away. From the optical reciprocity theorem it is clear that the antenna is most efficiently excited by an incident wave that is the time reversal of the emitted wave. Such excitation conditions can conveniently be realized by focusing a radially polarized laser-beam by an objective lens with high NA or by a PM, an important issue we will come back to in the instrumentation section.

9.2.1 *Instrumentation*

9.2.1.1 Optics layout

The combination of spectroscopic (Raman fingerprint and PL) imaging and scanning probe microscopy both with high spatial resolutions on the nanometer scale is one of the topmost goals that near-field gap-mode optical microscopy has targeted ever since its invention [5]. A satisfactory combination of both techniques is not trivial. Besides the stringent instrumentation requirements regarding the stability, the accuracy and the reproducibility, at least two more technical problems have to be considered.

The first problem concerns the capability of handling non-transparent sample substrates. Near-field gap-mode optical microscopy based on an *inverted confocal microscope* which is the most commonly adopted configuration nowadays, has been restricted to transparent sample substrates since its invention [6, 7]. The intrinsic reason is due to the position of the objective lens used for laser focusing and optical signal collection that is placed below the sample. Transparent substrates, such as mica [8] and glass [9] or ultra-thin metal films with a thickness allowing enough laser power of a certain frequency to penetrate through, are strictly required. Side-illumination proved to be an effective solution regarding this problem [10]. However, the far-field background from the large projection of the far-field focus and the limited signal collection angle have to be taken in account, especially during near-field optical imaging.

The second problem is how to selectively image the optical signals of interest. A near-field-enhanced Raman spectrum normally exhibits well-defined Raman peaks superimposed on a broad and strong background. This background mainly consists of the PL from the excited tip antenna, the sample, and the substrate. Since both the Raman and PL information provide characteristic insights into the properties of the tested species, it is helpful to extract these two types of information from one spectrum. Recently, there were several reports about tip-enhanced Raman and PL imaging based on band pass filters to select photons within a certain wavelength range [11, 12]. The usual FWHM of the commercially available band pass filters are at least 5 nm, which is much broader than the average FWHM of a Raman peak. Hence, a band pass filter is liable of providing false Raman imaging pictures especially for systems consisting of multi-Raman active components that exhibit similar Raman lines. In addition, intrinsically band pass filtering allows only for wavelength selectivity while being incapable of separating, e.g. the inelastically scattered photons of a Raman process from the emitted PL photons. For true Raman imaging, this method is vulnerable to the artifacts induced by the sample or substrate.

To tackle the above problems, we converted the conventional design concept of a *tip-enhanced optical microscope* in developing a new type of optical layout using a PM as the laser focusing and signal collection element [13, 14]. We fix the position of the tip apex in the laser focus, while scanning the sample *versus* the tip apex by using a piezo-controlled scanning table for spectroscopic imaging. The tip-enhanced optical images are recorded by collecting the whole spectrum consisting of sharp Raman peaks and a dispersive PL background from each data point. Spectroscopic imaging pictures can be plotted either with the intensity integration of a certain Raman peak or the PL from a certain wavelength range via spectral fitting [14].

Figure 9.2 shows an illustration of the optics layout. The optical setup follows the basic design of a confocal microscope, whereby

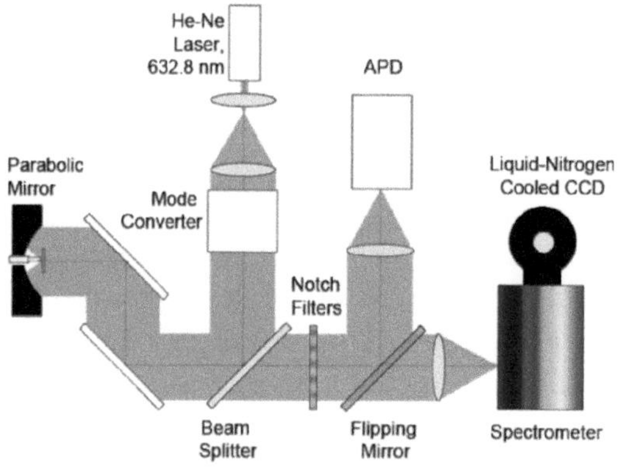

Figure 9.2 Optics layout of a PM-assisted near-field optical microscope. Adapted from Ref. [13]. Copyright (2008) the Royal Microscopical Society.

the microscope objective is replaced by a PM. As excitation source we use a single-mode 4 mW He-Ne laser operated at 632.8 nm. High-order laser modes are generated using a polarization converter containing four quarters of differently-oriented half-wave plates, as described in [15]. Higher spatial frequencies are removed by a spatial filter. The collimated beam, expanded to match the 2 cm aperture of our PM, is then reflected by a 50:50 nonpolarizing laser line beam-splitter and directed onto the PM. Special care must be taken for the precise alignment of the beam with the optical axis of the PM, since the slightest misalignment leads to severe coma aberrations in the field distribution around the focal region [16]. The sample is then scanned through the focus of the PM. The maximum excitation power focused onto the sample in our experiments is 250 μW. The mirror, with a NA close to unity, serves as the focusing as well as the collecting element for the scattered/back-reflected light. After passing through two spectral notch-filters to remove the laser radiation, the inelastically scattered light is either focused onto an avalanche photodiode for image recording or redirected towards a spectrometer equipped with a liquid-nitrogen cooled CCD camera for spectroscopic investigations.

9.2.2 *PM Optics*

PMs with small NA are very common in astronomic telescope design and in telecommunication antennas. High NA PMs have been used as efficient light collectors in single molecule spectroscopy [17–21] however, for imaging they have been avoided due to their poor off-axis imaging properties. The classical imaging devices for microscopy and imaging are objective lenses, because of their wide field of view and excellent resolution resulting from the high degree of correction and the high achievable NA. A PM focuses a parallel beam, traveling along the optical axis, to one single point without any aberrations in the geometrical optics approximation and in this sense it is an ideal focusing device. The drawback is that slight deviations of the incident beam from the optical axis or from parallelism would lead to a deteriorated focus.

Confocal microscopy is a point to point imaging technique and hence has to correct aberrations only for one single object point.

This purpose is perfectly achieved by PM in a stage scanning microscope where the focus rests on the optical axis. In addition it is fairly easy for a PM to achieve a high aperture without chromatic aberrations since the deflecting surface can be made highly reflective over a large spectral range [13, 16].

A PM configuration used for a near-field optical microscope is illustrated in Fig. 9.3a. The PM is placed above the sample stage and a Au tip is approached through an axial hole into the focus of the PM. The mirror is illuminated symmetrically with a radially polarized laser beam creating via reflection a diffraction-limited focal spot. Several advantages are obtained with this configuration:

(1) The PM focuses the laser beam and collects optical signals from above the sample. Hence the bottleneck problem of the opaque substrates is no longer an issue of concern.
(2) PM focusing is based on reflection and is free from chromatic aberration; it can therefore be easily adapted to work with different laser wavelengths.

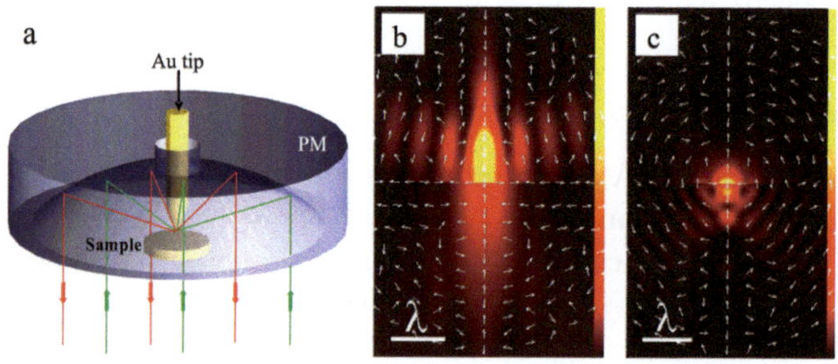

Figure 9.3 (a) Sketch of a PM situated above a sample and the tip antenna approached through an axial hole into the focus of a radially polarized laser beam. Green arrows indicate the incoming laser beam and the red arrows indicate the outgoing laser beam and optical signals (Raman and PL). Snap-shots of the electric field intensity distribution created by a radially polarized laser beam in the focal volume of (b) a high NA (0.99) PM incident from above and (c) a high NA (1.4) objective lens at an air–glass interface incident from below. The white arrows indicate the polarization directions at every individual point. λ in this case is corresponding to 632.8 nm. The colour-code in (b) and (c) is normalized in each individual case. Adapted from Ref. [14]. Copyright (2009) John Wiley & Sons, Inc.

(3) A NA of 0.99 can be achieved by a PM due to its geometric properties and its focusing principle. Benefiting from this large NA, an efficient and symmetric signal collection over a solid angle of π above the sample surface can be easily realized, which is challenging for an air objective lens. Only a minor fraction of the signal is lost due to the small axial hole opened at the top of the PM and the shadowing by the tip shaft.

(4) As shown in Fig. 9.3b, when illuminated by a radially polarized laser beam, the large NA creates in the diffraction-limited focal volume of an intensity maximum in which the field is almost completely polarized along the longitudinal direction (14.3 times stronger than that of the transverse direction) with a decay constant on the order of λ (see Fig. 9.3b). In contrast, when the laser illuminates in transmission configuration through the objective lens (see Fig. 9.3c), the converging waves form an intensity maximum at the glass–air interface with the longitudinally polarized field component (3.5 times stronger than that of the transverse component) decaying within a small fraction of $\lambda/2$ into the air filled half space. Therefore, excitation of the longitudinal plasmonic oscillation of the tip antenna is much more efficient when the tip is positioned in the focus of a radially polarized laser beam in a PM [14].

(5) The focusing capabilities of a PM go beyond the possibilities of the more common aplanatic lenses. To determine the focus spot size, we used a Si substrate covered with a thin PVA-film with a low concentration of Nile Blue molecules embedded with random orientation and gold cones (100 nm in height and 120 nm in diameter at the base) fabricated on a Si wafer. A single, fixed absorption dipole in both systems provides an ideal system for probing the local field distribution. Due to the dominant longitudinal field distribution in the focus of the radially polarized mode, molecules with the transition dipole moment oriented perpendicular to the surface are most efficiently excited. An example of a fluorescence image from such a single molecule is shown in Fig. 9.4a. The dark stripe in the scan, interrupting the profile, comes from the characteristic blinking behavior typical for single molecules. Figure 9.4b shows a scan line through the center of the pattern (circles),

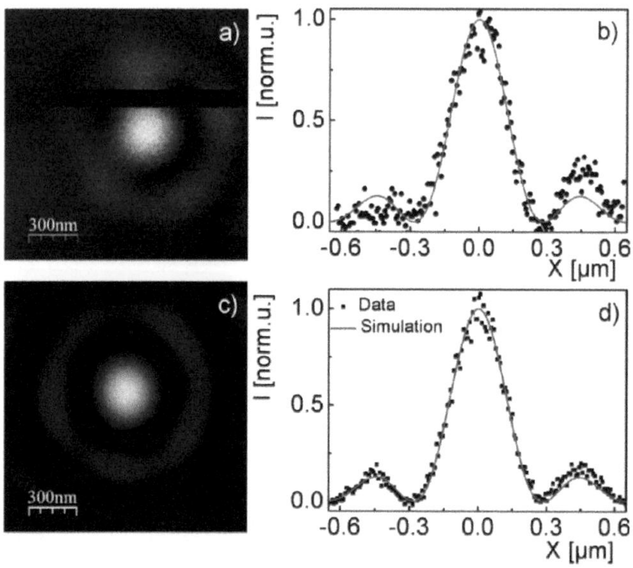

Figure 9.4 (a) Confocal image of a single Nile Blue molecule with a transition dipole moment oriented almost perpendicular to the sample surface scanned through the focus of a radially polarized laser beam in a PM. (b) Line section through the center of the focus, compared with the calculated $\mathbf{E}_z^2$ distribution. (c) PL pattern of a gold cone scanned through a radial mode. (d) Comparison between the experimental PL signal and the calculated $|\mathbf{E}_z|^2$ distribution created by a radially polarized laser beam in the focal volume of (b) a high NA (0.99) PM and (c) a high NA (1.4) objective lens at an air–glass interface. In this case λ corresponds to 632.8 nm. Adapted from Ref. [23]. Copyright (2008) Optical Society of America.

compared to the calculated field strength $\left|\left|\tilde{\mathbf{E}}\right|\right|^2$. The calculated profile fits the experimental data very well, in particular over the central maximum. The focal spot of a radial beam can be confined to an area of 0.134 λ^2 with a radius of the Airy disk of r = 0.45 λ as compared to r = 0.61 λ as obtained from the scalar diffraction theory for NA~1 [22]. This is the smallest diffraction-limited focal spot reported for a given wavelength under ambient conditions so far [23]. Figure 9.4c and d are the results obtained from one Au nanocone, which demonstrate the same results.

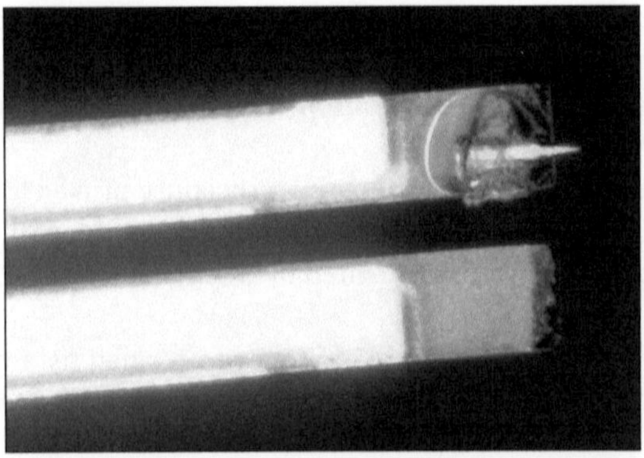

Figure 9.5 Quartz tuning fork with a gold-tip glued at one of the prongs used shear-force for tip-sample distance control. Typical vibration amplitudes are on the order of 0.1 nm. Adapted from PhD-Thesis of Marcus Sackrow, University of Tübingen.

9.2.3 *Tip-Sample Distance Control and Image Recording*

Commonly used methods to control the nanometer distance between the tip-apex and the sample surface are based on the *shear-force feedback* [24] and the electron tunneling [25] for conducting samples.

For shear-force distance control the tip is placed on one prong of a quartz tuning fork (see Fig. 9.5) which is excited mechanically to vibrate near its resonance frequency. The accumulated charge at the electrodes is recorded and analyzed via a phase locked loop (PLL) to measure the amplitude and the phase shift of the oscillation. The tuning fork is placed onto a piezoelectric tube to provide fine adjustment in all three axes and to excite the vibration. The tip can be brought to the focus and to the sample surface through an axial hole in the PM. To move the tip near the focus it is possible to look from the side into the PM to see the scattering at the tip; for fine adjustment the back reflection into the microscope is used. When the tip is brought as close as a few nanometres to the sample surface, the interaction between the tip and the sample/substrate results in a shift of the resonance frequency and/or a damping of the amplitude. The relation of this two parts strongly depends on the tip and sample materials [26]. Often the phase is used to control the distance to the sample via a feedback loop. Similarly the tunneling current can be used for tip-sample distance control. The feedback loop as well as the scanning and data acquisition is usually done by a commercial AFM/STM controller.

For producing tips the following procedure proved to be reliable. The tips are produced by electrochemical etching the end of a 125 μm gold wire in concentrated hydrochloric acid (HCl, 37%), see Fig. 9.6.

As counter electrode a platinum wire formed as a ring is used to generate a circular constant field on the gold wire. A frequency

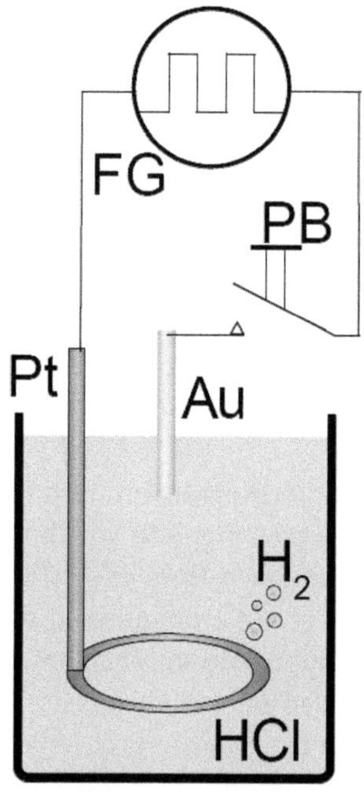

Figure 9.6 Sketch of the setup for tip etching. A frequency generator (FG) produces voltage pulses which are brought to the gold wire (Au) and the platinum counter electrode. Both are placed in hydrochloric acid (HCl), during the etching process hydrogen gas is produced. The tip forms at the air–liquid interface. Etching stops itself when the part of the Au wire dipping into the HCl solution is used up and the contact between the tip and the interface breaks. Voltage pulses are used to reduce shaking of the solution surface by strong bubbling push button (PB). Adapted from PhD-Thesis of Marcus Sackrow, University of Tübingen, Germany.

generator provides rectangular pulses with 8 V and a duration of 30 μs at a repetition frequency of 3 kHz which can be brought to the gold wire by a push button. With a wide-field microscope the tip can be observed from the side to stop the etching process. The tip is cleaned with distilled water and characterized with a SEM before using.

9.3 Different Types of Gap-Modes

9.3.1 *Gap-Mode of Metallic System*

9.3.1.1 Au tip and Au substrate

Together the discovery of surface-enhanced Raman scattering [27–31] and the raise of nano-optics in the past several years, the plasmonic properties of Au- and Ag-nanostructures have been a hot topic since assemblies of metal nanoparticles find a variety of applications in the fields of biosensors, photonic waveguiding, optical switching, and so on. The process of metal nanosphere assembly or aggregation is known to alter the optical properties of the nanostructure due to the interparticle electromagnetic coupling (see Chapter 6), which will especially causes variations for Au nanoparticles in the PL intensity and plasmonic resonance shifts. This electromagnetic coupling strongly depends on the excitation wavelength, the particle material, size, orientation, shape and the interparticle distances [32–36], addressing this issue precisely is a demanding task. The plasmon resonance is often determined from the white light spectrum of a metallic nanoparticle. Recently it was shown that the PL emission spectrum of aggregated Au particles also closely follows the resonance curve obtained from the white light scattering spectrum regardless of the specific geometry of an individual nanoparticle [37] PL emission of gold covers a large spectral range from the visible to the infrared and hence is an important criterion to evaluate the efficiency of the plasmonic excitation. For Au-nanoparticles plasmon excited PL-emission is well explained by the interband transitions of *d*-band electrons into the conduction band and subsequent radiative recombination [38] while for the infrared emission intraband transitions mediated by the strongly confined fields near metal nanostructures have been suggested [39]. Since the creation of an electron–hole pair is proportional to the square of the local optical field, strong emission of PL photons by their recombination points at locations with high field strength. In order to relate the plasmon resonance to the PL emission with a precisely tuneable geometrical arrangement under well-defined polarization conditions, we have investigated the plasmon excitation in smooth Au thin film and Au nanocones as well as plasmon coupling between a sharp Au nanocone and a sharp Au tip as a function of their tip-to-tip position [14].

Flat Au-surfaces are often used as ideal substrates for tip-enhanced Raman scattering (TERS) [10, 40, 41] and often a broad and substantial PL background is observed along with the Raman spectrum of the adsorbed molecules particularly when the tip-sample configuration is excited with 632.8 nm radiation. This background shows a red-shift when the Au-tip approaches to the surface which is interpreted as an increased electromagnetic coupling of the tip- and localized-sample surface plasmon [42]. Figure 9.7a shows a near-field PL image recorded with a sharp Au tip in the focus of a PM from a clean 100 nm gold film under 632.8 nm illumination [43]. Figure 9.7b represents the simultaneously recorded topography, while Fig. 9.7c depicts the respective error signal of the feed-back loop that keeps the gap width constant. In the tip-enhanced PL image we observe bright spots with a count rate about twice as high than the average signal on the sample (FWHM: 12 nm). The optical features correlate with small topographic protrusions with a height of about 1 nm, as seen in the line cuts presented in Figs. 9.7d and e. The lateral extension of these protrusions has a similar FWHM as in the optical image, indicating that in both topographic and optical measurements the lateral resolution is only limited by the end radius of the tip. The final shape of the features is the result of the convolution between the tip and the actual structure on the surface. By comparing measurements taken with different tips having known end-radii, we concluded that the real lateral extension of the protrusions is limited to dimensions comparable to their height, in the range of 1–2 nm [43].

Figure 9.8 shows optical images of a Au nanocone recorded with our PM-assisted optical microscope in confocal configuration using

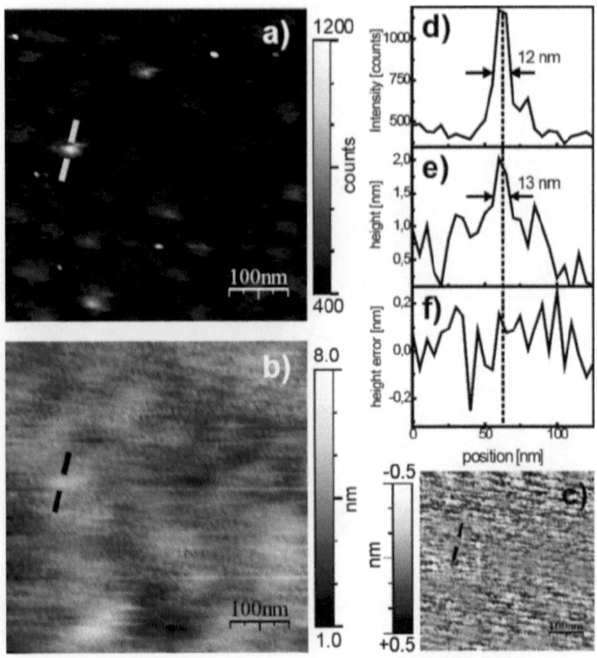

Figure 9.7 High-resolution image of the gold PL excited in the ~2 nm gap between a gold-tip and a flat gold surface (a) on a planar 100 nm thick gold film deposited on silicon. The topography (b) and error-signal (c) images were recorded simultaneously with the near-field optical image. (d), (e) and (f) are cross-sections taken trough the bright spot in the near-field optical image, the topographic image, and image of the error signal marked by dashed lines in the corresponding figures. Adapted from Ref. [43]. Copyright (2008) Wiley-VCH Verlag GmbH & Co, KGaA, Weinheim.

radially (a) and azimuthally (b) polarized higher order laser modes. As described in section 2, in the focal volume of a radially polarized laser beam a dominantly stronger electric field distribution exists in the longitudinal direction than that of the transverse direction; while for an azimuthally polarized laser beam the electric field in the focal volume is exclusively distributed in the transverse direction [16]. Such a distinctive difference in the electric field distribution is helpful for determining the orientation of the principal dipole moment oscillation in a plasmonic nanostructure. The nanocone is more efficiently excited by the longitudinal electric field than

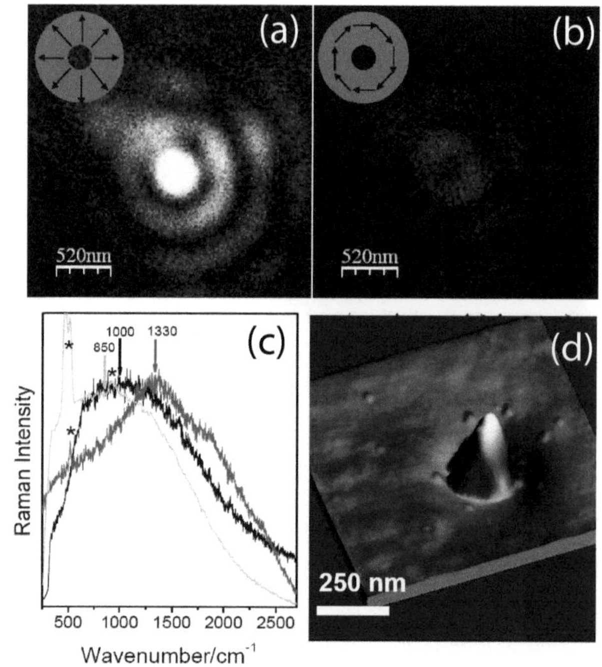

Figure 9.8 PL scans of a Au cone excited with (a) radially and (b) azimuthally polarized laser mode. Insets of figures (a, upper left) and (b, upper right) illustrate the polarization directions of the respective laser modes. (c, lower left) Plasmon-enhanced spectra (normalized) collected from one Au cone (light gray), the apex of the Au tip (black) and the small gap between the Au tip apex and the top of the cone (dark gray). The peaks labeled with * come from the Si substrate. (d, lower right) A superposition of the 3D topography of the Au cone color-coded with the PL intensity obtained when the cone is scanned beneath the tip. The bright color indicates a strong optical signal. Adapted from Ref. [14]. Copyright (2009) John Wiley & Sons, Inc.

the transverse electric field resulting in the 10 times stronger PL emission signals observed in Fig. 9.8a as compared to the Fig. 9.8b. This indicates that the principal dipole moment is orientated vertically *versus* the substrate, i.e. parallel to the cone axis [14].

By precisely positioning the tip above a nanocone in the well-polarized field we can use tip-enhanced PL imaging as a sensitive probe to study plasmonic dipole–dipole coupling as a function of the tip-to-tip position. Figure 9.8c shows three plasmon-enhanced spectra obtained from a Au nanocone, from the tip apex and from a configuration where the tip apex is positioned in close proximity above of the nanocone. Both the nanocone and the tip apex exhibit strong PL emission with the intensity maxima (obtained by spectral fitting) at 850 cm^{-1} (1.854 eV) and 1000 cm^{-1} (1.835 eV) red-shifted from the excitation wavelength (632.8 nm). When the tip is positioned right above the nanocone, a sharper peak with a small shoulder appears in the spectrum exhibiting an intensity maximum at 1330 cm^{-1} (1.794 eV), which is 0.060 eV red-shifted *versus* that of the nanocone and 0.041 eV red-shifted *versus* that of the Au tip alone [14]. While the red-shifted PL emission energy reflects the coupling between the plasmonic dipole of the tip apex and the plasmonic dipole of the Au cone [44], the values of the red-shifted PL emission energy can be used for the evaluation of the dipole coupling strength.

Figure 9.8d shows a three dimensional topographic image recorded by scanning a nanocone below the tip at constant distance of several nanometers with the PL signal intensity as a colour code. It can clearly be seen that the strongest PL appears when the tip is positioned right above the top of the cone. Together with the red-shifted PL emission observed in Fig. 9.8c, we conclude that the strongest dipole–dipole coupling can be achieved by aligning the plasmonic dipole moment of the tip apex right on top of the dipole moment of a Au cone. The much weaker optical signal observed when the tip scans above the Si substrate shows that the PL from the Si can be neglected. In order to estimate the field enhancement in the gap between the tip and the cone, we compare the maximum near-field PL intensity $I_{NF} \approx 450$ kcounts/s from the distance curves with the PL intensity $I_{\mathrm{conf}} \approx 3$ kcounts/s measured in confocal far-field mode from a flat Au film, normalized to the effective area emitting the PL signal [3]. The effective area in the confocal far-field mode corresponds to the focus of the PM, which was determined experimentally as $0.134\lambda^2$. The effective area in the near-field gap-mode configuration is essentially determined by the tip size, which is around 30 nm in diameter. The tip-cone gap size is held as 2 nm. Based on the above parameters, the local electric field enhancement of 105 is obtained, which corresponds to an $||\tilde{\mathbf{E}}||^4$ enhancement of ca. 1.2×10^8 [44].

9.3.1.2 Au tip and monolayer adsorbates/Au substrate

If we now consider a layer of molecules adsorbed on the gold surface, one should obtain a better optical contrast over the topographic islands, due to the forth power dependence of the Raman signal of the molecules on the local field strength. In order to verify this expectation, we prepared a layer of *benzotriazole* (BTA) molecules on gold films by immersing the samples in a 5% solution of BTA in methanol. Molecules that were not chemisorbed on the metal were carefully rinsed off with methanol. BTA is known to adsorb on metals via a partial chemical bond of the para-nitrogen, forming closed monolayer films [45, 46]. Raman spectra of these molecules are well known in the literature [45, 47]. An advantage in using BTA as field-strength sensors is that the molecules are not resonant at the excitation wavelength of 632.8 nm, so that experimental complications due to bleaching effects can be avoided.

An optical near-field image of such a sample, together with the respective topographic and feedback error images, is presented in Fig. 9.9.

The overall contrast in Fig. 9.9a is higher than in the PL near-field image from Fig. 9.7a, with the intensity increased up to a factor of 4 over the most intense protrusions. Besides these sharp optical signals over the protrusions, illustrated in the line cuts d–f, we find additional features in the near-field Raman image of BTA on gold. These features are circular patterns with local intensity minima at the core correlating with the centre of the protrusion (line cuts d and e). The FWHM of the optical signal is ~12 nm. The intensity of the minimum in the centre is still higher than the background signal, while the diameter of the circular optical features scales with the diameter of the islands in the topographical image. We conclude that the molecules accumulate at the steps of the islands, and the optical signal is in this case modulated due to the increased number of scatterers in the gap [43].

It is rather interesting to note that very small protrusions give a significantly stronger optical near-field signal, both for PL and Raman scattering, than higher (~5 nm) and much broader topographic features for which we do not find their near-field optical correspondent. The increase of the tip-enhanced optical

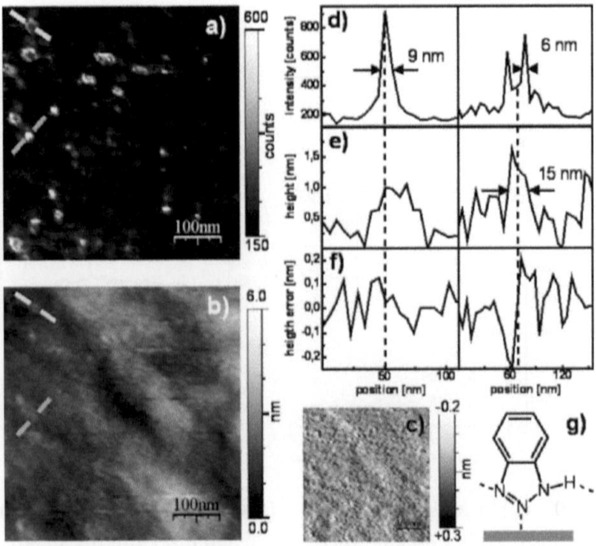

Figure 9.9 (a) Near-field optical image of a smooth Au film covered with a layer of BTA molecules. Simultaneously recorded topography (b) and error-signal (c) images. The graphs on the right side show line cuts over two small gold islands marked with gray (d)-(f) and white (g)-(i) lines in the respective images. (g) BTA molecule with a partial bond to the gold surface. Adapted from Ref. [43]. Copyright (2008) Wiley-VCH Verlag GmbH & Co, KGaA, Weinheim.

signal can not be explained as a topographic artifact, since the height variation over the protrusions, visible in the error signal (see Fig. 9.9c), is not higher than the overall noise of the respective scans. From a careful spectroscopic analysis of these surfaces, we must also exclude impurities, such as carbon nanoparticles or other luminescent features. However, it is well known that the peak of the PL spectrum closely follows the plasmon resonance of the nanostructure [36, 48, 49]. Indeed, typical near-field PL spectra (see Fig. 9.10a) taken with the tip positioned over a bright spot (dash-dotted line) show a slight red-shift over the protrusion as compared to a smooth region with weaker optical signal (dashed line), indicating plasmonic coupling of the protrusion to the tip, while no Raman lines or other luminescent peaks can be observed. Furthermore, we observed this strong enhancement over small

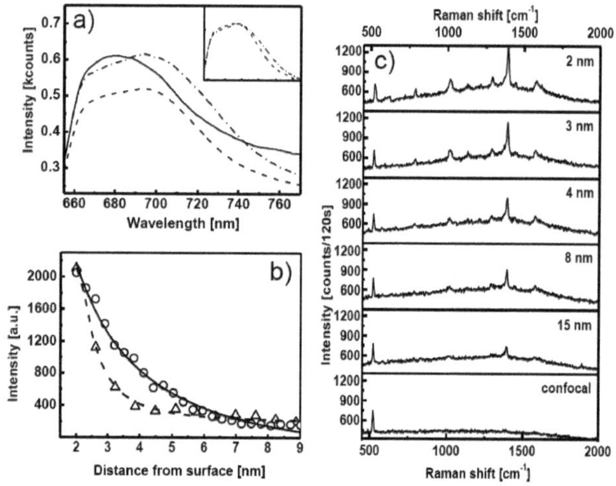

Figure 9.10 (a) PL spectrum of the tip (full line), the PL spectrum measured with the tip over a flat area of the film (dashed line) and the spectrum of the tip over a small protrusion (dash-dotted line). The respective normalized spectra are shown in the inset. The gap width was in both cases held at ~2 nm. (b) Approach curves of the gold tip over a pure gold film showing enhanced gold luminescence (circles) and over a BTA-covered gold film (triangles) cumulating Raman signal plus luminescence background as a function of the tip-sample distance. (c) Respective distance dependent Raman spectra of BTA on a flat gold film. No Raman lines can be observed from the same sample area in confocal mode. Adapted from Ref. [43]. Copyright (2008) Wiley-VCH Verlag GmbH & Co, KGaA, Weinheim.

protrusions for different combinations of film thicknesses and tip radii. We therefore conclude that small gold islands modulate the plasmonic coupling and the optical field in the gap between the apex of the tip and the otherwise planar gold surface.

The longitudinal electric field component in the focal volume induces a localized plasmon in the front end of the tip, which couples to its mirror image in the gold film. This is consistent with the observation that the PL spectrum of the tip-surface system is red-shifted by ~20 nm (dashed line in Fig. 9.10a as compared to the spectrum of the free tip (full line in Fig. 9.10a), similar to a dipole coupling to its mirror image in a gold surface [50]. Since the gap width, *d*, between the tip and surface is much smaller than the incident wavelength, we can neglect retardation and describe the

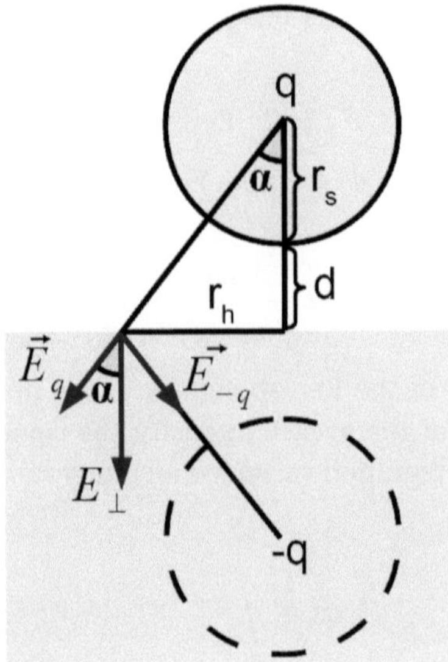

Figure 9.11 A sketch illustrates a charged sphere and its image charge inside a Au substrate. Adapted from the Ph.D thesis of Marcus Sakrow, University of Tübingen.

local electric field in the gap by the superposition of the external field and the field defined by the oscillating surface charge on the tip apex, associated with the surface plasmon and the respective mirror image in the sample surface (see Fig. 9.11). To understand how the tip-sample distance influences the gap field, we introduced below a simple analytical model describing the coupling of a localized plasmon at the end of the tip and its mirror image in the gold surface by a point charge q and its mirror charge $-q$ (see Fig. 9.11).

A tip with radius r_s is situated at the distance d above a plain gold surface. The tip is regarded as a charged gold sphere. The electric field $\mathbf{E}_q$ is given by:

$$\mathbf{E}_q = \frac{q}{r^2}\hat{\mathbf{r}}, \tag{9.1}$$

where r is the distance measured from the point where the tip-axis intersects the sample surface.

The perpendicular part of the electric field (the main contribution to the gap-mode) which is formed by the interaction between the charged sphere and its image charge inside the metal is given by:

$$E_{\perp} = 2\left|\left|\mathbf{E}_q\right|\right| \cos(\alpha)\,, \tag{9.2}$$

where $\cos(\alpha) = (r_s + d)/r$. From Eq. (9.1) and Eq. (9.2), we get the following equation:

$$E_{\perp} = 2\frac{q\,(r_s + d)}{\left((r_s + d)^2 + r_h^2\right)^{3/2}}\,. \tag{9.3}$$

The intensity of the Raman signals in the near-field is proportional to $||\mathbf{E}||^4$. For the overall intensity, the signal can be obtained using r_h as the integration variable and integrate over circular areas around the tip:

$$I_{\mathrm{TERS}} \propto \int_0^{\infty} E_{\perp}^4 2\pi r_h \mathrm{d}r_h\,. \tag{9.4}$$

Considering Eq. (9.3), we derived the following relationship between the near-field-enhanced Raman intensity and the tip-sample distances:

$$I_{\mathrm{TERS}} \propto \frac{16\pi q^4}{5\,(r_s + d)^6}\,. \tag{9.5}$$

Assuming that the Au-PL of a surface element of the sample is proportional to $|E_{local}\,(d)|^2$, the relative near-field-excited PL intensity $I_{\mathrm{PL}}\,(d)$, emerging from the gap region for a flat sample, is obtained by integrating over the whole focal area:

$$I_{\mathrm{PL}} \propto \int_0^{\infty} E_{\perp}^2 2\pi r_h \mathrm{d}r_h = b\frac{2\pi q^2}{(r_s + d)^2} + c\,, \tag{9.6}$$

where b and c are fit parameters for the relative intensity and the background. Similarly, for the BTA-covered sample the gap width dependent near-field optical signal must take into account both the enhanced Raman signal, being proportional to $|E_{local}\,(r, d)|^4$ and the luminescence background $|E_{local}\,(r, d)|^2$. Integrating over the focal area yields:

$$I_{Raman}\,(d) = a\frac{16\pi q^4}{5\,(r_s + d)^6} + b\frac{2\pi q^2}{(r_s + d)^2} + c\,, \tag{9.7}$$

the first term describing the Raman-contribution, the second term describing the background.

As shown in Figs. 9.10b and c, Eq. (9.6) and Eq. (9.7) can be fitted quite well to the spectrally integrated near-field optical approach. The approach curves are consistent with the gap field–enhanced spectra recorded from a BTA covered sample region, which show a considerably sharper gap width–dependent enhancement for the Raman lines than for the luminescence background. The line at $\sim$520 cm^{-1} originates from the Si substrate below the Au film and experiences no near-field enhancement; its constant intensity proves that the sample stays centered in the focus while the spectra were recorded. Since the protrusions are small compared to the tip radius, they have only little influence on the overall field distribution in the gap. We may therefore model the effect of the small protrusions by a half-sphere emerging from the surface, which is polarized by the field in the gap. The additional field of such a protrusion can be described by the field of a small dipole being located on the sample surface with a magnitude based on the polarizability of a gold sphere and a radius equal to the height of the protrusion. For a standard protrusion with a radius of 1 nm, this model predicts a PL enhancement of $\sim$1.4 and a Raman enhancement of $\sim$2 over the gold island for a tip radius of 10 nm and a gap width of 2 nm, which is in very good agreement with our experimental findings (see Figs. 9.9 and 9.10).

9.3.1.3 Au tip and single molecule/Au substrate

The Pettinger group has recently presented a tip-enhanced Raman spectroscopic and microscopic study on *brilliant cresyl blue* (BCB) adsorbed on gold single-crystal surfaces employing a PM-assisted gap-mode microscope operating in an ultrahigh vacuum (UHV). Because of substantially reduced photobleaching under UHV conditions, they were able not only to record TER spectra of single dye molecules, but also Raman images of a single dye molecule with a resolution of a few nm [37].

BCB is an organic dye with electronic transitions near to the resonance of the localized surface plasmon modes of the tunneling gap between the gold tip and gold sample, making it an ideal model

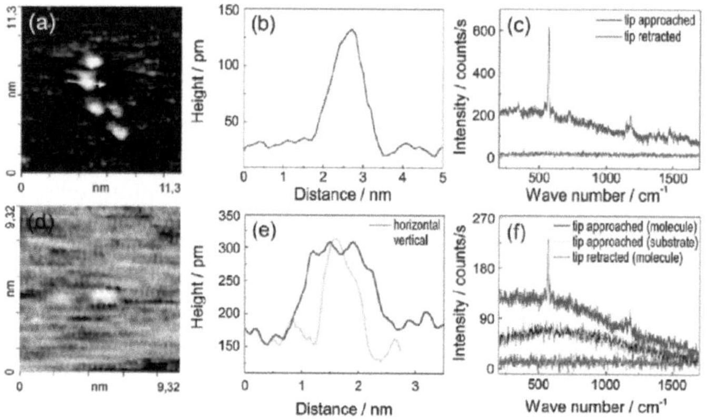

Figure 9.12 STM images of five (a) and a single BCB molecule (d) adsorbed on Au(111), the corresponding height cross-sections (b), (e) and the UHV-RR and UHV-TERR spectra (c), (f). Reprinted with permission from Ref. [50]. Copyright (2010) American Physical Society.

system. Its resonant excitation with a 633 nm wavelength laser produces a weak, but detectable resonance Raman signal (far-field), while its adsorption at a metal surface effectively quenches the dye fluorescence. In the experiments shown in Figs. 9.12a and d, regions of a gold surface covered with either five BCB molecules or only a single, well separated one, were imaged by scanning tunneling microscopy. The STM height profiles (see Fig. 9.12b and e) display BCB molecules with a lateral size of 1.4×0.7 nm^2 and an effective height of 120–140 pm. TER spectra were recorded with the molecules located exactly in the gap between gold tip and substrate Fig. 9.12c and f. For 90 molecules a Raman intensity of 7800 counts/s is observed at 570 cm^{-1}, which corresponds to an intensity of ~87 counts/s for one molecule. Thus, the signal intensities of 410 counts/s for five and 110 counts/ s for a single molecule are in good agreement with the number of probed molecules. After retraction of the tip by 1 μm a far-field Raman signal of the investigated surface area could not be detected for the low BCB coverages. Since single-molecule spectra of BCB were recorded with an integration time of only 1 s, the sensitivity also allows Raman imaging of single molecules on a time scale in which the probed surface section is stable [50].

9.3.2 *Gap-Mode of Metal-Organic Semiconductor System*

9.3.2.1 Au tip and diindenoperylene molecule

Semiconducting organic materials have been studied intensively due to their interesting opto-electronic properties, giving rise to a broad range of applications such as organic field-effect transistors, organic light emitting diodes or organic solar cells. The electronic properties of thin semiconductor films are usually different from the monomer or bulk crystal properties; moreover, they depend strongly on the microscopic film morphology, such as grain boundaries.

We had reported the near-field imaging of a *diindenoperylene* (DIP) thin film based on the plasmon–polariton coupling between the excited gold tip and the semiconductor film. A near-field optical image of the DIP film and the simultaneously recorded topographic image are shown in Fig. 9.13 a and b. A sharp optical resolution of ca. 17 nm at one molecular boundary is demonstrated. The DIP molecular domain boundaries being one to four molecular layers (1.5 $\sim$ 6 nm) high are resolved topographically by a shear-force scanning tip and optically by the 6×10^5 times enhanced PL (see Fig. 9.13c). The excitation is four magnitudes enhanced and the intrinsically weak PL quantum yield of the DIP-film is 15-fold enhanced by the tip. The enhanced PL contrast results from the local film morphology via stronger coupling between the tip plasmon and the exciton-polariton in the DIP film. The increasing and red-shifted PL intensity with the tip approaching closer to the DIP sample indicate a stronger plasmon–polariton coupling (see Fig. 9.13d). The boundary of the DIP thin film appears to be brighter due to a stronger tip-sample coupling as compared to that of inside the thin film (see Fig. 9.13e) [51].

9.3.2.2 Au tip and organic solar cell blends

Based on the different plasmon–polariton coupling, near-field gap-mode microscopy can also spectroscopically imaging the material contrast in blended semiconductor thin film, such as the organic solar cell blend films. Organic solar cell composed of conjugated semiconductor polymers is a promising candidate for new generation of low-cost and flexible photovoltaic devices. With blends of

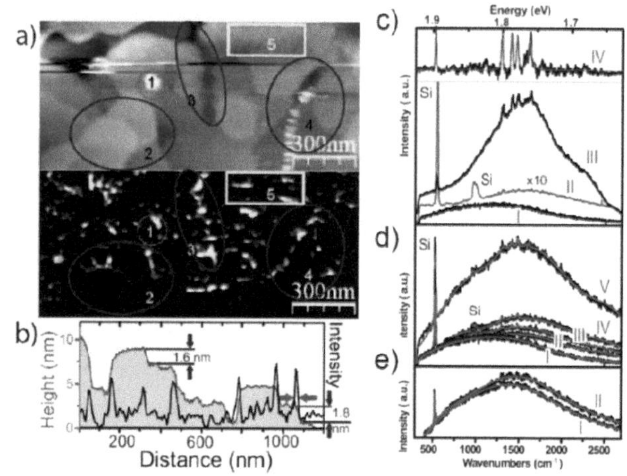

Figure 9.13 (a) Simultaneously recorded topographic (upper panel) and near-field optical images (lower panel). Scan area 1.5 × 0.7 μm^2, laser power at the focus: 170 μW, tip-sample distance: 3 nm. (b) Line profile through the topographic image with its correlated optical intensity. (c) Confocal far-field PL-spectrum (I) of a tip alone integrated for 10 s, (II) of DIP film integrated for 900 s, and (III) tip-enhanced spectrum integrated for 10 s at a tip–DIP distance of 3 nm. Excitation power: 170 μW. (IV) Tip-enhanced Raman spectrum obtained from (III) after subtraction of PL-background with fitted spectrum (gray curve). (d) Spectra collected at different tip-sample distance. II, 3 nm; III, 2.5 nm; IV, 2 nm; V, 1.5 nm and I, spectrum of tip alone. Integration time: 10 s. (e) PL spectra, (I) from inside a domain and (II) from a domain edge. Tip-sample distance: 3nm. Excitation power: 170 μW. Acquisition time: 30s. Adapted from Ref. [51]. Copyright (2010) the American Physical Society.

poly(3-hexylthiophene)) and *[6,6]-penyl-C61 butyric acid methyl ester* (P3HT:PCBM) which have been intensely used as electron donor and acceptor in organic solar cells devices, 5% energy conversion efficiencies were achieved [52, 53]. One of the main factors affecting the power conversion efficiency of P3HT:PCBM solar cells is the nanoscale morphology issue.

Wang *et al.* had reported the correlated topographical and near-field spectroscopic mapping images obtained from the blend film, which has been thermally annealed for 30 min. (see Fig. 9.14). The PL intensity variation reflects the efficiency of charge generation at the P3HT:PCBM interface. Raman spectroscopy, which is very sensitive to the degree of π-electron delocalization along the chain axis of conjugated polymers, studies nanoscale changes in morphology by monitoring the vibrational modes of P3HT polymer chains that are involved in electronic transitions.

Figure 9.14a shows the AFM image of a sample area. The corresponding near-field mapping image (see Fig. 9.14b), obtained by summing the total spectral intensity from 283 cm^{-1} to 2926 cm^{-1} shows intense optical signals at the places where the island is located. Gradually decreasing optical intensity from the maximum height of the island to the flat area is clearly resolved. Figure 9.14c stacks 32 Raman spectra taken from a one-line near-field spectroscopic map (1 μm) along the arrow indicated in Fig. 9.14b. While the P3HT Raman peak at 1450 cm^{-1}, represented by the red line across Fig. 9.14c, indicates the presence of P3HT molecules throughout this region, a clear optical signal evolution in the range from 1600 to 2600 cm^{-1}, corresponding to PCBM PL, is observed following the one-line scan. Figure 9.14c shows that from spectra No. 7 to 12, PCBM tends to aggregate more and more severely as supported by the PCBM PL increase and the topographically obvious clustering. The FWHM of the C=C stretching Raman peaks is broadened where the PL ratio of PCBM against P3HT is larger (Fig. 9.14e). From spectra No. 1 to 12, the PL ratio of PCBM against P3HT increases 6 times. Correspondingly, the FWHM of the P3HT C=C stretching Raman vibration peak increases by ca. 10 cm^{-1}, indicating that the increasing clustering of PCBM induces reduced crystallinity of the P3HT domain [54].The above fine optical information cannot be revealed by the conventional far-field microscopy due to its resolution limitation (see Fig. 9.14f) [55].

9.3.2.3 Gap-mode of metal-inorganic semiconductor system

Dye sensitized TiO_2 solar cell is one of the excitonic solar cells. Interfacial *electron transfer* (ET) plays a central role in dye-sensitized solar cells, photocatalysis, environmental chemistry, surface chemistry, and molecular electronics [56]. The *forward electron transfer* (FET) kinetics in various dye-TiO_2 systems has typical half-times ranging from femtoseconds to several hundred picoseconds. When the adsorbed molecule is strongly bound close to the TiO_2

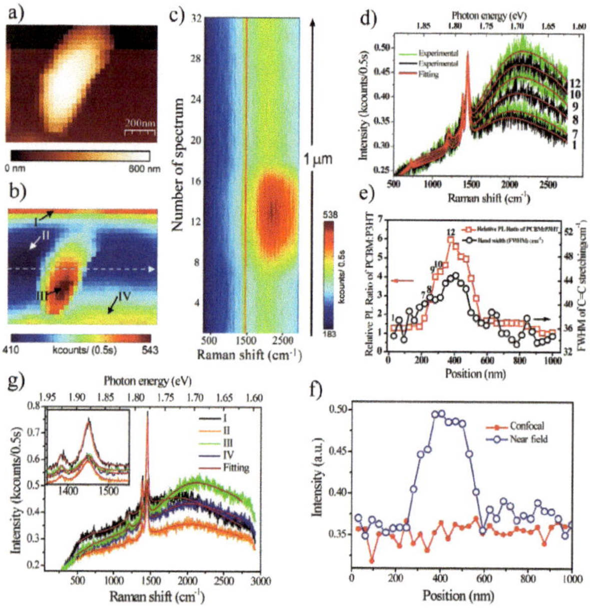

Figure 9.14 (a) Topographical image. (b) near-field spectroscopic mapping images obtained by summing the total spectral pixel intensity from 283 cm^{-1} to 2926 cm^{-1}. (c) 32 Spectra extracted from a line scan (1 μm) the position of which is indicated in (b) by an arrow. (d) Comparisons of several spectra extracted from the Raman line mapping sequence shown in figure (c). The green and black spectra are the experimental data while the red curves are the fitted spectra. (e) The relative PL ratios between PCBM and P3HT (curves labelled with red open rectangles) and the FWHM changes of the P3HT C=C stretching Raman peaks (curves labeled with black open circles) along the line indicated in figure (b). (f) Comparison of the near-field and confocal spectroscopic mapping along the same line indicated in figure (b). The optical signals plotted here are derived from the PCBM PL intensities. (g) Near-field spectra collected at positions I, II, III and IV as indicated in figure (b). The red curves are the fitting spectra. The inset figure shows the enlarged spectra from 1350 cm^{-1} to 1550 cm^{-1}. Sample is thermally annealed at 140°C for 30 min. Scan area: 1 × 0.75 μm. Laser power: 250 μW. Acquisition time: 0.5 s per spectrum. Adapted from Ref. [55]. Copyright (2010) Wiley-VCH Verlag GmbH & Co. KGaA, Weinheim.

surface, then rapid FET is expected. In the so-called wide-band limit, Franck-Condon factors do not matter for FET and the transfer rate is primarily controlled by the electronic coupling [57]. Following FET, the injected electron relaxes and is localized to sub-band states. *Backward electron transfer* (BET) from the semiconductor to the oxidized dye molecules often follows. Compared to ultrafast FET processes, BET processes take longer, with lifetimes ranging from sub-nanoseconds to several milliseconds [58]. The wide timescale variation is most likely associated to the inhomogeneous molecular surface bonding and vibrational relaxation energy barriers for the BET although, other contributions can be from the existence of trap states and non-Brownian electron diffusion in semiconductors. Furthermore, the dynamics of the BET processes are also often multiexponential and stretched exponential [59].

Interfacial ET in TiO_2-based systems is important in artificial solar energy harvesting systems, catalysis, and in advanced oxidative waste water treatment. The fundamental importance of ET processes and impending applications make the study of interfacial ET a promising research area. Photoexcitation of dye molecules adsorbed on the surface of wide band-gap semiconductors, such as TiO_2, results in the injection of electrons from the dye molecules to the conduction band of the semiconductor or energetically accessible surface electronic states. Using Raman spectroscopy and ensemble–averaging approaches, the chemical bonding and vibrational relaxation of the ET processes have been extensively studied. However, due to the complexity of the interfacial ET energetics and dynamics, significant questions remain on characterizing the source of the observed complexities. To address these important issues, we have applied advanced spectroscopic and imaging techniques such as confocal and tip-enhanced near-field Raman as well as PL spectroscopic and topographic imaging. Here we explore single surface states on TiO_2 as well as the interfacial electronic coupling of *alizarin* to TiO_2 single crystalline surfaces.

We have used confocal and tip-enhanced high-resolution near-field luminescence and Raman imaging spectroscopy to analyze the interfacial charge-transfer energetics down to the single site and single-molecule level, well beyond the spatial resolution of the optical diffraction limit [60].

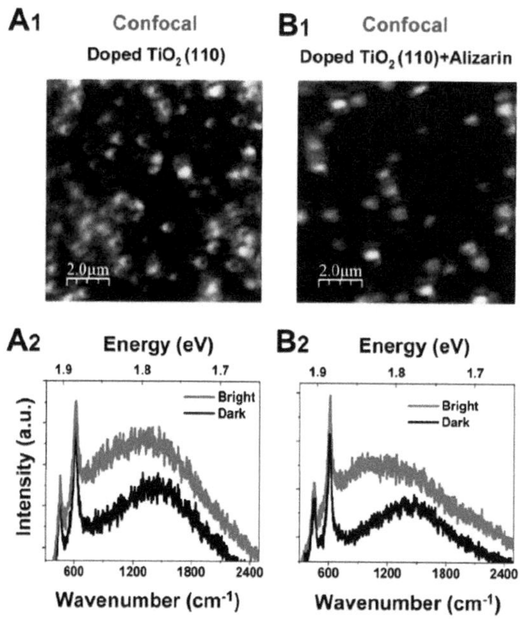

Figure 9.15 Optical response of the abundant surface states of the Nb-doped rutile TiO_2 (110) surface. (A1) Confocal imaging and (A2) Raman/luminescence spectra of TiO_2 (110) surface. (B1) Confocal imaging and (B2) Raman/luminescence spectra of TiO_2 (110) surface with alizarin (1 μM). Adapted from Ref. [60]. Copyright (2011) American Chemical Society.

Panels A1 and B1 in Fig. 9.15 show confocal images of the Nb-doped rutile TiO_2 (110) surface under the conditions of without (A) and with alizarin (B), respectively. Most of the hot spots in the confocal image (without the tip) in Fig. 9.15A1 show as donut-shape features, and only a small number of them show as circular or ellipsoidal-shaped spots. Such different diffraction limited fluorescence excitation patterns can only be observed when single quantum systems are raster scanned through the field distribution of a tightly focused radially polarized laser beam. Since the excitation rate is proportional to the square of the projection of the transition dipole moment onto the electric field.

Distinct excitation patterns are observed if single quantum systems, such as a single molecule, defects, or surface states are imaged, revealing their orientation and the dimensionality of their optical transition. We have calculated the optical patterns of quantum systems with one dipole moment, two perpendicular dipole moments, or three perpendicular dipole moments excited by radially polarized laser beams. Only for quantum systems with a two dimensional transition dipole moment as recently found for excitons in quantum dots [61] we can observe circular patterns. Here the ring patterns suggest that the transition dipole moment of the surface states can assume two perpendicular orientations with the same probability and both lying in the surface plane [61]. This is consistent with recent quantum-chemical calculations which suggest that O-vacancy formation in rutile Ti(110) surface results in two excess electrons occupying 3d orbitals on Ti atoms neighboring the vacancy [62].

Surface states have been demonstrated to play important roles in interfacial ET process [63]. Their density and energy distributions are the possible parameters that affect the charge-transfer pathways. We have used TiO_2/alizarin, a typical model system with strong electronic coupling, to probe these parameters. However, as shown in Fig. 9.15B1 and B2, we observed less hot spots and inhomogeneous imaging patterns (ellipsoidal shaped and donut-shape) and clear blue-shifts in the PL background of the spectra comparing with the TiO_2 crystal surface. These observations can be interpreted by the intensity difference of the electronic coupling between TiO_2 and alizarin molecules [60].

The nanoscale optical images of single surface states are visualized by the strong exciton plasmon–polariton coupling localized at the sub-domain boundaries with a spatial resolution of $\sim$15 nm. We suggest that the abundant surface states in the doped TiO_2 generate excitons under laser excitation which are strongly coupled to the surface plasmon of the Au tip. Bright luminescent spots arise from localized surface states which are distributed mainly along the sub-domain boundaries as evidenced in the superimposed topographic (in blue) and near-field optical (in red) image (see Fig. 9.16A). They are expected to play a crucial role in the charge transport process

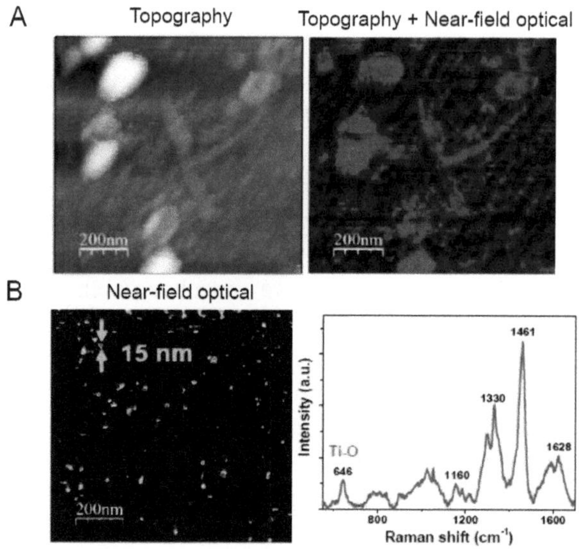

Figure 9.16 (A) Simultaneously obtained topographic and near field optical images of a clean Niobium-doped rutile TiO_2 (110) surface. Single surface states localized at the sub-domain boundaries can be clearly observed. (B) For the Alizarin adsorbed Nb-doped TiO_2 (110) surface, Raman spectrum with the Ti-O peak at 646 cm^{-1} provides the evidence of the strong electronic coupling at the interface. Adapted from Ref. [60]. Copyright (2011) American Chemical Society.

on the semiconductor surfaces. The interfacial electronic molecule-substrate coupling has been characterized by probing the molecule-perturbed surface states distribution and the associated specific Raman vibrational modes.

Figure 9.16 B shows the high resolution (15 nm) surface states imaging and the Raman spectrum of the Alizarin adsorbed Nb-doped TiO_2 (110) surface. We observed a direct evidence of the formation of the Alizarin–TiO_2 charge-transfer complex: besides the normal vibrational modes from Alizarin, we also observed a new Raman peak at 646 cm^{-1}, which is a typical Ti-O stretching mode from a bridging Ti-O-C bond. The strong C-O stretching at 1330 cm^{-1} reveals that the Alizarin-TiO_2 charge-transfer complex forms primarily through the hydroxyl groups of Alizarin and a Ti atom of TiO_2 surface.

9.4 Conclusion

In the above sections, we have systematically explained the principle, the concept of design, the instrumentation and the applications of PM-assisted gap-mode optical microscopy.

As has been well-demonstrated in the basic principle of gap-mode, the tip antenna and the sample system, be it a metallic structure or a semiconductor molecule, are an integrated unit. Hence inevitably the tip-sample gap distance and the polarizability of the substrate have to be carefully taking into account. The coupling strength between the oscillating dipole in the tip antenna and its image dipole inside the metallic substrate decreases sharply with the increase of the gap distance, leading to weaker enhancements in both the Raman and the PL processes. Due to the lower polarizability of dielectric material, the oscillating surface charge at the tip apex induces weaker mirror charge as compared to the metallic substrate that can be described by the free electron model. A direct result is a much weaker gap-mode formed between the tip antenna and the dielectric substrate. The near-field enhancement for this case is largely determined by the electric field confinement only by the tip antenna. Logically, semiconducting molecules, especially when forming a crystalline structure, can be polarized by the oscillating electric field at the apex of the tip and lead to a moderate gap-mode via the tip-plasmon and molecular-polariton coupling. Finite-difference time domain (FDTD) simulations described well the above analysis (see Fig. 9.17). The different coupling strength in the gap-mode, influences directly the sensitivity and the resolution of gap-mode optical microscopy. So far the single molecule detections using near-field optical microscopy are mostly reported with a gap-mode formed between a noble metallic tip and noble metallic substrate. In addition, the strong confinement of the electric field inside the gap-mode (by decreasing the tip-sample distance and using noble metallic tip and substrate) further decreases the waist of the gap-mode, leading to a possible improvement of the optical resolution to a scale of below 10 nm, even smaller than the tip apex size. In such a way, high-resolution gap-mode optical microscopy has found a way around the diffraction limited resolution in classic optical microscopy.

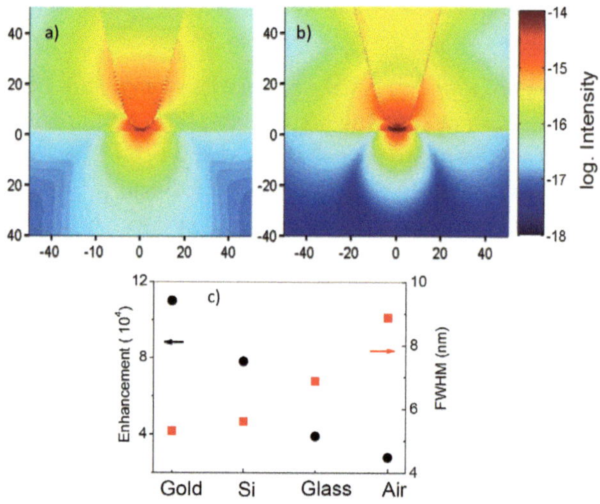

Figure 9.17 The FDTD calculation of the focus field intensity distribution. (a) a gold tip approached 2 nm away from a glass surface; and (b) a gold tip approached 2 nm away from a gold surface. (c) Comparisons of the enhancement factors and the FWHM of the gap-modes comprised of gold tips and different substrates (gold, Si, glass) The enhancement and FWHM in air are evaluated considering a gold tip alone. The FDTD calculations were performed with a commercial FDTD-package from Lumerical. The shape of the gold tip was modeled from typical SEM images and its length is set as 200 nm. The thicknesses of the substrates are 150 nm. Parameters for the calculation volume are the following: height: 300 nm, width: 200 nm, grid: 5 nm. The high precision volume part is around the tip apex, which has been modeled with a radius of 25 nm and a grid of 0.5 nm.

The success of gap-mode optical microscopy is further strengthened by the integration of PM optics in the microscope design. In another word, the PM optics facilitates to bring the fundamental scientific concept to applications in reality. It makes the choice of sample system more flexible, the signal collection more efficient, and the tip antenna excitation more optimized. Though several factors, such as the reproducibility of tip fabrication, the imaging speed, the application to aqueous condition, still need to be improved, high-resolution PM-assisted gap-mode optical microscopy is finding its way to a broad acceptance and application.

References

1. Drechsler, A., Lieb, M. A., Debus, C., Meixner, A. J., and Tarrach, G. (2001) Confocal microscopy with a high numerical aperture parabolic mirror, *Opt. Express*, **9**, 637–644.
2. Muhlschlegel, P., Eisler, H. J., Martin, O. J. F., Hecht, B., and Pohl, D. W. (2005) Resonant optical antennas, *Science*, **308**, 1607–1609.
3. Schuck, P. J., Fromm, D. P., Sundaramurthy, A., Kino, G. S., and Moerner, W. E. (2005) Improving the mismatch between light and nanoscale objects with gold bowtie nanoantennas, *Phys. Rev. Lett.*, **94**, 017402/1–4.
4. Talley, C. E., Jackson, J. B., Oubre, C., Grady, N. K., Hollars, C. W., Lane, S. M., Huser, T. R., Nordlander, P., and Halas, N. J. (2005) Surface-enhanced Raman scattering from individual Au nanoparticles and nanoparticle dimer substrates, *Nano Lett.*, **5**, 1569–1574.
5. Stockle, R. M., Suh, Y. D., Deckert, V., and Zenobi, R. (2000) Nanoscale chemical analysis by tip-enhanced Raman spectroscopy, *Chem. Phys. Lett.*, **318**, 131–136.
6. Hayazawa, N., Inouye, Y., Sekkat, Z., and Kawata, S. (2001) Near-field Raman scattering enhanced by a metallized tip, *Chem. Phys. Lett.*, **335**, 369–374.
7. Hartschuh, A., Anderson, N., and Novotny, L. (2003) Near-field Raman spectroscopy using a sharp metal tip, *J. Microsc.-Oxford*, **210**, 234–240.
8. Deckert, V., and Bailo, E. (2008) Tip-enhanced Raman spectroscopy of single RNA strands: Towards a novel direct-sequencing method, *Angew. Chem. Int. Edit.*, **47**, 1658–1661.
9. Hartschuh, A., Sanchez, E. J., Xie, X. S., and Novotny, L. (2003) High-resolution near-field Raman microscopy of single-walled carbon nanotubes, *Phys. Rev. Lett.*, **90**, 095503/1–4.
10. Pettinger, B., Ren, B., Picardi, G., Schuster, R., and Ertl, G. (2004) Nanoscale probing of adsorbed species by tip-enhanced Raman spectroscopy, *Phys. Rev. Lett.*, **92**, 096101/1–4.
11. Hartschuh, A., Qian, H. H., Meixner, A. J., Anderson, N., and Novotny, L. (2005) Nanoscale optical imaging of excitons in single-walled carbon nanotubes, *Nano Lett.*, **5**, 2310–2313.
12. Hartschuh, A., Qian, H. H., Georgi, C., Anderson, N., Green, A. A., Hersam, M. C., and Novotny, L. (2008) Exciton energy transfer in pairs of single-walled carbon nanotubes, *Nano Lett.*, **8**, 1363–1367.

13. Stanciu, C., Sackrow, M., Meixner, A. J. (2008) High NA particle- and tip-enhanced nanoscale Raman spectroscopy with a parabolic-mirror microscope, *J. Microsc.*, **229**, 247–253.
14. Zhang, D., Wang, X., Braun, K., Egelhaaf, H. J., Fleischer, M., Hennemann, L., Hintz, H., Stanciu, C., Brabec, C. J., Kern, D. P., and Meixner, A. J. (2009) Parabolic mirror-assisted tip-enhanced spectroscopic imaging for non-transparent materials, *J. Raman Spectrosc.*, **40**, 1371–1376.
15. Dorn, R., Quabis, S., and Leuchs, G. (2003) Sharper focus for a radially polarized light beam, *Phys. Rev. Lett.*, **91**, 2339011/1-4.
16. Lieb, M. A., and Meixner, A. J. (2001) A high numerical aperture parabolic mirror as imaging device for confocal microscopy, *Opt. Express*, **8**, 458–474.
17. Ambrose, W. P., Basche, T., and Moerner, W. E. (1991) Detection and spectroscopy of single pentacene molecules in a para-terphenyl crystal by means of fluorescence excitation, *J. Chem.Phys.*, **95**, 7150–7163.
18. Fleury, L., Tamarat, P., Lounis, B., Bernard, J., and Orrit, M. (1995) Fluorescence-spectra of single pentacene molecules in p-terphenyl at 1.7-K, *Chem. Phys. Lett.*, **236**, 87–95.
19. Vandermeer, H., Disselhorst, J. A. J. M., Kohler, J., Brouwer, A. C. J., Groenen, E. J. J., and Schmidt, J. (1995) An insert for single-molecule magnetic-resonance spectroscopy an external magnetic-field, *Rev. Sci. Instrum.*, **66**, 4853–4856.
20. Durand, Y., Woehl, J. C., Viellerobe, B., Gohde, W., and Orrit, M. (1999) New design of a cryostat-mounted scanning near-field optical microscope for single molecule spectroscopy, *Rev. Sci. Instrum.*, **70**, 1318–1325.
21. Enderlein, J., Ruckstuhl, T., and Seeger, S. (1999) Highly efficient optical detection of surface-generated fluorescence, *Appl. Optics*, **38**, 724–732.
22. Born, M., and Wolf, E. (eds) (1999) *Principles of Optics: Electromagnetic Theory of Propagation, Interference and Diffraction of Light*, 7th edn, Cambridge U. Press.
23. Stadler, J., Stanciu, C., Stupperich, C., and Meixner, A. J. (2008) Tighter focusing with a parabolic mirror, *Opt. Lett.*, **33**, 681–683.
24. Karrai, K., and Grober, R. D. (1995) Piezoelectric tip-sample distance control for near-field optical microscopes, *Appl. Phys. Lett.*, **66**, 1842–1844.
25. Binnig, G., Rohrer, H., Gerber, C., and Weibel, E. (1982) Surface studies by scanning tunneling microscopy, *Phys. Rev. Lett.*, **49**, 57–61.

26. Hoppe, S., Ctistis, G., Paggel, J. J., and Fumagalli, P. (2005) Spectroscopy of the shear force interaction in scanning near-field optical microscopy, *Ultramicroscopy*, **102**, 221–226.

27. Fleischmann, M., Hendra, P. J., and McQuillan, A. J. (1974) Raman spectra of pyridine adsorbed at a silver electrode, *Chem. Phys. Lett.*, **26**, 163–166.

28. Jeanmaire, D. L., and Van Duyne, R. P. (1977) Surface Raman spectroelectrochemistry: Part I. Heterocyclic, aromatic, and aliphatic amines adsorbed on the anodized silver electrode, *J. Electroanal. Chem.*, **84**, 1–20.

29. Albrecht, G. M., and Creighton, A. J. (1977) Anomalously intense Raman spectra of pyridine at a silver electrode, *J. Am. Chem. Soc.*, **99**, 5215–5217.

30. Moskovits, M. (1978) Surface roughness and the enhanced intensity of Raman scattering by molecules adsorbed on metals, *J. Chem. Phys.*, **69**, 4159–4161.

31. Kneipp, K., Moskovits, M., and Kneipp, H. (eds) (2006) *Surface Enhanced Raman Scattering: Physics and Applications*, Springer, Berlin.

32. Jain, P. K., Eustis, S., and El-Sayed, M. A. (2006) Plasmon coupling in nanorod assemblies optical absorption, discrete dipole approximation simulation, and exciton-coupling mode, *J. Phys. Chem. B*, **110**, 18243–18253.

33. Mirin, N. A., Bao, K., and Nordlander, P. (2009) Fano resonances in plasmonic nanoparticle aggregates, *J. Phys. Chem. A*, **113**, 4028–4034.

34. Lal, S., Grady, N. K., Kundu, J., Levin, C. S., Lassiter, J. B., and Halas, N. J. (2008) Tailoring plasmonic substrates for surface enhanced spectroscopies, *Chem. Soc. Rev.*, **37**, 898–911.

35. Kelly, K. L., Coronado, E., Zhao, L. L., and Schatz, G. C. (2003) The optical properties of metal nanoparticles: The influence of size, shape, and dielectric environment, *J. Phys. Chem. B*, **107**, 668–677.

36. Eustis, S., and El-Sayed, M. (2005) Aspect ratio dependence of the enhanced fluorescence intensity of gold nanorods: Experimental and simulation study, *J. Phys. Chem. B*, **109**, 16350–16356.

37. Steiner, M., Debus, C., Failla, A. V., and Meixner, A. J. (2008) Plasmon-enhanced emission in gold nanoparticle aggregates, *J. Phys. Chem. C*, **112**, 3103–3108.

38. Mooradian, A. (1969) Photoluminescence of Metals, *Phys. Rev. Lett.*, **22**, 185–187.

39. Beversluis, M. R., Bouhelier, A., and Novotny, L. (2003) Continuum generation from single gold nanostructures through near-field mediated intraband transitions, *Phys. Rev B*, **68**, 115433–115442.
40. Neacsu, C. C., Dreyer, J., Behr, N., and Raschke, M. B. (2006) Scanning-probe Raman spectroscopy with single-molecule sensitivity, *Phys. Rev. B*, **73**, 193406/1–4.
41. Deckert-Gaudig, T., and Deckert, V. (2009) Ultraflat transparent gold nanoplates—ideal substrates for tip-enhanced Raman scattering experiments, *Small*, **5**, 432–436.
42. Pettinger, B., Domke, K. F., Zhang, D., Schuster, R., and Ertl, G. (2007) Direct monitoring of plasmon resonances in a tip-surface gap of varying width, *Phys. Rev. B*, **76**, 113409/1–4.
43. Sackrow, M., Stanciu, C., Lieb, M., and Meixner, A. J. (2008) Imaging nanometre-sized hot spots on smooth Au films with high-resolution tip-enhanced luminescence and Raman near-field optical microscopy, *Chem. Phys. Chem*, **9**, 316–320.
44. Fleischer, M., Stanciu, C., Stade, F., Stadler, J., Braun, K., Heeren, A., Haffner, M., Kern, D. P., and Meixner, A. J. (2008) Three-dimensional optical antennas: Nanocones in an apertureless scanning near-field microscope, *Appl. Phys. Lett.*, **93**, 111114/1–3.
45. Cao, P. G., Gu, R. N., and Tian, Z. Q. (2002) Electrochemical and surface-enhanced Raman spectroscopy studies on inhibition of iron corrosion by benzotriazole, *Langmuir*, **18**, 7609–7615.
46. Nilsson, J. O., Tornkvist, C., and Liedberg, B. (1989) Photoelectron and infrared reflection absorption-spectroscopy of benzotriazole adsorbed on copper and cuprous-oxide surfaces, *Appl. Surf. Sci.*, **37**, 306–326.
47. Chant, H. Y. H., and Weaver, M. J. (1999) A vibrational structural analysis of benzotriazole adsorption and phase film formation on copper using surface-enhanced Raman spectroscopy, *Langmuir*, **15**, 3348–3355.
48. Dulkeith, E., Niedereichholz, T., Klar, T. A., Feldmann, J., von Plessen, G., Gittins, D. I., Mayya, K. S., and Caruso, F. (2004) Plasmon emission in photoexcited gold nanoparticles, *Phys. Rev. B*, **70**, 205424/1–4.
49. Bouhelier, A., Bachelot, R., Lerondel, G., Kostcheev, S., Royer, P., and Wiederrecht, G. P. (2005) Surface plasmon characteristics of tunable photoluminescence in single gold nanorods, *Phys. Rev. Lett.*, **95**, 267405–267409.
50. Steidtner, J., and Pettinger, B. (2008) Tip-enhanced Raman spectroscopy and microscopy on single dye molecules with 15 nm resolution, *Phys. Rev. Lett.*, **100**, 236101/1–4.

51. Zhang, D., Heinemeyer, U., Stanciu, C., Sackrow, M., Braun, K., Hennemann, L. E., Wang, X., Scholz, R., Schreiber, F., and Meixner, A. J. (2010) Nanoscale spectroscopic imaging of organic semiconductor films by plasmon-polariton coupling, *Phys. Rev. Lett.*, **104**, 056601/1–4.
52. Ma, W. L., Yang, C. Y., Gong, X., Lee, K., and Heeger, A. J. (2005) Thermally stable, efficient polymer solar cells with nanoscale control of the interpenetrating network morphology, *Adv. Funct. Mater.*, **15**, 1617–1622.
53. Li, G., Shrotriya, V., Huang, J. S., Yao, Y., Moriarty, T., Emery, K., and Yang, Y. (2005) High-efficiency solution processable polymer photovoltaic cells by self-organization of polymer blends, *Nat. Mater.*, **4**, 864–868.
54. Yun, J. J., Peet, J., Cho, N. S., Bazan, G. C., Lee, S. J., and Moskovits, M. (2008) Insight into the Raman shifts and optical absorption changes upon annealing polymer/fullerene solar cells, *Appl. Phys. Lett.*, **92**, 251912/1–3.
55. Wang, X., Zhang, D., Braun, K., Egelhaaf, H. J., Brabec, C. J., and Meixner, A. J. (2010) High-resolution spectroscopic mapping of the chemical contrast from nanometer domains in P3HT:PCBM organic blend films for solar-cell applications, *Adv. Funct. Mater.*, **20**, 492–499.
56. Oregan, B., and Gratzel, M. (1991) A low-cost, high-efficiency solar-cell based on dye-sensitized colloidal TiO_2 films, *Nature*, **353**, 737–740.
57. Ernstorfer, R., Felber, S., Storck, W., Galoppini, E., Wei, Q., and Willig, F. (2005) Distance dependence of heterogeneous electron transfer probed in ultra-high vacuum with femtosecond transient absorption, *Res. Chem. Intermediat.*, **31**, 643–647.
58. Biju, V., Micic, M., Hu, D. H., and Lu, H. P. (2004) Intermittent single-molecule interfacial electron transfer dynamics, *J. Am. Chem. Soc.*, **126**, 9374–9381.
59. Guo, L., Wang, Y., and Lu, H. P. (2010) Combined single-molecule photon-stamping spectro scopy and femtosecond transient absorption spectroscopy studies of interfacial electron transfer dynamics, *J. Am. Chem. Soc.*, **132**, 1999–2004.
60. Sevinc, P. C., Wang, X., Wang, Y., Zhang, D., Meixner, A. J., and Lu, H. P. (2011) Simultaneous spectroscopic and topographic near-field imaging of TiO_2 single surface states and interfacial electronic coupling, *Nano Lett.*, **11**, 1490–1494.
61. Chizhik, A. I., Chizhik, A. M., Khoptyar, D., Bär, S., Meixner, A. J., and Enderlein, J. R. (2011) Probing the radiative transition of single molecules with a tunable microresonator, *Nano Lett.*, **11**, 1700–1703.

62. Morgan, B. J., and Watson, G. W. (2010) Intrinsicn-type defect formation in TiO_2: A comparison of rutile and anatase from GGA plus U calculations, *J. Phys. Chem. C*, **114**, 2321–2328.

63. Enache-Pommer, E., Liu, B., and Aydil, E. S. (2009) Electron transport and recombination in dye-sensitized solar cells made from single-crystal rutile TiO_2 nanowires, *Phys. Chem. Chem. Phys.*, **11**, 9648–9652.

Chapter 10

Wet-Chemical Synthesis Techniques for Colloidal Plasmonic Nanostructures Assisted by Convective or Microwave Dielectric Heating

L. Carbone
IPCF-CNR UOS Pisa, Via G. Moruzzi 1 - 56124 Pisa, Italy
carbone@ipcf.cnr.it

In this chapter we review different synthesis methodologies of colloidal metal-based single and hybrid nanoarchitectures accomplished in phase solution in the presence of coordinating ligands and the assistance of traditional convective heating.

10.1 Introduction

Collective oscillations of conduction electrons known as plasmons, have a characteristic resonance frequency ω_p which depends on the mass, density and charge of the carriers. If the incident light frequency matches the plasmon resonance frequency, a strong absorption and scattering cross-section is obtained. Metal

Handbook of Molecular Plasmonics
Edited by Fabio Della Sala and Stefania D'Agostino

ISBN 978-981-4303-20-0 (Hardcover), 978-981-4303-21-7 (eBook)
www.panstanford.com

nanocrystals (NCs) show strong optical responses because of their large free-electron density, drastic boundary and surface effects. In this case the absorption and scattering cross-section depend not only on the nanostructure size and shape but also on the relative permittivity of the metal and surrounding dielectric constant (see Chapter 1). By varying nanoparticle morphology and/or particle local enclosing environment the resonance frequency can be tuned over a broad spectral range [1, 2], see Chapter 3.

The possibility to manipulate the strong light confinement associated with plasmon resonances has been translated in the design and development of metal NCs with tailored dimensions and sizes. The synthesis of metal NCs has thus become an interdisciplinary subject that is attracting a profound interest of the scientific community owing to their broad range of potential applications. For instance metal NCs have showed enhanced selective catalytic and sensing activity in chemical reaction and fuel cells depending on their morphology [3–5], have been tested as optical contrast agents [6], as tools for medical diagnostic [7], biosensing [8, 9] and when coupled with periodic lattice, led to the development of diverse subwavelength photonic components [10].

In order to improve applicative effectiveness and likely identify new technological advantages, the synthetic chemistry of metallic NCs has created nanoarchitectures exhibiting morphologies as disparate as spheres [11–13], rods [14], wires [15], cubes [16], nanoshells and nanoboxes [17–19], nanorattles [17, 20], and many others for a variety of materials. In addition, with the aim to benefit from two synergically-operating physico-chemical properties or investigate their mutual effects, more elaborated hybrid nanoarchitectures based on two or more combined components have been created [3, 21].

In the first part of the present chapter we review different synthesis methodologies of colloidal metal-based single and hybrid nanoarchitectures accomplished in phase solution in the presence of coordinating ligands and the assistance of traditional convective heating. All the methodologies hereafter described, share the same classical basic principles of crystal formation: indeed a first synthetic stage known as nucleation phase, consisting with the precipitation from the liquid of a nascent crystalline nucleus is followed by the growth stage. This last step is heavily governed by coordinating molecules dispersed in the liquid which commonly control the particle crystallographic evolvement determining the final morphology and regulate the solubility in organic or inorganic solvents. A diversity in the experimental approaches to the here-discussed growth methods must be searched in the physico-chemical phenomena which promote the nucleation step therefore inevitably affecting the future particle shape.

An interesting helpful support on the development and interpretation of further possible synthetic procedures will come from the second part of the chapter, whereby *wet-chemical synthesis* of *colloidal* metallic single or hybrid nanostructures carried out under microwave dielectric heating will be reported. This is described as another way to selectively deliver thermal energy inside the reaction domain frequently dominated by a series of "*secondary effects*" and it will be analyzed with the aim to evidence, whether exist, possible mechanistic differences with the convectional heating.

10.1.1 *Wet-Chemical Synthesis: Basic Principles*

Generally, state-of-art colloidal NCs synthesis in liquid phase, regardless of the insulating, semiconducting or metallic material character, consists in decomposing or reducing appropriate molecular reactive precursors in polar or non-polar solvents, frequently coordinating, and the contemporary support of selected ligand molecules (namely surfactants) which act at the same time as reactivity moderators, NC growth-addressing reagents, and in solvent-dispersing and -stabilising media [22, 23]. Shape-homogeneous and size-monodisperse inorganic NCs of different materials with different morphologies can be produced [22–26]. According to *classical nucleation theory* [22, 24] two virtual consecutive synthesis stages can be individuated: (1) seed nucleation and (2) growth of NCs thereof (Fig. 10.1a).

(1) Nucleation is based on the preliminary formation of small seeds of the nascent material, these are adopted as the structural basements onto which the NCs will be developed, thereby, seeds formed in a sudden burst would ensure a uniform size distribution of the forthcoming NCs (homogeneous nucleation). Two main

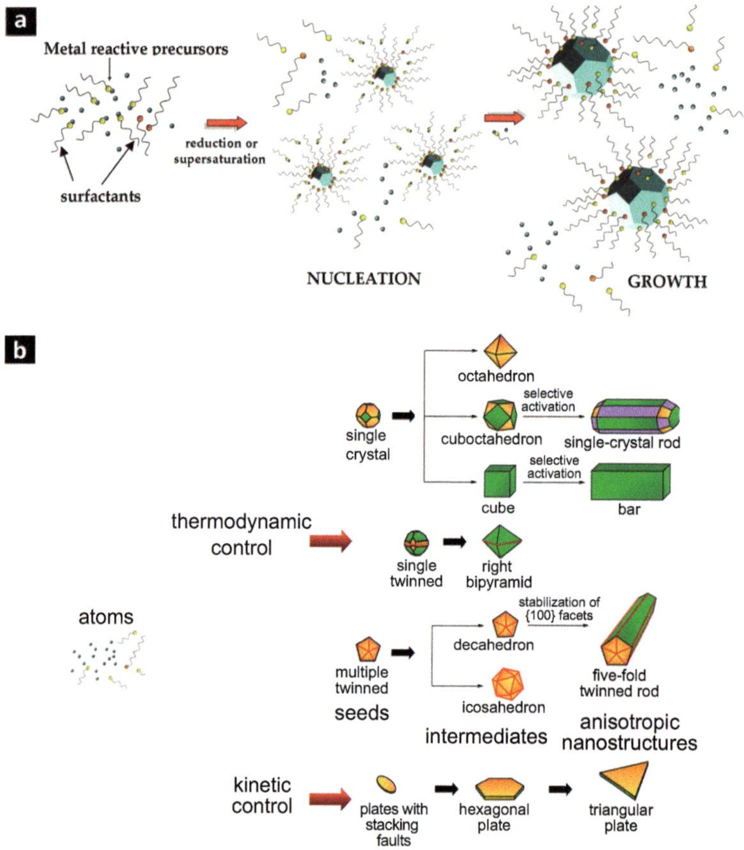

Figure 10.1 (a) The drawing in the picture summarizes the mechanistic stages of the synthesis: the system reaches the physico-chemical conditions to start the nucleation, at this step small seeds are in equilibrium with unreacted metal materials dispersed in solution which during the growth stage will feed the nuclei. (b) Schematic illustration of the synthetic pathways which determine the final shapes. Thermodynamic or kinetic control of the reaction defines the energetic stability of the facets exposed. The picture has been adapted from Ref. [34] with permission, copyright Wiley-VCH Verlag Gmbh & Co. KGaA. The authors established the following color map: as green they indicated the {100} facets, as orange the {111}, and as purple the {110}.

driving forces can behave as initiators of nucleation, one involving a charge transfer as in the cases of growth promoted by reduction routes, electrochemical methods, photochemical methods and galvanic replacements, and the other associated to supersaturation commonly attained in the growth methodologies based on precursor decomposition. In the first case is very likely that ions are reduced and later aggregated to form small seeds wherefrom NCs will be originated. For instance Turkevich and coworkers reported as early as 1963 the synthesis of 10 nm Au nanospheres upon reducing a $HAuCl_4$ solution with sodium citrate at water boiling temperature [27]. In a similar manner, by using $NaBH_4$ as strong reducing agent, have been prepared sphere-like Au [14, 28], and Co NCs [29], or Ag spheres with N_2H_4 [30]. Differently, in the case of supersaturation-promoted nucleation, when the solution precursor concentration results to be very high, a critical level is reached at which nucleation begins. Increasingly supersaturation conditions positively affect the free formation energy of new crystal phase, thermodynamically favoring the nuclei formation [22]. It appears evident that foreign coordinating molecular species, impurities and/or any further additives dissolved in the reaction environment, wherein seeds are formed, will actively alter the crystalline seed formation and most of all govern the kinetic of subsequent NC growth step [31]. An interesting evidence is represented by Au nanospheres either produced via $NaBH_4$-promoted reduction in the presence of sodium citrate ligands or via ascorbic acid in the presence of *cetyltrimethylammonium bromide* (CTAB): multiple-twinned and single crystalline NCs were respectively obtained by the two experimental approaches [32]. Generally, thermodynamics mainly regulates the formation of seeds and specifically, their shape evolution is dominated by the tendency to minimize the surface energy. In the case of face-centered-cubic (fcc) structured metal nanoparticles, the differences in the densities of metal atoms allocated on the $\{111\}$, $\{100\}$ and $\{110\}$ crystallographic facets result in an order of surface energies as follows: $\gamma_{\{111\}} < \gamma_{\{100\}} < \gamma_{\{110\}}$ [33, 34]. Due to energetic reasons, shapes with highly developed $\{111\}$ and $\{100\}$ facets should be mainly exhibited, therefore, single-crystal seeds will prefer cuboctahedral and octahedral shapes [33, 34]. Whereas, if single- or multiple-

Another possible route that the hybrid system could adopt to release the lattice interfacial strain consists in a progressive change of the chemical composition whereby atoms interdiffuse across the interface thus quietening the lattice mismatch, this is what happens in the case of alloy formation [17, 19].

The growth regimes above-described are heavily influenced by the selective coordinating ability played by surfactant molecules unevenly overcoating the NCs' surfaces thus differently affording or restricting the deposition on the seed surface facets.

10.2 Synthesis under Conventional Convective Heating

10.2.1 *Hard-Templated Growth*

Usually the term *templated-growth* means a growth methodology adopted to develop nanostructures of different materials within confining hard-nano(micro)-containers in the form of porous membranes or organic and ceramic mesoporous templates promoting nano(micro)-sized unidirectional evolution [43]. The most significant experimental steps of metal growth occurring in those shape-restricted environments, consist in the use of membranes showing cylindrical-shaped pores followed by the material deposition through an either electrochemical or chemical (electroless) reduction of metal ions and then by the dissolution of the template itself with the release of the metal nanorods, wires or fibrils originated [44–47]. Nowadays, several types of membranes are available either sold commercially as polycarbonate and polyester membranes [47, 48] or lab-fabricated as alumina [46, 49], silicate/aluminosilicate [50, 51] and glass[52] membranes whose pores' diameters and density can be widely tailored in the range of 0.005–10 μm and 108–1011 pores/cm^2 respectively as well as engineered in the pore arrays distribution [43]. By means of electrochemical deposition several examples have been reported about the fabrication of monometallic and single-crystalline nanowires based on materials as different as Cd [49], Zn and ZnO [53], Au, Ag and Cu [54, 55], Sn [56], Fe [57], poly-crystalline Co, Ni and

Rh [54], multisegmented bimetallic rod-shaped crystals of Au/Pt [46], Pb/Cu [58] (Fig. 10.2a), Au/Sn [59] (Fig. 10.2b,c), Fe/Cu [57, 60], by twos alternatively incorporated among the following species Pd, Pt, Ni, Co, Ag, Cu, Au [61] (see Fig. 10.2d), Co/Cu [62] and alloyed Co_xPt_{1-x} nanowires [63]. Many other experimental cases involve an electroless deposition generally consisting with chemically functionalized pore host walls which thereby catalyze the metal reduction [64, 65]. For instance wires [64, 66, 67] or spherical islands [68] of gold, nanowires of Ni and Cu [69], Pt [70, 71], Ag [72] and Pd [71] and nanotubules of Pb and Sn [73] (Fig. 10.2e,f) have been delivered trough this methodology.

The hard-templating growth methods described before should not be confused, on the contrary, with soft-templating approaches which usually are accomplished in solution phase in the presence of free surfactant molecules which severely attend the nucleation of a new particle and the further shape development selectively and dynamically interacting with the evolving particle surface facets.

10.2.2 *Precipitation-Promoted Growth*

A simple precipitation technique is applied to produce hybrid nanostructures in a single synthetic step. The two solid portions of the hybrids co-nucleate at the same time either segregated individually as in the case of Ag/AgBr dimers [74] (Fig. 10.2g,h) or compositionally alloyed as in the case of Au–Ag nanoparticles [75] (Fig. 10.2i). The co-precipitation is induced because of a reduced solubility. The addition in the reactive domain of a weak reducing agent in all likelihood disturbs ion solubility firstly modifying solution polarity and secondly favoring ion reduction. In the first case $AgNO_3$ is solubilized in toluene by means of didodecyldimethyl ammonium bromide complexing agent and the support of dodecanthiol (DDT) which will strongly passivate the NC surface formed and further precipitated by methanol. Whereas in the second case AuAg alloyed NCs, have been produced by reducing a refluxing water mixture of $HAuCl_4$ and $AgNO_3$ via sodium citrate.

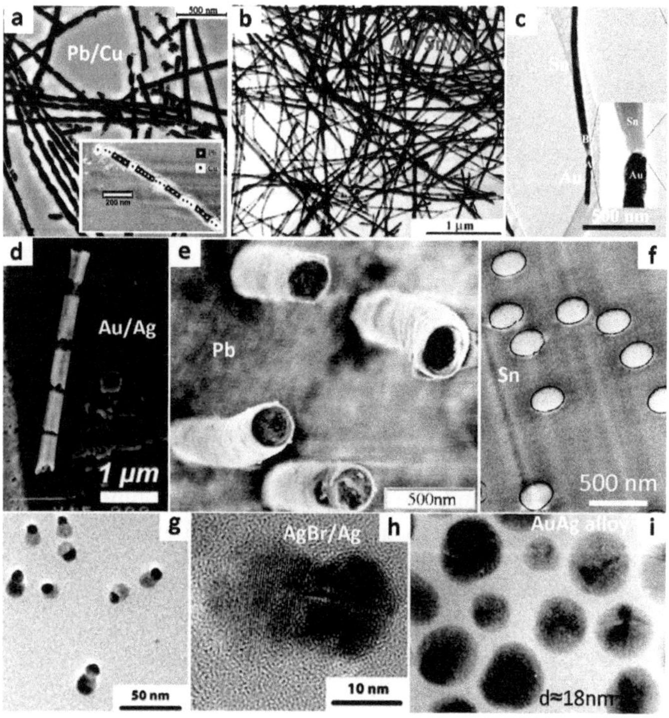

Figure 10.2 Panel of TEM (transmission electron microscope) and SEM (scanning electron microscope) images of templated grown NCs. (a) TEM image of Pb/Cu nanowires. Inset below indicates schematically the spatial points where the sample have been spotted by electron beam during EDX (energy dispersive X-ray spectroscopy) analysis in order to assess the chemical composition. Adapted from Ref. [58] with permission, copyright AIP. (b) TEMs of Au/Sn/Au nanowires and (c) of the two-metal junction Sn/Au. Adapted from Ref. [59] with permission, copyright American Chemical Society. (d) Field emission SEM image of multi segmented bimetallic Au/Ag nanorods. Adapted from Ref. [61] with permission, copyright AAAS. SEM images of (e) Pb nanotubules and (f) ultramicrotomized slice of membrane pores filled with Sn elongated crystals. The images have been adapted from Ref. [73] with permission, copyright IOP publishing. TEM images of lower (g) and higher (h) magnification relative to AgBr/Ag heterodimers. Images adapted from Ref. [74] with permission, copyright Wiley-VCH Verlag Gmbh & Co. KGaA. (i) HR-TEM image of alloyed AuAg nanospheres. Adapted from Ref. [75] with permission, copyright American Chemical Society.

10.2.3 *Electrochemical and Shape-Controlled Growth*

The electrochemical route for the synthesis of metal nanomaterials foresees the electrochemical oxidation-reduction of metal complexes accomplished in a simple two- [76] or three- [77] electrode type cell. The electrodes are immersed in an electrolytic solution basically composed of soft-templating molecules which operate in the reaction domain both as shape-inducing reagents stabilizing and delineating the nanoparticles shape and size and furthermore as supporting electrolyte [76, 78, 79]. A two-electrode setup is sketched in Fig. 10.3a.

Au nanoparticles have been reported by two-step electrolysis in a three component reactive environment based on CTAB, acetone and water, thereafter assisted by ultrasonic treatment. At the first electrolytic step small Au NCs are produced as seeds, then, at the second step, the Au growth is promoted by the preformed seed [80]. Thiol-monolayer coated Au spherical nanostructures have been observed upon electrochemical dissolution of gold in a bi-component mixture of H_2O and ethanol. The particles were formed both with and without the presence of a strong reducing agent [81] (Fig. 10.3b,c). *Tetradodecylammonium bromide* (TTAB) surfactant is employed to synthesize and stabilize Au NCs. Upon tuning experimental parameters as the amount of TTAB, the growth temperature and the current density is possible to control the NC size in the average diameter range of 8–60 nm. Particularly, the crystal size decreases by increasing the amount of TTAB [82]. The fabrication of suspended Au nanorods was similarly achieved in aqueous solution of CTAB and TTAB, at temperature of 36°C and the addition of cyclohexane in order to tune the rod aspect ratio [83] (see Fig. 10.3d). Bulk gold-based anode is oxidized and then newly reduced forming nanoparticles within the electrolytic solution at platinum cathode proximity. Interestingly the authors have found out that a critical role in rod formation is played by the presence in the reactive environment of a silver plate, in all likelihood due to generation of Ag ions. Dumbbells-like gold nanostructures have been electrochemically fabricated by using micelle templates, the latter attained by the coassistance of one main surfactant, the CTAB, and a further co-surfactant being tetradecylammonium bromide

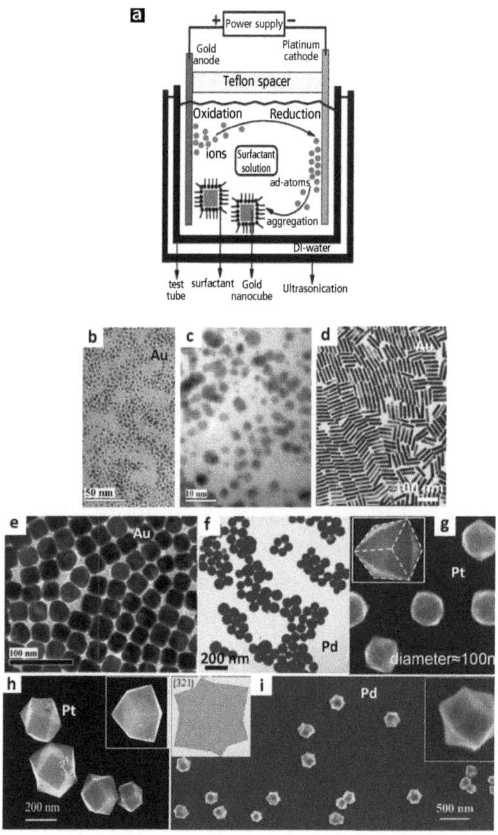

Figure 10.3 (a) Two-electrode setup for solution-phase nanocrystal synthesis in the presence of shape and size directing surfactant molecules. The image has been adapted from Ref. [86] with permission, copyright Elsevier. TEM images at lower (b) and higher (c) magnification of thiol-coated gold round-shaped nanoparticles. The image has been adapted from Ref. [81] with permission, copyright American Chemical Society. (d) Micrograph of shape-monodispersed Au nanorods, showing a tendency to self-align on TEM grid. The image has been adapted from Ref. [83] with permission, copyright American Chemical Society. (e) TEM image of gold nanocubes. The image has been adapted from Ref. [86] with permission copyright Elsevier. (f) PVP-coated Pd colloidal nanospheres electrochemically produced. The image has been adapted from Ref. [87] with permission, copyright American Chemical Society. Sequence of SEM images respectively of (g) Pt tetrahexahedrons, (h) Pt and (i) Pd concave hexoctahedral NCs. The insets display the specimens at higher magnification. The images have been adapted from Ref. [77] with permission, copyright Royal Society of Chemistry.

(TOAB). The addition of acetone solvent as third component affected the Au nanoparticle shape. X-ray diffraction (XRD) investigation revealed that the gold nanodumbbells consisted of single-crystalline particles showing fcc structure [84]. Single-crystalline and size monodisperse truncated Au nanocubes have been electrochemically obtained, using a CTAB surfactant solution and acetone. The nanocube edge was about 30 nm long [85, 86] (Fig. 10.3e). Electrochemical synthesis of spherical size-monodisperse Pd NCs has been accomplished in the presence of *poly(vinyl-pyrrolidone)* (PVP) which, according to the authors, beyond playing the role of solution electrolyte and shape-regulating ligand, prevents the electrodeposition of as-formed metal NCs onto the Pt cathode. Compared to bulk Pd, the as-synthesized nanosized Pd particles displayed high electrocatalytic activity towards CH_3OH oxidation [87] (Fig. 10.3f). In the shadow of doubt because of debautable sample characterization we mention Au–Ag alloy nanoparticles synthesized at different metal compositions in aqueous solution by an electrochemical co-reduction of $HAuCl_4$ and $AgNO_3$ in the presence of PVP. XRD analysis and UV-visible absorption spectra of the bimetallic systems seem suggesting the homogeneous alloy nanoparticle formation [88]. Further examples of colloidal NCs synthesized via electrochemical method included slightly elongated Ag nanospheres showing two peaks at 410 and 470 nm in the absorption spectrum features resulting from transverse and longitudinal surface plasmon resonances, respectively [89], as well as Cu nanorods [90]. Tian *et al.* have presented a solution electrochemical treatment of preformed Pt nanospheres supported on glassy carbon and subjected to a square-wave potential at 10 Hz with the assistance of ascorbic acid [91, 92]. Upon controlling the growth time, single-crystalline tetrahexahedral Pt NCs could be observed, whereas, when ascorbic acid is replaced by sodium citrate, concave Pt hexoctahedrons are obtained [77]. In a similar manner trapezohedral and concave hexoctahedral Pd NCs have been illustrated as well as. When the glassy carbon supported Pt seeds were exposed to air, prior to the potential appliance, fivefold twinned Pt nanorods originated [77] (Fig. 10.3g–i).

10.2.4 *Photochemical and Shape-Controlled Growth*

In the photochemical approach a direct or sensitizer-mediated [93] metal source reduction to form metal NCs, is promoted by high-power light commonly in the UV range of frequencies; the shape-controlled syntheses are carried out in solution phase and in the presence of surfactant molecules working as stabilizing and shape-tailoring agents. In situ UV light irradiation promotes the initial formation of metal clusters which behave as seeds for the next growth step as well as they exhibit interesting photochemical reactivity in the excited states depending on the cluster atoms' number [94].

Photochemical-approached examples of growth of metal NCs include nonaqueous synthesis of Au spheres and cubic Pd NCs in formamide in the presence of PVP [95]. Colloidal Au nanorods have been prepared by UV irradiation of $HAuCl_4$ aqueous solution containing hexadecyltrimethylammonium chloride in a concentration range to ensure rod-like micelles formation [96, 97]. In a similar way Kim *et al.* reported Au nanorod growth in the presence of CTAB, TTAB and acetone, the synthesis resulted kinetically favored by Ag^+ ions dispersed in solution [98, 99] whereas, the irradiation time could be drastically reduced from several hours to a few tens of minutes by introducing in the reaction batch a radical photoinitiator [100]. Au nanowires have been produced in aqueous solution of block copolymer poly(ethylene oxide)-poly(propylene oxide)-poly(ethylene oxide) [101]. Hongjun *et al.* have described a simple method to fabricate Ag and Au nanofilaments at the air-H_2O interface through the reduction of metal salt with UV light in the presence of polyacrilic acid [102]. Zheng *et al.* focused on the effect of laser excitation wavelength and power on the final size and shape of Ag NCs: planar triangles, disks pyramids and pentagons could be produced irradiating with monochromatic light at different wavelengths a water solution containing Ag^+ ions, silver clusters and sodium citrate [103]. The addition of PVP in the reaction domain and sample excitation with a blue light-emitting diode centered at 465 nm promoted the growth of Ag nanodecahedra [104]. Several other examples have reported for instance polygonal Au NCs observed UV-irradiating a solution containing gold salt in the presence of sodium oxalate [105]; multiple-shaped Pt NCs have been observed by irradiating for 3h a Pt(IV) solution containing CTAB and 2,7-dihydroxynaphthalene (2,7-DHN) in alkaline conditions: the CTAB to metal ion molar ratios, and the 2,7-dihydroxynaphthalene concentration affected the particle size and shape, in general promoting transition from spherical- to cubic- to bacteria-like-shaped Pt NCs upon increasing the absolute amount of CTAB [106] (Fig. 10.4a–c). In the same growth conditions the authors controlled Rh NC shape transition from spherical to flower-like to cubic [107]. Flat Au nanorods have been developed upon irradiation of $AuCl_4$-based ethanolic solution containing TiO_2 sols and any other surfactant molecule, different morphologies of NCs evolved depending on the solution composition [93]. During light-irradiation charge excitons generated in the TiO_2 particles played a catalytic effect leading electrons to reduce free gold ions whereas the corresponding holes were scavenged by the ethanol. When the reaction is performed in the presence of already formed seed of gold, six-star Au NCs were produced (Fig. 10.4d–i). Light irradiation is also utilized to effect the particle size and shape as in the cases described by Jin *et al.* who reported a light-induced ripening process responsible for the shape evolution of Ag nanospheres into triangular nanoprisms in the form of colloidal suspension [108] (Fig. 10.4j–l), and El Sayed *et al.* who, described a pulsed laser-mediated (100 fs-7 ns and 800 nm) transformation of Au nanorods into nanospheres as due to laser-generated multiphotons ionization processes which led to rod particle melting in order to minimize the electrostatic charge repulsions [109] (Fig. 10.4m,n).

10.2.5 *Galvanic Displacement-Mediated Growth*

In the experimental strategies to develop fully metal-based single or hetero-nanoarchitectures, one main procedure, often unintentionally employed, is represented by *galvanic displacement* which consists in the deposition of a solution-dispersed less active metal (a noble metal as Au, Pd, and Pt) onto a more active metal core (a less noble metal as Ag). Via red-ox permutation additional metal ions, after precipitation onto formerly nucleated metal nanoparticle,

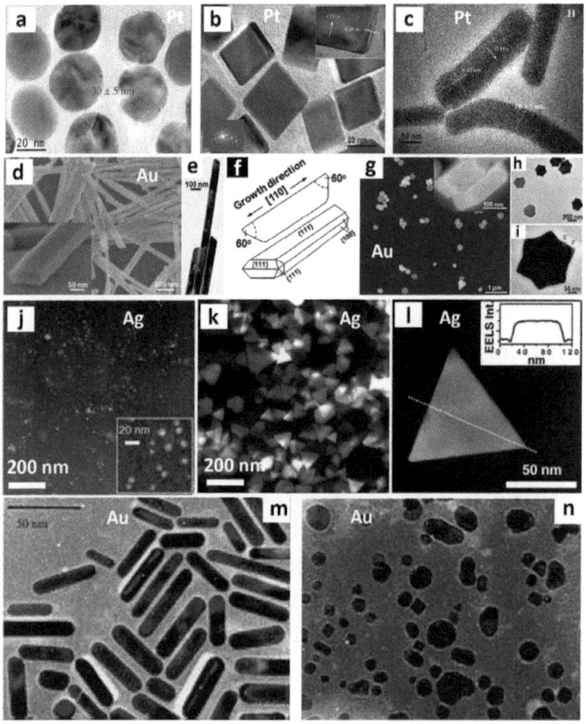

Figure 10.4 The panel reports TEM images of samples subjected to photoirradiation. (a) Sperical, (b) cubic and (c) bacteria-like size and shape monodispersed Pt nanocrytals fabricated irradiating, in alkaline conditions, a Pt-containing solution in the presence of different amount of CTAB. The images have been adapted from Ref. [106] with permission, copyright American Chemical Society. Sequence of images relative to TiO_2-sensitized growth of Au nanorods. (d) SEM image of flat Au nanorods (see the cross-section in the inset below) and (e) corresponding TEM image of the same sample. (f) Crystallographic drawing showing the plate-like morphology. (g) SEM image of star-shaped Au NCs produced in the same growth conditions of the nanorods however in the presence of small gold seeds. (h) TEM image of more particles and (i) single particle evidencing the star-like cross-section. The images have been adapted from Ref. [93] with permission, copyright American Chemical Society. TEM images of showing the morphology change from (j) silver shere to (k) triangle prisms whose triangular flat facet is reported in the picture (l). The images have been adapted from Ref. [108] with permission, copyright AAAS. (m) TEM image of Au nanorods grown in micellar solution before and (n) after laser photoirradiation. The images have been adapted from Ref. [109] with permission, copyright American Chemical Society.

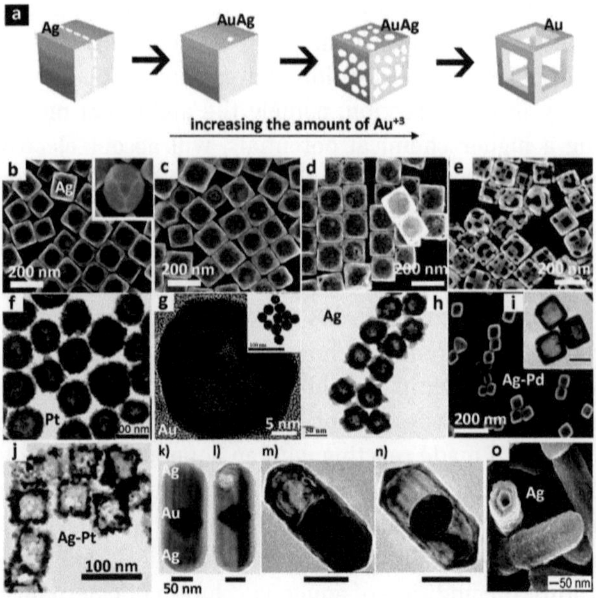

Figure 10.5 (a) Sketch of the entire sequential Au for Ag galvanic displacement adapted from Ref. [19]. (b–e) Panel of four TEM images summarizing the Au for Ag galvanic process. (b) Ag original nanocubes; (c) AuAg alloyed nanocubes showing very few pinholes; (d) AuAg alloyed nanocubes with higher density of pinholes; (e) porous Au nanoframes. All the images have been adapted from Ref. [112] with permission, copyright American Chemical Society. (f) TEM image of hollow urchin-like Pt nanospheres adapted from Ref. [21] with permission, copyright Wiley-VCH Verlag Gmbh & Co. KGaA. (g) High-resolution TEM image of nanogold shell adapted from Ref. [115] with permission, copyright American Chemical Society. Inset in the figure reports low magnification TEM image of a group of particles. (h) TEM picture of Ag nanoshell adapted from Ref. [117] with permission, copyright American Chemical Society. (i) SEM image of Ag-Pd alloy empty nanoboxes, in the inset is displayed the corresponding TEM image of the sample. (j) TEM picture of Pt-decorated Ag nanocubes, both the micrographs have been adapted from Ref. [118] with permission, copyright American Chemical Society. (k–n) TEM details of the sequential steps of the galvanic replacement in Ag–Au–Ag heterometal nanorod. Images adapted from Ref. [123] with permission, copyright American Chemical Society. (o) Multi-walled Ag nanorattles, the figure evidences the pentagonal cross-section of the nanotube and the different families of coaxial walls. Picture adapted from Ref. [125] with permission, copyright Wiley-VCH Verlag Gmbh & Co. KGaA.

are reduced oxidizing the original metal atoms thus partially or completely replacing them [19, 43, 110] (Fig. 10.5a). In the course of selective substitution the galvanic replacement occurs exclusively in one direction, namely the additional metal species, showing a higher chemical potentials, will accept electrons from the original metal which, on the other hand, owns lower redox potentials. In fact, due to size-dependence of nanostructure redox potentials metal-to-metal exchange is not always straightforwardly predictable. Based on the Faraday second law of electrolysis, one mole of transferred electrons involves one gram-equivalent of material oxidized/reduced, however, being different the electric charges of metallic ions in solution the mass balance of the cation exchange is not always compensated as evidenced in the case of original Ag NCs made reacting with an aqueous solution of $HAuCl_4$, whereby three silver atoms must be oxidized to reduce one gold ion. This non-conterbalanced atom inter-exchanging behaviour tends to generate hollow metallic nanostructures [19, 111, 112]. As detailed in literature the addition of aqueous solution of chloroauric acid at 100°C to a solution of silver NCs promotes the oxidation of silver and its partial dissolution, whereas gold epitaxially precipitates at the beginning decorating silver surface as small islands then forming a nanoscale crystalline shell, perhaps incomplete, around the nascent hollow nanostructures [111, 112]. At this step silver atoms diffuse from the inner core into the lattice of the Au to form a compositionally alloyed continuous AuAg shell [112]. Upon increasing the amount of $HAuCl_4$ in solution the volume of the hollow part grows up originating AuAg internally-empty NCs and this tendency is supported by the high reaction temperature as well as higher reactivity of pure silver than alloyed silver. Afterwards, still persisting the replacement reaction, the system experiences a dealloying process due to Ag removal out of the shell during which, few pinholes are generated inside the shell: the interior of core-shell is now opened to the outer reactive environment giving rise to a porous hetero-nanostructure. In this regard Lu *et al.* successfully demonstrated the selective ablation of silver from Au–Ag alloy nanoboxes employing water dispersed inorganic etchant as $Fe(NO)_3$ or NH_4OH [19]. Reasonably the pinhole formation can be as well interpreted as a gold-promoted displacement reaction occurring locally rather than over the entire shell. Supplemental addition of $HAuCl_4$ in solution promotes further etching of silver from the shell giving rise to cage-like and then frame-like single-metal Au nanostructures whose shape perfectly match the original Ag consumed templates, whereas the size is slightly higher as a result of outer layer-on-layer gold deposition (Fig. 10.5b–e). This mechanism has been widely observed for different cases whereby transmetallation is carried out on Ag nanospheres [111] or nanocubes [16, 19, 112] to produce Au nanocages/boxes, Au truncated octahedrons [113], and empty Pt spheres [114] when the galvanic displacement reaction is performed in o-dichlorobenzene and using *oleylamine* (OLAM) as surfactant and inorganic to organic transferring agent. Guo *et al.* have reported urchin-like metallic and bimetallic NCs based on Au/Pt starting from roughly spherical Ag NCs at 100°C in aqueous domain; in a similar fashion where obtained Pd, Pd/Pt Au/Pd and Au/Pt hollow nanospheres [21] (Fig. 10.5f). The same authors reported the fabrication of Pt nanotubes from Ag nanowires. Via two-step reduction reactions, the former exploiting metal redox potentials and the second supported by the employment of a reducing reagent, hollow Au [115, 116] (Fig. 10.5g) and Ag [117] nanospheres (Fig. 10.5h) have been reported starting from Co NCs. Under water refluxing experimental conditions Au for Ag polyol-coated nanowires have been presented by Sun *et al.* [112] as well as AuAg alloyed nanoboxes produced in similar chemical conditions however lower reaction temperature. Upon addition of Na_2PdCl_4 or Na_2PtCl_4 to a refluxing aqueous solution of Ag nanocubes, Pd-Ag alloy single crystals (Fig. 10.5i) and hetero-nanoboxes of Pt-Ag (Fig. 10.5j) have been respectively produced [118]. Pd atoms rapidly diffuse into Ag thus forming a homogeneous alloy without following a further dealloying process, whereas Pt and Ag do not interdiffuse even at refluxing conditions resulting in two distinc metallic identities. PVP- and sodium citrate-coated Ag triangular nanoplates could be similarly converted to Au nanorings [119]. Interestingly Mirkin *et al.* described a back-filling process consisting in selectively etching the flat facets of Ag triangular nanoprisms (length side $\approx$70 nm) with $HAuCl_4$ in order to form frame-like Au–Ag nanotriangles: in all likelihood the gold precipitation onto large flat facets produces larger increase of surface energy thus accounting for axial etching without producing any empty prism. Then the triangular frames could be refilled by precipitating solution-dispersed Ag^{+} and Au^{+3} ions with ascorbic acid giving rise to porous triangular AuAg nanoprisms [120]. Nano-Galvanic cell reaction of DDT-coated clusters of Ag, Pd or Cu with Au(I) or Pd(II) thiolate complexes in toluene produced a wide variety of bimetallic clusters (3–4 nm in diameter) [121, 122]. Asymmetric and later symmetric hollow formation has been accomplished by Seo *et al.* by reaction at 100°C with $HAuCl_4$ of an aqueous solution of PVP- and *diethylene glycol* (DEG)-coated Ag–Au–Ag heterometal nanorods [123]. Heteronanorods have been previously synthesized trough a seeded-growth approach depositing at 260°C a DEG-based solution containing $AgNO_3$ and PVP onto PVP-coated Au decahedral seeds [124] (Fig. 10.5k). Surprisingly, during the $HAuCl_4$-driven redox replacement, a pore is initially formed inside the silver domain (Fig. 10.5l) merely on one rod end broadening until that rod tip is completely empty (Fig. 10.5m); further gold addition promotes the etching of the remaining heterostructure side (Fig. 10.5n). The authors developed two main hypothesis to explain the asymmetric etching tendency, the former based on the fact that after generation of one pore this works as active gate accelerating the transferring of Ag^{+} ions outside and causing expansion of the hole, and the latter considering that electrons produced by silver oxidation in the region surrounding the pore are transferred via rod conductive metal surface making the Au^{+3} ions evenly precipitating on the surface. This behavior will reduce the activity of another concurring pore.

When displacement reactions are associated with electroless deposition of silver, exotic coaxially-organized heterostructures, characterized by hybrid chemical compositions, are generated [17, 125]. The heterostructures present a concentric topology of the various metals, a sort of Matrioshka arrangement known as nanorattles as defined by the same authors [17]. Different shape and material configurations have been reported in literature, for instance onto spherical AuAg alloy core NCs, Ag could be deposited by chemical reduction promoted by ascorbic acid at high temperatures and water environment, afterwards, a redox displacement $HAuCl_4$-promoted altered the silver shell composition in alloy shell further creating a partial hollow ring between the core and the shell. As habit in the galvanic replacement a further increment of gold concentration in solution produced a dealloying process of the shell generating pores [17]. The procedure could be alternatively extended to fabricate multiple-walled nanoshells with different morphology, the experimental condition in this case is to start from already empty shell and to build up onto it several alternative layers of voids and shells [17]. Thus, a wealth of similar examples can be found in literature as cube-shaped double-walled AuAg alloys shells [17], triple-walled nanotubes [125], (Fig. 10.5o) spherical nanorattles entirely made of Au in the core and in the shell [126], Au nanorods surrounded by pierced AuAg shell [20], single- and double-walled Cu_2O core-$Fe(OH)_x$ shell nanorattles in the shape of cubes, octahedrons or nanoframes as well as hollow cages of Fe_2O_3 and Fe_3O_4 [127]. Very interestingly hollow NCs show efficiency of sensitivity higher than for solid metals, resulting from the coupling between external and internal surface plasmon fields, the shift of plasmon extinction peak position with respect to changes in the dielectric environment is decisively higher than for solids [20, 128].

An interesting O_2-mediated redox displacement has been shown occurring in aqueous domains and very long reaction times (t = 20 hr) as reported in the case of Pd nanocubes synthesized in a solution containing ethylene glycol (EG), water and PVP and heated at at 100°C in air. After several hours of reaction the O_2-initiated corrosion process became competitive with the Pd reduction step creating holes in the cube surface which sustained the inner NC etching until PVP-coated nanocages are formed. Afterwards the cages started to dissolve from the corner of the nanoboxes inducing a reshaping of the squared cubes into rounded cubes [18]. Critical in the corrosion process resulted the presence of Cl^- ions in solution and/or adsorbed on the particle surface in facilitating the surface diffusion massively accelerating the NC surface hole formation [18]. Octgonal Pd nanorods have been grown by exploiting the O_2-supported oxidative etching and in particular, $PdCl_4^{2-}$ ions could be reduced in solution of EG and water and in the presence of PVP giving rise to cuboctahedral Pd NCs, then the presence of oxygen and Cl^- favoured the corrosion leading to the formation of nanorods with octagonal section. Thereafter the addition of $AuCl_4^-$ ions generated an asymmetric galvanic replacement reaction between the two metals meaning that Au is chemically reduced on one tip of the Pd nanorod forming tadpole-like heterostructures [129]. In a similar fashion a single Au sphere was selectively reduced either on one end of Te nanorods or decorating the rod surface depending on the pH conditions and of CTAB concentration [130, 131].

10.2.6 *Seed-Promoted Growth*

The above discussed basic principles of wet-chemical synthesis have evidenced the fundamental issues related to the nucleation phase as inevitable staging post towards the definitive NC morphology growth. Chemical reduction or thermal decomposition of the molecular species containing the ions/atoms that will build the inorganic nucleated seeds and later supply the growth are the two main physico-chemical forces promoting the nucleation. The crystalline phase of the as-formed nuclei, their interplay with the surfactant molecules or additives free in solution and temperature-regulated diffusion rates of the active species which will feed nuclei are the key factors that will dictate a thermodynamic- or kinetic-controlled growth [31, 39]. Shape and composition manipulation of possible nanoarchitectures can be accomplished via homogeneous or heterogeneous nucleation. The former definitely consists in one–pot reaction whereby the seeds are *in situ* nucleated and immediately experience the growth without any possibility to separate the two stages. The latter is a synthetic procedure, nowadays overused, to develop metal hybrid NCs. It founds on the injection in the reaction mixture, of previously formed NCs which will therefore behave as "seeds" in the reactive environment. Compatibly with the thermodynamic principles of the hybrid NCs formation [23, 41, 42], the second material will nucleate on the foreign seeds allowing the system to by-pass the nucleation stage. It is evident the efficacy of this method to design all the possible metal associations and likely create different topologies of hybrid NCs depending on the original seed shapes [23].

10.2.6.1 Homogeneous nucleation

This paragraph will review a set of examples of single-metal colloidal NCs synthesized in phase solution, without dealing with all the theoretical basics of the *in situ* nucleation and growth already detailed. Basically, the controlled development of crystal shapes is regulated by physico-chemical factors as temperature, solution composition and ability of surfactant molecules (or mixture of the them) free in solution to selectively adsorb onto crystalline surface facets enhancing isotropycal or anisotropycal particle growth or development of defect planes [34].

The first example of shape-controlled synthesis of spherical Au NCs goes back to 1963 when Turkevich *et al.* reported the sodium citrate-promoted chemical reduction of aqueous solution of $HAuCl_4$ at 100°C to produce 10 nm Au nanospheres [27]. In this case sodium citrate behaved as reducing and strongly stabilizing agent. In a similar coordinating environment and using $NaBH_4$ as strong reducing agent, sphere-like Au [14, 28] have been prepared. Jana *et al.* reported the synthesis of Au, Ag and Pt NCs by transferring trough didodecyldimethylammonium bromide inorganic metal salts in organic solvent (toluene) and reducing them with tetrabutylammonium borohydride or N_2H_4. Fatty acids or amines employed in the synthesis, allowed to control the NC size in the range of 1–7 nm [13]. Very similarly, Brust *et al.* described the synthesis of small Au spheres (1–3 nm) via a two-phase method. A gold salt was reduced in toluene with $NaBH_4$ in the presence of a long alkyl chain ammonium bromide-based transferring agent [132, 133]. Fièvet *et al.* [134] and then the group of Xia [16, 135], developed the polyol synthesis based on the reduction of metal salts in a mixture of alcohol and PVP polymer at high temperatures (196°C in the case of EG). The polyol reducing power is considerably enhanced at high temperatures. Size-monodisperse metal NCs showing various compositions, sizes and shapes, as in the case of Ag, Pd, Au, and Pt have been fabricated via polyol method. Upon injection of a solution of $HAuCl_4$, EG and PVP in boiling EG, Au NCs could be produced. The key role of PVP in controlling the particle shape was here evident. Depending on the PVP to Au^{3+} molar ratios Au tetrahedrons or octahedrons (~200 nm edge length) could be produced [136]. As we have already described, the final shape of fcc NCs is possibly dictated by a large exposure of the most stable $\{111\}$ and $\{100\}$ crystal facets, as in the case of tetrahedrons and icosahedrons. The boundary effect played by surfaces in dictating NC shape has been further demonstrated by adopting the same synthesis conditions and introducing a certain amount of Ag^+ ions which better stabilized the $\{100\}$ crystal facets: cubic NCs (150 nm edge length) were thus produced. Similarly, separate injections of a silver salt and PVP to boiling pentanediol produced Ag polyhedral shapes, whereas traces of $CuCl_2$ in the reaction batch facilitated the Ag nanocubes growth [137]. Ag nanorices were synthesized following a polyol process however decreasing the reaction temperature to 100°C and prolonging the heating time to 8 h [138, 139] (Fig. 10.6a). Au nanosheets with planar dimension of several micrometers, were also equivalently addressed trough a polyol approach at temperatures as high as 120°C; a reaction temperature lower than ordinarily and appropriate conditions of precursor concentration facilitated the production of triangles, hexagons, nanosheets and nanobelts [140]. Lim *et al.* reported the synthesis of anisotropic Au nanostructures by reduction of $HAuCl_4$ with PVP in water-based environment. Upon changing the Au^{3+} concentration the system evolved from nanotadpoles, to nanokites, to triangular and hexagonal microplates [35]. Au octahedral NCs were obtained from thermal decomposition (250°C) of $HAuCl_4$ carried out in block copolymers [141]. Further interesting shapes as Au nanorods or less symmetric morphologies have been largely described, we will deal with them in the following paragraph as they are the product of a seeded-growth approach [14, 40, 142].

Fully size monodisperse Ag nanospheres (diameter ~10 nm) have been attained both by thermal reduction of a single source silver precursor in isoamyl ether and oleic acid (OLAC)[143], and, by means of a polyol-based approach, employing PVP to $AgNO_3$ molar ratios particularly high in order to emphasize the PVP polymer binding ability thus erasing any selectivity [135]. The polyol-based methodology applied to cobalt acetate in a reaction atmosphere made of 1,2 DDT, OLAC and trioctylphosphine (TOP) or tributylphosphine (TBP) allowed to synthesize Co nanospheres with a little control of their diameters (5–10 nm) [144, 145]. Xia and coworkers have successfully produced a broad collection of

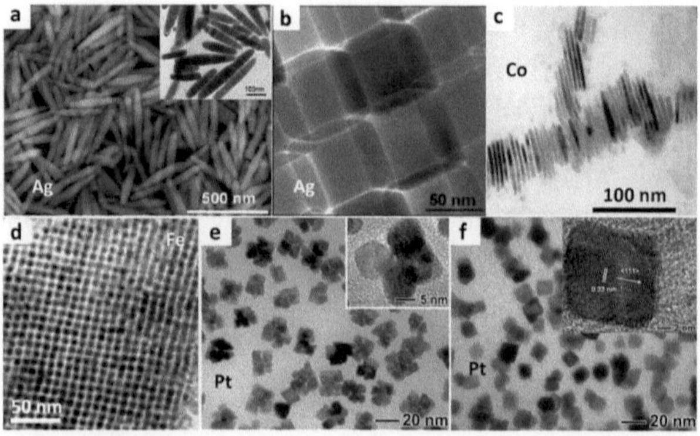

Figure 10.6 Panel of sample resulting from *in situ* homogeneous nucleation mechanism. (a) SEM image of Ag nanorices synthesized following the so-called polyol method. Inset in the picture displays TEM of the sample whereby are evident alternate stripes of light and dark areas along its [111] growth direction. Images have been adapted from Ref. [138] with permission, copyright American Chemical Society. (b) TEM micrograph of Ag nanocubes. The 3D effect is a consequence of the tilting of the TEM grid. The image has been adapted from Ref. [146] with permission, copyright American Chemical Society. (c) TEM photo of Co nanorods showing hexagonal close packed crystal structures. The image has been adapted from Ref. [153] with permission, copyright AAAS. (d) TEM micrograph of assembly of Fe nanocubes adapted from Ref. [155] with permission, copyright AAAS. Picture (e) and (f) reports TEM photos of (e) multioctahedral Pt NCs grown under the effect of Fe_3^+/Fe_2^+ ions dispersed in the solvent, and (f) of the same sample grown without Fe_3^+/Fe_2^+ ion influence. Insets respectively display magnified view of the samples. The images have been adapted from Ref. [157] with permission, copyright American Chemical Society.

single metal NCs in different shapes as for instance Ag cubes [16, 112, 135] mainly using the so-called polyol process carried out in EG and PVP at 160°C. Ethylene glycol behaved as reducing agent whereas PVP interacted with silver surface trough its polar functional groups therefore preventing the particle agglomeration and precipitation. Traces in the solution of Br^- ions further contributed to stabilize the {100} facets strongly promoting the formation of cubes [146] (Fig. 10.6b). Constant values of PVP to $AgNO_3$ molar ratios and a gradual decrease of the silver concentration in solution facilitated the formation of the most thermodynamically stable seeds, namely multiple-twinned nuclei showing decahedral shapes (Fig. 10.1b). Elongation of the decahedral seeds produced rods with pentagonal cross-section, enclosed in five lateral {100} facets. PVP polymer assisted rod elongation strongly and selectively binding {100} facilitating the formation of Ag nanowires [112, 135, 147]. Alternatively, in order to energetically stabilize Ag {111} facets, PVP could be experimentally replaced, adopting less coordinating ligands for Ag^+ ions as sodium citrate [135, 148], or polyacrylamide [36]; in this cases nanodisks and nanobelts were observed. Thermal annealing of spherical PVP-coated Ag colloids (diameter <5 nm) produced flat triangles [37, 148]. Disk-shaped particles were also obtained upon sonication of Ag^+ ions and N_2H_4 performed in reverse micelles based on H_2O, isooctane and (2-ethyl-hexyl)sulfosuccinate, in those experiments N_2H_4 played the role of reducing agent easily accelerating the solid phase formation [30]. In similar experimental conditions however in the presence of Cl^- ions as additive, Cu nanorods were reported [149]. Thermal treatment of dimethylformamide (DMF)-based $AgNO_3$ solution with the assistance of PVP polymer produced nanoprisms [150]. Thermal decomposition of metallorganic Cu-acetylacetonate carried out in OLAM and Ni-acetylacetonate in OLAM and hot TOP (or TBP) produced respectively sphere-like paricles of Cu [151] and Ni [152]. Puntes *et al.* reported the synthesis of magnetic Co nanorods and spheres by decomposition of $Co_2(CO)_8$ in o-dichlorobenzene at 182°C in the presence of OLAC and trictylphosphinoxide. Short times of reaction promoted nanorods showing hexagonal-close-packed crystal structure, whereas longer times favored a more thermodynamically stable ϵ phase exhibiting sphere-like shape [153] (Fig. 10.6c). The support of linear alkylamine on the other hand, promoted disk-like shapes [154].

H_2-mediated decomposition of a metallorganic precursor of iron and the reaction assistance of OLAC and hexadecylamine produced size monodisperse Fe nanocubes ($\sim$7 nm edge length) with high tendency to self-assemby in 3-D superlattices [155]. (Fig. 10.6d). As far as 1941 Rampino *et al.* exploited the effect of H_2 in generating a Pt precursor and the coassistance of polymers as polyvinyl alcohol (PVA) or sodium polyacrilate to form Pt NCs [156]. Lim *et al.* reported the synthesis of multiarmed Pt NCs synthesized in aqueous solution containing H_2PtCl_6, PVP and additives as Fe^{3+}/Fe^{2+} at 100°C for 24 h. The concentration of Fe^{3+}/Fe^{2+} ions controlled the reduction kinetic and the corresponding morphologies: upon increasing their amount a reduction of the system kinetic was evident and multioctahedral-shaped particles were observed (Fig. 10.6e). On the other hand, an acceleration of the reaction kinetic could be detected removing completely Fe^{3+}/Fe^{2+} ions from the solution generating octahedral-shaped NCs [157] (Fig. 10.6f). A few years earlier, the critical kinetic roles of Fe^{3+}/Fe^{2+} species and of O_2 in dictating particle morphologies had been addressed: Pt NCs in the form of spheres, stars, tetrapod- and wire-shaped NCs could be observed [158, 159]. Similarly, further kinetic-affecting additives were tested as for instance Na^+ ions liable to change Pt NC morphology from irregular spheroids to tetrahedra and octahedra [160] or Ag^+ ions which enhanced the growth in the $< 100 >$ direction promoting the sequential transformation cubes-cuboctahedra-octahedra [161]. Very recently Lim *et al.* supported the idea of NC shape tunability additive-promoted by introducing in the reaction surroundings small metallic Co NCs as additive; a change of the relative amounts of OLAM and Co NCs induced a Pt NC shape evolution from small spheres to cubes, while polypods could be obtained without the presence of metallic Co seeds [162].

Pd nanospheres (1–3 nm) were reported by Teranishi *et al.* upon reduction of $PdCl_4^{2-}$ precursor in alcoholic environment in the presence of PVP [163]. Xia and coworkers gave a fundamental contribution in detailing the shape-controlled synthesis of metal NCs and particularly in the case of Pd [34]. Based on experimental results they discussed the thermodynamic and kinetic pathways occurring at the synthesis nucleation stage monitoring a series of experimental parameters which can force the system in one direction or the other (see Fig. 10.1b). They could predict and experimentally verify all the possible Pd shapes as cuboctahedrons [164], nanocubes, bars and rods [18, 165], right bipyramids [166] and nanoplates [36].

10.2.6.2 Heterogeneous nucleation

Heterogeneous nucleation often reported as *seeded growth* represents one solution-phase route to build hybrid nanoarchitectures by pre-forming one nanocrystalline component, hereafter individuated as seed, and afterwards employing it as solid-substrate to facilitate the precipitation onto it of a second component according to the thermodynamic principles of hybrid NCs formation. The activation energy for heteronucleation is significantly lower than for homogeneous nucleation; the presence of seeds allows the system to save the formation energy otherwise necessary to create a new surface from scratch [23]. Key factors in establishing the topological configuration of the hybrid system, will be: well-defined seed geometry; the existence in solution of ligands/additives, both originally carried by the seeds and those dispersed in the solvent which selectively inhibit or favour the access to crystal facets; the interfacial energy gain in the heteroepitaxial growth, namely the facet-dependent wettability between seeds and deposited material (see Sec. 10.1.1.1), and finally the temperature. A large assortment of possible configurations can be individuated as for instance: onion-like core-shell , phase-segregated heterodimer, or cores decorated by shell islands [23].

Jana *et al.* have evidently corroborated the use of heterogeneous nucleation in energetically facilitating heteroepitaxial growth [167–169]. In a rod-like micellar template based on CTAB and using preformed citrate-capped Au seeds, in the average diameter range of 3–4 nm, they reported the synthesis of Au nanorods and wires [167, 169]. In analogous fashion they reported the synthesis of Ag nanorods and wires employing citrate-capped Ag seeds ($\sim$4 nm) [168]. Nikookakht *et al.* reproduced the same synthesis with a few changes, they reported the growth of Au nanorods using CTAB-coated seeds and apparently non-considerable amounts of Ag^+ ions as additives in solution [14]. As evidenced by structural analysis, citrate-capped Au seeds promoted the growth of Au nanorods, showing a pentagonal cross-section, surrounded by $\{100\}$ and $\{110\}$ lateral facets developed along the rod axis [40, 170], whereas when CTAB-capped Au seeds were employed and in the presence of Ag^+ ions, nanorods uniquely surrounded by $\{110\}$ lateral facets were observed [171]. Based on the assumption that the addition of ascorbic acid allows the partial Au reduction up to Au(I), the total rod formation mechanism is not yet completely evident. Furthermore, the preparation of 1-D nanostructures is a challenging task especially for structurally isotropic metals such as gold showing fcc symmetry. Several possible experimental hypothesis have been drawn, as for instance: a seed-mediated electron transfer form ascorbic acid to Au(I) [172], the presence of negatively or positively charged seeds which could affect the electron transfer [28, 173], the selective adsorption of CTAB as responsible of elongated growth [172, 174, 175], the presence of Ag^+ ions which could induce AgBr precipitation (Br^- from CTAB) on the rod surface thus attracting CTAB molecules [14], or possible selective Ag precipitation on specific gold facets (preferentially the {100}) assisted by underpotential deposition mechanism [171].

Liu *et al.* reported different outcomes obtained by reducing $HAuCl_4$ trough ascorbic acid in aqueous solution containing CTAB and Ag^+ ions and adopting as seeds, in one case single crystalline CTAB-coated Au seeds and in the second case multiple-twinned citrate-capped Au seeds. According to the first procedure single crystalline Au nanorods were observed (Fig. 10.7a), whereas penta-twinned bipyramids were produced in the second case [32] (Fig. 10.7b). Grzelczak *et al.* reported the growth of Pt NCs seeded by CTAB-coated Au nanorods in an aqueous solution containing K_2PtCl_4, CTAB, and eventually Ag^+ at 40°C. The addition of ascorbic acid produced homogeneous Au-Pt core-shell nanorods at low concentration of Ag^+ and dumbbell-shaped NCs at higher concentrations [176–178] (Fig. 10.7c,d). The same group reported quasi-epitaxial growth of Ni shell around elongated Au cores [179]. In a similar way Au–Ag core-shell sphere-like [180] and rod-like [181] NCs were observed. In those examples an increase of Ag^+ concentration promoted a uniform shell growth until a threshold silver amount above which a lateral thickening was promoted. Relatively to this specific bimetallic association, Cho *et al.* showed an interesting system based on anisotropic Au core and isotropic Ag shell to produce final octahedral shapes [182] (Fig. 10.7e). Sun *et al.* reported the synthesis of Pt-Ag nanowires obtained by reduction of a silver salt in EG at 160°C and the assistance of PVP polymer in the presence of EG-coated Pt seeds [183]. Habas *et al.* described heterogeneous nucleation seeded by well-faceted crystals [184]. They prepared cubic NCs of Pt capped with tetradecylammonium bromide in order to seed the reduction of K_2PdCl_4 in aqueuos solution in the presence of the same surfactant. Pt-Pd core-shell NCs in a cube-like morphology were observed. The addition in

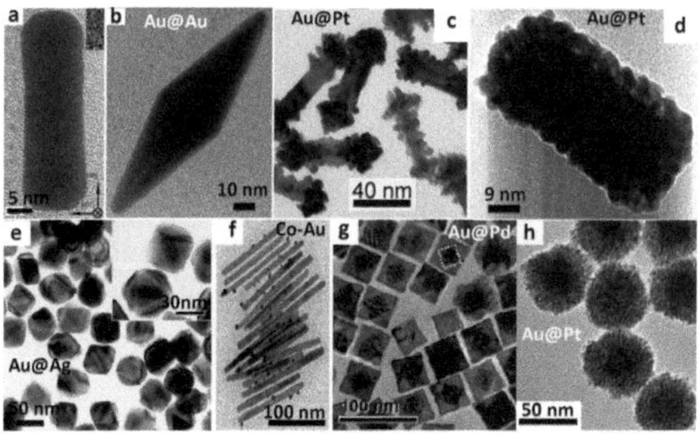

Figure 10.7 High-resolution TEM pictures of a single Au nanorod seeded by (a) single crystalline CTAB-coated and (b) multiple-twinned citrate-coated Au seeds. Images adapted from Ref. [32] with permission, copyright American Chemical Society. (c) TEM image of Pt-decorated Au nanorods showing high Pt deposition on the rod tips. The image has been adapted from Ref. [177] with permission, copyright American Chemical Society. (d) TEM image of a single Au-Pt core-shell nanorod. Image adapted from Ref. [178] with permission, copyright Wiley-VCH Verlag Gmbh & Co. KGaA. (e) TEM picture of Au–Ag core-shell NCs, inset evidences the elongated shape of the Au core. Image adapted from Ref. [182] with permission, copyright Wiley-VCH Verlag Gmbh & Co. KGaA. Picture (f–n) report the different epitaxial growth morphologies for the binary system Pt-Pd core-shell. (f) TEM image of gold-tipped Co nanorods. Image adapted from Ref. [185] with permission, copyright Wiley-VCH Verlag Gmbh & Co. KGaA. (g–h) TEM micrographs of Au-Pd nanocubes and Au-Pt nanoparticles respectively. Images are adapted from Ref. [42] with permission, copyright American Chemical Society.

the reactive environment of increasing amount of NO_2 allowed different growth directions with respect to the original Pt cubes thus producing Pt-Pd cuboctahedra and octahedra. An enhancement of the lattice mismatch between the seeds and the deposited material produced segregated hybrid configuration as in the case of Au growth seeded by Pt nanocubes; gold nanorods uniquely decorated with one Pt particles were evident [184]. Au-decorated Co nanorods were observed by reaction in organic solvent of amine-capped Co elongated seeds in the presence of a particular Au precursor (Au-tetrahydrothiophene). Metal reduction was assisted by lauric acid and hexadecylamine [185] (Fig. 10.7g). Fan *et al.* delineated the guide lines for epitaxial and non-epitaxial core-shell growth based on the lattice mismatch differences between substrate and deposited metal material. Trough a seeded-growth methodology, they showed how the shell epitaxially adapt to the core crystal structure in the case of Au-Pd (see Fig. 10.7g) and Au–Ag metallic combinations producing core-shell nanocubes and non-epitaxially as in the case of Pt-decorated Au spheres [42] (see Fig. 10.7h). Hydrophobic $CoPt_3$ and FePt nanoparticles were transferred in aqueous domain with CTAB and afterwards used as seeds to promote Au or Ag deposition thus giving rise to bimetallic heterodimers [186, 187]. Segmented Ag–Au–Ag rod-like heterostructures have been also shown. Upon seeding Ag growth in DEG at 260°C with Au multiple-twinned dodecahedral NCs with the support of PVP polymer heterometallic nanorods could be generated [124] (see Fig. 10.5k–l). An interesting trimetallic Au-Pd-Pt core-shell-shell hybrid system with metal materials arranged according to a spherical onion-like configuration has been recently reported. Mediated by ascorbic acid, a few Pd monolayers were created on Au nanospheres and then in the same reducing conditions Pt clusters deposited onto the bimetallic Au-Pd hybrid system [3].

10.3 Synthesis under Microwave Dielectric Heating

10.3.1 *Basic Principles of Microwave-Promoted Growth*

Traditional heating methodologies rely on the transfer of energy from hotter to colder in-contact interfaces ultimately employing one or more of the known heat transfer mechanisms: conduction, convection and/or radiation.

Microwaves (MWs) correspond to a fraction of the electromagnetic spectrum including a range of frequencies from approximately 0.3 (L-band of the microwave spectrum) to 300 GHz (D-band). As a rule, with the aim to avoid band interferences, microwave apparatus for domestic or chemical applications are allowed to operate at an established frequency window either of 0.915 GHz or 2.45 GHz (S-band of microwave spectrum).

When electric fields of electromagnetic waves interact with a polarizable medium (dielectric) three main effects may be

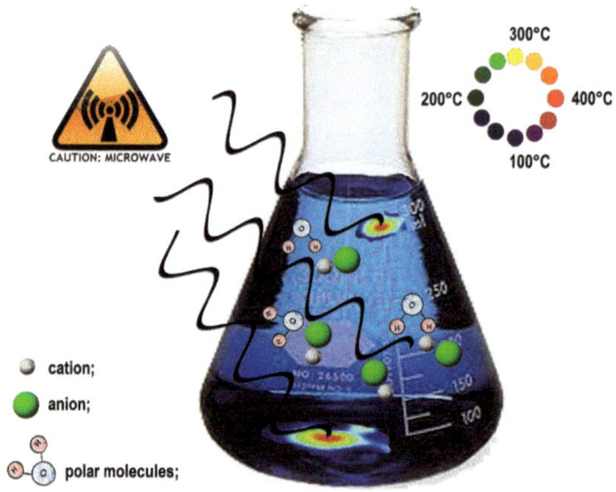

Figure 10.8 Microwave-promoted dielectric heating. Dipolar and conductive polarizations have been highlighted in the drawing. Furthermore two spots in the solution evidence an interior localized superheating and subsequent temperature homogenization. A color map has been reported in order to describe a possible temperature distribution.

promoted: electronic polarization during which the applied field causes a distortion of the electron cloud and nucleus in opposite directions; dipolar polarization activating rotations of weakly bonded dipoles (molecular specimens) and/or conduction polarization which actuates alternate motions of free charges (ionic specimens), see Fig. 10.8. In fact, in the event of microwave regime, electronic polarization is not a mechanism as significative as dipolar and/or conductive polarization, the latter being responsible for dielectric heating [188].

Dielectric polarizability depends on the facility of a material to polarize in response to electric fields: it is expressed by the permittivity ϵ which, generally speaking, represents a measure of the resistance that is encountered in the dipole or charge reorganization. In fact, the permittivity (resistivity) will almost always be dependent on the frequency of the applied field and on the viscosity of the reactive environment to some extent, thus it has the form of a complex value. When the frequency irradiation is low, dipole/charges' re-orientation is regularly in phase with the applied field coherently re-displacing themselves; alternatively in the field of very high applied frequencies dipole/charge organization falls: no polar movement is promoted thus the material can be considered as non-polar. Vice versa, in the MW frequency region (10^{+9} cycles/s), corresponding to rotational excitation energies in molecules, the polarization doesn't react instantaneously to the applied electric field causing a phase difference in the response: a delay between the cause and the effect is created, this causes energy loss by dipolar frictions and/or ionic collisions therefore promoting resistive heating in the medium. By means of prolonged polar re-orientations, electromagnetic energy is straightly and selectively delivered to microwave-active species and, due to frictional forces, mainly dissipated as heat [188–192].

Whether from one side the existence of permanent/induced dipoles or ions ensures the capability of a material to absorb microwave energy, the ability to convert it into heat is expressed by the dielectric loss $tan\ \delta$, which has to be considered as a parameter quantifying the dissipation of the electromagnetic energy in dielectric media. It correlates the imaginary part of the complex permittivity to the real part providing, to some extent, information of the dielectric heating event efficiency. As materials exhibit different values of $tan\ \delta$, chemicals do not interact equivalently with MWs, thus, a convenient choice of one dielectric reactive medium may promptly generate heat inside the reaction mixture and a spatial distribution of temperatures diffusively created thereof [192].

Different temperature regimes promoted by MWs engender singular microwave effects in the reaction environment: the most critically debated in literature are hereafter mentioned. In the cases of molecular/ionic species thoroughly MW-absorbing, localized superheatings up to 26°C above reactive medium conventional boiling points have been observed at atmospheric pressure. Reasonably, this behavior is attributed to a scarce formation of boiling nuclei during MW dielectric heating [193–195]. Surface effects and hot spots have been also described, the former originated in the case of heterogeneous systems meaning in the presence of MW-active solid supports, the latter due to non-homogeneous heat transfer [194, 196–198]. Finally, when operating in suitably-fabricated sealed vessels, a build up of pressure is achieved, thus mimicking an autoclave settings with a fast heating rate however [190, 194, 197].

All these factors reveal actively significative from the point of view of fast and mass heating, reaction rates 1–2 orders higher than through traditional heating approaches and moreover for energy saving [194, 199].

Further considerable experimental features indicated by scientific literature as crucial for MW effectiveness are for instance: the volume and geometry of reaction vessels, the material which they are made of, applied MW irradiation powers, the design of resonant cavities whether single mode or multi-mode as well as the accuracy of temperature and pressure measurement equipments [190–192, 200–202].

Since the introduction of MWs in organic and inorganic synthesis, the main question delineating the scientific research has been whether it is worth comparing conventional heating approaches with microwave-promoted heating methodologies. In fact, the lack of a sole prevailing microwave effect has rendered challenging to interpret the physico-chemical arguments that form the basis of synthesis MW efficacy, specially to discriminate whether a direct molecular activation or simply a thermal effect is involved. Although purposive or not, reaction conditions as homogeneous or heterogeneous reactive mixtures, the use of specific chemical reagents or vessel-cavity geometry may advantage one of the disparate MW effects giving rise to unexpected outcomes; aged and recent studies agree with assessing the influence of MW strictly in selective thermal activation [190, 196, 203, 204]. Thus, except in the profile heating rates, no evident differences between conventional and MW heating procedures have been observed when two experiments are performed at the same temperature [203, 204].

Application of MWs has been proved extremely suitable in the field of wet-chemical nanoparticle synthesis particularly for pure metal NCs [205]. Unless few rare exceptions of MW-assisted syntheses of metal NCs set up in polar non-protic [206, 207] or wholly non-polar environment [208], most of the metal synthetic pathways presented in literature are based on hydrothermal routes namely polar and protic reaction media. In details these latter approaches found on hot reduction either of water-dispersed metal salts supported by the active presence of particle-stabilizing reductants, [209–212] or in alcoholic media operating as both solvents (or chelating ligands in the case of polyols) and reducing agents, and further assisted by some size/shape-regulating additives (polymers or surfactants) [213–219]. As expected, according to the nature of the reaction context the paths of heat-delivery are different: syntheses carried out in non-polar dispersing media are completely transparent to MWs, therefore in these growth circumstances, uniquely metal complexes are selectively involved in the MW interactions and then transfer of energy to the non-active solvent occurs. Vice-versa, in the case of polar growth domain, it entirely couples to MWs owing to simultaneous solvent and metal salt involvement in dielectric interactions. A rapid heating of the reaction mixture (a few seconds) up to the solvent boiling point or even to higher temperatures under solvothermal conditions performed in sealed vessels is thus promoted, leading to shortening of the time required for completing molecular precursor conversion and furthermore favoring higher reaction selectivity [209–219].

This section specifically reviews wet-chemical hydrothermally synthesized single- and bi-component metal NCs whose growth is promoted by microwave irradiation. Ordinarily, dealing with metallic NCs is mainly synonymous of Ag or Au NC growth, two widespread materials which share a realistic technical interest due to their widely effective applicability in nanotechnology [135, 220]. The two noble metals show similar fcc crystal structure as well as crystallographic lattice constants. Further examples of metallic nanomaterials will be presented and discussed in detail.

10.3.2 *Mono-Metallic Nanostructures*

Size monodisperse Ag NCs in the diameter range of 5–7 nm were prepared at 140–155°C and ambient pressure, by MW-promoted reduction of fatty acid silver salts in alcoholic media [213]. Tuning of NC diameters was promoted by selecting carboxylate anions with different alkyl chain lengths, as well as changing the reaction temperature and/or heating time. The size tuning indicated that carboxylate anions not only acted as surface stabilizers against premature aggregation, but also as moderators of the alcohol-driven reduction reaction, regulating the extent of precursor consumption in the nucleation stage. On the other side, the reaction temperature, which was dictated by the boiling point of the alcohol solvent, controlled the rate at which subsequent growth of nuclei took place [213]. The same size evolution trend observed for further Ag-based systems over largely varying dimensional ranges, suggested similar molecular dynamics [209, 211, 217, 218].

A limited solubility of inorganic metal salts in many organic solvents results in heterogeneous reaction mixtures, in this case a non uniform heating pattern is likely expected. Moreover, as in experiments based on MW heating organic reagent (polymer) and inorganic metal salts are combined at first and afterwards exposed to irradiation, metal complexes formation seems to be favored prior to reduction. Ag NCs produced in heterogeneous systems were based on short alkyl chain alcohols and therein insoluble PVA working as both reductant and stabilizer [214, 221] PVP-stabilized spherical Ag NCs, 3–12 nm of diameter range, have been achieved by Ag^+ ions reduction either in DMF [206] or ethanol [219, 221–223] or methanol [224]. A mixture of Ag nanowires (diameters $= 40 \div 120$ nm and several microns of lengths) and nanorods were described upon MW-assisted and poly 1,2-ethanedithiol-promoted reduction of Ag_2O at temperatures between 80–140°C [225]. Interestingly, EG as an alternative to poly 1,2-ethanedithiol produced only spherical particles. Similarly, uniform Ag nanowires with average dimensions of 45 nm × 4–12 μm, were aerobically synthesized by dissolution of $AgNO_3$ in EG and the contemporaneous presence of PVP and NaCl [15]. Cl^- ions aimed at inducing the precipitation of AgCl thus contributing to decrease the Ag^+ concentration in solution and preventing the conversion of Ag^+ to Ag^0: this effect markedly concurred to control 1-D growth. Absence or very low concentrations of NaCl produced only roughly rounded nanoparticles. The authors highlighted a tendency of the reaction system to form networks of nanowires due to interparticle structural fusion of metal tips, both promoted by O_2 and Cl^- co-operating as etching reagents and by MW hot spots very pronounced at the ends of growing wires and responsible for local temperatures higher in proximity of wire vertices than at the midsections [15].

Based on a two-phase system, an interesting growth procedure of size monodisperse spherical Ag NCs has been reported by Gao *et al.* [12]. In a sealed vessel $AgNO_3$ salt is reduced by EG in a mixture of DDT and toluene at 160–170°C for 3 h under MW radiation; EG and toluene immiscibility creates a net interface inducing the solubilization of silver salt in the glycolic polar phase and DDT molecules packing across the interface pointing the long alkyl chains in the organic phase and the thiol groups bound to silver ions. Ag NCs growth is thus promoted at the interface by means of DDT molecules being at the same time NC capping agents and assembly promoters. Upon variating the EG:DDT molar ratio, size monodispersed spherical or cubic NCs can be observed along with close-packed either hexagonal or orthogonal superlattices therefrom (Fig. 10.9a–d) [12]. Roughly spherical Ag NCs in a size range of 5–10 nm have been similarly described by Varma and coworkers upon swift MW irradiation (30–60 s) at moderately low temperatures (30–60°C) of a water-based $AgNO_3$ solution. Glutathione, a tripetide existing in concentrations up to 10 mM in cell cytosol and considered a powerful endogenous antioxidant [226], has been used as reducing agent of silver salt as well as NC capping molecule in aqueous environment. The system reactivity showed consistent dependency on the MW irradiation power in particular showing a decreasing of the reaction times as well as particle sizes with the power increase. By comparison, equivalent reactions performed under conventional convective heating at 60°C did not yield any particle growth even after 24 h [227]. The authors successfully extended the synthetic procedure to other noble metals such as Pd, Pt and Au, in these cases producing spherical NCs larger than for Ag case (~100 nm) [227]. A few examples of nanophase Pt powders in sphere- and rod-like morphologies have been furthermore detailed by MW-assisted solvothermal techniques [228–230]. Based on the use of H_2PtCl_6 as metal precursor the growth methods account on methanol or EG to promote the metal reduction, whereas the sample growth is assisted by PVP molecules. Temperatures and irradiation time changes allowed to modulate the NC size.

Accurately as for silver-based nanostructured crystals, MW-assisted solution phase growth methods have been diffusively used to produce Au NCs with different sizes and shapes. Spherical particles with average diameters below 10 nm were differently achieved, for instance upon PVP-assisted hydrothermal reduction of

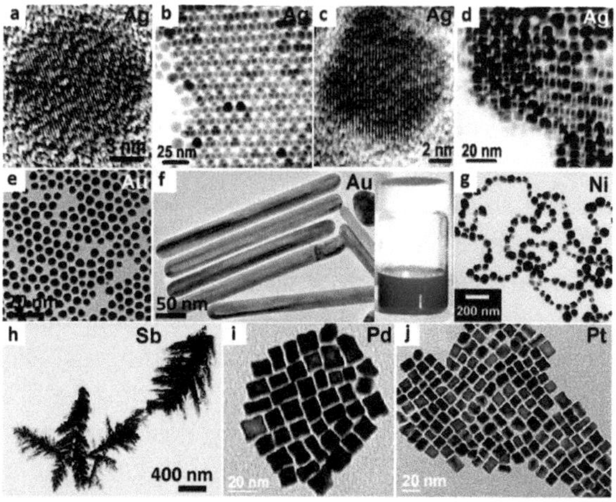

Figure 10.9 TEM picture collection and sketches mimicking the respective shape of mono-metallic NCs produced by MW irradiation. (a) HR-TEM and (b) low-magnification TEM images of Ag spherical NCs the latter showing hexagonal-like ordered superstructures. (c) HR-TEM image of cubic Ag NC prepared in EG and DDT. (d) Low-magnification TEM picture of two layers of superlattices of a cubic silver sample. (a–d) pictures are reproduced and adapted from Ref. [12] with permission, copyright American Chemical Society. (e) TEM image of spherical Au NCs obtained in a mixture of OLAC:OLAM = 1:1. The image is reproduced from Ref. [238] with permission, copyright American Chemical Society. (f) Au nanorod TEM image and photograph of corresponding gold nanorod solution. Pictures adapted from Ref. [241], copyright American Chemical Society. (g) TEM micrograph of Ni particles showing a chain-like arrangements obtained by MW-polyol synthesis in the presence of PVP. The image is reproduced from Ref. [250] with permission, copyright Wiley-VCH Verlag Gmbh & Co. KGaA. (h) Low-magnification TEM picture of Sb NCs showing dendritic shape and dimensions in the range of hundreds of nm. The picture has been adapted from Ref. [245] with permission, copyright Elsevier. (i,j) Monodisperse cubic NCs respectively of Pd and Pt prepared under chemical conditions of OLAM:OLAC = 2. The images have been adapted from Ref. [246] with permission, copyright American Chemical Society.

$HAuCl_4$ in methanol [231], ethanol [232], supported and promoted by poly(propyleneimine) dendrimer [233], in alcoholic medium in the presence of PVA molecules [214] or an amphiphilic macrocyclic polyamine (TNMR) [234] or in DMF-based polar environment [206]. In a similar fashion, Tsuji and coworkers succeded in preparation of

PVP-supported Au nanoplates with shapes ranging from hexagonal to pentagonal, to regular or truncated-triangular, by reducing chloroauric acid in EG [235, 236]: dissimilar solvent boiling points, namely dissimilar temperature profiles in the reaction batch, might be one pivotal argument responsible for discrepancy in the shape effect. As well as $HAuCl_4$ concentration was found to be another important factor in NC shape and size regulation [236]. Further studies, focused on MW polyol synthesis methods, evidenced a shape selective oxidative etching behaviour played by $AuCl_4^-$ and Cl^- anions, hence addition of a MCl salt (M = Na, K,H) resulted strongly effective to enhance the chloride shape selectivity [236]. Mallikarjuna *et al.* reported controllable adjustement of Au NC shapes, all the time showing a fcc crystal structure, via MW-assisted reduction of aqueous sugar solutions of $HAuCl_4$ [237]. Molecular features and concentration of sugar molecules heavily biased the particle size and shape: in particular, high sugar concentrations selectively regulated the reaction kinetic giving raise to spherical nanostructures when sucrose or α-D-glucose were used, whereas hexagons, cubes and prisms were favored with maltose; vice versa in low sugar concentration condition no selectivity was reported and mixtures of polygonal shapes were observed. When further supported by PVP presence, the method could be successfully extended to Ag, Pd, Pt NCs [237]. More evidences of the kinetic role played by ligands in modulating the NC growth rate and morphology was provided by Mohamed *et al.* who reported the shape tailoring of Au NCs as depending on the OLAM and OLAC concentration. Pure amine content or molecular ratio OLAM:OLAC = 1:1 led to uniform and size monodisperse $\sim$10 nm Au spheres, whereas a remarkable increase of oleic acid into the reaction mixture, apart from enhancing the growth rates, favored the development of non-isotropic shapes as faceted prisms or hexagons [238] (see Fig. 10.9e). Alkylamine-capped Au NCs were synthesized by ethanol reduction under MW irradiation in reverse micelles. Size monodispersity and stability of the amine-capped NCs were highly dependent on concentration and alkylamine chain length: generally the NC size tended to shrink upon increasing the alkylamine chain length [239].

In addition to the aforementioned round- and polygonal-shaped Au NCs, a few examples of single-material gold nanorods and wires were produced. Mixtures of different shapes with a fraction of elongated particles tunable between 6 and 31% were assessed by hydrothermal synthesis of TOAB-stabilized gold nanorods. High concentration of TOAB and MW reaction temperature rise, markedly influenced the nanorod formation yield and morphology [240]. In CTAB micellar media and alkaline conditions due to 2,7-DHN the reduction of gold salt was accomplished to generate gold nanoparticles. By varying metal salt to surfactant molar ratio and the 2,7-DHN solution concentration a strict size and shape control could be achieved from spheres to rods to prisms up to 85% of yield [241] (Fig. 10.9f). Zhu *et al.* reported single crystalline sodium dodecyl sulphate- and PVP-stabilized Au nanorods and wires upon reduction of $HAuCl_4$ in EG. [242]. Conductive gold nanochains, having electrical resistivity comparable to pure metals, were grown in solution under microwave using DNA as a reducing and capping agent [243].

Microwave-promoted polyol synthesis of Ni NCs in the size range of 30–100 nm in the presence of PVP and dodecylamine (DDA) was described by Li *et al.* Detailed experimental studies indicated that DDA concentration was the main factor affecting the Ni morphology, whereas the particle size distribution was controlled by the concentration of metal source and the DDA to PVP molar ratio (Fig. 10.9g) [223, 244]. Further interesting examples of single metal NCs developed with the contribution of MWs in solution-phase growth procedures are: dendrites of Sb composed of units of nanoflakes [245] (Fig. 10.9h), pure metallic Pt and Pd cubes as well as Rh and Ru faceted particles were prepared in DMF by using a mixture of OLAM and OLAC [246] (Fig. 10.9i,j), PVP-overcoated Co [223, 247], Fe [223], Cu [248] and Pt, Pd, Ru, Rh, Ir [231, 249] nanospheres could be synthesized in EG environment.

10.3.3 *Bi-Metallic Hetero-Nanostructures*

Hetero-nanostructures composed of two metal segments and organized according to either concentric or segregated architectures have been included in the following paragraph.

Controlling and modulating shape of larger nanoparticles have been most successfully attained using seeded approaches in order to grow templateless and polymerless larger metal NCs from smaller

[212, 251]. Addition of preformed seeds has the advantage to increase the overall growth rate and further catalyze surface growth of secondary ions onto former particles [167].

Non-spherical-shaped Ag nanostructures could be synthesized under MW heating in EG under the synergic assistance of PVP and 2 nm preformed Pt seeds [215, 216]. Whether from one side metal seeds catalyze the nucleation of a new phase onto their surface and drive the initial architecture of the core-shell, PVP concentration and reactive ion supply on the other, governs the shape evolvement suppressing the development of some crystal facets and favoring the growth of others [216, 252] (Fig. 10.10). Seed existence in the reactive environment and adequate MW heating time unequivocally favored the further formation of anisotropic shapes as wires, rods or films contrarily to spherical particles which were the only feasible thermodynamic shape in seedless syntheses. By using a two-step MW-polyol method, low concentration of PVP and relatively medium amount of seeds promoted a symmetrical elongation of five-twinned decahedral seeds, having only ten $\{111\}$ facets, thus giving rise to core-shell nanorods/wires in turn exhibiting longitudinal PVP-stabilized $\{100\}$ facets and active uncovered $\{111\}$ on the tips (see Fig. 10.10). At higher PVP concentration and low Pt seeds, PVP uniformly covered the seeds resulting into isotropic shapes as spheres and cubes now exposing only $\{100\}$ facets. Low PVP concentration but short heating time generated core-shell nanoplates having on the top and the bottom two flat $\{111\}$ faces [215, 216].

In order to basically interpret the aforementioned seed-promoted core-shell NC growth mechanism, Tsuji *et al.* examined the two-step synthesis of Au–Ag core-shell nanostructures [207, 253]. Based on the fact that Au and Ag show the same fcc crystal structure with similar lattice constants, nanostructures exhibiting comparable shapes and sizes are expected to evolve. Furthermore Au–Ag core-shell systems are ideal candidates to be investigated at TEM structural analyses as they own different mass contrast therefore easily distinguishable in TEM images [33, 253].

In the first synthetic step MW-polyol reduction of $HAuCl_4$ with the assistance of PVP produced a mixture of Au cores in the shape of triangular/hexagonal twin plates, octahedrons and multiple-

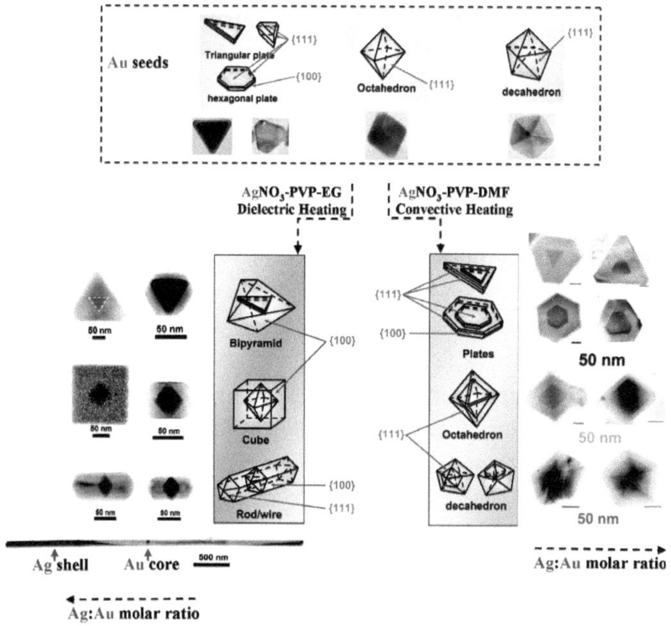

Figure 10.10 Schematic picture summarizing the Ag shell growth primarily governed by preformed Au seeds, under different heating and chemical conditions. Top line of the picture reports sketches and corresponding TEM images of triangular/hexagonal twin plates, octahedral and multi-twinned decahedral Au seeds formed from very initial metal clusters. The bottom line displays the two reaction paths: the first one performed in EG and supported by microwave irradiation (left side), the second one carried out in dimethylformamide using traditional oil-bath heating approach (right side). The picture reports drawings and TEM images which highlight the shapes of core-shell particles differently evolved by initial Au seed shape. Each single case is supported by two TEM images at different Ag:Au molar ratios. Thicker shells correspond to higher Ag:Au molar ratios. The whole picture has been readapted from Refs. [207, 216, 253] with permissions, copyrights American Chemical Society and Elsevier, respectively.

twinned decahedrons all uniquely exposing {111} crystal surfaces. Later when Ag^+ ions were reduced, the particle morphology developed according to the shape of pristine Au seeds although a non-negligible kinetic role is played by PVP molecules in selectively stabilizing one NC surface with respect to another at different Ag/Au molar ratios. Au cores having the shape of triangular/hexagonal twin plates favored the overgrown of truncated bipyramidal crystals at low Ag/Au molar ratios and fully developed triangular bipyramids at higher precursor molar ratios further showing thickening of the shell. (Fig. 10.10) Chemical analysis performed by energy-dispersive X-ray spectroscopy (TEM-EDS) confirmed the Au–Ag core-shell geometry and ruled the alloy existence completely out. In the case of octahedral cores, low Ag/Au molar ratios promoted square cubic core-shell crystals further growing at higher precursor molar ratios up to reach a cube edge length of 40 nm (Fig. 10.10). Finally, pentagonal multiple-twinned decahedral Au cores facilitated the growth of rods or wires with pentagonal section depending on the Ag/Au molar ratios; Ag growth occurred from the {111} facets of seeds in the two <110> opposite directions at the same rate [207, 253] (Fig. 10.10). In all these mentioned cases core-shell NCs were uniquely enclosed by {100} facets. On the contrary when Au^{+3} and Ag^{+} ions were mixed and reduced together in one pot reaction only spherical Au–Ag alloys were observed without any core-shell architecture.

Different physico-chemical reaction conditions catalyzed a distinctive kinetic path when similar Au seeds were used as shell-growth-promoters in DMF and subjected to traditional heating: in this case core-shell nanostructures exposing only {111} surface facets were observed [207]. A DMF surrounding provided epitaxial silver growth fully preserving the seed geometry thereof from triangular/hexagonal plate-like Au seeds triangular/hexagonal plate-like Au–Ag core-shell particles were formed otherwise from octahedral/decahedral Au seeds, octahedrons/decahedrons core-shell originated [207] (Fig. 10.10).

Based on their studies Tsuji *et al.* could ultimately outline two main reaction pathways: a first metal seed-independent route which provides homogeneous nucleation of Ag^{+} ions forming small clusters whose subsequent growth was uniquely assisted by PVP to produce isotropic particles and a second way based on preformation of metal seeds which then catalyze heterogeneous silver nucleation in the extent above mentioned [252]. The authors reported further mechanistic details extending their reaction studies both to longer MW irradiation times and to a three-steps synthesis procedure namely employing the core-shell particles as growth-promoters themselves for another seeded reaction. In the first case, very long irradiation times induced additional nucleation of Ag nanospheres on the originally formed Au–Ag wires/rods in all likelihood promoted by local heating of the preformed metallic surfaces whereas, in the second case, Ag elongated particles thicker and with higher aspect ratio could be grew seeding with core-shell elongated particles a new $AgNO_3$ reduction in EG under MWs, in this case bent rods/wires were evident as well [252]. The existence in the reaction environment of chloride anions played a competitive role in the growth of Au–Ag core-shell NCs; their interfering behavior was discussed both in terms of a decrease of the free Ag^+ in solution due to precipitation of low soluble salt AgCl and shape selective oxidative etching caused by Cl^-/O_2 [18]. Interestingly, unlike spherical NCs which underwent an oxidative etching effect, cubic, twinned bi-pyramidal and penta-twinned rods/wires were unexpectedly resistant against chemical corrosion [207, 254].

Harpeness *et al.* reported the synthesis of roughly spherical Au-Pd core-shell NCs by the MW-polyol-assisted simultaneous reduction of gold and palladium salts. In this case the largely different Au^{3+} or Au^+ and Pd_2^+ redox potentials firstly ensured gold reduction and the formation of metal seeds which in turn promoted heterogeneous nucleation of Pd [255] (Fig. 10.11a). In a similar fashion Pd-Pt core-shell NCs consisting of cubic Pd NC evenly decorated by Pt pellets was very recently described [256] (Fig. 10.11b).

Hybrid synthesis approaches based on first convective and second dielectric heating allowed to fabricate core-shell NCs composed of 50 nm Ag cores and 5 nm of polydiacetylene (PADA) shell. Ag is initially reduced in aqueous sodium citrate solution then in a second step uniform and concentric PADA precipitation on metal seed is MW-promoted (Fig. 10.11c).

Bimetallic alloys consisting in mixtures of two metallic elements giving rise to a single homogeneous solid phase have been under investigations in the last years, as for example iron-platinum-based alloys due to their magnetic properties [260]. MW-assisted synthesis of spherical monodisperse fcc FePt and FePd nanoparticles, respectively produced by the reduction in organic solvents (octylether and nonadecane) of corresponding metal acetylacetonate precursors

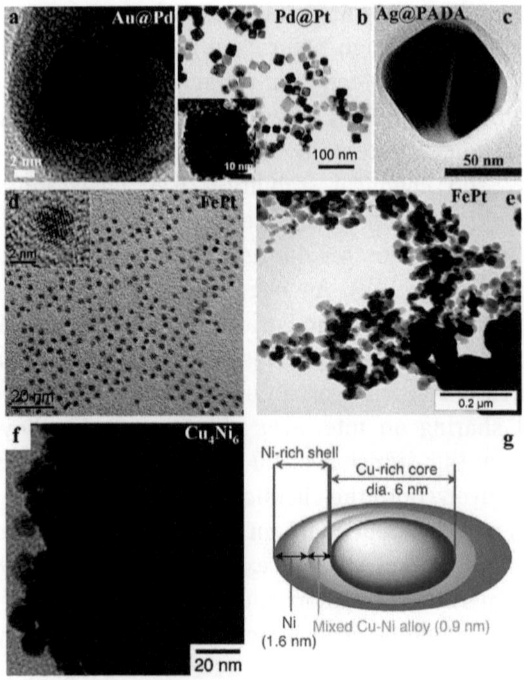

Figure 10.11 (a) HRTEM image of core-shell Au-Pd nanoparticle. Image reproduced from Ref. [255] with permission copyright American Chemical Society. (b) TEM image of Pd-Pt core-shell nanostructures synthesized at 3:1 $PdCl_2$:K_2PtCl_4 molar ratio. Reproduced from Ref. [256] with permission copyright American Chemical Society. (c) TEM image of Ag-polydiacetylene core-shell NCs reproduced from Ref. [257] with permission copyright Elsevier. (d) TEM picture of monodisperse fcc FePt nanoparticles reproduced from Ref. [208] with permission, copyright Royal Society of Chemistry. Inset shows alloy core-shell particle with a fringe spacing consistent with the (200) plane of fcc structure. (e) Low resolution TEM image of FePt alloy NCs after annealing at 600°C in Ar atmosphere. Reproduced from Ref. [258] with permission, copyright Royal Society of Chemistry. (f) TEM image of Cu_4Ni_6 alloy NCs drop-casted from hexane-based solution. (g) Schematic Cu_4Ni_6 nanostructure as determined by HAADF-STEM analysis. (f) and (g) pictures are adapted from Ref. [259] with permission, copyright Royal Society of Chemistry.

and the support of OLAM and/or OLAC ligands was described. When performed at high pressure a mixture of NCs showing fcc and ferromagnetic face-centered-tetragonal (fct) crystal phases was shown [208] (Fig. 10.11d). The fct phase of FePt binary alloys has received significant attention due to its high coercitivity, magnetic anisotropy and chemical stability [261]. Based on a different reduction efficiency of metal precursors under MWs, Harpeness *et al.* obtained similar results using platinum chloride and iron acetate [258] (Fig. 11e). An interesting core-shell system of Cu-Ni nanoparticles was demonstrated upon the reduction of Cu_2^+ and Ni_2^+ formate complexes sustained by OLAM ligands; each single particle consisted of a Cu-rich core surrounded by a Ni-rich shell sharing an interlayer of mixed Cu-Ni alloy the latter generated by the diffusion of Cu or Ni atoms into Ni shell or Cu core respectively. Both the chemical composition and the magnetic properties resulted dependent on the Cu:Ni precursor molar ratios [259] (Fig. 10.11f,g). Additional examples of binary alloys comprise spherical Au–Ag alloy nanoparticles in the 5–50 nm size range obtained in aqueous environment by using polyacrylamide as stabilizing agent and hydrazine as reducing agent [262]; PVP-stabilized PtNi nanoparticles (diameter 3–6 nm) prepared by hydrazine-mediated reduction in the presence of carbon [263]; alloyed and PVP-coated size-monodisperse PtRu nanoparticles with 2–3 nm of average diameters grown in EG [264] and finally a wide collection of bimetallic alloyed NCs, up to 25 nm in size, created by all the possible combinations of the following metals Au, Ag, Pt, Pd, Ru, Rh, Cu, Ni [246].

10.4 Conclusions

The technologycal scenario offered by hybrid NCs, namely the opportunity to develop solid platforms of which physico-chemical properties can be manipulated by the morphology and/or composition of the single component unit, has invested colloidal metal NCs with a key role in diverse areas of material science.

Generally, synthetic chemistry and, particularly experimental approaches based on liquid phase chemistry intensively aim at meeting the unique academic and technological requests of disparate arrays of differently-shaped colloidal metal NCs topologically-set according to disparate simple and hybrid configurations and morphologies.

In this chapter we have reviewed all the advancements of the wet-chemical synthesis approaches to colloidal metal NCs highlighting the basic principles which govern this part of chemistry and specifically, how their elegant manipulation influences the synthesis outcomes. Beyond this, we have tried to delineate the synthetic effectiveness of thermal energy when delivered via traditional heat transfer modes or via microwaves.

Although the physico-chemical differences between the two heating methodologies, no discordant outcomes are evident, except an effective influence of the profile heating rates up to reduce the growth times from hours, in the case of convective heating, to minutes or seconds for MW irradiation. Despite of this, unexpected *microwave effects* often emerging in different temporal phases of the synthesis experiment may strongly behave as ungovernable kinetic factors. On the other hand, MW-based methodologies show a large experimental flexibility, as they do not require complete solubility of chemical precursors, thermal decomposition can be promoted in the course of selective heating.

A wealth of theoretical basics, experimental cases and the straightforward reading of the heating effects, not always unequivocally allowable in the MW-mediated syntheses, have made procedures promoted by traditional convective heating widely privileged.

References

1. Murray, W. A., and Barnes, W. L. (2007) Plasmonic materials, *Adv. Mater.*, **19**(22), 3771–3782.
2. Liz-Marzán, L. M., (2005) Tailoring surface plasmons through the morphology and assembly of metal nanoparticles, *Langmuir*, **22**(1), 32–41.
3. Fang, P. P., Duan, S., Lin, X. D., Anema, J. R., Li, J. F., Buriez, O., Ding, Y., Fan, F. R., Wu, D. Y., Ren, B., Wang, Z. L., Amatore, C., and Tian, Z. Q. (2011) Tailoring Au-core Pd-shell Pt-cluster nanoparticles for enhanced electrocatalytic activity, *Chem. Sci.*, **2**(3), 531–539.

4. Alayoglu, S., Nilekar, A. U., Mavrikakis, M., and Eichhorn, B.(2008) Ru-Pt core-shell nanoparticles for preferential oxidation of carbon monoxide in hydrogen, *Nat. Mater.*, **7**(4), 333–338.
5. Tao, F., Grass, M. E., Zhang, Y., Butcher, D. R., Renzas, J. R., Liu, Z., Chung, J. Y., Mun, B. S., Salmeron, M., and Somorjai, G. A. (2008) Reaction-driven restructuring of Rh-Pd and Pt-Pd core-shell nanoparticles, *Science*, **322**(5903), 932–934.
6. Murphy, C. J., Gole, A. M., Stone, J. W., Sisco, P. N., Alkilany, A. M., Goldsmith, E. C., and Baxter, S. C. (2008) Gold nanoparticles in biology: Beyond toxicity to cellular imaging, *Accounts Chem. Res.*, **41**(12), 1721–1730.
7. Rosi, N. L., and Mirkin, C. A. (2005) Nanostructures in biodiagnostics, *Chem. Rev.*, **105**(4), 1547–1562.
8. Jain, P. K., Huang, X., El-Sayed, I. H., and El-Sayed, M. A. (2008) Noble metals on the nanoscale: Optical and photothermal properties and some applications in imaging, sensing, biology, and medicine, *Accounts Chem. Res.*, **41**(12), 1578–1586.
9. Haes, A. J., and Van Duyne, R. P. (2002) A nanoscale optical biosensor: Sensitivity and selectivity of an approach based on the localized surface plasmon resonance spectroscopy of triangular silver nanoparticles, *J. Am. Chem. Soc.*, **124**(35), 10596–10604.
10. Lal, S., Link, S., and Halas, N. J. (2007) Nano-optics from sensing to waveguiding, *Nat. Photonics*, **1**(11), 641–648.
11. Jackson, J. B., and Halas, N. J. (2001) Silver nanoshells variations in morphologies and optical properties, *J. Phys. Chem. B*, **105**(14), 2743–2746.
12. Gao, F., Lu, Q. Y., and Komarneni, S. (2005) Interface reaction for the self-assembly of silver nanocrystals under microwave-assisted solvothermal conditions, *Chem. Mater.*, **17**(4), 856–860.
13. Jana, N.R., and Peng, X. (2003) Single-phase and gram-scale routes toward nearly monodisperse Au and other noble metal nanocrystals, *J. Am. Chem. Soc.*, **125**(47), 14280–14281.
14. Nikoobakht, B., and El-Sayed, M. A. (2003) Preparation and growth mechanism of gold nanorods (NRs) using seed-mediated growth method, *Chem. Mater.*, **15**(10), 1957–1962.
15. Gou, L., Chipara, M., and Zaleski, J. M. (2007) Convenient, rapid synthesis of Ag nanowires, *Chem. Mater.*, **19**(7), 1755–1760.
16. Sun, Y., and Xia, Y. (2002) Shape-controlled synthesis of gold and silver nanoparticles, *Science*, **298**(5601), 2176–2179.

17. Sun, Y., Wiley, B., Li, Z.-Y., and Xia, Y. (2004) Synthesis and optical properties of nanorattles and multiple-walled nanoshells/nanotubes made of metal alloys, *J. Am. Chem. Soc.*, **126**(30), 9399–9406.
18. Xiong, Y., Wiley, B., Chen, J., Li, Z.-Y., Yin, Y., and Xia, Y. (2005) Corrosion-based synthesis of single-crystal Pd nanoboxes and nanocages and their surface plasmon properties, *Angew. Chem. Int. Edit.*, **44**(48), 7913–7917.
19. Lu, X., Au, L., McLellan, J., Li, Z.-Y., Marquez, M., and Xia, Y. (2007) Fabrication of cubic nanocages and nanoframes by dealloying Au/Ag alloy nanoboxes with an aqueous etchant based on $Fe(NO_3)_3$ or NH_4OH, *Nano Lett.*, **7**(6), 1764–1769.
20. Khalavka, Y., Becker, J., and Sönnichsen, C. (2009) Synthesis of rod-shaped gold nanorattles with improved plasmon sensitivity and catalytic activity, *J. Am. Chem. Soc.*, **131**(5), 1871–1875.
21. Guo, S., Dong, S., and Wang, E. (2008) A general method for the rapid synthesis of hollow metallic or bimetallic nanoelectrocatalysts with urchinlike morphology, *Chem. Eur. J.*, **14**(15), 4689–4695.
22. Park, J., Joo, J., Kwon, S. G., Jang, Y., and Hyeon, T. (2007) Synthesis of monodisperse spherical nanocrystals, *Angew. Chem. Int. Edit.*, **46**(25), 4630–4660.
23. Carbone, L., and Cozzoli, P. D. (2010) Colloidal heterostructured nanocrystals: Synthesis and growth mechanisms, *Nano Today*, **5**(5), 449–493.
24. Cozzoli, P. D., Pellegrino, T., and Manna, L. (2006) Synthesis, properties and perspectives of hybrid nanocrystal structures, *Chem. Soc. Rev.*, **35**(11), 1195–1208.
25. Tao, A. R., Habas, S., and Yang, P. (2008) Shape control of colloidal metal nanocrystals, *Small*, **4**(3), 310–325.
26. Xia, Y., Xiong, Y., Lim, B., and Skrabalak, S. E. (2009) Shape-controlled synthesis of metal nanocrystals: Simple chemistry meets complex physics, *Angew. Chem. Int. Edit.*, **48**(1), 60–103.
27. Enüstün, B. V., and Turkevich, J. (1963) Coagulation of colloidal gold, *J. Am. Chem. Soc.*, **85**(21), 3317–3328.
28. Gole, A., and Murphy, C. J. (2004) Seed-mediated synthesis of gold nanorods role of the size and nature of the seed, *Chem. Mater.*, **16**(19), 3633–3640.
29. Petit, C., Taleb, A., and Pileni, M.-P. (1998) Self-organization of magnetic nanosized cobalt particles, *Adv. Mater.*, **10**(3), 259–261.

30. Maillard, M., Giorgio, S., and Pileni, M.-P. (2002) Silver nanodisks, *Adv. Mater.*, **14**(15), 1084–1086.
31. Pileni, M.-P. (2007) Control of the size and shape of inorganic nanocrystals at various scales from nano to macrodomains, *J. Phys. Chem. C*, **111**(26), 9019–9038.
32. Liu, M., and Guyot-Sionnest, P. (2005) Mechanism of silver(I)-assisted growth of gold nanorods and bipyramids, *J. Phys. Chem. B*, **109**(47), 22192–22200.
33. Wang, Z. L. (2000) Transmission electron microscopy of shape-controlled nanocrystals and their assemblies, *J. Phys. Chem. B*, **104**(6), 1153–1175.
34. Xiong, Y., and Xia, Y. (2007) Shape-controlled synthesis of metal nanostructures: The case of palladium, *Adv. Mater.*, **19**(20), 3385–3391.
35. Lim, B., Camargo, P. H. C., and Xia, Y. (2008) Mechanistic study of the synthesis of Au nanotadpoles, nanokites, and microplates by reducing aqueous $HAuCl_4$ with poly(vinyl pyrrolidone), *Langmuir*, **24**(18), 10437–10442.
36. Xiong, Y., Siekkinen, A. R., Wang, J., Yin, Y., Kim, M. J., and Xia, Y. (2007) Synthesis of silver nanoplates at high yields by slowing down the polyol reduction of silver nitrate with polyacrylamide, *J. Mater. Chem.*, **17**(25), 2600–2602.
37. Germain, V., Li, J., Ingert, D., Wang, Z. L., and Pileni, M. P. (2003) Stacking faults in formation of silver nanodisks, *J. Phys. Chem. B*, **107**(34), 8717–8720.
38. Jun, Y.-W., Lee, S.-M., Kang, N.-J., and Cheon, J. (2001) Controlled synthesis of multi-armed CdS nanorod architectures using monosurfactant system, *J. Am. Chem. Soc.*, **123**(21), 5150–5151.
39. Lee, S.M., Cho, S. N., and Cheon, J. (2003) Anisotropic shape control of colloidal inorganic nanocrystals, *Adv. Mater.*, **15**(5), 441–444.
40. Wang, Z. L., Gao, R. P., Nikoobakht, B., and El-Sayed, M. A. (2000) Surface reconstruction of the unstable {110} surface in gold nanorods, *J. Phys. Chem. B*, **104**(23), 5417–5420.
41. Ledentsov, N., Ustinov, V., Shchukin, V., Kop'ev, P., Alferov, Z., and Bimberg, D. (1998) Quantum dot heterostructures: Fabrication, properties, lasers (Review), *Semiconductors*, **32**(4), 343–365.
42. Fan, F.-R., Liu, D.-Y., Wu, Y.-F., Duan, S., Xie, Z.-X., Jiang, Z.-Y., and Tian, Z.-Q. (2008) Epitaxial growth of heterogeneous metal nanocrystals: From gold nano-octahedra to palladium and silver nanocubes, *J. Am. Chem. Soc.*, **130**(22), 6949–6951.

43. Jones, M. R., Osberg, K. D., Macfarlane, R. J., Langille, M. R., and Mirkin, C. A. (2011) Templated techniques for the synthesis and assembly of plasmonic nanostructures, *Chem. Rev.*, **111**(6), 3736–3827.
44. Martin, C. R., (1996) Membrane-based synthesis of nanomaterials, *Chem. Mater.*, **8**(8), 1739–1746.
45. Kline, T. R., Tian, M., Wang, J., Sen, A., Chan, M. W. H., and Mallouk, T. E. (2006) Template-grown metal nanowires, *Inorg. Chem.*, **45**(19), 7555–7565.
46. Martin, B. R., Dermody, D. J., Reiss, B. D., Fang, M., Lyon, L. A., Natan, M. J., and Mallouk, T. E. (1999) Orthogonal self-assembly on colloidal gold-platinum nanorods, *Adv. Mater.*, **11**(12), 1021–1025.
47. Hurst, S. J., Payne, E. K., Qin, L., and Mirkin, C. A. (2006) Multisegmented one-dimensional nanorods prepared by hard-template synthetic methods, *Angew. Chem. Int. Edit.*, **45**(17), 2672–2692.
48. Penner, R. M., and Martin, C. R. (1986) Controlling the morphology of electronically conductive polymers, *J. Electrochem. Soc.*, **133**(10), 2206–2207.
49. Al-Mawlawi, D., Liu, C. Z., and Moskovits, M. (1994) Nanowires formed in anodic oxide nanotemplates, *J. Mater. Res.*, **9**(04), 1014–1018.
50. Beck, J. S., Vartuli, J. C., Roth, W. J., Leonowicz, M. E., Kresge, C. T., Schmitt, K. D., Chu, C. T. W., Olson, D. H., and Sheppard, E. W. (1992) A new family of mesoporous molecular sieves prepared with liquid crystal templates, *J. Am. Chem. Soc.*, **114**(27), 10834–10843.
51. Wu, C.-G., and Bein, T. (1994) Conducting polyaniline filaments in a mesoporous channel host, *Science*, **264**(5166), 1757–1759.
52. Tonucci, R. J., Justus, B. L., Campillo, A. J., and Ford, C. E. (1992) Nanochannel array glass, *Science*, **258**(5083), 783–785.
53. Wang, J.-G., Tian, M.-L., Kumar, N., and Mallouk, T. E. (2005) Controllable template synthesis of superconducting Zn nanowires with different microstructures by electrochemical deposition, *Nano Lett.*, **5**(7), 1247–1253.
54. Tian, M., Wang, J., Kurtz, J., Mallouk, T. E., and Chan, M. H. W. (2003) Electrochemical growth of single-crystal metal nanowires via a two-dimensional nucleation and growth mechanism, *Nano Lett.*, **3**(7), 919–923.
55. Wang, J., Tian, M., Mallouk, T. E., and Chan, M. H. W. (2003) Microtwinning in template-synthesized single-crystal metal nanowires, *J. Phys. Chem. B*, **108**(3), 841–845.

56. Tian, M., Wang, J., Snyder, J., Kurtz, J., Liu, Y., Schiffer, P., Mallouk, T. E., and Chan, M. H. W. (2003) Synthesis and characterization of superconducting single-crystal Sn nanowires, *Appl. Phys. Lett.*, **83**, 1620–1622.
57. Dubois, S., Chassaing, E., Duvail, J. L., Piraux, L., and Waals, M. G. (1999) Preparation and characterization of electrodeposited Fe and Fe/Cu nanowires, *J. Chim. Phys. Chim. Biol.*, **96**(8), 1316–1331.
58. de Horne, F. M., Piraux, L., and Michotte, S. (2005) Fabrication and physical properties of Pb/Cu multilayered superconducting nanowires, *Appl. Phys. Lett.*, **86**, 152510/1-3.
59. Wang, J.-G., Tian, M.-L., Mallouk, T. E., and Chan, M. H. W. (2004) Microstructure and interdiffusion of template-synthesized Au/Sn/Au junction nanowires, *Nano Lett.*, **4**(7), 1313–1318.
60. Corredor, J. T., Rojas, C. E., and Jorge, G. A. (2010) Characterization of Fe/Cu multilayers prepared by electrodeposition, *Acta Microsc.*, **19**, 23–31.
61. Nicewarner-Peña, S. R., Freeman, R. G., Reiss, B. D., He, L., Peã, D. J., Walton, I. D., Cromer, R., Keating, C. D., and Natan, M. J. (2001) Submicrometer metallic barcodes, *Science*, **294**(5540), 137–141.
62. Liu, K., Nagodawithana, K., Searson, P. C., and Chien, C. L. (1995) Perpendicular giant magnetoresistance of multilayered Co/Cu nanowires, *Phys. Rev. B*, **51**(11), 7381–7384.
63. Mallet, J., Yu-Zhang, K., Chien, C. L., Eagleton, T. S., and Searson, P. C. (2004) Fabrication and magnetic properties of fcc Co_xPt_{1-x} nanowires, *Appl. Phys. Lett.*, **84**(19), 3900–3902.
64. Asefa, T., and Lennox, R. B. (2005) Synthesis of gold nanoparticles via electroless deposition in SBA-15, *Chem. Mater.*, **17**(10), 2481–2483.
65. Dubrovsky, T. B., Hou, Z., Stroeve, P., and Abbott, N. L. (1998) Self-assembled monolayers formed on electroless gold deposited on silica gel: A potential stationary phase for biological assays, *Anal. Chem.*, **71**(2), 327–332.
66. Menon, V. P., and Martin, C. R. (1995) Fabrication and evaluation of nanoelectrode ensembles, *Anal. Chem.*, **67**(13), 1920–1928.
67. Gu, J., Shi, J., Xiong, L., Chen, H., Li, L., and Ruan, M. (2004) A new strategy to incorporate high density gold nanowires into the channels of mesoporous silica thin films by electroless deposition, *Solid State Sci.*, **6**(7), 747–752.

68. Byeon, J. H., and Kim, J. W. (2011) Fabrication of a gold island film on porous polymeric substrates by a strategic electroless deposition, *J. Electrochem. Soc.*, **158**(1), D15-D20.
69. Zhang, Z., Dai, S., Blom, D. A., and Shen, J. (2002) Synthesis of ordered metallic nanowires inside ordered mesoporous materials through electroless deposition, *Chem. Mater.*, **14**(3), 965–968.
70. Ko, C. H., and Ryoo, R. (1996) Imaging the channels in mesoporous molecular sieves with platinum, *Chem. Commun.*, (21), 2467–2468.
71. Kijima, T. (2010) Synthesis and applications of noble-metal nanotubes, in *Inorganic and Metallic Nanotubular Materials* (ed. Kijima, T.), Springer-Verlag, Berlin.
72. Brumlik, C. J., Menon, V. P., and Martin, C. R. (1994) Template synthesis of metal microtubule ensembles utilizing chemical, electrochemical, and vacuum deposition techniques, *J. Mater. Res.*, **9**(5), 1174–1183.
73. de Horne, F. D., Piraux, L., and Michotte, S. (2009) Electroless template grown superconducting lead and tin nanotubes, *Nanotechnology*, **20**(38), 385603–385608.
74. Saunders, A. E., Popov, I., and Banin, U. (2007) Synthesis and characterization of organic-soluble Ag/AgBr dimer nanocrystals, *Z. Anorg. Allg. Chem.*, **633**(13–14), 2414–2419.
75. Link, S., Wang, Z. L., and El-Sayed, M. A. (1999) Alloy formation of gold-silver nanoparticles and the dependence of the plasmon absorption on their composition, *J. Phys. Chem. B*, **103**(18), 3529–3533.
76. Yu, Y. Y., Chang, S.-S., Lee, C.-L., and Wang, C. R. C. (1997) Gold nanorods electrochemical synthesis and optical properties, *J. Phys. Chem. B*, **101**(34), 6661–6664.
77. Zhou, Z. Y., Tian, N., Huang, Z. Z., Chen, D. J., and Sun, S. G. (2008) Nanoparticle catalysts with high energy surfaces and enhanced activity synthesized by electrochemical method, *Faraday Discuss.*, **140**, 81–92.
78. Pérez-Juste, J., Pastoriza-Santos, I., Liz-Marzán, L. M., and Mulvaney, P. (2005) Gold nanorods: Synthesis, characterization and applications, *Coordin. Chem. Rev.*, **249**(17–18), 1870–1901.
79. Yin, B. S., Ma, H. Y., and Chen, S. H. (2004) New progress in electrochemical preparation of nanomaterials, *Prog. Chem.*, **16**(2), 196–203.
80. Shen, L. M., Yao, J. L., and Gu, R. A. (2007) Synthesis of gold nanoparticles by two step electrolysis and investigation on their SERS activities, *Acta Chim. Sinica*, **65**(3), 203–206.

81. Nagaraju, D. H., and Lakshminarayanan, V. (2008) Electrochemical synthesis of thiol-monolayer-protected clusters of gold, *Langmuir*, **24**(24), 13855–13857.
82. Huang, C. J., Chiu, P. H., Wang, Y. H., Chen, K. L., Linn, J. J., and Yang, C. F. (2006) Electrochemically controlling the size of gold nanoparticles, *J. Electrochem. Soc.*, **153**(12), D193–D198.
83. Chang, S.-S., Shih, C.-W., Chen, C.-D., Lai, W.-C., and Wang, C. R. C. (1998) The shape transition of gold nanorods, *Langmuir*, **15**(3), 701–709.
84. Huang, C. J., Chiu, P. H., Wang, Y. H., and Yang, C. F. (2006) Synthesis of the gold nanodumbbells by electrochemical method, *J. Colloid Interf. Sci.*, **303**(2), 430–436.
85. Huang, C. J., Chiu, P. H., Wang, Y. H., Chen, R. W., and Meen, T.-H. (2006) Synthesis of the gold nanocubes by electrochemical technique, *J. Electrochem. Soc.*, **153**(8), D129-D133.
86. Huang, C.-J., Wang, Y.-H., Chiu, P.-H., Shih, M.-C., and Meen, T.-H. (2006) Electrochemical synthesis of gold nanocubes, *Mater. Lett.*, **60**(15), 1896–1900.
87. Pan, W., Zhang, X. K., Ma, H. Y., and Zhang, J. T. (2008) Electrochemical synthesis, voltammetric behavior, and electrocatalytic activity of Pd nanoparticles, *J. Phys. Chem. C*, **112**(7), 2456–2461.
88. Zhou, M., Chen, S. H., Zhao, S. Y., and Ma, H. Y. (2006) One-step synthesis of Au–Ag alloy nanoparticles by a convenient electrochemical method, *Physica E*, **33**(1), 28–34.
89. Zhu, J., Zhu, X., and Wang, Y. C. (2005) Electrochemical synthesis and fluorescence spectrum properties of silver nanospheres, *Micro. Eng.*, **77**(1), 58–62.
90. Yang, X. G., Chen, S. H., Zhao, S. Y., Li, D. G., and Ma, H. Y. (2003) Synthesis of copper nanorods using electrochemical methods, *J. Serb. Chem. Soc.*, **68**(11), 843–847.
91. Tian, N., Zhou, Z.-Y., Sun, S.-G., Ding, Y., and Wang, Z. L. (2007) Synthesis of tetrahexahedral platinum nanocrystals with high-index facets and high electro-oxidation activity, *Science*, **316**(5825), 732–735.
92. Ding, Y., Gao, Y., Wang, Z. L., Tian, N., Zhou, Z.-Y., and Sun, S.-G. (2007), Facets and surface relaxation of tetrahexahedral platinum nanocrystals, *Appl. Phys. Lett.*, **91**, 121901/1-3.
93. Huang, X., Qi, X. Y., Huang, Y. Z., Li, S. Z., Xue, C., Gan, C. L., Boey, F., and Zhang, H. (2010) Photochemically controlled synthesis of anisotropic Au nanostructures: Platelet-like Au nanorods and six-star Au nanoparticles, *ACS Nano*, **4**(10), 6196–6202.

94. Sakamoto, M., and Majima, T. (2010) Photochemistry for the synthesis of noble metal nanoparticles, *B. Chem. Soc. Jpn*, **83**(10), 1133–1154.

95. Han, M.Y., and Quek, C. H. (1999) Photochemical synthesis in formamide and room-temperature Coulomb staircase behavior of size-controlled gold nanoparticles, *Langmuir*, **16**(2), 362–367.

96. Esumi, K., Matsuhisa, K., and Torigoe, K. (1995) Preparation of rodlike gold particles by UV irradiation using cationic micelles as a template, *Langmuir*, **11**(9), 3285–3287.

97. Dong, S.A., and Zhou, S. P. (2007) Photochemical synthesis of colloidal gold nanoparticles, *Mater. Sci. Eng. B*, **140**(3), 153–159.

98. Kim, F., Song, J. H., and Yang, P. (2002) Photochemical synthesis of gold nanorods, *J. Am. Chem. Soc.*, **124**(48), 14316–14317.

99. Placido, T., Comparelli, R., Giannici, F., Cozzoli, P. D., Capitani, G., Striccoli, M., Agostiano, A., and Curri, M. L. (2009) Photochemical synthesis of water-soluble gold nanorods: The role of silver in assisting anisotropic growth, *Chem. Mater.*, **21**(18), 4192–4202.

100. Ahmed, M., and Narain, R. (2010) Rapid synthesis of gold nanorods using a one-step photochemical strategy, *Langmuir*, **26**(23), 18392–18399.

101. Kim, J. U., Cha, S. H., Shin, K., Jho, J. Y., and Lee, J. C. (2004) Preparation of gold nanowires and nanosheets in bulk block copolymer phases under mild conditions, *Adv. Mater.*, **16**(5), 459–464.

102. Chen, H., Jia, J., and Dong, S. (2007) Photochemical formation of silver and gold nanostructures at the air-water interface and their electrocatalytic properties, *Nanotechnology*, **18**(24), 245601/1–8.

103. Zheng, X., Xu, W., Corredor, C., Xu, S., An, J., Zhao, B., and Lombardi, J. R. (2007) Laser-induced growth of monodisperse silver nanoparticles with tunable surface plasmon resonance properties and a wavelength self-limiting effect, *J. Phys. Chem. C*, **111**(41), 14962–14967.

104. Zheng, X. L., Zhao, X. J., Guo, D. W., Tang, B., Xu, S. P., Zhao, B., Xu, W. Q., and Lombardi, J. R. (2009) Photochemical formation of silver nanodecahedra: Structural selection by the excitation wavelength, *Langmuir*, **25**(6), 3802–3807.

105. Huang, W. C., and Chen, Y. C. (2008) Photochemical synthesis of polygonal gold nanoparticles, *J. Nanopart. Res.*, **10**(4), 697–702.

106. Kundu, S., and Liang, H. (2009) Photoinduced formation of shape-selective Pt nanoparticles, *Langmuir*, **26**(9), 6720–6727.

107. Kundu, S., Wang, K., and Liang, H. (2009) Photochemical generation of catalytically active shape selective rhodium nanocubes, *J. Phys. Chem. C*, **113**(43), 18570–18577.
108. Jin, R., Cao, Y., Mirkin, C. A., Kelly, K. L., Schatz, G. C., and Zheng, J. G. (2001) Photoinduced conversion of silver nanospheres to nanoprisms, *Science*, **294**(5548), 1901–1903.
109. Link, S., Burda, C., Mohamed, M. B., Nikoobakht, B., and El-Sayed, M. A. (1999) Laser photothermal melting and fragmentation of gold nanorods: Energy and laser pulse-width dependence, *J. Phys. Chem. A*, **103**(9), 1165–1170.
110. Cortie, M. B., and McDonagh, A. M. (2011) Synthesis and optical properties of hybrid and alloy plasmonic nanoparticles, *Chem. Rev.*, **111**(6), 3713–3735.
111. Sun, Y., Mayers, B. T., and Xia, Y. (2002) Template-engaged replacement reaction: A one-step approach to the large-scale synthesis of metal nanostructures with hollow interiors, *Nano Lett.*, **2**(5), 481–485.
112. Sun, Y., and Xia, Y. (2004) Mechanistic study on the replacement reaction between silver nanostructures and chloroauric acid in aqueous medium, *J. Am. Chem. Soc.*, **126**(12), 3892–3901.
113. Yin, Y., Erdonmez, C., Aloni, S., and Alivisatos, A. P. (2006) Faceting of nanocrystals during chemical transformation from solid silver spheres to hollow gold octahedra, *J. Am. Chem. Soc.*, **128**(39), 12671–12673.
114. Selvakannan, P. R., and Sastry, M. (2005) Hollow gold and platinum nanoparticles by a transmetallation reaction in an organic solution, *Chem. Commun.*, 13, 1684–1686.
115. Schwartzberg, A. M., Olson, T. Y., Talley, C. E., and Zhang, J. Z. (2006) Synthesis, characterization, and tunable optical properties of hollow gold nanospheres, *J. Phys. Chem. B*, **110**(40), 19935–19944.
116. Liang, H.-P., Wan, L.-J., Bai, C.-L., and Jiang, L. (2005) Gold hollow nanospheres: Tunable surface plasmon resonance controlled by interior-cavity sizes, *J. Phys. Chem. B*, **109**(16), 7795–7800.
117. Chen, M., and Gao, L. (2006) Synthesis and characterization of Ag nanoshells by a facile sacrificial template route through *in situ* replacement reaction, *Inorg. Chem.*, **45**(13), 5145–5149.
118. Chen, J., Wiley, B., McLellan, J., Xiong, Y., Li, Z.-Y.,and Xia, Y. (2005) Optical properties of Pd-Ag and Pt-Ag nanoboxes synthesized via galvanic replacement reactions, *Nano Lett.*, **5**(10), 2058–2062.

119. Sun, Y., and Xia, Y. (2003) Triangular nanoplates of silver: Synthesis, characterization, and use as sacrificial templates for generating triangular nanorings of gold, *Adv. Mater.*, **15**(9), 695–699.

120. Métraux, G. S., Cao, Y. C., Jin, R., and Mirkin, C. A. (2003) Triangular nanoframes made of gold and silver, *Nano Lett.*, **3**(4), 519–522.

121. Shon, Y.-S., Dawson, G. B., Porter, M., and Murray, R. W. (2002) Monolayer-protected bimetal cluster synthesis by core metal galvanic exchange reaction, *Langmuir*, **18**(10), 3880–3885.

122. Huang, T., and Murray, R. W. (2003) Luminescence of tiopronin monolayer-protected silver clusters changes to that of gold clusters upon galvanic core metal exchange, *J. Phys. Chem. B*, **107**(30), 7434–7440.

123. Seo, D., and Song, H. (2009) Asymmetric hollow nanorod formation through a partial galvanic replacement reaction, *J. Am. Chem. Soc.*, **131**(51), 18210–18211.

124. Seo, D., Yoo, C. I., Jung, J., and Song, H. (2008) Ag–Au–Ag heterometallic nanorods formed through directed anisotropic growth, *J. Am. Chem. Soc.*, **130**(10), 2940–2941.

125. Sun, Y., and Xia, Y. (2004) Multiple-walled nanotubes made of metals, *Adv. Mater.*, **16**(3), 264–268.

126. Chen, H. M., Liu, R.-S., Asakura, K., Lee, J.-F., Jang, L.-Y., and Hu, S.-F. (2006) Fabrication of nanorattles with passive shell, *J. Phys. Chem. B*, **110**(39), 19162–19167.

127. Wang, Z., Luan, D., Li, C. M., Su, F., Madhavi, S., Boey, F. Y. C., and Lou, X. W. (2010) Engineering nonspherical hollow structures with complex interiors by template-engaged redox etching, *J. Am. Chem. Soc.*, **132**(45), 16271–16277.

128. Mahmoud, M. A., and El-Sayed, M. A. (2010) Gold nanoframes: Very high surface plasmon fields and excellent near-infrared sensors, *J. Am. Chem. Soc.*, **132**(36), 12704–12710.

129. Camargo, P. H. C., Xiong, Y., Ji, L., Zuo, J. M., and Xia, Y. (2007) Facile synthesis of tadpole-like nanostructures consisting of Au heads and Pd tails, *J. Am. Chem. Soc.*, **129**(50), 15452–15453.

130. Vinod, T. P., Yang, M., Kim, J., and Kotov, N. A. (2009) Self-guided one-sided metal reduction in Te nanowires leading to Au-Te matchsticks, *Langmuir*, **25**(23), 13545–13550.

131. Lin, Z.-H., Lin, Y.-W., Lee, K.-H., and Chang, H.-T. (2008) Selective growth of gold nanoparticles onto tellurium nanowires via a green chemical route, *J. Mater. Chem.*, **18**(22), 2569–2572.

132. Brust, M., Walker, M., Bethell, D., Schiffrin, D. J., and Whyman, R. (1994) Synthesis of thiol-derivatised gold nanoparticles in a two-phase liquid-liquid system, *J. Chem. Soc., Chem. Commun.*, 7, 801–802.
133. Brust, M., Fink, J., Bethell, D., Schiffrin, D. J., and Kiely, C. (1995) Synthesis and reactions of functionalised gold nanoparticles, *J. Chem. Soc., Chem. Commun.*, 16, 1655–1656.
134. Kurihara, L. K., Chow, G. M., and Schoen, P. E. (1995) Nanocrystalline metallic powders and films produced by the polyol method, *Nanostruct. Mater.*, **5**(6), 607–613.
135. Wiley, B., Sun, Y., Mayers, B., and Xia, Y. (2005) Shape-controlled synthesis of metal nanostructures the case of silver, *Chem. Eur. J.*, **11**(2), 454–463.
136. Kim, F., Connor, S., Song, H., Kuykendall, T., and Yang, P. (2004) Platonic gold nanocrystals, *Angew. Chem. Int. Edit.*, **43**(28), 3673–3677.
137. Tao, A., Sinsermsuksakul, P., and Yang, P. (2006) Polyhedral silver nanocrystals with distinct scattering signatures, *Angew. Chem. Int. Edit.*, **45**(28), 4597–4601.
138. Liang, H., Yang, H., Wang, W., Li, J., and Xu, H. (2009) High-yield uniform synthesis and microstructure-determination of rice-shaped silver nanocrystals, *J. Am. Chem. Soc.*, **131**(17), 6068–6069.
139. Wei, H., Reyes-Coronado, A., Nordlander, P., Aizpurua, J., and Xu, H. (2010) Multipolar plasmon resonances in individual Ag nanorice, *ACS Nano*, **4**(5), 2649–2654.
140. Li, C. C., Cai, W. P., Cao, B. Q., Sun, F. Q., Li, Y., Kan, C. X., and Zhang, L. D. (2006) Mass synthesis of large, single-crystal Au nanosheets based on a polyol process, *Adv. Func. Mater.*, **16**(1), 83–90.
141. Zhang, J., Gao, Y., Alvarez-Puebla, R. A., Buriak, J. M., and Fenniri, H. (2006) Synthesis and SERS properties of nanocrystalline gold octahedra generated from thermal decomposition of $HAuCl_4$ in block copolymers, *Adv. Mater.*, **18**(24), 3233–3237.
142. Sau, T. K., and Murphy, C. J. (2004) Room temperature, high-yield synthesis of multiple shapes of gold nanoparticles in aqueous solution, *J. Am. Chem. Soc.*, **126**(28), 8648–8649.
143. Lin, X. Z., Teng, X., and Yang, H. (2003) Direct synthesis of narrowly dispersed silver nanoparticles using a single-source precursor, *Langmuir*, **19**(24), 10081–10085.
144. Murray, C. B., Sun, S. H., Doyle, H., and Betley, T. (2001) Monodisperse 3d transition-metal (Co, Ni, Fe) nanoparticles and their assembly into nanoparticle superlattices, *MRS Bull.*, **26**(12), 985–991.

145. Sun, S., and Murray, C. B. (1999) Synthesis of monodisperse cobalt nanocrystals and their assembly into magnetic superlattices, *J. Appl. Phys.*, **85**, 4325–4330.

146. Yu, D., and Yam, V. W.-W. (2004) Controlled synthesis of monodisperse silver nanocubes in water, *J. Am. Chem. Soc.*, **126**(41), 13200–13201.

147. Sun, Y., Yin, Y., Mayers, B. T., Herricks, T., and Xia, Y. (2002) Uniform silver nanowires synthesis by reducing $AgNO_3$ with ethylene glycol in the presence of seeds and poly(vinyl pyrrolidone), *Chem. Mater.*, **14**(11), 4736–4745.

148. Sun, Y., Mayers, B., and Xia, Y. (2003) Transformation of silver nanospheres into nanobelts and triangular nanoplates through a Thermal Process, *Nano Lett.*, **3**(5), 675–679.

149. Tanori, J., and Pileni M.-P. (1995) Change in the shape of copper nanoparticles in ordered phases, *Adv. Mater.*, **7**(10), 862–864.

150. Pastoriza-Santos, I., and Liz-Marzán, L. M. (2002) Synthesis of silver nanoprisms in DMF, *Nano Lett.*, **2**(8), 903–905.

151. Uk Son, S., Kyu Park, I., Park, J., and Hyeon, T. (2004) Synthesis of Cu_2O coated Cu nanoparticles and their successful applications to Ullmann-type amination coupling reactions of aryl chlorides, *Chem. Commun.*, 7, 778–779.

152. Park, J., Kang, E., Son, S. U., Park, H. M., Lee, M. K., Kim, J., Kim, K. W., Noh, H. J., Park, J. H., Bae, C. J., Park, J. G., and Hyeon, T. (2005) Monodisperse nanoparticles of Ni and NiO: Synthesis, characterization, self-assembled superlattices, and catalytic applications in the suzuki coupling reaction, *Adv. Mater.*, **17**(4), 429–434.

153. Puntes, V. F., Krishnan, K. M., and Alivisatos, A. P. (2001) Colloidal nanocrystal shape and size control: The case of cobalt, *Science*, **291**(5511), 2115–2117.

154. Puntes, V. F., Zanchet, D., Erdonmez, C. K., and Alivisatos, A. P. (2002) Synthesis of hcp-Co nanodisks, *J. Am. Chem. Soc.*, **124**(43), 12874–12880.

155. Dumestre, F., Chaudret, B., Amiens, C., Renaud, P., and Fejes, P. (2004) Superlattices of iron nanocubes synthesized from $Fe[N(SiMe_3)_2]_2$, *Science*, **303**(5659), 821–823.

156. Rampino, L. D., and Nord, F. F. (1941) Preparation of palladium and platinum synthetic high polymer catalysts and the relationship between particle size and rate of hydrogenation, *J. Am. Chem. Soc.*, **63**(10), 2745–2749.

157. Lim, B., Lu, X., Jiang, M., Camargo, P. H. C., Cho, E. C., Lee, E. P., and Xia, Y. (2008) Facile synthesis of highly faceted multioctahedral Pt

nanocrystals through controlled overgrowth, *Nano Lett.*, **8**(11), 4043–4047.

158. Chen, J., Herricks, T., and Xia, Y. (2005) Polyol synthesis of platinum nanostructures: Control of morphology through the manipulation of reduction kinetics, *Angew. Chem. Int. Edit.*, **44**(17), 2589–2592.
159. Chen, J., Herricks, T., Geissler, M., and Xia, Y. (2004) Single-crystal nanowires of platinum can be synthesized by controlling the reaction rate of a polyol process, *J. Am. Chem. Soc.*, **126**(35), 10854–10855.
160. Herricks, T., Chen, J., and Xia, Y. (2004) Polyol synthesis of platinum nanoparticles control of morphology with sodium nitrate, *Nano Lett.*, **4**(12), 2367–2371.
161. Song, H., Kim, F., Connor, S., Somorjai, G. A., and Yang, P. (2004) Pt nanocrystals: Shape control and langmuir-blodgett monolayer formation, *J. Phys. Chem. B*, **109**(1), 188–193.
162. Lim, S. I., Ojea-Jimeénez, I., Varon, M., Casals, E., Arbiol, J., and Puntes, V. (2010) Synthesis of platinum cubes, polypods, cuboctahedrons, and raspberries assisted by cobalt nanocrystals, *Nano Lett.*, **10**(3), 964–973.
163. Teranishi, T., and Miyake, M. (1998) Size control of palladium nanoparticles and their crystal structures, *Chem. Mater.*, **10**(2), 594–600.
164. Xiong, Y., Chen, J., Wiley, B., Xia, Y., Aloni, S., and Yin, Y. (2005) Understanding the role of oxidative etching in the polyol synthesis of Pd nanoparticles with uniform shape and size, *J. Am. Chem. Soc.*, **127**(20), 7332–7333.
165. Xiong, Y., Cai, H., Wiley, B. J., Wang, J., Kim, M. J., and Xia, Y. (2007) Synthesis and mechanistic study of palladium nanobars and nanorods, *J. Am. Chem. Soc.*, **129**(12), 3665–3675.
166. Xiong, Y., Cai, H., Yin, Y., and Xia, Y. (2007) Synthesis and characterization of fivefold twinned nanorods and right bipyramids of palladium, *Chem. Phys. Lett.*, **440**(4–6), 273–278.
167. Jana, N. R., Gearheart, L., and Murphy, C. J. (2001) Seed-mediated growth approach for shape-controlled synthesis of spheroidal and rod-like gold nanoparticles using a surfactant template, *Adv. Mater.*, **13**(18), 1389–1393.
168. Jana, N. R., Gearheart, L., and Murphy, C. J. (2001) Wet chemical synthesis of silver nanorods and nanowires of controllable aspect ratio, *Chem. Commun.*, 7, 617–618.

169. Jana, N. R., Gearheart, L., and Murphy, C. J. (2001) Wet chemical synthesis of high aspect ratio cylindrical gold nanorods, *J. Phys. Chem. B*, **105**(19), 4065–4067.
170. Wang, Z. L., Mohamed, M. B., Link, S., and El-Sayed, M. A. (1999) Crystallographic facets and shapes of gold nanorods of different aspect ratios, *Surf. Sci.*, **440**(1–2), L809–L814.
171. Orendorff, C. J., and Murphy, C. J. (2006) Quantitation of metal content in the silver-assisted growth of gold nanorods, *J. Phys. Chem. B*, **110**(9), 3990–3994.
172. Murphy, C. J., Sau, T. K., Gole, A. M., Orendorff, C. J., Gao, J., Gou, L., Hunyadi, S. E., and Li, T. (2005) Anisotropic metal nanoparticles synthesis, assembly, and optical applications, *J. Phys. Chem. B*, **109**(29), 13857–13870.
173. Pérez-Juste, J., Liz-Marzán, L. M., Carnie, S., Chan, D. Y. C., and Mulvaney, P. (2004) Electric-field-directed growth of gold nanorods in aqueous surfactant solutions, *Adv. Func. Mater.*, **14**(6), 571–579.
174. Johnson, C. J., Dujardin, E., Davis, S. A., Murphy, C. J., and Mann, S. (2002) Growth and form of gold nanorods prepared by seed-mediated, surfactant-directed synthesis, *J. Mater. Chem.*, **12**(6), 1765–1770.
175. Gao, J., Bender, C. M., and Murphy, C. J. (2003) Dependence of the gold nanorod aspect ratio on the nature of the directing surfactant in aqueous solution, *Langmuir*, **19**(21), 9065–9070.
176. Grzelczak, M., Péerez-Juste, J., Rodríguez-González, B., and Liz-Marzán, L. M. (2006) Influence of silver ions on the growth mode of platinum on gold nanorods, *J. Mater. Chem.*, **16**(40), 3946–3951.
177. Grzelczak, M., Pérez-Juste, J., Garcia de Abajo, F. J., and Liz-Marzán, L. M. (2007) Optical properties of platinum-coated gold nanorods, *J. Phys. Chem. C*, **111**(17), 6183–6188.
178. Feng, L., Wu, X., Ren, L., Xiang, Y., He, W., Zhang, K., Zhou, W., and Xie, S. (2008) Well-controlled synthesis of Au@Pt nanostructures by gold-nanorod-seeded growth, *Chem. Eur. J.*, **14**(31), 9764–9771.
179. Grzelczak, M., Rodríguez-González, B., Pérez-Juste, J., and Liz-Marzán, L. M. (2007) Quasi-epitaxial growth of Ni nanoshells on Au nanorods, *Adv. Mater.*, **19**(17), 2262–2266.
180. Rodriguez-Gonzalez, B., Burrows, A., Watanabe, M., Kiely, C. J., and Liz-Marzán, L. M. (2005) Multishell bimetallic AuAg nanoparticles synthesis, structure and optical properties, *J. Mater. Chem.*, **15**(17), 1755–1759.

181. Becker, J., Zins, I., Jakab, A., Khalavka, Y., Schubert, O., and Sönnichsen, C. (2008) Plasmonic focusing reduces ensemble linewidth of silver-coated gold nanorods, *Nano Lett.*, **8**(6), 1719–1723.

182. Cho, E. C., Camargo, P. H. C., and Xia, Y. (2010) Synthesis and characterization of noble-metal nanostructures containing gold nanorods in the center, *Adv. Mater.*, **22**(6), 744–748.

183. Sun, Y., Gates, B., Mayers, B., and Xia, Y. (2002) Crystalline silver nanowires by soft solution processing, *Nano Lett.*, **2**(2), 165–168.

184. Habas, S. E., Lee, H., Radmilovic, V., Somorjai, G. A., and Yang, P. (2007) Shaping binary metal nanocrystals through epitaxial seeded growth, *Nat. Mater.*, **6**(9), 692–697.

185. Wetz, F., Soulantica, K., Falqui, A., Respaud, M., Snoeck, E., and Chaudret, B. (2007) Hybrid Co-Au nanorods: Controlling Au nucleation and location, *Angew. Chem. Int. Edit.*, **46**(37), 7079–7081.

186. Pazos-Pèrez, N., Gao, Y., Hilgendorff, M., Irsen, S., Pérez-Juste, J., Spasova, M., Farle, M., Liz-Marzán, L. M., and Giersig, M. (2007) Magnetic-noble metal nanocomposites with morphology-dependent optical response, *Chem. Mater.*, **19**(18), 4415–4422.

187. Pellegrino, T., Fiore, A., Carlino, E., Giannini, C., Cozzoli, P. D., Ciccarella, G., Respaud, M., Palmirotta, L., Cingolani, R., and Manna, L. (2006) Heterodimers based on CoPt3-Au nanocrystals with tunable domain size, *J. Am. Chem. Soc.*, **128**(20), 6690–6698.

188. Gabriel, C., Gabriel, S., Grant, E. H., Halstead, B. S. J., Michael, D., and Mingos, P. (1998) Dielectric parameters relevant to microwave dielectric heating, *Chem. Soc. Rev.*, **27**(3), 213–224.

189. Rao, K. J., Vaidhyanathan, B., Ganguli, M., and Ramakrishnan, P. A. (1999) Synthesis of inorganic solids using microwaves, *Chem. Mater.*, **11**(4), 882–895.

190. Lidström, P., Tierney, J., Wathey, B., and Westman, J. (2001) Microwave assisted organic synthesis–a review, *Tetrahedron*, **57**(45), 9225–9283.

191. Thostenson, E. T., and Chou, T. W. (1999) Microwave processing: Fundamentals and applications, *Compos. Part A: Appl. Sci. Manuf.*, **30**(9), 1055–1071.

192. Bilecka, I., and Niederberger, M. (2010) Microwave chemistry for inorganic nanomaterials synthesis, *Nanoscale*, **2**(8), 1358–1374.

193. Baghurst, D. R., and Mingos, D. M. P. (1992) Superheating effects associated with microwave dielectric heating, *J. Chem. Soc., Chem. Commun.*, 9, 674–677.

194. Galema, S. A. (1997) Microwave chemistry, *Chem. Soc. Rev.*, **26**(3), 233–238.

195. Pol, V. G., Langzam, Y., and Zaban, A. (2007) Application of microwave superheating for the synthesis of TiO_2 rods, *Langmuir*, **23**(22), 11211–11216.

196. Berlan, J. (1995) Microwaves in chemistry: Another way of heating reaction mixtures, *Rad. Phys. Chem.*, **45**(4), 581–589.

197. Mingos, D. M. P., and Baghurst, D. R. (1991) Applications of microwave dielectric heating effects to synthetic problems in chemistry (Tilden Lecture), *Chem. Soc. Rev.*, **20**(1), 1–47.

198. Whittaker, A. G., and Mingos, D. M. P. (2000) Arcing and other microwave characteristics of metal powders in liquid systems, *J. Chem. Soc. Dalton*, **2000**(9), 1521–1526.

199. Komarneni, S. (2003) Nanophase materials by hydrothermal, microwave hydrothermal and microwave-solvothermal methods, *Curr. Sci.(India)*, **85**(12), 1730–1734.

200. Kappe, C. O., Dallinger, D., and Murphree, S. S. (eds) (2009) *Practical Microwave Synthesis for Organic Chemists*, Wiley-VCH, Weinheim.

201. Stadler, A., Yousefi, B. H., Dallinger, D., Walla, P., Van der Eycken, E., Kaval, N., and Kappe, C. O. (2003) Scalability of microwave-assisted organic synthesis. From single-mode to multimode parallel batch reactors, *Org. Process. Res. Dev.*, **7**(5), 707–716.

202. Kappe, C. O. (2006) The use of microwave irradiation in organic synthesis. From laboratory curiosity to standard practice in twenty years, *Chimia*, **60**, 308–312.

203. Pein, A., Baghbanzadeh, M., Rath, T., Haas, W., Maier, E., Amenitsch, H., Hofer, F., Kappe, C. O., and Trimmel, G. (2010) Investigation of the formation of $CuInS_2$ nanoparticles by the oleylamine route: Comparison of microwave-assisted and conventional syntheses, *Inorg. Chem.*, **50**(1), 193–200.

204. Baghbanzadeh, M., Carbone, L., Cozzoli, P. D., and Kappe, C. O. (2011) Microwave-assisted synthesis of colloidal inorganic nanocrystals, *Angew. Chem. Int. Edit.*, **50**(48), 11312–11359.

205. Tsuji, M., Hashimoto, M., Nishizawa, Y., Kubokawa, M., and Tsuji, T. (2005) Microwave-assisted synthesis of metallic nanostructures in solution, *Chem. Eur. J.*, **11**(2), 440–452.

206. Pastoriza-Santos, I., and Liz-Marzán, L. M. (2002) Formation of PVP-protected metal nanoparticles in DMF, *Langmuir*, **18**(7), 2888–2894.

207. Tsuji, M., Matsuo, R., Jiang, P., Miyamae, N., Ueyama, D., Nishio, M., Hikino, S., Kumagae, H., Kamarudin, K. S. N., and Tang, X. L. (2008) Shape-dependent evolution of Au@Ag core-shell nanocrystals by PVP-assisted N,N-dimethylformamide reduction, *Cryst. Growth Des.*, **8**(7), 2528–2536.

208. Nguyen, H. L., Howard, L. E. M., Giblin, S. R., Tanner, B. K., Terry, I., Hughes, A. K., Ross, I. M., Serres, A., Burckstummer, H., and Evans, J. S. O. (2005) Synthesis of monodispersed fcc and fct FePt/FePd nanoparticles by microwave irradiation, *J. Mater. Chem.*, **15**(48), 5136–5143.

209. Yamamoto, T., Yin, H., Wada, Y., Kitamura, T., Sakata, T., Mori, H., and Yanagida, S. (2004) Morphology-control in microwave-assisted synthesis of silver particles in aqueous solutions, *B. Chem. Soc. Jpn*, **77**(4), 757–761.

210. Liu, F.-K., Huang, P.-W., Chang, Y.-C., Ko, C.-J., Ko, F.-H., and Chu, T.-C. (2005) Formation of silver nanorods by microwave heating in the presence of gold seeds, *J. Cryst. Growth*, **273**(3–4), 439–445.

211. Yin, H., Yamamoto, T., Wada, Y., and Yanagida, S. (2004) Large-scale and size-controlled synthesis of silver nanoparticles under microwave irradiation, *Mater. Chem. Phys.*, **83**(1), 66–70.

212. Liu, F.-K., Huang, P.-W., Chu, T.-C., and Ko, F.-H. (2005) Gold seed-assisted synthesis of silver nanomaterials under microwave heating, *Mater. Lett.*, **59**(8–9), 940–944.

213. Yamamoto, T., Wada, Y., Sakata, T., Mori, H., Goto, M., Hibino, S., and Yanagida, S. (2004) Microwave-assisted preparation of silver nanoparticles, *Chem. Lett.*, **33**(2), 158–159.

214. Abargues, R., Gradess, R., Canet-Ferrer, J., Abderrafi, K., Valdes, J. L., and Martinez-Pastor, J. (2009) Scalable heterogeneous synthesis of metallic nanoparticles and aggregates with polyvinyl alcohol, *New J. Chem.*, **33**(4), 913–917.

215. Tsuji, M., Nishizawa, Y., Hashimoto, M., and Tsuji, T. (2004) Syntheses of silver nanofilms, nanorods, and nanowires by a microwave-polyol method in the presence of Pt seeds and polyvinylpyrrolidone, *Chem. Lett.*, **33**(4), 370–371.

216. Tsuji, M., Nishizawa, Y., Matsumoto, K., Miyamae, N., Tsuji, T., and Zhang, X. (2007) Rapid synthesis of silver nanostructures by using microwave-polyol method with the assistance of Pt seeds and polyvinylpyrrolidone, *Colloid. Surf. A*, **293**(1–3), 185–194.

217. Katsuki, H., and Komarneni, S. (2003) Microwave-assisted polyol synthesis of Ag powders, *J. Mater. Res.*, **18**(4), 747–750.

218. Katsuki, H., and Komarneni, S. (2003) Nano-and micro-meter sized silver metal powders by microwave-polyol process, *J. Jpn. Soc. Powder Powder Metallurgy*, **50**(10), 745–750.

219. Pal, A., Shah, S., and Devi, S. (2009) Microwave-assisted synthesis of silver nanoparticles using ethanol as a reducing agent, *Mater. Chem. Phys.*, **114**(2–3), 530–532.

220. Daniel, M.-C., and Astruc, D. (2003) Gold nanoparticles: Assembly, supramolecular chemistry, quantum-size-related properties, and applications toward biology, catalysis, and nanotechnology, *Chem. Rev.*, **104**(1), 293–346.

221. Groisman, Y., and Gedanken, A. (2008) Continuous flow, circulating microwave system and its application in nanoparticle fabrication and biodiesel synthesis, *J. Phys. Chem. C*, **112**(24), 8802–8808.

222. Singh, A. K., and Raykar, V. S. (2008) Microwave synthesis of silver nanofluids with polyvinylpyrrolidone (PVP) and their transport properties, *Colloid. Polym. Sci.*, **286**(14–15), 1667–1673.

223. Komarneni, S., Katsuki, H., Li, D., and Bhalla, A. S. (2004) Microwave-polyol process for metal nanophases, *J. Phys.: Condens. Matter*, **16**(14), S1305–S1312.

224. He, B. L., Tan, J. J., Kong, Y. L., and Liu, H. F. (2004) Synthesis of size controlled Ag nanoparticles, *J. Mol. Catal. A-Chem.*, **221**(1–2), 121–126.

225. Zhu, Y. J., and Hu, X. L. (2004) Microwave-assisted polythiol reduction method a new solid-liquid route to fast preparation of silver nanowires, *Mater. Lett.*, **58**(9), 1517–1519.

226. Procopio, A., Alcaro, S., Cundari, S., De Nino, A., Ortuso, F., Sacchetta, P., Pennelli, A., and Sindona, G. (2005) Molecular modeling, synthesis, and preliminary biological evaluation of glutathione-s-transferase inhibitors as potential therapeutic agents, *J. Med. Chem.*, **48**(19), 6084–6089.

227. Baruwati, B., Polshettiwar, V., and Varma, R. S. (2009) Glutathione promoted expeditious green synthesis of silver nanoparticles in water using microwaves, *Green Chem.*, **11**(7), 926–930.

228. Li, D. S., and Komarneni, S. (2006) Synthesis of Pt nanoparticles and nanorods by microwave-assisted solvothermal technique, *Z Naturforsch B*, **61**(12), 1566–1572.

229. Yu, W., Tu, W., and Liu, H. (1998) Synthesis of nanoscale platinum colloids by microwave dielectric heating, *Langmuir*, **15**(1), 6–9.

230. Komarneni, S., Li, D., Newalkar, B., Katsuki, H., and Bhalla, A. S. (2002) Microwave-polyol process for Pt and Ag nanoparticles, *Langmuir*, **18**(15), 5959–5962.

231. Tu, W., and Liu, H. (2000) Rapid synthesis of nanoscale colloidal metal clusters by microwave irradiation, *J. Mater. Chem.*, **10**(9), 2207–2211.

232. Jiang, Z. L., Feng, Z. W., and Shen, X. C. (2001) Microwave synthesis of Au nanoparticles with the system of $AuCl_4$-CH_3CH_2OH, *Chinese Chem. Lett.*, **12**(6), 551–554.

233. Sun, X., and Luo, Y. (2005) Size-controlled synthesis of dendrimer-protected gold nanoparticles by microwave radiation, *Mater. Lett.*, **59**(29–30), 4048–4050.

234. Shen, M., Sun, Y., Han, Y., Yao, R., and Yan, C. (2008) Strong deaggregating effect of a novel polyamino resorcinarene surfactant on gold nanoaggregates under microwave irradiation, *Langmuir*, **24**(22), 13161–13167.

235. Tsuji, M., Hashimoto, M., Nishizawa, Y., and Tsuji, T. (2003) Preparation of gold nanoplates by a microwave-polyol method, *Chem. Lett.*, **32**(12), 1114–1115.

236. Tsuji, M., Miyamae, N., Hashimoto, M., Nishio, M., Hikino, S., Ishigami, N., and Tanaka, I. (2007) Shape and size controlled synthesis of gold nanocrystals using oxidative etching by $AuCl_4^-$ and Cl^- anions in microwave-polyol process, *Colloid. Surface A*, **302**(1–3), 587–598.

237. Mallikarjuna, N. N., and Varma, R. S. (2007) Microwave-assisted shape-controlled bulk synthesis of noble nanocrystals and their catalytic properties, *Cryst. Growth Des.*, **7**(4), 686–690.

238. Mohamed, M. B., AbouZeid, K. M., Abdelsayed, V., Aljarash, A. A., and El-Shall, M. S. (2010) Growth mechanism of anisotropic gold nanocrystals via microwave synthesis formation of dioleamide by gold nanocatalysis, *ACS Nano*, **4**(5), 2766–2772.

239. Shen, M., Du, Y. K., Hua, N. P., and Yang, P. (2006) Microwave irradiation synthesis and self-assembly of alkylamine-stabilized gold nanoparticles, *Powder Technol.*, **162**(1), 64–72.

240. Liu, F.-K., Chang, Y.-C., Ko, F.-H., and Chu, T.-C. (2004) Microwave rapid heating for the synthesis of gold nanorods, *Mater. Lett.*, **58**(3–4), 373–377.

241. Kundu, S., Peng, L., and Liang, H. (2008) A new route to obtain high-yield multiple-shaped gold nanoparticles in aqueous solution using microwave irradiation, *Inorg. Chem.*, **47**(14), 6344–6352.

242. Zhu, Y.-J., and Hu, X. L. (2003) Microwave-polyol preparation of single-crystalline gold nanorods and nanowires, *Chem. Lett.*, **32**(12), 1140–1141.

243. Kundu, S., and Liang, H. (2008) Microwave synthesis of electrically conductive gold nanowires on DNA scaffolds, *Langmuir*, **24**(17), 9668–9674.

244. Dongsheng, L., and Sridhar, K. (2006) Microwave-assisted polyol process for synthesis of Ni nanoparticles, *J. Am. Ceram. Soc.*, **89**(5), 1510–1517.

245. Zhou, B., Hong, J.-M., and Zhu, J.-J. (2005) Microwave-assisted rapid synthesis of antimony dendrites, *Mater. Lett.*, **59**(24–25), 3081–3084.

246. Abdelsayed, V., Aljarash, A., El-Shall, M. S., Al Othman, Z. A., and Alghamdi, A. H. (2009) Microwave synthesis of bimetallic nanoalloys and CO oxidation on ceria-supported nanoalloys, *Chem. Mater.*, **21**(13), 2825–2834.

247. Cheng, W. T., and Cheng, H. W. (2009) Synthesis and characterization of cobalt nanoparticles through microwave polyol process, *Aiche J.*, **55**(6), 1383–1389.

248. Zhu, H.-T., Zhang, C.-Y., and Yin, Y.-S. (2004) Rapid synthesis of copper nanoparticles by sodium hypophosphite reduction in ethylene glycol under microwave irradiation, *J. Cryst. Growth*, **270**(3–4), 722–728.

249. Tu, W., and Liu, H. (2000) Continuous synthesis of colloidal metal nanoclusters by microwave irradiation, *Chem. Mater.*, **12**(2), 564–567.

250. Li, D. S., and Komarneni, S. (2006) Microwave-assisted polyol process for synthesis of Ni nanoparticles, *J. Am. Ceram. Soc.*, **89**(5), 1510–1517.

251. Liu, F.-K., Huang, P.-W., Chang, Y.-C., Ko, F.-H., and Chu, T.-C. (2005) Combining optical lithography with rapid microwave heating for the selective growth of Au/Ag bimetallic core/shell structures on patterned silicon wafers, *Langmuir*, **21**(6), 2519–2525.

252. Tsuji, M., Matsumoto, K., Miyamae, N., Tsuji, T., and Zhang, X. (2006) Rapid preparation of silver nanorods and nanowires by a microwave-polyol method in the presence of Pt catalyst and polyvinylpyrrolidone, *Cryst. Growth Des.*, **7**(2), 311–320.

253. Tsuji, M., Miyamae, N., Lim, S., Kimura, K., Zhang, X., Hikino, S., and Nishio, M. (2006) Crystal structures and growth mechanisms of Au@Ag core-shell nanoparticles prepared by the microwave-polyol method, *Cryst. Growth Des.*, **6**(8), 1801–1807.

254. Tsuji, M., Nishio, M., Jiang, P., Miyamae, N., Lim, S., Matsumoto, K., Ueyama, D., and Tang, X. L. (2008) Role of chloride ions in the formation of Au@Ag core-shell nanocrystal structures by using a microwave-polyol method, *Colloid. Surface A*, **317**(1–3), 247–255.

255. Harpeness, R., and Gedanken, A. (2004) Microwave synthesis of core-shell gold/palladium bimetallic nanoparticles, *Langmuir*, **20**(8), 3431–3434.

256. Zhang, H., Yin, Y., Hu, Y., Li, C., Wu, P., Wei, S., and Cai, C. (2010) Pd@Pt core-shell nanostructures with controllable composition synthesized by a microwave method and their enhanced electrocatalytic activity toward oxygen reduction and methanol oxidation, *J. Phys. Chem. C*, **114**(27), 11861–11867.

257. Yokoyama, T., Masuhara, A., Onodera, T., Kasai, H., and Oikawa, H. (2009) Development of fabrication process for Ag/polydiacetylene (core/shell) hybridized nanocrystals, *Synth. Met.*, **159**(9–10), 897–899.

258. Harpeness, R., and Gedanken, A. (2005) The microwave-assisted polyol synthesis of nanosized hard magnetic material, FePt, *J. Mater. Chem.*, **15**(6), 698–702.

259. Yamauchi, T., Tsukahara, Y., Sakata, T., Mori, H., Yanagida, T., Kawai, T., and Wada, Y. (2010) Magnetic Cu-Ni (core-shell) nanoparticles in a one-pot reaction under microwave irradiation, *Nanoscale*, **2**(4), 515–523.

260. Zeng, H., Li, J., Liu, J. P., Wang, Z. L., and Sun, S. (2002) Exchange-coupled nanocomposite magnets by nanoparticle self-assembly, *Nature*, **420**(6914), 395–398.

261. Weller, D., and Doerner, M. F. (2000) Extremely high-density longitudinal magnetic recording media, *Annu. Rev. Mat. Sci.*, **30**(1), 611–644.

262. Pal, A., Shah, S., and Devi, S. (2007) Synthesis of Au, Ag and Au–Ag alloy nanoparticles in aqueous polymer solution, *Colloid. Surface A*, **302**(1–3), 51–57.

263. Deivaraj, T. C., Chen, W. X., and Lee, J. Y. (2003) Preparation of PtNi nanoparticles for the electrocatalytic oxidation of methanol, *J. Mater. Chem.*, **13**(10), 2555–2560.

264. Bensebaa, F., Patrito, N., Page, Y. L., L'Ecuyer, P., and Wang, D. (2004) Tunable platinum-ruthenium nanoparticle properties using microwave synthesis, *J. Mater. Chem.*, **14**(22), 3378–3384.

Index